DILUTED MAGNETIC SEMICONDUCTORS

DILUTED MAGNETIC SEMICONDUCTORS

MUKESH JAIN

The University of Western Australia
Australia

World Scientific
Singapore • New Jersey • London • Hong Kong

Published by

World Scientific Publishing Co. Pte. Ltd.
P O Box 128, Farrer Road, Singapore 9128
USA office: Suite 1B, 1060 Main Street, River Edge, NJ 07661
UK office: 73 Lynton Mead, Totteridge, London N20 8DH

Library of Congress Cataloging-in-Publication Data

Jain, Mukesh Kumar.
    Diluted magnetic semiconductors / Mukesh Jain.
        p.   cm.
    ISBN 9810201761
        1. Diluted magnetic semiconductors.   I. Title.
    QC611.8.M25J35   1991
    537.6'223--dc20   91-26394
                                                                              CIP

Printed in Singapore by General Printing Services Pte. Ltd.

Contributors

A. Abe Matsushita Electric Industrial Co. Ltd., 3-15 Yagumo-Nakamachi, Moriguchi, Osaka, Japan.

E. Anastassakis National Technical University, Physics Department, Zografou Campus, Athens 157 73, Greece.

G. Bauer Semiconductor Physics Group, Johannes-Kepler-Universität Linz, A-4040 Linz, Austria.

C. Benecke Central Research Units, F. Hoffmann-La Roche & Co. Ltd., Basel, Switzerland.

D. Coquillat Groupe d'Etude des Semiconducteurs, Université des Sciences et Techniques du Languedoc, Montpellier II, Place E. Bataillon, F-34060-MONTPELLIER-Cédex 1, France.

T. Dietl Institute of Physics, Polish Academy of Sciences, Al. Lotników 32/46 Pl 02-668 Warszawa, Poland.

A. Fujimori Department of Physics, University of Tokyo, Hongo 7-3-1, Bunkyo-ku, Tokyo 113, Japan.

J. K. Furdyna Department of Physics, University of Notre Dame, Notre Dame, Indiana 46556, USA.

T. M. Giebultowicz National Institute of Standards and Technology, Gaithersburg, MD 20899, USA.

M. Godlewski Institute of Physics, Polish Academy of Sciences, Al. Lotników 32/46, Pl 02-668 Warsaw, Poland.

O. Goede Humboldt-Universität zu Berlin, Sektion Physik, Invalidenstrasse 110, 1040 Berlin, Germany.

H. -E. Gumlich Institut für Festkorperphysik der Technischen Universität, Berlin, Germany.

W. Heimbrodt Humboldt-Universität zu Berlin, Sektion Physik, Invaldienstrasse 110, 1040 Berlin, Germany.

M. Jain   *Department of Electrical and Electronic Engineering & Department of Physics, The Univertsity of Western Australia, Nedlands, WA 6009, Australia.*

A. Lewicki   *Department of Physics, Purdue University, West Lafayette, Indiana 47907, USA & Department of Solid State Physics, Akademia Górniczo-Hutnicza (AGH), PL-30-059 Kraków, Poland.*

G. E. Marques   *Departamento de Física, Universidade Federal de São Carlos, 13560 São Carlos SP, Brazil.*

G. N. Pain   *Telecom Australia Research Laboratories, 770 Blackburn Road, Clayton, Victoria 3168, Australia.*

H. Pascher   *Experimentalphysik I, Universität Bayreuth, D-8580 Bayreuth, Germany.*

J. M. Pawlikowski   *Institute of Physics, TUW, Wyspianskiego 27, 50-370 Wroclaw, Poland.*

J. L. Robins   *Department of Physics, The Univertsity of Western Australia, Nedlands, WA 6009, Australia.*

A. I. Schindler   *Department of Physics, Purdue University, West Lafayette, Indiana 47907, USA.*

R. K. Singh   *School of Physics, Barkatullah University, Bhopal 462026, India.*

K. Swiatek   *Institute of Physics, Polish Academy of Sciences, Al. Lotników 32/46, Pl 02-668 Warsaw, Poland.*

A. Twardowski   *Institute of Experimental Physics, Warsaw University, Hoza 69, 00681 Warsaw, Poland & Physics Department, Eindhoven University of Technology, Eindhoven, The Netherlands.*

Introduction

In recent years, diluted magnetic semiconductors (DMS), also known as semimagnetic semiconductors (SMSC), have attracted the attention of the scientific and industrial community. These materials have some unique properties which enhance their potential for use in a wide range of opto-electronic devices. DMS are semiconductors with a fraction of their constituent ions replaced by transition metal ions. These materials behave in a similar manner to their non-magnetic counterpart semiconductors in the absence of any external magnetic field. The band gap- and lattice-parameters can be tuned by changing appropriately the concentration of the added magnetic ions. The possibility for band gap engineering in these materials makes them useful for various device applications. The tunability of the lattice parameter with magnetic ion content also extends their utility into the field of heterostructures where lattice matching is important.

The presence of substituted magnetic ions in the DMS results in magnetic properties which distinguish them from ordinary semiconductors. In contrast to normal magnetic semiconductors for which the application of an external magnetic field does not produce a marked response by the magnetic ions, the magnetic ions in DMS do respond to an applied magnetic field and change the energy band and impurity level parameters. Much of the interest in DMS is related to the wide range of their magnetic properties such as paramagnetic, antiferromagnetic.

A unique and important feature of DMS is the spin-spin exchange interaction between the localized magnetic moments of the magnetic ions and the conduction and/or valence band electrons. This interaction affects the energy band, electronic structure and impurity level parameters of the semiconductors, resulting in new physical effects particularly when in the presence of strong magnetic fields. The consequences are quite dramatic: g-factors are effectively enhanced by as much as two orders of magnitude, Faraday rotations become very large and the magneto-resistance can become negative, reaching exceptionally large values and leading to an insulator to metal transition induced by increasing magnetic field.

The most common and extensively studied DMS are those in which Mn^{2+} ions are introduced into A^{II}-B^{VI} compound semiconductors. These also have the privilege of being the first to be described as DMS. Until today, these are the DMS which can accommodate the highest possible magnetic ion concentration without the crystallographic quality and structure of the host materials being destroyed. These alloys have the potential to be used in various device applications such as IR detectors and solar cells. The properties of these alloys have been studied thoroughly on a worldwide basis since 1977 when, for the first time giant enhancement of magnetic-optical effects in CdMnTe were reported. The importance of these materials is evidenced by the large number of invited talks and research papers presented at international conferences and by the fact that Vol. 25 of the Semiconductors and Semimetals series has been dedicated to these materials.

The next stages of DMS development involved the replacement of Mn^{2+} by other transition metal ions such as Fe^{2+} and Co^{2+}, and the replacement of group II elements by group IV elements, resulting in DMS alloys such as $Pb_{1-x}Mn_xTe$. These new DMS are not simply an extension of the $A^{II}MnB^{VI}$ type DMS because they show some additional unique and important properties which are quite different to those of the $A^{II}MnB^{VI}$. These developments led to applications involving new types of opto-electronic device. More recently, additional new types of DMS, for example $A^{III}MB^{V}$ and $(A_{1-x}M_x)_2As_3$, have been developed which further enlarge the DMS family. Attention has also been extended towards the development of various devices based on these materials and to the development of epitaxial films grown by a variety of growth techniques.

Work on $A^{II}MB^{VI}$ DMS has already been well presented in some earlier reviews and in Vol. 25 of the Semiconductors and Semimetals series. The research presented in this book is thus additional to this and also includes some updating of information in the field. This book contains sixteen chapters beginning with a description of the fundamental aspects of these alloys, followed by details of the new materials, superlattices and heterostructures. It concludes with a look at different possible applications using these DMS materials. By way of an introduction for readers who are new to this field, the first chapter contains a description of fundamental concepts such as the structure, energy gap and preparation techniques of DMS. The next six chapters describe the various properties of the most common materials including, optical, luminescent, electronic, transport, phase transition and light scattering properties. These will give a clear understanding of the current state of knowledge of these various phenomena. Chapters 8 to 10 are dedicated to describing the special properties of the more recently developed DMS alloys in which either ions other than Mn^{2+} are used as the magnetic element or the IV-VI compounds are used as hosts. Chapters 11 and 12 contain work on DMS superlattices and heterostructures. The last four chapters of the book give detailed information about some possible applications of these DMS alloys.

Whilst considerable effort has been made to present the most up to date information and the latest developments in DMS materials, the field is expanding so rapidly that total coverage is no longer possible in a book such as this. Further, it should be noted that some degree of overlap of material between chapters has been permitted so that each chapter is relatively self contained.

Contents

Chapter 16 On Eu Activated II-VI Semiconducting Compounds

M. Godlewski and K. Świątek

DILUTED MAGNETIC SEMICONDUCTORS

MATERIAL PREPARATION, CRYSTAL STRUCTURE AND ENERGY GAP OF DILUTED MAGNETIC SEMICONDUCTORS

MUKESH JAIN[*+] and JOHN L. ROBINS[*]

[*]Department of Physics &

[+]Department of Electrical and Electronic Engineering,
The University of Western Australia,
Perth, Australia

CONTENTS

1. INTRODUCTION

During recent years, there has been increasing interest and considerable experimental and theoretical activity focussed on a new group of semiconductors[1-10] which have some unique properties that enhance their potential for use in a wide range of opto-electronic device applications. Because of these features, which have both defence and commercial applications, worldwide experimental and theoretical efforts have been directed towards understanding the underlying physics of the unusual phenomena associated with these special semiconductors. Recognition of the importance of these materials has been increasing rapidly since 1977 when Kamarov et al[11] first reported the giant enhancement of magnetic-optical effects in CdMnTe. As a result all major conferences on semiconductors now include these materials either in invited talks or special sessions. These materials are known either as Diluted Magnetic Semiconductors (DMS) or Semimagnetic Semiconductors (SMSC).

The DMS or SMSC are semiconductors formed by replacing a fraction of the cations in a range of compound semiconductors by the transition metal ions. Some examples of DMS are $Cd_{1-x}Mn_xTe$, $Zn_{1-x}Fe_xSe$, $Pb_{1-x}Mn_xTe$ and $In_{1-x}Mn_xAs$. These can be expressed generally in a number of different ways including $A_{1-x}M_xB$ or $AB(M)$ or $(AB)_{1-x}(MB)_x$. Here x indicates the fraction of the non-magnetic cations (A) of the compound semiconductors (AB), which is randomly replaced by magnetic '3d' or '4f' ions (M). Formation of the DMS can also be described as alloying an ordinary semiconductor AB with a magnetic semiconductor MB in the required proportion. This leads to an alternative definition of DMS as mixed crystals of ordinary and magnetic semiconductors. A consequence of their ternary nature is that the lattice constants, energy gap and other band parameters of the DMS can be tuned between the values applicable to the primary binaries AB and MB by varying the value of x. The possibility thus exists to tailor the properties of these DMS materials and hence they loom large as important materials in the field of device technology.

A wide range of devices can be produced using these DMS materials by exploiting their various semiconducting and magnetic properties[12-20]. In zero magnetic field, DMS behave in a similar fashion to normal nonmagnetic ternary semiconductors. For example, the energy gap, lattice parameters, effective mass etc can be varied in a controlled fashion[7-8] by varying the composition x. This property underlies their applications as lattice matched substrates, specially for Hg based alloys, and as devices such as solar cells, IR detectors and lasers. For example, the CdTe-MnTe system with its zinc-blende structure represents a versatile substrate for opto-electronic devices because its lattice constant can be varied[21] between 6.37Å and 6.48Å by changing the x from 0 to 0.77. Other types of device may be developed by exploiting the presence of the substituted transition metal ion, e.g. Mn^{2+}, in the DMS. For example, optical transitions within the Mn^{2+} can produce some electroluminescent properties in the DMS, and $Zn_{1-x}Mn_xSe$ and $Zn_{1-x}Mn_xS$ are already recognised as potentially good materials for flat panel display devices[13-15,22]. With an applied external magnetic field, the presence of magnetic ions in the DMS systems affects the free charge carrier behaviour and modifies the electronic properties of the semiconductors through the sp-d exchange interaction between the localized magnetic moments and the spins of band electrons. In contrast to magnetic semiconductors for which the application of an external magnetic field does not produce a marked response by the magnetic ions, the magnetic ions in DMS do respond to an applied magnetic field and change the energy band and impurity level parameters. Also sensitive to an applied external magnetic field are the spin-spin interactions in DMS which result in a large increase in Faraday rotation, an enhancement by as much as two orders of magnitude in g-factors, and giant negative magneto-resistance[5-8,23-24]. These various unique characteristics of DMS make them special and quite different from the ordinary and magnetic semiconductors and hence make possible their fabrication into special devices such as magnetic field sensors and optical isolators[25].

In the early stages of DMS development, it was the II-VI compound semiconductors containing substitutional Mn^{2+} ions which received major attention and consequently these are the most common and thoroughly studied DMS[26-28]. The replacement of Mn^{2+} by other transition metal ions[29-30] such as Fe^{2+}, and then the replacement of the group II elements by the group IV elements[31-32] such as replacing the Cd by Pb in CdMnTe type DMS were the next stages of DMS development. More recently, attention has been given to the development of new types of DMS. One of the different types is based on the III-V compound semiconductors[33-34], resulting in DMS such as $In_{1-x}Mn_xAs$. Another type is produced by introducing new transition and rare earth ions[35-38] such as Co and Eu instead of Mn and Fe. Much experimental and theoretical research work is now being carried out on these new materials but compared to the II-VI based DMS, they are still at an early stage and very little literature is available on them. The successful preparation of epitaxial thin films of some of these DMS by using techniques such as molecular beam epitaxy (MBE) or metal organic chemical vapour deposition (MOCVD) is also introducing new possibilities for superlattices[39-40], quantum well heterostructures[41-42] and special device technology. These techniques overcome some of the problems associated with large single crystal growth, such as limited composition range, and are adaptable to the production of large scale devices. Epitaxial films of some DMS grown on GaAs substrates are proving to be suitable for integrated optical circuit structures.

This book contains sixteen chapters and most of them are dedicated to new work on DMS materials. The work includes some important devices developed using these DMS, some of the new DMS now being produced and DMS superlattices and heterostructures. By way of an introduction for readers who are new to this field, this first chapter is devoted to describing the fundamental concepts of DMS in a simple manner. The discussion is thus limited to the fundamentals of structure, energy gap and preparation techniques. In general, it is also restricted to the most fully investigated II-VI and IV-VI DMS although some of the more recent developments are included where possible. This treatment begins in section 2 with a description of the fundamental properties of the host materials as an understanding of these is basic to an understanding of the DMS developed from them. Then in section 3, the structure and composition of the II-VI and IV-VI DMS are presented. Methods of preparation of both bulk crystals of DMS and thin films are given in section 4, together with details of material preparation and other parameters required for crystal growth. Finally, in section 5, a brief discussion is given of some recent developments in the field of DMS including epitaxial films and superlattices.

2. GENERAL PROPERTIES OF HOST MATERIALS

The properties and compositions of the binary host materials influence the properties of the DMS ternary materials. The crystal structure and most of the optical and electrical properties of the DMS are qualitatively similar to those of their host material compounds, for example CdMnTe resembles CdTe and CdMnSe resembles CdSe, especially in crystal structure. It is therefore advisable to begin by reviewing briefly the properties of the host binaries.

2.1 The A^{II}-B^{VI} Host Binaries

Almost all the A^{II}-B^{VI} compounds crystallise either in the zinc blende or wurtzite structures[43-44], as shown in Table 1. The common, and dominant, feature of these structures is that each atom of one element is tetrahedrally bonded to four atoms of the other element. In zinc blende these tetrahedra are arranged in a cubic type structure whilst in wurtzite they are in a hexagonal type structure. Indeed, the centres of similar tetrahedra are arranged in a face-centred cubic (fcc) array in the former and a hexagonal close-packed (hcp) array in the latter. The similarity of these two structures is best recognised by noting that the only difference between fcc and hcp structures, when viewed as stacked close-packed planes (the (111) in fcc), is that the repeat sequence in fcc is abcabc whilst in hcp it is ababab. It should be noted however that it has been reported that some of the compounds listed in Table 1 also occur in layered form consisting of repeat sequences of the close-packed layers of tetrahedra other than the abcabc and ababab which correspond to pure zinc blende and wurtzite respectively.

Table 1

Crystal structure of A^{II}-B^{VI} compounds[43-46]

Compound	Crystal structure type	Lattice parameter Å	
		a	c
ZnS	Zinc blende	5.41	
	Wurtzite	3.82	6.26
ZnSe	Zinc blende	5.67	
	Wurtzite	4.01	6.54
ZnTe	Zinc blende	6.10	
	Wurtzite	4.27	6.99
CdS	Zinc blende	5.83	
	Wurtzite	4.13	6.75
CdSe	Zinc blende	6.08	
	Wurtzite	4.31	7.02
CdTe	Zinc blende	6.48	
HgS	Zinc blende	5.85	
	Cinnabar	4.15	9.50
HgSe	Zinc blende	6.07	
HgTe	Zinc blende	6.46	

The zinc blende structure, otherwise known as sphalerite, may be visualised as two interlocking fcc structures each comprised of one type of atom, with one shifted by one fourth of the body diagonal of the fcc unit cell of the other. The wurtzite structure may be visualised as two interlocking hcp structures, again with each comprised of one type of atom and with one shifted along the c axis of the other. In both of these structures each atom has four atoms of the other type in its first coordination sphere and twelve atoms of its own type in the second coordination sphere. For example, Zn is surrounded by four Te atoms and twelve Zn atoms in the wurtzite structure and also by four Te atoms and twelve Zn atoms in the zinc blende structure. With this common feature, it is often difficult to distinguish between the two crystal structures. The basic difference is the position of, and relative distance to, atoms in the third coordination sphere, with this distance being slightly greater in the zinc blende than in the wurtzite structure relative to the nearest neighbours spacing.

Where these compounds form in both zinc blende and wurtzite structures, the interatomic distance within the tetrahedra of both forms is very similar. This can be seen in Table 2 which shows some of the properties of the A^{II}-B^{VI} semiconductors. For example, in ZnS these interatomic distances are 2.36Å and 2.35Å in the zinc blende and wurtzite forms respectively. The distance between like atoms (the nearest cation-cation distance, d_{cc}), in the second coordination sphere, is also very similar for these compounds, an example being the 3.81Å and 3.82Å in the zinc blende and wurtzite forms of ZnS[47], respectively. This indicates that the distances between like atoms and unlike atoms (bond length) are similarly related in the two structures.

For group II elements the atoms have two electrons in the outer shell, in configuration s^2. Correspondingly, atoms of group VI elements have six electrons, in configuration s^2p^4. Where the relative electro-negativity of the metal (group II) atom is sufficiently strong that these atoms can give up the two electrons to the group VI atoms, doubly charged ions are formed and the structure takes the octahedral configuration of NaCl type structures. This occurs in the alkali metal (group IIA) chalcogenides. However, where the ionisation potential is sufficiently high that the metals do not fully give up but rather just share their electrons with their non-metal neighbours, tetrahedrally oriented covalent bonds are formed. This is the case with group IIB and group VI atoms where the tetrahedrally based zinc blende and wurtzite structures are formed. These bonds are in the form of four elongated two-electron clouds arranged around each atom in positions pointing to the corners of a tetrahedron centered on the atom. However, in the case of these A^{II}-B^{VI} compounds, the group VI elements do have a

Table 2

Some important physical properties of A^{II}-B^{VI} compounds[48-49]

Compound	Crystal structure	Distances between unlike atoms in tetraheda Å	Optical forbidden bandwidth eV	Melting point °C
ZnS	Zinc blende	2.36	3.6	1830
	Wurtzite	2.35		
ZnSe	Zinc blende	2.45	2.6	1515
	Wurtzite	2.45		
ZnTe	Zinc blende	2.64	2.2	1295
CdS	Zinc blende	2.52	2.4	1750
	Wurtzite	2.52		
CdSe	Zinc blende	2.62	1.7	1258
	Wurtzite	2.63		
CdTe	Zinc blende	2.79	1.5	1098
HgS	Zinc blende	2.53	2.0	1450
HgSe	Zinc blende	2.63	0.6	800
HgTe	Zinc blende	2.80	0.02	670

sufficiently strong relative electro-negativity that the electron clouds are displaced from their centrally located position between dissimilar atoms and are shifted closer to the group VI atoms than the group II atoms. Thus these compounds show properties of both covalent and ionic bonding.

If the A^{II}-B^{VI} compounds are grouped according to the element which forms the cation, as in Table 2, then within these subgroups the properties change as the atomic number of the chalcogen changes. As this atomic number increases, the optical forbidden bandwidth and the melting point decrease. At the same time, as the overall atomic number of the elements in these compounds increases, the ionicity of the bonds decreases and the bonds become more metallic as the electron clouds broaden without actually allowing the electrons to be shared by the lattice as a whole. Relative variation of the metallic and ionic composition of the bonds in this group of compounds produces a range of material types from those which are almost insulators (ZnS) to those with narrow forbidden bands (HgSe) and even semimetals (HgTe).

2.2 The A^{IV}-B^{VI} Host Binaries

The A^{IV}-B^{VI} compounds are the chalcogenides of the elements in subgroup IVB of the periodic table. These A^{IV}-B^{VI} compounds can be divided into two further subgroups on the basis of their crystal structure. One group contains the compounds PbS, PbSe, PbTe, SnTe and a high temperature modification of GeTe(I), which all have an NaCl type lattice[43]. The other group includes GeS, GeSe, SnS and SnSe which have an orthorhombic structure[43]. The crystal structure and lattice parameters of these A^{IV}-B^{VI} compounds are shown in Table 3. In general it is only compounds of the former group which are used as a basis for DMS and only those will be discussed in detail here.

Table 3

Crystal structure of A^{IV}-B^{VI} compounds[43,45,50-51]

| Compound | Crystal structure | Lattice parameter Å | | |
		a	b	c
GeS	Orthorhombic	4.30	3.65	10.44
GeSe	Orthorhombic	4.38	3.82	10.79
GeTe(I)	NaCl type (distorted)	6.00		
SnS	Orthorhombic	4.34	3.99	11.20
SnSe	Orthorhombic	4.46	4.19	11.57
SnTe	NaCl type	6.32		
PbS	NaCl type	5.94		
PbSe	NaCl type	6.15		
PbTe	NaCl type	6.45		

The crystal structure for the compounds which form in the NaCl type lattice can be visualised as two interlocking fcc lattices, each containing one type of atom, with one lattice shifted by a distance a/2 along any one of the three cubic axes of the other. The bonds are between dissimilar atoms and the experimental values for their interatomic spacings (determined from the measured lattice parameters) are listed in Table 4. Also listed in that table are the sums of the ionic radii, and the sums of the covalent radii, of the pairs of elements involved. In all cases the experimental values lie between the related summed values, indicating the binding to be neither purely ionic nor purely covalent, but rather a mixture of the two types. Some of the important properties such as forbidden bandwidth of the A^{IV}-B^{VI} compounds are shown in Table 5.

Table 4

Interatomic spacings in NaCl type compounds[52]

Compounds	Exp.interatomic spacings Å	Sum of covalent radii (Å)	Sum of ionic radii(Å)
PbS	2.97	2.50	3.04
PbSe	3.07	2.63	3.18
PbTe	3.23	2.83	3.41
SnTe	3.16	2.77	3.33
GeTe	3.00	2.59	3.14

An explanation of the origin of this combined ionic-covalent bonding in these compounds has been offered by Krebs[53] and may be summarised as follows, in terms of PbS. The outer shell electron configurations for Pb and S are s^2p^2 and s^2p^4 respectively. If the compound was purely ionic, there would be Pb^{2+} ions, each with three empty p orbitals and S^{2-} ions, each with three filled p orbitals containing two electrons in each. The p-type orbitals associated with each ion would be directed along the regular orthogonal coordinate axes towards adjacent ions of dissimilar type. Thus between each adjacent pair there would be an empty p-type orbital associated with the Pb^{2+} ion and an occupied p-type orbital associated with the S^{2-} ion. Because of this uneven filling of the orbitals the electrons would be drawn away from the sulphur and closer to the lead ion. This in turn would tend to partly neutralise the charges on the ions, reducing the ionicity of the bond and adding a covalent (homopolar) component. As each pair of electrons associated with one p state of the S^{2-} ion would be shared between the bonding to the two Pb^{2+} ions on either side along a coordinate direction, there would be a resonant binding of the sulphur ion to the two lead ions.

Table 5

Some important physical properties of
A^{IV}-B^{VI} compounds[8, 43]

Compound	Optical forbidden bandwidth eV	Melting point ^{o}C
GeS	1.80	665
GeSe	1.16	670
GeTe	0.50	725
SnS	1.08	881
SnSe	0.90	860
SnTe	0.26	806
PbS	0.30	1119
PbSe	0.22	1076
PbTe	0.25	917

This added covalent component would reduce the interatomic spacing in the PbS to a value less than that expected for pure ionic bonding, thus explaining the observations reported in Table 4. Indeed, as the covalent component increases from PbS through PbSe to PbTe, so the relative shift away from the pure ionic value towards the pure covalent value increases. The relatively narrow forbidden band of these compounds can also be explained in terms of Krebs' model in that the resonant binding between electrons in the p states reduces the energy required for an electron transition to the conduction band.

3. CRYSTAL STRUCTURE, LATTICE PARAMETER AND THE BAND GAP OF DMS

3.1 The $A_{1-x}^{II}M_xB^{VI}$ DMS

$A_{1-x}^{II}M_xB^{VI}$ are the DMS in which a fraction of the cations of the II-VI compound semiconductors are randomly substituted by transition metal ions[1-10] such as Mn^{2+}, Co^{2+} etc. Most of the research carried out on DMS so far has concentrated on this family of alloys with specific attention to those in which Mn^{2+} is the added magnetic ion[8]. The possible composition range and crystal structure of $A_{1-x}^{II}M_xB^{VI}$ DMS are shown in Table 6. It may be seen from the table that none of the $A_{1-x}^{II}M_xB^{VI}$ can be grown in single crystal form over the complete composition range $0 \leq x \leq 1$.

The $A_{1-x}^{II}M_xB^{VI}$ DMS can be divide into three categories based on the different substituted transition metal ions: $A_{1-x}^{II}Mn_xB^{VI}$, $A_{1-x}^{II}Fe_xB^{VI}$ and $A_{1-x}^{II}Co_xB^{VI}$. Each category may be divided further into sub-categories: the wide-gap DMS and the narrow-gap DMS. The wide-gap DMS are those in which transition metal ions replace Zn or Cd cations ($Zn_{1-x}Fe_xS$, $Cd_{1-x}Fe_xSe$ $Zn_{1-x}Mn_xS$ etc.)[54-57], whereas the narrow gap DMS are those in which Hg ions are replaced ($Hg_{1-x}Mn_xSe$, $Hg_{1-x}Fe_xTe$ etc.)[58-59]. As Mn based II-VI DMS are the most commonly and widely studied, therefore, we start by reviewing the properties of these DMS, including their lattice parameters and band gaps, followed by a review of the properties of the less studied Fe and Co based DMS.

Table 6

Crystal structure of $A_{1-x}^{II}M_xB^{VI}$ diluted magnetic
semiconductors

Alloy	Composition range	Crystal structure
$Zn_{1-x}Mn_xS$	$0.00 < x \leq 0.10$	Zinc blende
	$0.10 < x \leq 0.45$	Wurtzite
$Zn_{1-x}Mn_xSe$	$0.00 < x \leq 0.30$	Zinc blende
	$0.30 < x \leq 0.57$	Wurtzite
$Zn_{1-x}Mn_xTe$	$0.00 < x \leq 0.86$	Zinc blende
$Cd_{1-x}Mn_xS$	$0.00 < x \leq 0.50$	Wurtzite
	$0.94 < x \leq 1.00$	NaCl type
$Cd_{1-x}Mn_xSe$	$0.00 < x \leq 0.50$	Wurtzite
	$0.94 < x \leq 1.00$	NaCl type
$Cd_{1-x}Mn_xTe$	$0.00 < x \leq 0.77$	Zinc blende
	$0.96 < x \leq 1.00$	NiAs type
$Hg_{1-x}Mn_xS$	$0.00 < x \leq 0.37$	Zinc blende
$Hg_{1-x}Mn_xSe$	$0.00 < x \leq 0.38$	Zinc blende
$Hg_{1-x}Mn_xTe$	$0.00 < x \leq 0.75$	Zinc blende
$Zn_{1-x}Fe_xS$	$0.00 < x \leq 0.10$	Zinc blende
$Zn_{1-x}Fe_xSe$	$0.00 < x \leq 0.30$	Zinc blende
$Zn_{1-x}Fe_xTe$	$0.00 < x \leq 0.01$	Zinc blende
$Cd_{1-x}Fe_xSe$	$0.00 < x \leq 0.03$	Wurtzite
$Cd_{1-x}Fe_xTe$	$0.00 < x \leq 0.15$	Zinc blende
$Hg_{1-x}Fe_xSe$	$0.00 < x \leq 0.15$	Zinc blende
$Hg_{1-x}Fe_xTe$	$0.00 < x \leq 0.12$	Zinc blende
$Zn_{1-x}Co_xS$	$0.00 < x \leq 0.14$	Zinc blende
$Zn_{1-x}Co_xSe$	$0.00 < x \leq 0.05$	Zinc blende
$Cd_{1-x}Co_xSe$	$0.00 < x \leq 0.22$	Wurtzite

3.1.1 Composition Dependence of Crystal Structure and Lattice Parameter

As seen in Table 6, the Mn based DMS can be grown over a wider composition range than Fe and Co based DMS. It is quite remarkable that such a wide composition range is possible in these DMS, considering the difference in the different crystal structures of the component binary semiconductors; MnS, MnSe and MnTe crystallise in either rocksalt or NiAs type structures whereas the II-VI compounds crystallise in zinc blende or wurtzite structures.

Possible solubility limits of Mn in II-VI compounds to form the $A_{1-x}^{II}M_xB^{VI}$ DMS are shown in Fig.1. $Hg_{1-x}Mn_xTe$ crystallises in the zinc blende structure with x between 0 and 0.75. However, whilst single crystal structure is observed for x up to 0.35, for higher values of x, besides the zinc blende structure of $Hg_{1-x}Mn_xTe$, other phases (e.g. MnTe and $MnTe_2$) also appear[60]. The lattice parameter decreases linearly[5] from 6.46Å to 6.38Å as the value of x varies from 0 to 0.75.

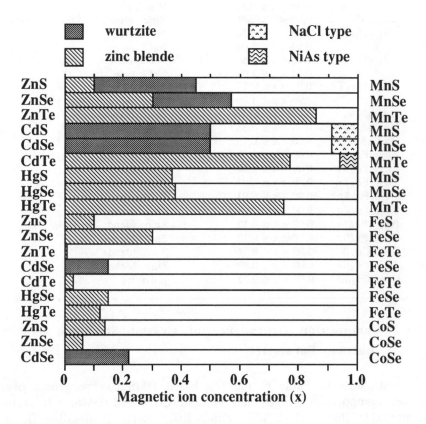

Fig.1. Crystal structures of the $A_{1-x}^{II}Mn_xB^{VI}$ DMS as a function of magnetic ion concentration.

$Hg_{1-x}Mn_xSe$ and $Hg_{1-x}Mn_xS$ also crystallise in the zinc blende structure but for only half the range of x values applicable to their counterpart $Hg_{1-x}Mn_xTe$ DMS. The zinc blende phase is stable up to x = 0.38 and 0.37 for $Hg_{1-x}Mn_xSe$[61] and $Hg_{1-x}Mn_xS$[62] respectively. In both cases, the lattice parameter also decreases with increase in Mn concentration in the zinc blende phase. It is reported that the lattice parameter in all three Hg based DMS decreases linearly with Mn concentration whilst the stable phases exist, but beyond this concentration limit no significant changes occur in the lattice parameter values.

Whereas the mercury manganese chalcogenide DMS show only one crystal structure over the possible composition range, the cadmium manganese chalcogenide DMS behave quite differently in that they show two different structures in the cadmium and manganese rich regions. $Cd_{1-x}Mn_xTe$ crystallises in the zinc blende (x ≤ 0.77) and the NiAs structures (x ≥ 0.96)[8], whilst $Cd_{1-x}Mn_xSe$ and $Cd_{1-x}Mn_xS$ crystallise in the wurtzite (x ≤ 0.50) and the rocksalt structure (x ≥ 0.94)[63-64]. The zinc blende structure of $Cd_{1-x}Mn_xTe$ and the wurtzite structure of $Cd_{1-x}Mn_xSe$ and $Cd_{1-x}Mn_xS$ in cadmium rich regions resemble the respective structures of CdTe, CdSe and CdS. This supports the concept that the structures of host binaries effect the structures of the DMS based on them. In these DMS, the lattice parameters decrease with increase in the Mn concentration[64-65].

Like other telluride based DMS, $Zn_{1-x}Mn_xTe$ also crystallises in the zinc blende structure[43,66] with the widest range of manganese concentration (x ≤ 0.86) found among any of the $A_{1-x}^{II}Mn_xB^{VI}$ DMS. The lattice parameter increases with increase in Mn concentration in $Zn_{1-x}Mn_xTe$. Both $Zn_{1-x}Mn_xSe$ and $Zn_{1-x}Mn_xS$ show two different structures with Mn concentration[66-67] but their behaviour is different to that of $Cd_{1-x}Mn_xSe$ and $Cd_{1-x}Mn_xS$ despite the fact that they also showed two different structures. $Cd_{1-x}Mn_xSe$ and $Cd_{1-x}Mn_xS$ show different structures for crystals with compositions near the two ends of the composition range, corresponding to dilution of the two separate binaries. Conversely, as can be seen in Fig. 1, in $Zn_{1-x}Mn_xSe$ and $Zn_{1-x}Mn_xS$, the structure is different either side of a particular composition. $Zn_{1-x}Mn_xSe$ and $Zn_{1-x}Mn_xS$ show the zinc blende structure for x up to 0.30 and 0.10 respectively, whilst beyond this compositions the zinc blende phase changes to the wurtzite with x extending to 0.57 for $Zn_{1-x}Mn_xSe$ and 0.45 for $Zn_{1-x}Mn_xS$.

With increasing concentration of magnetic ions, the lattice parameter decreases in $Cd_{1-x}Mn_xTe$ and $Hg_{1-x}Mn_xTe$, whilst it increases in $Zn_{1-x}Mn_xTe$. This variation is shown in Fig. 2 where it can be seen that the relationships are linear over the full range of Mn concentrations for which the zinc blende structure persists (full lines). It is clear that if these linear dependencies are extrapolated (dashed lines) to x = 1, they meet at a single point corresponding to a = 6.334Å. It can be concluded that this is the lattice parameter that MnTe would have if it crystallised in the zinc blende structure rather than its usual NiAs structure[6,8]. These DMS can thus be regarded as true pseudobinary alloys of CdTe (or ZnTe or HgTe) and MnTe.

The right hand scale in Fig.2 is the corresponding cation-cation distance d_{cc} in the alloys. Clearly the relationship between this parameter d_{cc} and composition x is the same as that between the lattice parameter a and composition x. This is because of the fixed relationship between d_{cc} and a (namely a = $(4/\sqrt{3})b$) in the zinc blende structure. It will be recalled from Section 2 that in both the zinc blende and wurtzite structures of the A^{II}-B^{VI} compounds, the basic structural unit is a tetrahedron in which each A (or B) ion is tetrahedrally bonded, by s-p^3 bonds, to four nearest neighbour B (or A) ions. This spacing is the bond length b = $r_{II} + r_{VI}$, where the r values are the covalent radii of the II and VI group atoms. These tetrahedra are arranged in an fcc configuration to form the zinc blende structure or in an hcp configuration to form the wurtzite structure. As the fcc and hcp structures are both close-packed structures, each A (or B) ion is surrounded by twelve second nearest neighbours of the same type A (or B) at a "cation-cation distance" d_{cc} where $d_{cc} = \sqrt{8/3}\, b$ for both structures.

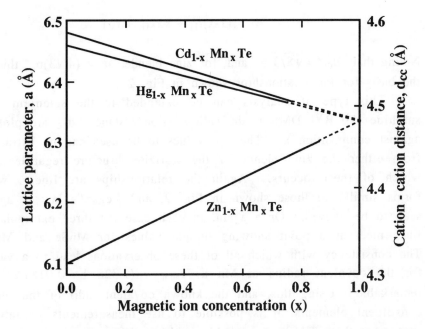

Fig. 2. Composition dependence of the lattice parameters (a) in the $A_{1-x}^{II}Mn_xTe$ DMS. Extrapolation (dashed lines) of the lattice parameters (full lines) to x = 1 gives the lattice parameter value for the zinc blende phase (hypothetical) of MnTe. The corresponding cation-cation distances d_{cc} are shown in the right hand side.

The relationships between the cation-cation distance d_{cc} and the lattice parameters (a for zinc blende and a and c for wurtzite) are:

for zinc blende $\quad a = \sqrt{2}d_{cc} \qquad = (4/\sqrt{3}) \, b$

for wurtzite $\qquad a = d_{cc} \qquad\quad = \sqrt{8/3} \, b$

$\qquad\qquad\qquad c = \sqrt{8/3} \, d_{cc} = (8/3) \, b$

Clearly, the bond length b is basic to both structures, and in both cases it is equal to the sum of the covalent radii of the constituents.

In a ternary DMS alloy where Mn ions replace a fraction x of the group II cations, the average "cation" radius will be $(1-x)r_{II} + xr_{Mn}$ and the average bond length b will be given by

$$b = (1-x)r_{II} + x.r_{Mn} + r_{VI}$$

Noting that $d_{cc} = \sqrt{8/3}\ b$ and, for zinc blende, $a = (4/\sqrt{3})b$, this is the basis for the relationships shown in Fig. 2

This type of analysis can be extended to the selenium and sulphide A^{II}-B^{VI} DMS if the cation-cation distances d_{cc} are plotted against composition x. The d_{cc} values to be used can be extracted from either the zinc blende or the wurtzite structure regardless of which of these occurs. Again the relationships are linear, with forms similar to those shown in Fig. 2, and Vegard's law is again seen to be obeyed. Once again, in each case the three extrapolated lines meet at a point showing unique values for MnSe and MnS. The consistency with which all of these observations lead to a value for the covalent radius of Mn as $r_{Mn} = (1.326 \pm 0.018)Å$[67] is remarkable. Using this, and the known covalent radii of the other constituent elements, it is possible to use measurements of lattice spacing in these DMS as a reliable guide to their composition.

Of course, if these structures are examined on an atomic scale, such as by extended X-ray absorption fine structure (EXAFS), it will be seen that the bond lengths are not the average values but are specific to the elements involved (e.g. a Cd-Te bond will be 2.180Å and a Mn-Te bond will be 2.731Å). Thus on this scale the zinc blende and wurtzite structures will be slightly distorted, depending on the actual Mn distribution. However such fine structure will not be detected during normal X-ray lattice parameter determinations.

II-VI compounds containing Fe such as $Cd_{1-x}Fe_xSe$, $Hg_{1-x}Fe_xTe$ and $Zn_{1-x}Fe_xSe$ are relatively new DMS[68-73] and are not a simple extension of the Mn based DMS family. The different behaviour is related to the different physical situation within these Fe based DMS. Although these materials are still relatively unexplored, it is recognised that the Fe^{2+} ion contains one more electron than the Mn^{2+} ion, so that the Fe^{2+} $3d^6$ state is located above the Mn^{2+} $3d^5$ state in the semiconductor band structure. The main physical properties of the $A^{II}_{1-x}Fe_xB^{VI}$ DMS depend on the location of the Fe^{2+}

$3d^6$ states relative to the top of the valence band and the bottom of the conduction band. The magnetism of the Fe^{2+} sub-system is quite different from that of the Mn^{2+} sub-system. The magnetic properties of these DMS are explained on the basis of Fe-Fe exchange interaction. Fe ions reveal only a field induced magnetic moment and at low temperatures, those alloys with low Fe concentration exhibit Van Vleck paramagnetism while in those with higher Fe content ($x \geq 0.1$), a spin glass phase is found[73-74]. The permanent magnetic moment of the Fe^{2+} ion vanishes at low temperatures. At high temperature, for all achievable values of x, the exchange interaction between Fe^{2+} and their nearest neighbour is antiferromagnetic.

The best known DMS of the $A_{1-x}^{II}Fe_xB^{VI}$ type are the narrow gap DMS such as $Hg_{1-x}Fe_xSe$ and $Hg_{1-x}Fe_xTe$, and the wide gap DMS such as $Cd_{1-x}Fe_xSe$ and $Cd_{1-x}Fe_xTe$ DMS. Table 6 and Fig. 1 show the possible range of x in the Fe based DMS together with their crystal structures. As can be seen from the table, the solubility of Fe is about 2-15% in the $A_{1-x}^{II}Fe_xTe$ DMS[75-76], and about 15-30% in the $A_{1-x}^{II}Fe_xSe$ DMS[76-77] which are far below the solubilities in the Mn based DMS. The narrow gap $Hg_{1-x}Fe_xSe$ and $Hg_{1-x}Fe_xTe$ crystalise in the zinc blende structure with x varying from 0 to 0.15 in $Hg_{1-x}Fe_xSe$ and from 0 to 0.12 in $Hg_{1-x}Fe_xTe$, as shown in Fig. 1. The lattice constant decreases with decreasing Fe concentration in both cases. The zinc iron chalcogenides also crystallise in the zinc blende structure but with different Fe concentration limits; value of x varies from 0 to 0.10 in $Zn_{1-x}Fe_xS$, from 0 to 0.30 in $Zn_{1-x}Fe_xSe$ and from 0 to 0.01 in $Zn_{1-x}Fe_xTe$. This low maximum solubility limit of Fe in $Zn_{1-x}Fe_xTe$ is also seen in $Cd_{1-x}Fe_xTe$ in which x varies only from 0 to 0.03. $Cd_{1-x}Fe_xTe$ crystalises in the zinc blende structure, whereas $Cd_{1-x}Fe_xSe$ shows an hexagonal structure with x varying from 0 to 0.15. Only a limited amount of data is available on these materials. The lattice parameters decrease linearly with increasing Fe content.

Recently, some DMS alloys in which a fraction of the cations has been replaced by Co^{2+} ions[35-36,78-81] have been successfully synthesized. These DMS alloys are relatively new and very little information is available on them; for example, there is no literature available on the narrow gap $Hg_{1-x}Co_xB^{VI}$ DMS although these have been prepared experimentally. Crystals of $Zn_{1-x}Co_xS$ and $Zn_{1-x}Co_xSe$ have been prepared by a chemical vapor transport technique with x varying[80] from 0 to 0.145 in the $Zn_{1-x}Co_xS$ and from 0 to 0.048 in the $Zn_{1-x}Co_xSe$. The behaviour of the lattice parameters in these DMS is quite different to that in DMS alloys containing Mn. In contrast to $Zn_{1-x}Mn_xB^{VI}$ DMS, in which the lattice parameters increase with increasing Mn concentration and Vegard's law is obeyed, the values of the lattice parameters in these alloys do not change with changes in the Co^{2+} concentration. This behaviour of the lattice parameter can be explained on the basis of the covalent radii of the constituent elements. The covalent radius of Mn is larger than that of Zn in the $Zn_{1-x}Mn_xB^{VI}$ alloys, whereas in $Zn_{1-x}Co_xB^{VI}$ alloys the covalent radius of Co is smaller than that of Mn and is similar to that of Zn. The modified Bridgman method has been used to grow the $Cd_{1-x}Co_xS$ and $Cd_{1-x}Co_xSe$ DMS, both of which crystallise in the wurtzite structure. In contrast to the wurtzite structured DMS containing Mn, in which the c/a ratio is very close to the ratio $\sqrt{8/3} = 1.633$ as predicted by the ideal close packing of spheres, the c/a ratio in these DMS is not exactly 1.633. This is associated with a small axial distortion along the crystal c axis, and because of this distortion the magnetic susceptibility becomes anistropic at low temperatures.

3.1.2 Composition Dependence of Energy Gap

The $A_{1-x}^{II}Mn_xB^{VI}$, in which manganese ions substitute for the cations in the A^{II}-B^{VI} compound semiconductors, are direct gap semiconductors with the band extrema occurring at the Γ-point. In the A^{II}-B^{VI} compounds, the two valence electrons of the group II element and the six electrons of the group VI element are distributed accordingly to the s-p^3 orbital bonding configuration. A small concentration of Mn, below 1 atomic percent, does not make any significant change in the band structure, whereas higher concentrations, of the order of few atomic percent and more, lead to significant changes in the energy gap of these materials, which in turn affect their electrical and optical properties. The main change occurs because of the presence of d-levels contributed by the Mn 3d shell. Through the process of p-d hybridisation, the Mn d-levels are slightly broadened into relatively flat and narrow bands. The other effect of adding Mn to II-VI semiconductors is an opening of energy gap.

Substitution of Mn ions into the semimetal mercury chalcogenides produces a particularly interesting phenomenon. In the case of $Hg_{1-x}Mn_xTe$, where the parent semiconductor (HgTe) is a semimetal with the inverted Grove-Paul band structure[82] and has the Γ_8 and Γ_6 energy levels with $E(\Gamma_6) > E(\Gamma_8)$, the band gap can be tuned from zero to positive by changing the Mn concentration. For the lowest values of x, $Hg_{1-x}Mn_xTe$ behaves as a zero gap semiconductor but with slightly larger values of x ($x \leq 0.1$), the energy gap opens, see Fig.1 of Chapter 12. As the x value increases further, the alloy becomes a positive gap semiconductor. The energy gap varies linearly with increase in the Mn concentration[83-84]. This type of behavior is similar to that of $Hg_{1-x}Cd_xTe$ in which substitution of Cd into HgTe opens the energy gap, transforming the zero gap HgTe into the open gap CdTe crystal. This type of band gap engineering near the zero gap condition is very interesting and useful for making certain opto-electronic devices such as infrared detectors. The main features of the variation of energy gap with composition as described above for $Hg_{1-x}Mn_xTe$ are also valid for the other mercury based DMS, notably $Hg_{1-x}Mn_xSe$ and $Hg_{1-x}Fe_xTe$.

Substitution of Mn for Cd in the CdTe lattice increases the gap to the point that the originally opaque material becomes transparent to visible light[85-86]. The energy gap varies linearly[86-88] with x in $Cd_{1-x}Mn_xTe$ and $Cd_{1-x}Mn_xSe$ and nonlinearly[89] with x in $Cd_{1-x}Mn_xS$ where the energy gap initially decreases and then increases with x. The energy gap in the $Zn_{1-x}Mn_xTe$ varies linearly[90-91] from 2.2eV for x = 0 to 2.8eV for x = 2.8, whereas in $Zn_{1-x}Mn_xSe$, the energy gap decreases rapidly[92] as x increases from low Mn content up to x = 0.5 then increases almost linearly with Mn content there after. As mentioned above, the energy gap varies linearly with Mn concentration in the $Cd_{1-x}Mn_xTe$, $Hg_{1-x}Mn_xTe$ and $Zn_{1-x}Mn_xTe$ DMS. Extrapolation of these curves should give the value for the hypothetical zinc blende $MnTe^9$. At 300K, there is some small variation among the values obtained by extrapolation but at lower temperature, i.e. 4.2K, all three curves give a single value of 3.18eV for the hypothetical zinc blende MnTe. This is shown in Fig. 3. The value for the MnTe energy gap, obtained by extending the curve for any one of these DMS to x = 1, can then be used in the spirit of virtual crystal approximation to obtained energy gap values for the full range of x for the two other corresponding DMS.

The nature of the energy gap in Fe based DMS differs from that of Mn based compounds. Because the substitutional Fe^{2+} possesses both spin and orbital momentum and contains one more electron than the Mn^{2+} so the $Fe^{2+}3d^6$ level is located above the $Mn^{2+}3d^5$ state in the semiconductor band structure. Narrow gap $Hg_{1-x}Fe_xSe$ and wide gap $Cd_{1-x}Fe_xTe$ and $Cd_{1-x}Fe_xSe$ are the best known DMS of the $A^{II}_{1-x}Fe_xB^{VI}$ type. The main physical properties of these materials depend on the location of the $Fe^{2+}3d^6$ state relative to the top of the valence band and bottom of the conduction band; the $Fe^{2+}3d^6$ level being located[69,93-94] either in the forbidden gap as a deep donor state, for example as in $Cd_{1-x}Fe_xTe$ and $Cd_{1-x}Fe_xSe$, or in the conduction band as a resonant donor state as in $Hg_{1-x}Fe_xSe$. For low values of x, the Fe ions act as impurities in the band gap causing the pinning of the Fermi level, whereas for higher values of x, the position of the $Fe^{2+}3d^6$ level will correspond to the valence band edge of the alloy.

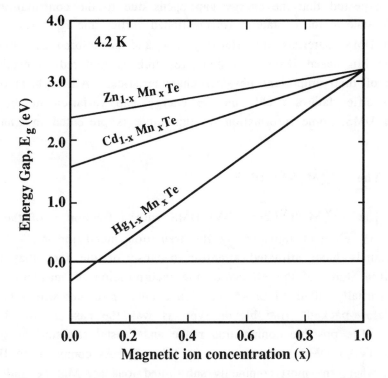

Fig.3. Variation of energy gap E_g as a function of M n concentration in $A_{1-x}^{II} Mn_x Te$ DMS at 4.2K. Extrapolation of E_g to x = 1 gives value of energy gap for the hypothetical zinc blende phase of MnTe.

The energy gap in $Cd_{1-x}Fe_xSe$ varies[68,94] from 1.8eV to 3.3eV with an increase in the Fe content from 0 to 0.15. The variation of energy gap in this range of x is found to be almost linear. The value of the energy gap in $Cd_{1-x}Fe_xTe$ does not change with an increase in Fe content through the very limited possible range of x (x ≤ 0.03). Thus substitution of Fe ions into CdTe either does not effect the band gap or the effect is so small as to be not recordable. The energy gap variation due to the Γ_6-Γ_8 direct transition, as a function of Fe content, is not well understood because the transition is too weak to be seen in the reflectivity spectrum. In spite of this it is expected that the energy gap opens due to the contribution of the Fe state to the Γ states. As mentioned earlier, the Fe^{2+} and Co^{2+} based DMS materials are relatively new and unexplored as compare to the Mn based DMS, and more research is required to establish some of their important physical characteristics. Although there is no specific data available on the energy gap variation in the Co based DMS, some information on this topic is presented in Chapter 10.

3.2 The $A_{1-x}^{IV}M_xB^{VI}$ DMS

The $A_{1-x}^{IV}M_xB^{VI}$ are IV-VI DMS with a fraction x of the the group IV element replaced by the transition metal ion M. These compounds have attracted attention because of the fact that they permit a study of the effects of the incorporation of magnetic ions with partially filled 3d or 4f states in a cubic structure which is not the "close-packed" zinc blende type as was the case in the II-VI hosts. The possible composition range and crystal structure of some of the IV-VI DMS are summarise in Table 7. As compared to II-VI DMS, where the most commonly substituted ions are Mn, Fe and Co, the IV-VI DMS contain mainly Mn, Eu and Gd as the substituted magnetic ions. As mentioned in section 2.2, the common IV-VI DMS are those based on the pseudobinary lead compounds and little is known about the other possible IV-VI DMS. As shown in Fig. 4, in comparision to their II-VI counterparts, the solubility limits of the magnetic ions in these IV-VI DMS are generally small and vary markedly from alloy to alloy. Materials in this group of DMS show interesting properties which make them useful for certain device applications[95-97]. Variation of the band gap within these IV-VI DMS is strongly dependent on the substituted magnetic ion; substitution of Eu producing a much more rapid change in band gap than Mn. Most of the IV-VI DMS crystallise in the rocksalt structure with the minimum direct gap located at the L-point of the Brillouin zone. This is to be compared to the II-VI DMS which crystallise in either zinc blende or wurtzite structures with the direct gap located at the Γ point of the Brillouin zone.

3.2.1 Composition Dependence of Crystal Structure and Lattice Parameter

The lead chalcogenide based DMS are useful for various device applications and have been investigated in more detail than the other IV-VI DMS. In $Pb_{1-x}Mn_xB^{VI}$ DMS, the magnetic ions incorporated into the cubic rocksalt structure are randomly distributed in the Pb-fcc sub-lattice. $Pb_{1-x}Mn_xTe$ crystallises in the cubic rocksalt structure for x in the range from 0 to 0.12[98-99] and with the lattice constant decreasing linearly with increasing x throughout this composition range. Crystals with $x \cong 0.15$ show deformation of the cubic structure towards the hexagonal structure of MnTe whereas the crystals with $x \cong 0.20$ reveal a very poor crystallographic quality. Lattice parameter measurements for samples in which x varies from 0.15 to 0.20 are not reproducible, but indicate the presence of come cubic PbTe rich phase which shows the lack of homogeneity. Up to now, the highest composition of manganese in $Pb_{1-x}Mn_xTe$ alloys for which physical properties have been reported is x = 0.12. In agreement with this is the report by Vanuyarkho[98] that the solubility limit of MnTe in PbTe is 12%. Escorne et al[99] as well as Vinogradova[100] found a Vegard type behaviour for the weighted average of the bond length up to this possible composition with an empirial rule $a = (a_0 - 0.0491x)$, where a denotes the lattice constant of $Pb_{1-x}Mn_xTe$ and a_0 that of PbTe. In addition to the Bridgman technique, the hot-wall technique has also been used to grow $Pb_{1-x}Mn_xTe$ single crystals[101].

$Pb_{1-x}Mn_xSe$ and $Pb_{1-x}Mn_xS$ also crystallise in the rocksalt structure. The maximum possible composition ranges are with x up

Table 7

Crystal structure of $A_{1-x}^{IV}M_xB^{VI}$ diluted magnetic semiconductors

Alloy	Composition range	Crystal structure
$Pb_{1-x}Mn_xS$	$0.00 < x \leq 0.05$	Rocksalt
$Pb_{1-x}Mn_xSe$	$0.00 < x \leq 0.17$	Rocksalt
$Pb_{1-x}Mn_xTe$	$0.00 < x \leq 0.12$	Rocksalt
$Sn_{1-x}Mn_xTe$	$0.00 < x \leq 0.70$	Rocksalt
$Ge_{1-x}Mn_xTe$	$0.00 < x \leq 0.70$	Rocksalt
	$0.95 < x \leq 1.00$	NiAs

to 0.17 in $Pb_{1-x}Mn_xSe$ and up to 0.05 in $Pb_{1-x}Mn_xS$ solid solutions[31]. As with $Pb_{1-x}Mn_xTe$, the lattice constants in these two DMS decrease with increasing Mn content[102]. An important feature of these compounds is the fact that the addition of manganese to the diamagnetic lead salt hosts causes an antiferromagnetic exchange interaction which is rather weak in comparison to the corresponding II-VI DMS compounds.

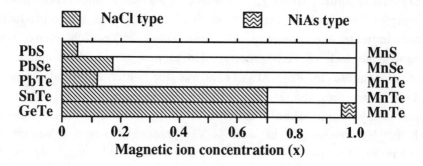

Fig. 4.   Crystal structures of the $A_{1-x}^{IV}Mn_xB^{VI}$ DMS as a function of magnetic ion concentration.

$Sn_{1-x}Mn_xTe$ crystallises in the rocksalt structure with manganese concentrations up to 0.70[103-104]. The lattice parameter decreases from 6.32Å to about 6.00Å as the Mn concentration increases from 0.0 to 0.70. $Ge_{1-x}Mn_xTe$ also crystallises in the rocksalt structure[43,105] with x ranging from 0.0 to 0.70. This is followed at higher concentrations by a two-phase region of alloys with NaCl and NiAs type structures until at x = 0.95 where a NiAs type single phase structure is observed. Literature is not available on selenides and sulphides of these two DMS systems.

3.2.2   Compositional Dependence of Energy Gaps

The IV-VI semiconductor compounds have their extrema of the conduction and valence bands at the L-point of the Brillouin zone. The replacement of the IV group element by Mn or Eu leads to a significant change of the energy gap in the IV-VI DMS. These compounds have attracted special attention because of the fact that they permit a study of the effects of the incorporation of magnetic ions with partially filled 3d or 4f states in the cubic but non-zinc blende type structure. As mentioned earlier, the lead chalcogenide based IV-VI DMS have been studied in-dept but not much is known about the other IV-VI DMS. This is one of the fields in which more research is needed to understand the different IV-VI DMS and to determine how to use them in the various device applications. The lead chalcogenides are known to have both a direct and an indirect band gap. The values of the direct and the indirect energy gaps[106] at 300K are 0.41 and 0.37eV for PbS, 0.29 and 0.26eV for PbSe and 0.32 and 0.29eV for PbTe.

The incorporation of a small amount of Mn into the lead chalcogenide semiconductors produces a significant change in their band gaps. The energy gap increases nonlinearly in $Pb_{1-x}Mn_xTe$ with an increase in Mn concentration[107]. The average value of dE_g/dx, the change of forbidden energy gap as a function of composition, is observed to be 1.3eV[108]. Similar to PbTe, PbMnTe also shows both a direct and an indirect transition. The values of E_g for the direct and the indirect transitions differ by 0.08eV which is considerably larger than the corresponding difference for PbTe. It is reported that the reason for having the higher difference between the values of direct and indirect energy gaps may be due either to an additional impurity absorption in the indirect transition range or to a departure from the dependence of the absorption coefficient on the energy gap difference, (E_g-E_{gi}), resulting from the more complex band structure of the $Pb_{1-x}Mn_xTe$ solid solution. It is possible to grow both p- and n-type PbMnTe and in both cases the band gap is found to be the same for a given x value.

As the energy band structure of PbSe is similar to that of PbTe, it is expected that the incorporation of Mn into PbSe will influence the energy gap in a similar fashion to that observed in the case of $Pb_{1-x}Mn_xTe$. However, the energy gap variation as a function of Mn content is not very well understood in $Pb_{1-x}Mn_xSe$. Photoconductivity measurements carried out on this DMS show a decrease in band gap with increasing Mn content, whilst Nel'son et al[109] have reported nonlinear increase in energy gap with x, similar to the $Pb_{1-x}Mn_xTe$ DMS. A detailed investigation is required to establish a relationship between the energy gap and Mn concentration. In contrast to the above mentioned behaviour in selenium and tellurium DMS, the energy gap decreases linearly with increasing Mn concentration in the $Pb_{1-x}Mn_xS$ DMS[5]. This difference between the sulphide and selenide DMS warrants further investigations.

4. PREPARATION OF DMS CRYSTALS

The Bridgman technique is both a convenient and commonly available technique which is widely used to grow bulk crystals of most of the DMS. There is a wide range of DMS materials, with a corresponding diversity of values for the parameters which are important to crystal growth. These include principally the melting points and vapour pressures of both the constituents and the final products and of course the individual equilibrium phase diagrams. This in turn leads to a range of conditions, often specific to individual DMS, which must be used when applying the Bridgman technique. In general we restrict ourselves here to presenting only the unique growth conditions in most cases, except where other unusal factors need to be emphasised. In some cases other techniques such as chemical transport and RF induction heating methods have been employed and these also are included below.

It is generally much more convenient to start with the binary compounds than the individual constituent elements, especially as these binaries are now commercially available in high purity (99.999%) form. However, if it is necessary to start with the constituent elements one should be very particular about their handling and purity. Even a very small amount of impurity in or oxidation of some of these elements can make them unsuitable for the purpose of crystal growth. In addition, special precautions should also be taken when handling these materials because some of the constituent elements are very poisonous. There are various ways, readily available in literature, to purify these chemical elements up to the limit of about 6N or 7N purity which is high enough to prepare good quality DMS crystals. Vacuum distillation is one of the best ways to purify many of these elements. Zinc, cadmium, lead, mercury and sulphur can be purified by double or triple vacuum distillation. Tellurium and selenium need zone refining after vacuum distillation to achieve the required purity. Manganese can be purified by vacuum sublimation carried out at 1000°C under a high dynamic vacuum.

The majority of the A^{II}-B^{VI} and A^{IV}-B^{VI} binary compounds can be prepared by melting suitable amounts of the constituent elements in evacuated quartz ampoules. To avoid contamination from the quartz tubing at high temperature, it is sometimes advisable to coat the inner-walls of the ampoules with graphite. This not only reduces the contamination but also results in a better quality material. In order to prepare the polycrystalline binary compounds, the constituent elements are mixed in stoichiometric proportion and sealed in a quartz ampoule under a vacuum of $\sim 10^{-6}$

Torr. The sealed ampoule is then placed in a furnace and heated. It is important to note that during the initial heating, one section of the ampoule must be kept at a temperature below the lower of the boiling points of the two constituents, whilst other sections of the ampoule will generally need to be at a higher temperature to cause the reaction to occur. After few hours of heating, when the reaction is complete, the temperature of the whole ampoule can be raise to the melting point of the compound.

CdTe is prepared by heating a quartz ampoule containing Cd and Te elements in stoichiometric proportion at 700°C for a few hours and then at 1100°C, the melting point of CdTe being 1098°C. To prepare CdSe and CdS, the low temperature region in the ampoule should be maintained below 685°C for CdSe and 440°C for CdS, as the boiling temperatures of Cd, Se and S are 767, 685, and 440°C respectively, with the high temperature region being maintained at about 700°C for CdSe and 1000°C for CdS. In these cases, the reaction occurs at temperatures which are lower than the melting points of the alloys. Mercury- and zinc-chalcogenides can be prepared in a similar fashion. The required lower and higher temperatures of the ampoule can be estimated in term of the boiling temperatures of constituent elements and the melting points of the binary compounds. In order to avoid an explosion, the heating must be slow. Binary sulphides can also be prepared by precipitation from aqueous solutions of their salts, by passing H_2S gas through the solution.

Lead sulphide can also be prepared by heating an evacuated ampoule containing Pb and S. Special precautions are required in the preparation of PbS because there is a high risk that the ampoules will explode at about 800-1000°C. The danger of explosions occuring shifts to higher temperatures if slow heating is employed. The explosions associated with rapid heating are usually attributable to the high vapour pressure of S and SO_2. Special precautions should be taken to avoid ampoule explosion during the preparation of all sulphides. For example, to avoid explosions, lead sulphide is prepared by heating the ampoule in steps, increasing the ampoule temperature from room temperature to 700°C gradually over 9-10 hours, then keeping the ampoule at this temperature for couple of hours before again gradually increasing the temperature in further slow steps. Generally, lead captures oxygen and forms a thin film of lead oxide on the top portion of the material. This oxide film can be removed by electrolytic etching. Other A^{IV}-B^{VI} binary compounds can be synthesized by melting the constituent elements in evacuated sealed ampoules. MnSe and MnS can be prepared the same way using 700°C the lower temperature for MnSe and 450°C for MnS whilst using 1000°C for the higher temperature in both cases. These materials take a few days to synthesize, whereas MnTe takes only a few minutes in preparation. MnTe can be prepared by heating a carbonised quartz ampoule containing Mn and Te at about 1000°C for approximately 5 minutes.

As mentioned earlier, the Bridgman technique is one of the best suited for growing bulk crystals of DMS alloys and indeed most DMS crystals have been grown using this method. The Bridgman technique is a melt technique in which material is heated to a temperature above its melting point and then the melt is cooled in a carefully controlled manner. Bridgman crucibles are available commercially in various sizes and different materials. The Bridgman crucible containing the melt materials moves downwards with a constant, usually extremely slow, speed through a furnace from a high to lower temperature zone. During this process, the molten material starts solidify from the lower end of the crucible and the growth of a single grain proceeds behind the liquid/solid interface as this moves through the melt, resulting in the formation of a single bulk crystal. The furnace temperature and lowering speed are the two main parameters which must be varied from material to material to obtain the best quality bulk crystal.

To prepare bulk crystals[110-112] of $Hg_{1-x}Mn_xTe$ by the Bridgman technique, the starting HgTe and MnTe binary compounds are mixed in appropriate proportion and placed in a crucible. The growth speed is recommended to be as low as 1 mm/hr and the furnace temperature to be used is nearly 850°C. Higher growth rates give a higher degree of polycrystallinity. Like $Hg_{1-x}Cd_xTe$, $Hg_{1-x}Mn_xTe$ shows a slight Hg-deficiency in the as-grown crystals[110]. However, stoichiometry can be achieved by annealing the as-grown crystals at a temperature between 180-220°C under mercury vapour for a very long period of about 8 days. By using this technique, $Hg_{1-x}Mn_xTe$ can be grow with x up to 0.75 whilst retaining the zinc blende structure of HgTe. Recently, a solid state recrystallisation technique has been successfully employed to get good quality crystals of $Hg_{1-x}Mn_xTe$[113]. Bulk single crystals of $Hg_{1-x}Mn_xSe$ and $Hg_{1-x}Mn_xS$ can also be prepared by the same technique with the same growth speed. The working temperature of the furnace should be maintained at about 600°C for $Hg_{1-x}Mn_xS$[114] and 900°C for $Hg_{1-x}Mn_xSe$[8].

As mentioned above in section 3.1.1, $Zn_{1-x}Mn_xTe$ forms in single phase solid solutions over the widest range of manganese concentration ($x \leq 0.86$)[62]. Appropriate quantities of ZnTe and MnTe are mixed and sealed in evacuated thick walled silica ampoules. The temperature of the Bridgman furnace should be maintain at about 1000°C. For low Mn concentrations, good quality large crystals of $Zn_{1-x}Mn_xTe$ can be grown, but with high contents of Mn the quality of the crystals is not good. As shown in Table 6, $Zn_{1-x}Mn_xSe$ shows two different structures for different values of x, namely the zinc blende structure for x in the range $0 \leq x \leq 0.30$ and the wurtzite structure for $0.30 < x \leq 0.57$. ZnSe and MnSe are first mixed in appropriate proportions, pressed into pellets and heated at about 1000°C for several days. The resulting material is then used as starting material for the growth. The melting point of these materials is expected to be little higher than 1500°C. Due to the high melting point of this system, it is difficult to grow crystals of $Zn_{1-x}Mn_xSe$[56] by the Bridgman technique because quartz starts softening at this temperature. This problem can be circumvented by the use of external counter pressures. Using this technique, crystals of $Zn_{1-x}Mn_xSe$ have been grown only for x up to 0.10, with argon at about 100 atmospherus being used as a pressurising gas.

RF induction heating with a self-sealing graphite crucible is another technique which is better suited to growing the crystals of $Zn_{1-x}Mn_xSe$[115]. The starting materials are loaded into a graphite crucible and the inverted graphite crucible is placed at the centre of an induction coil. After evacuating the system, the temperature is slowly increased beyond the melting point, with the centre of the crucible being heated to about 100°C above the material melting point. The graphite crucible is then lowered out of the stationary induction coil heating system at a rate of about 10mm/hr. Good quality single crystals of $Zn_{1-x}Mn_xSe$ have been grown over the complete possible range of x using this technique. $Zn_{1-x}Mn_xS$ also crystallizes in two different structures; the zinc blende structure for x up to 0.10 and wurtzite structure for x in the range 0.11 to 0.45. Because of the high melting points of ZnS and MnS, the Bridgman technique is again not suitable for this system. The chemical transport technique is the most suitable technique for growing single crystals of $Zn_{1-x}Mn_xS$. Firstly, ZnS and MnS are mixed in the desired proportion, pressed into pellets, and sintered at about 1000°C for several days. The sintered material is then placed in quartz tubes (12 to 15 cm long with an inner diameter of about 15mm) together with iodine ($10mg/cm^3$) as the transport agent. The tubes are sealed under vacuum and inserted into a furnace where the temperature is 1000°C at one end and 980°C at the other end. It takes about one week to produce good quality single crystals of $Zn_{1-x}Mn_xS$.

The Bridgman technique is the best suited technique to grow single crystals of $Cd_{1-x}Mn_xTe$. CdTe and MnTe are mixed in the desired proportion and sealed in a quartz ampoule under vacuum. A Bridgman furnace with a temperature gradient of 50°C/cm is used. The growth process starts with a temperature of 1120°C and the ampoule is lowered with a speed of 3mm/hr. It is possible to grow crystals of $Cd_{1-x}Mn_xTe$ with x up to 0.77 by this technique[27] and their macroscopic quality and homogeneity are extremely good. All of these crystals show the zinc blende structure. For $x \geq 0.77$, several phases are observed which shows nonstoichiometry in the alloy. Crystals of $Cd_{1-x}Mn_xSe$ can also be produced by the same procedure, except that a starting temperature of 1270°C must be used[8].

The optimum procedure for growing $Cd_{1-x}Mn_xS$ alloys has not yet been established. However, the following different procedures have been employed[8]. To prepare the sintered starting materials MnS and CdS powders were mixed in the desired molar ratio and pressed into pellets. These pellets were then sealed in a quartz tube under high vacuum and sequentially annealed at each of the following temperatures; 1000, 800, 700 and 600°C for approximately 100 hours in each case. The samples were then quenched in air after having been taken out of the furnace. For the Bridgman technique, because of the high melting points of CdS and MnS, carbon crucibles under high inert gas pressure were used in place of quartz to contain the melt. After annealing the pellets again at 1000°C in a nitrogen/H_2S atmosphere, the pellets were then powdered and placed for one hour in a furnace filled with argon gas at about 60kg/cm^2 pressure and at a temperature maintained slightly above the melting point of the alloy; 1450°C for the Cd-rich alloys and 1620°C for the Mn rich alloys[8]. Single crystals of $Cd_{1-x}Mn_xS$ have also been grown by using chemical transport and induction heating methods[8]. To grow crystals by the chemical transport technique, the binary compounds, CdS and MnS, were mixed in the desired ratio and placed in a furnace for one week at 950°C together with iodine (10mg/cm^3). In the induction heating technique, self sealing graphite crucibles were used similar to those used for $Zn_{1-x}Mn_xSe$. Crystals of 1cm^3 in size were obtained by this technique.

Crystals of PbMnTe alloys can be grown by the Bridgman method under tellurium vapor pressure in a three zone furnace similar to that used for the growth of $Cd_{1-x}Mn_xTe$ alloys[99]. The appropriate amount of PbTe, Mn and Te, or Pb, Mn and Te are mixed and used as starting materials for crystal growth. The growth rate used was 2.8 mm/hr. The temperature profile during crystallisation was of the order of 40°C/cm. The Bridgman technique can also be used to grow $Pb_{1-x}Mn_xS$ single crystals using PbS and MnS as starting binaries. The PbS and MnS were weighed in the appropriate proportions and sealed in quartz ampoules under high vacuum. The hot zone of the furnace was at 1180°C and the temperature gradient was about 20°C/cm[95]. Electron microprobe analysis of the these crystals showed good agreement between the crystal composition and starting material proportion.

5. SOME RECENT DEVELOPMENTS IN DMS

5.1 The $A_{1-x}^{III}Mn_xB^V$ DMS

The A^{III}-B^V compounds which act as host binaries are those formed from the group III elements gallium, indium and aluminium, and the group V elements arsenic, phosphorous and antimony. All of these A^{III}-B^V compound semiconductors crystallise in the zinc blende structure[116-117]. Doping these binaries with a range of transition-metal impurities is not a new phenomenon in III-V semiconductor technology. Indeed during the past few years, numerous review papers have been devoted exclusively to the study of deep impurities or defects in transition metal doped III-V semiconductors[118-124]. The solubility limits of the transition-metal impurities in the III-V semiconductors is rather low, generally being around 10^{17} cm^{-3}, beyond which surface segregation, and in extreme cases, phase segregation occurs[125]. Due to this low doping level limitation, these materials have been used mainly as p-type or current blocking layers. Preparation of III-V DMS is an extremely difficult task, which is the reason that not much work has yet been done on these materials.

Recently, however, $A_{1-x}^{III}Mn_xB^V$ DMS have been successfully prepared by molecular beam epitaxy (MBE).

Epitaxial films of $In_{1-x}Mn_xAs$ and $Ga_{1-x}Mn_xAs$ have been prepared by MBE[33,126]. Manganese compositions varing from $x = 0$ to 0.18 have been achieved in $In_{1-x}Mn_xAs$ epitaxial films and it has been observed that the homogeneity and magnetic properties of these films are sensitive to the growth conditions. Films deposited at low temperature (200°C) show homogeneous incorporation of Mn which results in the formation of a homogeneous alloy of InMnAs. This material shows a paramagnetic behaviour which obeys a Currie-Weiss law relationship. The average lattice constant decreases with increasing Mn concentration in the films whilst incorporation of Mn into InAs reduces the band gap from that of the pure InAs. Films grown at about 300°C exhibit ferromagnetic behaviour similar to that of MnAs which may exist in the films in the form of clusters. Films deposited at higher substrate temperature, ~400°C, were observed to be porous and did not adhered to the substrates. Hall measurements up to 6kOe show a linear dependence of the Hall voltage on the magnetic field and n-type conduction, the same as is observed in undoped InAs films. The carrier concentration and the mobility of the $x = 0.18$ material are reported to be $1.0 \times 10^{16} cm^{-3}$ and 540 $cm^2/Vsec$ respectively at 77K. Growth of epitaxial films of GaMnAs has also been achieved by MBE at a growth temperature of 400°C, but growth at a substrate temperature of 580°C resulted in an accumulation of Mn on the surface. GaMnAs films deposited at 400°C exhibit ferromagnetic behaviour. Research on III-V DMS has only been started very recently and as yet very little literature is available on these DMS.

5.2 The $A_{1-x}Eu_xB$ DMS

The $A_{1-x}Eu_xB$ are DMS in which a fraction x of the cations in the AB compound semiconductors has been randomly substituted[127-130] by Eu . Doping of $A^{II}B^{VI}$ compounds with rare-earths such as Eu leads to the appearance of several interesting properties in these semiconducting materials, some of which are particularly well suited for device applications. For example, ZnEuS is a suitable material for use in multicolour electroluminescent devices. HgEuSe and HgEuTe can be prepared by the Bridgman technique, whereas ZnEuS is prepared by using an iodine transport method[127]. For the samples studied, the concentration of the Eu in the HgTe and ZnS crystals was 0.46 and 0.57 wt% respectively. It was observed[127] that the temperature dependence of the magnetic susceptibility obeyed the Curie-Weiss law at $T \geq 80K$ for $Hg_{1-x}Eu_xTe$ and at $T \geq 10K$ for $Hg_{1-x}Eu_xSe$.

MBE growth of $Pb_{1-x}Eu_xTe$ has demonstrated the usefulness of this narrow gap materials for infra-red opto-electronic applications, specifically in the fabrication of $PbTe/Pb_{1-x}Eu_xTe$ heterojunction diode lasers[133-134]. $Pb_{1-x}Eu_xTe$ films, prepared by the co-evaporation of PbTe, Eu and Te in a vacuum of 1×10^{-6} Torr, were single-phase and x-ray reflection diffraction patterns could be indexed as NaCl type structures[136]. The band gap increased with an increase in the x value and the rate of increase of the band gap with x, i.e. the value for (dE_g/dx), is reported to be 3.1 eV/x for $0 \leq x < 0.1$ and 1.4 eV/x for $0.1 < x \leq 0.3$. $Pb_{1-x}Eu_xTe$ single crystal films have also be prepared by a modified hot wall technique with up to 5% Eu content[128]. The preparation and physical properties of these materials are discussed in detail elsewhere in this book (see Chapter 16).

5.3 DMS Quantum Wells and Superlattices

One of the most important developments in DMS is the successful preparation of DMS superlattices and multiple quantum well structures[126,135-140]. As mentioned earlier in section 3, the lattice constant and energy gap can be tuned in DMS by changing the concentration of the incorporated transition metal ions. The variation of energy gap and lattice parameter with composition make these materials ideally suited for band gap engineering and for lattice matching possibilities during growth. Very recently photo-assisted molecular beam epitaxy (PAMBE)[141] and atomic layer epitaxy (ALE)[7,142] have also been used to grow DMS epitaxial films. In PAMBE, the procedure is similar to normal MBE except that the substrate is illuminated during the entire film growth process. The added energy appears to increase the atomic mobility and leads to improved epitaxial growth. ALE, which is an equilibrium growth method, is used to grow thin films via a stepwise monoatomic layer by layer growth process. The number of layers deposited is determined solely by the number of operational cycles performed. Since all deposition is halted for a short period between each successive deposition pulse, a thermodynamic equilibrium is approached at the end of each reaction step. The successful preparation of epitaxial layers of DMS indicates a bright future for these materials, especially for opto-electronic and magneto-optic applications.

The strong exchange interaction which occurs in bulk DMS is also found to take place in the quasi-two-dimensional quantum well situation. The first superlattice and multiple quantum well structures were reported for II-VI DMS with the wells consisting of wide-gap non-DMS materials sandwiched between wide-gap DMS barriers, for example, $Cd_{1-x}Mn_xTe/CdTe/Cd_{1-x}Mn_xTe$. Usually (100) or (111) GaAs has been used as a substrate on which to grow these structures but recently other substrates have also been used for better lattice matching. Another achievement has been the successful growth of the binaries MnSe and MnTe in the zinc blende structure, which will help in understanding the structural behaviour in DMS crystals and will also help in the preparation of ZnTe/MnTe type structures[143]. Preparation and characterisation of superlattices and quantum wells are discussed in detail in Chapters 11 and 12.

Although the major activity has been concentrated on the $A^{II}MB^{VI}$, $A^{IV}MB^{VI}$ and $A^{III}MB^{V}$ type DMS and their superlattices and heterostructures, the research is now being extended to new types of DMS, such as quaternary DMS alloys and $(A_{1-x}^{II}M_x)_3As_2$[148]. In quaternary DMS alloys, it is possible to change the variables, such as energy gap and magnetic susceptibility, independently. For example, in $Hg_{1-x-y}Cd_xMn_yTe$, x and y together determine the energy gap whilst only the value of y determines the magnetic properties. The quaternary DMS alloys provide an opportunity where the lattice parameter and energy gap can be varied without disturbing the magnetic properties or vice versa. It has been reported that $(A_{1-x}^{II}M_x)_3As_2$ type alloys also belong to the family of DMS. In the absence of a magnetic field, $(Cd_{1-x}^{II}Mn_x)_3As_2$ resembles its parent Cd_3As_2 but substitution of Cd by Mn in Cd_3As_2 produces a spin-spin exchange interaction between the band electrons and localized magnetic moments. Incorporation of Mn into the chalcoperite lattice[149] such as $CuInSe_2$ is another interesting field of DMS.

REFERENCES

1. R. R. Galazka and J. Kossut in Narrow Gap Semiconductors: Physics and Applications, Lecture Notes in Physics, No. **133** (Springler-Verlag, Berlin), p-245.
2. R. R. Galazka, Proc. 14th Int. Conf. on the Physics of Semicon., Edinburgh, 1978, ed. by B. L. H. Wilson, (Institute of Physics, London, 1978), p-133.
3. J. K. Furdyna, J. Appl. Phys., **53**, 7637 (1982) .
4. N. B. Brandt and V. V. Moshchalkov, Advances in Physics, **33**, 193 (1984).
5. R. R. Galazka and J. Kossut in landolt-Bornstein Series, III/17B.
6. J. K. Furdyna, J. Vac. Sci. Tech., A4, 2002(1986).
7. Diluted Magnetic (Semimagnetic) Semiconductors edited by R. L. Aggarwal, J. K. Furdyna and S. von Molnar, Vol. 89, (Mat. Research Society Symposia Proceedings, Pittsburge, 1987).

40

8. Semiconductors and Semimetals, Treatise Editors, R. K. Willardson and A. C. Beer, Vol. Eds., J. K. Furdyna and J. Kossut, Vol. 25, (Academic Press, Boston, 1988).

9. J. K. Furdyna, J. Appl. Phys., **64**, R29 (1988).

10. N. Samarth and J. K. Furdyna, Proc. of the IEEE, **78**, 990 (1990).

11. A. V. Kamarov, S. M. Pjabchenko, O. V. Terletskii, I. I. Zheru and R. D. Ivanchuck, Zh. eksp. teor. Fiz, **73**, 608 (1977).

12. N. Mikami, C. Nagao, T. Sawada and H. Takahashi, J. Appl. Phys., **69**, 433 (1991).

13. Y. H. Lee, I. Chung and M. Oh, App. Phys. Lett., **58**, 962 (1991).

14. R. Mach and G. O. Muller, J. Cryst. Growth, **86**, 866 (1988).

15. Electroluminescence, ed. by S. Strionoya and H. Kobayashi, Proc. in Physics, Vol. 38, (Springler-Verlag, Berlin, 1989).

16. P. Becla, J. Vac. Sci. Tech., A**6**, 2725 (1988).

17. A. Wall, C. Caprite, A. Franciosi, M. Vaziri, R. Reifenberger and J. K. Furdyna, J. Vac. Sci. Tech., A**4**, 2010 (1986).

18. A. Rohatgi, S. A. Ringel, J. Welia, K. Pollard, A. Erbit, P. V. Meyers and C. H. Liu, Solar Cells, **24**, 185 (1988).

19. D. L. Dreifus, R. M. Kolbas, R. L. Harper, J. R. Tassitirio, J. Hwang and J. F. Schetzina, Appl. Phys. Lett., **53**, 1279 (1988) .

20. A. wall, C. Caprite, A. Franciosi, R. Reifenberger and U. Debska, J. Vac. Sci. Tech., A**4**, 818 (1986).

21. R. Triboulet, A. Heurtel and J. Rioiex, Private communication.

22. S. Wong and P. Becla, J. Vac. Sci. Tech., A**4** 2019 (1984) .

23. S. I. Gubarev, N. I. Vitrichovskii, A. V. Komarov and V. B. Timofeev, Pls'ma Za. eksp. teor. Fis, **33**, 202 (1981).

24. P. Byszewski, K. Szienk, J. Kossut and R. R. Galazka, Phys. Stat. Sol(b). **95**, 359 (1979).

25. G. N. Pain, Chapter 14 of this book.

26. A. K. Ramdas, J. Appl. Phys. **53**, 7649 (1982).

27. R. Triboulet and G. Didier, J. Cryst. Growth, **52**, 614 (1981).

28. J. J. Hauser, Phys. Rev. B, **27**, 3460 (1983).

29. M. Vaziri, U. Debska and R. Reifenberger, Appl. Phys. Lett., **47**, 407 (1985).

30. A. Hoffman, R. Heitz and I. Broser, Phys. Rev. B, **41**, 5806 (1990).

31. G. Bauer, Mat. Res. Soc. Proc., **89**, 107 (1988).

32. S. Z. Korczak, W. Korczak, M. Subotowicz and H. Wasiewicz, Phys. Stat. Sol(b), **153**, 361 (1989).

33. H. Munekata, H. Ohno, S. von Molnar, A. Segmuller, L. L. Chang and L. Esaki, Phys. Rev. Lett., **63**,1849 (1989).
34. K. Huang and B. W. Wessels, J. Appl. Phys. **67**, 6882 (1990).
35. A. Lewicki, A. I. Schindler, I. Miotkowski and J. K. Furdyna, Phys. Rev. B, **41**, 4653 (1990).
36. P. M. Shand, A. Lewicki, B. C. Crooker, W. Giriat and J. K. Furdyna, J. Appl. Phys., **67**, 5246 (1990).
37. G. Jasiolek, Z. Golacki and M. Godlewski, J. Phys. Chem. Solids, **50**, 277 (1989).
38. K. R. Krylov, A. I. Ponomarev, I. M. Tsidil'Kovskii, N. P. Gavaleshko and V. V. Khomyak, Sov. Phys. Semicon. **23**, 268 (1989).
39. O. Geode and W. Heimbrodt, Phys. Stat. Sol.(b) **146**, 11 (1988).
40. G. N. Pain, S. P. Russo, R. G. Elliman, L. S. Wielunski, D. Gao, S. R. Glanvill, C. J. Rossouw, A. W. Stevenson, R. S. Rowe, G. B. Deacon, R. S. Dicson and B. O. West, Materials Forum, **15**, 35 (1991).
41. B. Lunn and J. J. Davies, Semicon. Sci. Tech., **5**, 1155 (1990).
42. R. L. Harper, S. Hwang, N. C. Giles, R. N. Bicknell, J. F. Schetzina, Y. R. Lee and A. K. Ramdas, J. Vac. Sci. Tech., A**6**, 2627 (1988).
43. Semiconducting II-VI, IV-VI and V-VI Compounds, Ed. by N. Kh. Abrikosov, V. F. Bankina, L. V. Poretskaya, L. E. Sheliniova and E. V. Shudnova (Plenum Press, New York, 1969).
44. Physics and Chemistry of II-VI Compounds, Ed. by M. Aven and J. S. Prener (North-Holland, Amsterdam, 1967).
45. M. Hansen and K. Anderko, Constitution of Binary Alloys (McGraw Hill, New York, 1968).
46. R. P. Elliot, Constitution of Binary Alloys (McGraw Hill, New York, 1965).
47. J. L. Birinan, Phys. Rev., **109**, 810 (1958).
48. B. F. Ormont, Structures of Inorg. Substances, (Gestekhteoizdat, Moscow, 1950).
49. P. H. Hamilton, Semicond. Prod. and Solid State Technology, **7**, 15 (1964).
50. A. Okazaki, J. Phys. Soc. Japan, **13**, 1151 (1958).
51. A. Okazaki, J. Phys. Soc. Japan, **11**, 470 (1956).
52. L. Pauling, The Nature of the Chemical Bond and the Structure of Molecules and Crystals (Cornell Univ. Press, New York, 1960).
53. H. Krebs, Actacryst, **9**, 95 (1956).
54. T. K. Nguyen and J. A. Gaj, Phys. Stat. Sol.(b), **83**, K133 (1977).

55. M. P. Vechhi, W. Giriat and L. Videla, Appl. Phys. Lett., **38**, 9 (1981).

56. A. Twardowski, T. Dietl and M. Demianiuk, Sol. State Comm., **48**, 845 (1983).

57. Y. R. Lee, A. K. Ramdas and R. L. Aggarwal, Phys. Rev. B, **33**, 7383 (1986).

58. G. Bastard, C. Rigaux, Y. Guldner, A. Mycielski, J. K. Furdyna and D. P. Mallin, Phys. Rev. B, **24**, 1961 (1981).

59. A. B. Davydov, B. B. Ponikarov and I, M, Tsidilhorskii, Phys. Stat. Sol(b)., **101**, 127 (1981).

60. R. T. Delves and B. Lewis, J. Phys. Chem. Solids, **24**, 549 (1963).

61. A. Pajaczkowska and A. Rabenau, Mat. Res. Bull., **12**, 2 (1977).

62. A. Pajaczkowska, Prog. Cryst. Growth Char., **1**, 289 (1978).

63. W. C. Cook Jr, J. Am. Ceram. Soc., **51**, 518 (1968) .

64. H. Wiedemeier and A. G. Sigai, J. Chem Thermodyn., **6**, 981 (1970).

65. A. H. Sadeek, Ph. D. Thesis, Rensselaer Pol. Inst., Troy, New York, 1971.

66. R. Juza, A. Rabenace and G. Pascher, Z. Anorg. Allgem. Chem., **285**, 61 (1956).

67. D. R. Yoder-Short, U. Debska and J. K. Furdyna, J. Appl. Phys., **58**, 4056 (1985).

68. A. Sarem, B. J. Kowalski and B. A. Orlowski, J. Phys. Cond. Matt., **2**, 8173 (1990).

69. A. Mycielski, J. Appl. Phys., **63**, 3279 (1988).

70. A. Lewicki, J. Spalek and A. Mycielski, J. Phys. C, **20**, 2005 (1987).

71. A. Twardowski, J. Appl. Phys., **67**, 5108 (1990).

72. W. J. M de Jonge, H. J. M. Swagten, G. E. P. Gerrits and A. Twardowski, Semicon. Sci. Tech., **5**, S270 (1990).

73. A. Lewicki, A. Mycielski and J. Spalek, Acta Phys. Polon, A**69**, 1043 (1986) .

74. Y. Guldner, C. Rigaux, M. Menant, D. P. Mullin and J. K. Furdyna, Solid State Comm., **33**, 133 (1980).

75. N. V. Joshi and L. Mogollon, Prog. Crys. Growth and Charat., **10**, 65 (1985).

76. A. Twardowski, A. M. Hennel, M. von Ortenberg and M. Demianiuk, Proc. 17th Int. Conf. Phys. of Semicond. (Springler, New York, 1984).

77. A. Lewicki, Z. Tarnawski and A. Mycielski, Acta. Phys. Pol. A**67**, 357 (1985).
78. B. T. Jonker, S. B. Qadri, J. J. Krebs, G. A. Prinz & L. Salamonea-Young, J. Vac. Sci. Tech., A7, 1360 (1989).
79. T. M. Giebultowicz, P. Klosowski, J. J. Phyne, J. J. Udovic, J. K. Furdyna and W. Giriat, Phys. Rev. B, **41**, 504 (1990).
80. A. Lewicki, A. I. Schindler, J. K. Furdyna and W. Giriat, Phys. Rev. B, **40**, 2379 (1989).
81. A. Lewicki, A. I. Schindler, I. Miotkowski, B. C. Crooks and J. K. Furdyna, Phys. Rev. B, **43**, 5713 (1991).
82. S. Groves and W. Paul, Proc. Int. Conf. Semi. Phys., Paris, 1964, p41.
83. M. Jaczynski, J. Kossut, R. R. Galazka, Phys. Stat. Sol(b), **88**, 3 (1978).
84. M. Dobrowolska, W. Dobrowolski, M. Otto, T. Dietl and R. R. Galazka, J. Phys. Soc. Jpn. Suppl., **49**, 815 (1980).
85. N. Bottka, J. Stankiewicz and W. Giriat, J. Appl. Phys., **52**, 4189 (1982).
86. Y. R. Lee and A. K. Ramdas, Sol. Stat. Comm., **51**, 861 (1984).
87. T. Kendelewicz, Sol. Stat. Comm., **36**, 127 (1980) .
88. J. Antoszewski and E. Kierzek-Pecold, Sol. Stat. Comm., **34**, 733 (1980).
89. M. Ikeda, K. Itoh and H. Sato, Jpn. J. Phys. Soc.., **25**, 455 (1968).
90. J. K. Furdyna, J. Vac. Sci. Tech., **21**, 220 (1982).
91. J. E. M. Toro, W. M. Becker, B. I. Wang, U. Debska and J. W. Richardson, Sol. Stat. Comm., **52**, 41 (1984).
92. M. Ke-Jun and W. Giriat, Sol. Stat. Comm., **60**, 927 (1986).
93. A. Mycielski, P. Dzwonkowski, B. J. Kowalski, B. A. Oslowski, M. Dobrowolska, M. Arciszewska, W. Dobrowolski and J. M. Branowski, J. Phys. C., Sol. Stat. Phys., **19**, 3605 (1986).
94. A. Twardowski, Chapter 8 of this book.
95. G. Karczewski, L. Kowalczyk and A. Szcerbakow, Sol. Stat. Comm., **38**, 499 (1981).
96. J. Niewodniczanska-Zawadzka and A. Szczerbakow, Sol. Stat. Comm., **34**, 887 (1980).
97. G. Karczewski and L. Kowalczyk, Sol. Stat. Comm., **48**, 653 (1983).
98. V. G. Vanyarkho, V. P. Zlomanov and A. V. Novoselova, Izv. Akad. Nauk., SSSR Neorg. Mater., **6**, 534 (1970).

99. M. Escorne, A. Mauger, J. L. Tholence and R. Triboulet, Phys. Rev. B, **29**, 6306 (1984).

100. M. N. Vinogradova, N. V. Kolomoets and L. M. Sysoeva, Sov. Phys. Semicon., **5**, 186 (1971).

101. A. Lopez-Otero, Thin Solid Films, **49**, 3, (1978).

102. B. V. Izvozchikov and I. A. Taksami, Fiz. Tekh. Poluprov., **1**, 565 (1967).

103. U. Sondermann, J. Magn. Magn. Mater., **2**, 216 (1976).

104. M. P. Mathur, D. W. Deis, C. K. Jones, A. Patterson, W. J. Carr Jr and R. C. Miller, J. Appl. Phys., **41**, 1005 (1970).

105. W. D. Johnston and D. E. Sestrich, J. Inorg. Nucl. Chem., **19**, 229 (1961).

106. W. W. Scanlon, J. Phys. Chem. Solids, **8**, 423 (1959).

107. I. A. Drabkin, G. F. Zabhayugina and I. U. Nel'son, Sov. Phys. Semi. **5**, 277 (1971).

108. J. Tholence, A. Mauger, M. Escorne and R. Triboulet, J. Appl. Phys., **55**, 2313 (1984).

109. I. V. Nel'son, I. A. Drabkin, Y. Y. Eliseera & M. N. Vigogradova, Sov. Phys. Semi., **8**, 568 (1974).

110. R. T. Holm and J. K. Furdyna, Phys. Rev. B, **15**, 844 (1977).

111. J. M. Wrobel, L. C. Bassett, J. L. Aubel, S. Sundaram and P. Becla, J. Appl. Phys., **60**, 1135 (1986).

112. S. Nagata, R. R. Galazka, D. P. Mullin, H. Akbarzaden, G. D. Khattak, J. K. Furdyna and P. H. Keesom, Phys. Rev. B, **22**, 3331 (1980).

113. R. G. Mani, T. McNair, C. R. Lu and R. Grober, J. Cryst. Growth, **97**, 617 (1989).

114. A. Pajaczkowska and A. Rabenau, Solid State Chemistry, **21**, 1 (1977).

115. U. Debska, W. Giriat, H. R. Harrison and D. R. Yoder-Short, J. Cryst. growth, **70**, 399 (1984).

116. Crystals, III-V Semiconductors, edited by K. -W. Benz, E. Bauser, K. G. Plessen, A. J. Marshall, J. Hesse and K. Ploog (Springler-Verlag, New York, 1980) p-1.

118. V. F. Masterov, V. P. Maslov and G. N. Talalakin, Sov. Phys. Semi., **11**, 837 (1977).

119. D. G. Andrianov, Yu. A. Grigor'ev, S. O. Klimonskii, A. S. Savel'ev and S. M. Yakubenya, Sov. Phys. Semicon., **18**, 162 (1984).

120. E. S. Demidov, A. -A. Eznevskii and V. V. Karzanov, Sov. Phys. Semicon., **17**, 412 (1983).

121. B. Lambert, R. Coquille, M. Gauneau, G. Grandpierre and G. Moisan, Semicon. Sci. Tech., **5**, 616 (1990).

122. B. Clerjaud, J. Phys. C, **18**, 3615 (1985).

123. G. F. Neumark and K. Kosai, Semiconductors and Semimetals, Vol. 19, ed. by R. K. Willardson & A. C. Beer (Academic Press, New York, 1983) p-1.

124. U. Kaufmann and J. Schneider, Adv. Electr. Elect. Phys., **58**, 141 (1981).

125. Y. Umehara and S. Koda, Metallography, **7**, 313 (1974).

126. H. Murekata, H. Ohno, S. vonMolnar, A. Harvit, A. Segmuller and L. L. Chang, J. Vac. Sci. Tech., B8, 176 (1990).

127. K. R. Krylov, A. I. Ponomarev, I. M. Tsidil'kovskil, N. P. Gavaleshko and V. V. Khomyak, Sov. Phys. Semicon., **23**, 268 (1989).

128. A. Krost, B. Harbecke, R. Faymanville, H. Schlgel, E. J. Fantner, K. E. Ambrasch and G. Bauer, J. Phys. C. **18**, 2119 (1985).

129. Yu. S. Gromovol, S. D. Darchuk, V. N. Konovalov, V. M. Lakeenkov, S. V. Plyatsko ans F. F. Sizov, Sov. Phys. Semicon., **23**, 639 (1989).

130. R. Suryanarayanana and S. K. Das, J. Appl. Phys., **67**, 1612 (1990).

131. H. Ohnishi, Y. Yamakoto and Y. Katayama, Int. Display Research Conference, San Diego (IEEE Ney Yoyk, 1985) p-159.

132. H. Kobayashi, S. Tanaka, V. Shanker, M. Shiiki, T. Kunou, J. Mita and H. Sasakura, Phys. Stat. Sol(a), **88**, 713 (1985).

133. D. L. Partin, J. Elect. Mater. **13**, 493 (1984).

134. D. L. Partin, Appl. Phys. Lett., **45**, 487 (1984).

135. R. L. Harper, S. Hwang, and N. C. Giles, R. N. Bicknell, J. F. Schetzina, Y. R. Lee and A. K. Ramdas, J. Vac. Sci. Tech. **6**, 2627 (1988).

136. G. Bauer and H. Clemens, Semicon. Sci. Technol. **5**, 3122 (1990).

137. R. L. Gunshor, N. Otsuka, M. Yamanishi, L. A. Kolodziejski, T. C. Bonsell, R. B. Bylsma, S. Datta, W. M. Becker and J. K. Furdyna, J. Cryt. Growth, **72**, 294 (1985).

138. A. Partovi, A. M. Glass, D. H. Olson, R. D. Feldman, R. F. Austin, D. Lee, A. M. Johnson and D. A. B. Miller, Appl. Phys. Lett., **58**, 334 (1991).

139. R. N. Bicknell, R. W. Yanka, N. C. Giles-Taylor, D. E. Blanks, E. L. Buckland and J. F. Schetzina, Appl. Phys. Lett., **45**, 92 (1984).

140. L. A. Kolodziejski, T. C. Bonsett, R. L. Gunshor, S. Datta, R. B. Bylsma, W. M. Bwcker, N. Otsuka, Appl. Phys. Lett., **45**, 440 (1984).

141. R. N. Bicknell, N. G. Giles, J. F. Schetzina, Appl. Phys. Lett., **50**, 691 (1982).

142. M. A. Herman, O. Jylha and M. Pessa, J. Crys. Growth, **66**, 480 (1984).

143. N. Pelekanos, Q. Fu, S. Durkin, M. Kobayashi, R. Gunshor, A. V. Nurmikko, 4th Int. Conf. on II-VI Compounds, Berlin, 1989.

144. S. F. Chehab and J. C. Woolley, J. Less. Common. Metal, **106**, 13 (1985).

145. T. Piotrowski, J. of Crystal Growth, **72**, 117 (1985).

146. S. Takeyama and S. Narita, Solid State Comm., **60**, 285 (1986).

147. J. C. Woolley, A. Monoogian, R. J. W. Hodgson and G. Lamorche, J. Mangsm and Mag. Mats., **78**, 164 (1989).

148. J. J. Neve, C. J. R. Bouwens and F. A. P. Blom, Solid State Comm., **38**, 27 (1981).

149. L. H. Greene, J. Orenstein, J. H. Wernick, G. W. Hull and E. Berry, Bull. Am. Phys. Soc., **31**, 383 (1986).

EXCHANGE INTERACTION EFFECTS ON OPTICAL PROPERTIES
OF WIDE-GAP SEMIMAGNETIC SEMICONDUCTORS AT Γ AND L POINT
OF THE BRILLOUIN ZONE

D. COQUILLAT

Groupe d'Etude des Semiconducteurs, Université des Sciences et
Techniques du Languedoc, Montpellier II, Place E. Bataillon,
F-34060 - MONTPELLIER-Cédex 1, FRANCE

Contents :

1. INTRODUCTION

In this paper we focus our attention on the effect of exchange
interaction between the spin of the s-like (p-like) conduction -
(valence -) band electrons and localized moment of the d-like Mn^{2+}
electrons in the wide-gap semimagnetic semiconductors (SMSC)
$A^{II}_{1-x}Mn_xB^{VI}$ alloys . This large sp-d exchange interaction affects the

energy bands of these materials resulting in interesting effects like, for example, extremly large Zeeman splitting of electronic levels and the resultant giant Faraday rotation (see the review articles : Gaj, 1980 ; Furdyna 1982, 1988 ; Brandt 1984 ; Galazka 1985 ; Goede 1988 ; and Vol. 25 of the Semiconductors and Semimetals series devoted entirely to SMSC). Even in absence of magnetic field, stricking alteration in the properties of SMSC, as departure from Varshni variation-like behavior of the energy gap occurs with alloying and temperature. It have origin in the sp-d exchange interaction.

Only a few investigations have been devoted to the effect of exchange interaction on the band structure of the SMSC systems ,far from the Brillouin zone center. At the L point, magneto-optical measurements clearly indicate that the magnitude of effect is much less spectacular than at the Γ point. On the other hand, the band structure of narrow gap $A^{II}_{1-x}Mn_xB^{VI}$ alloys, near L is similar to that of the wide-gap Mn-based $A^{II}B^{VI}$ materials having the same crystal structure. One can compare, in these conditions the magneto-optical effects at L point for the two classes of alloys (wide and zero or narrow-gap).

The general organization of the present paper reflects these three aspects, having their common origin in the sp-d exchange interaction. Section 2 presents examples of exchange interaction effects at Γ point, formulates exchange interaction sp-d in wide-gap SMSC , and describes experimental results of Zeeman splitting obtained on ternary alloys with Mn composition x up to 0.77 . Section 3 is devoted to the dependence of energy gap with concentration and temperature in absence of external magnetic field, and presents models built in terms of second order perturbation theory in sp-d interaction. Section 4 deals with optical results of transitions E_1 and $E_1+\Delta_1$ away from the Brillouin zone center for various SMSC. The effect of blurring with Mn concentration , and the energy dependence of E_1 transition with temperature and alloying in zero magnetic field are described. We then discuss Zeeman splitting of both E_1 and $E_1+\Delta_1$ transitions obtained by magneto-circular-cichroïsm technique on both $Cd_{1-x}Mn_xTe$ and $Zn_{1-x}Mn_xTe$ alloys and on the zero or narrow-gap $Hg_{1-x}Mn_xTe$.

2. EXCHANGE INTERACTION IN WIDE-GAP SMSC AT Γ POINT

2.1. Zeeman splitting in wide-gap Mn-based SMSC

Magneto-optical effects such as Zeeman splitting of valence and conduction bands in wide-gap SMSC, associated with measurements of magnetic properties (magnetization) enable us to study exchange interaction between carriers and Mn^{2+} ions.
The Faraday rotation resulting from a large difference in the index of refraction of the two circular polarization of light is in general very large of the order of 10 deg/cm Gs in $Cd_{1-x}Mn_xTe$ at 77K. (Komarov et al.,1977 ; Gaj et al.,1978a ; Kett et al.,1981 ; Kierzek-Pecold et al.,1984 ; Kullendorff et al.,1985) with a sign opposite to that in nonmagnetic II-VI compounds. The giant size of the effect is illustrated on Fig. 1, where each successive peak in the spectrum represents an additional Faraday rotation of 180° (Bartholomew et al., 1986).
Magnetoreflectivity measurements are commonly used to determine Zeeman splitting of the free exciton in wide-gap SMSC such as $Cd_{1-x}Mn_xTe$ (Komarov et al.,1977; Gaj et al.,1978b,1978c,1979; Rebmann et al.,1983; Lascaray et al.,1988) $Zn_{1-x}Mn_xTe$ (Komarov et al.,1978 ; Twardowski et al.,1983,1984 ; Barilero et al.,1984 ; Desjardins et al., 1986 ;

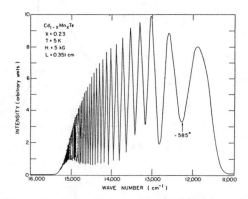

FIGURE 1 : Transmission spectrum through $Cd_{0.77}Mn_{0.23}Te$ in Faraday geometry at T = 5K and applied field H = 5T as function of wave number of initially linearly polarized light. Intensity of Faraday rotation is several order of magnitude larger than those in CdTe.[After Bartholomew et al.,1986].

50

Lascaray et al.,1987) $Zn_{1-x}Mn_xSe$ (Komarov et al.,1980b ; Twardowski et al.,1983). Magnetotransmission measurements on samples with a thickness below 1 μm allow obtained more detailed information on optical transitions (observation of ground state 1s and excited states, Twardowski et al.,1979). In wide-gap SMSC, the exciton ground state splits into six components, two visible in the σ^+ polarization, two in the σ^- polarization in Faraday geometry (one weak and one strong component for each σ polarization) and two in the π polarization, in Voigt geometry. Fig. 2 illustrates an example of magnetoreflectivity measurements of Zeeman splitting for $Cd_{1-x}Mn_xTe$ in the dilute regime and for a high concentrated alloy x = 0.73. For this latter, weak and strong component in each polarization are mixed.The size of the effect is about 100 times greater in SMSC and the selection rules are the same than in nonmagnetic semiconductor. Fig. 3 show the position of the six Zeeman components in $Cd_{0.9}Mn_{0.1}Te$ versus magnetic field. For compounds

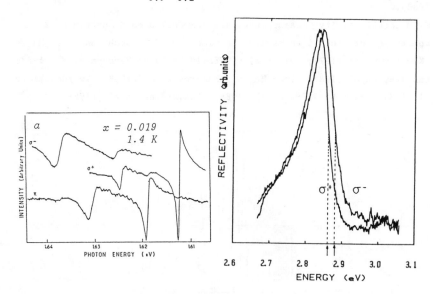

FIGURE 2 : a) reflectivity spectra of $Cd_{0.98}Mn_{0.02}Te$ at T = 1.4K and H = 1.1T in Faraday geometry (σ^+, σ^- configuration) and Voigt geometry (π configuration). [After Gaj et al.,1978b].
b) Reflectivity spectra of $Cd_{0.27}Mn_{0.73}Te$ at T =1.8K and H = 5.5 T in Faraday geometry. The middle point of the decreasing portion indicated by arrows in used for measuring ΔE. [After Lascaray et al.,1988].

with low Mn concentration (Fig.4), the splitting between strong
components ΔE increases with x and the saturation seams to be reach for
magnetic field weaker than 4T. In reality, exciton splitting, as we
shall see below, is proportional to magnetization of the crystal, and
details of magnetization such as high field magnetization steps are
observable. For higher concentration the values of splitting become
smaller and saturation is not reached.

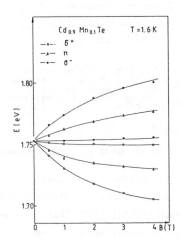

FIGURE 3 : Energy of Zeeman components of exciton versus
magnetic field in $Cd_{0.9}Mn_{0.1}Te$ in Faraday geometry.
Continuous lines are added for clarity. [After Twardowski
1981.See Gaj ,1988].

In wurtzite $A^{II}_{1-x}Mn_xB^{VI}$ compounds such as $Cd_{1-x}Mn_xSe$ (Arciszewska et
al.,1982,1986; Aggarwal et al.,1983; Komarov et al.,1980) and $Cd_{1-x}Mn_xS$
(Gubarev,1981; Abramishvili et al.,1984; Nawrocki et al.,1986; Gubarev
et al.,1986,1988) the results are more complicated than in cubic case.
In the "quasi-cubic" model, the top of the valence band, Γ_8 state in a
cubic case, is splitted by the hexagonal crystal field into a Γ_9 and Γ_7
state. However experimental results of Arciszewska et al., (1982).
show that the quasi-cubic approximation is inadequate in $Cd_{1-x}Mn_xSe$, an
anisotropy of the spin-orbit interaction, must be introduced.

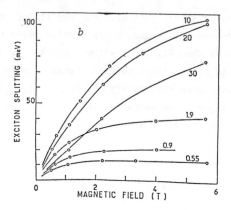

FIGURE 4 : Zeeman splitting between strong components versus magnetic field in $Cd_{1-x}Mn_xTe$, observed at 1.4K in magnetoreflectivity and Faraday geometry . Continuous lines are added to guide the eye. Composition values are given in mole % of Mn. [After Gaj et al.,1978b].

2.2 Exchange effects in new wide-gap SMSC compounds

In recent years the already broad family of SMSC based on Manganese was extended : SMSC systems containing transition-metal such as iron and cobalt have been crystallized. They possess the interesting magneto-optical properties of their Manganese-based counterparts, but in addition, have new physical properties . Indeed, exchange interaction between magnetic ion and bands electron is strongly influenced by the atomic ground state of the substitutional magnetic ion used.

The $^6S_{5/2}$ free-ion ground state of Mn^{2+} ($3d^5$) is not split by crystal field or spin-orbit, resulting in a simple $^6A_1$ orbital singlet ground state configuration. Unlike Mn^{2+}, magnetic situation is more complicated for iron, because Fe^{2+} ($3d^6$) ion possesses both spin and orbital momenta (S=2 and L=2). In cubic material like $Zn_{1-x}Fe_xSe$ or $Cd_{1-x}Fe_xTe$ the ground state of a free Fe^{2+} ion is split by tetrahedral crystal field into a ^5E orbital doublet and an upper ^5T orbital triplet. In addition spin-orbit interaction split ^5E in second order into five levels with a singlet A_1 ground state which has non permanent magnetic moment. This configuration leads to Van Vleck paramagnet behavior rather than the Brillouin paramagnetism which characterizes

the II-VI manganese-based SMSC. Results on magnetic and optical
properties of $Cd_{1-x}Fe_xSe$, (Petrou et al., 1987 ; Heiman et al., 1988)
$Zn_{1-x}Fe_xSe$ (Komarov et al.,1980 ; Twardowski et al.,1985,1987 ; Liu et
al.,1988) have been published in the last four years.
The $^4F_{9/2}$ free-ion ground level of Co^{2+} $(3d^7)$ is split by tetrahedral
crystal field into a lower $^4A_2$ orbital singlet ground-state, a $^4T_2$
triplet and an upper $^4A_1$ orbital triplet. The high value of Landé g
factor is due to a mixing of higher excited states with the ground
state by spin orbit coupling. Study of $Zn_{1-x}Co_xSe$ (Jonker et al.,1988;
Liu et al.,1989 ; Lewicki et al.,1989) and $Zn_{1-x}Co_xS$ (Lewicki et al.,
1989) indicates a Brillouin paramagnetism . Values of exchange
constants obtained by Zeeman splitting on $Zn_{1-x}Co_xSe$ and spin-flip
Raman scattering on $Cd_{1-x}Co_xSe$ (Bartholomew et al., 1989) show that
the exchange interaction are larger in new Co-based SCSM alloys than
that found in Mn or Fe-based counterparts (Table I).

2.3 Theoretical model of Exchange Interaction and approximations
in wide-gap SMSC

The exchange interaction between Manganese-$3d^5$ electrons and
delocalized electrons of conduction and valence bands may be described
in the vicinity of the Γ point (Kossut, 1976 ; Bastard et al., 1978) by
an Heisenberg type Hamiltonian :

$$H_{ex} = - \sum_{R_i} J(r-R_i) \ S_i \ \sigma$$

where S_i is manganese spin operator at the site R_i $(S_i = 5/2)$, σ the
spin operator for band electron, $J(r-R_i)$ represents the exchange
integral centred at R_i. The summation is only over the lattice sites
occupied by the Mn^{2+} ions. Two approximations are made to simplify H_{ex}.
First, since the mobile electron has a spacially extented wave
function, it interacts simultaneously with a large number of Mn^{2+} ions.
Using the molecular field approximation, this allows one to remplace S_i
by the thermal average $<S>$. In absence of magnetic field $<S>$ vanishes.
If the applied field is in the z direction, $<S>$ is equal to $<S_z>$ the
average spin per Mn site. $<S_z>$ is directly related to the magnetization
in the usual notation :

$$M = N_o \ x \ <S_z>$$

where N_o is the number of unit cells per unit volume. Second, $J(r-Ri)$ can be remplace by $xJ(r-R)$ where R denotes the coordinate of every site of the cation sublattice (virtual crystal approximation). With these approximations,

$$H_{ex} = - x \, \sigma_z \, <S_z> \, \sum_R J(r-R)$$

where the summation is extending over all cation sites. H_{ex} has the periodicity of the lattice.

In the case of parabolic conduction band and for wide-gap materials, further approximations are allowed. In wide-gap SMSC, the effective mass $m*$ is large, both Landau and spin splittings of the band states are considerably smaller. Consequently, it is justified to neglect the terms concerning these two splittings in the Hamiltonian. In this approximation, the magnetic splitting is due entirely to the exchange term H_{ex}. For a cubic zinc blende crystal, matrice of H_{ex} are diagonal:

$$< \psi_{\Gamma 6} \, |H_{ex}| \, \psi_{\Gamma 6} > = \begin{vmatrix} 3 \, A & 0 \\ 0 & - \, 3 \, A \end{vmatrix}$$

$$< \psi_{\Gamma 8} \, |H_{ex}| \, \psi_{\Gamma 8} > = \begin{vmatrix} 3 \, B & 0 & 0 & 0 \\ 0 & B & 0 & 0 \\ 0 & 0 & - \, B & 0 \\ 0 & 0 & 0 & - \, 3 \, B \end{vmatrix}$$

and :

$$< \psi_{\Gamma 7} \, |H_{ex}| \, \psi_{\Gamma 7} > = \begin{vmatrix} - \, B & 0 \\ 0 & B \end{vmatrix}$$

where $A = - \dfrac{1}{6} \, N_o \alpha \, x \, <S_z>$

$B = - \dfrac{1}{6} \, N_o \beta \, x \, <S_z>$

$\alpha = <s|J|s>$ and $\beta = <x|J|x>$ are exchange integrals for conduction and valence band respectively ($\alpha > 0$, $\beta < 0$), and $<S_z>$ is negative.

Describing excitonic magneto-optical splitting in terms of interband transitions, the selection rules governing transitions between the levels are shown in Fig. 5.

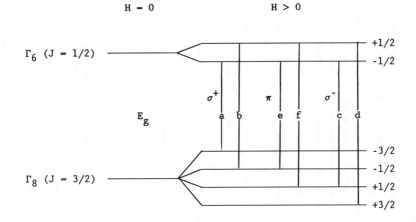

FIGURE 5 : Schematic illustration of the transitions between the spin-split valence band maxima and conduction band minima for zinc-blende wide-gap SMSC in magnetic field. Allowed transitions are indicated by arrows. In the Faraday geometry, transitions marked a and d are strong components whereas transitions b and c are weak components. In Voigt geometry, observed transitions are e and f.

In this mean field model, the Zeeman splitting ΔE between the two strong components ($|-3/2>_v \to |-1/2>_c$ and $|+3/2>_v \to |+1/2>_c$) is given by :

$$\Delta E = - N_0(\alpha-\beta) \times <S_z>. \qquad \text{Eq. 1}$$

The values of β/α, $N_0(\alpha-\beta)$, $N_0\alpha$ and $N_0\beta$ deduced from differents experiments, for several Mn-based compounds of various compositions and for Fe and Co based SMSC alloys, are collected in Table I.

2.4 Origin of the sp-d exchange

As it can be noted from table I, the exchange constants α and β are of opposite sign ($\alpha > 0$, $\beta < 0$) and the magnitude of β is larger than α. For a given Mn-based material in dilute regime, α and β appear to be insensitive to the Mn concentration. A physical mechanism common for Mn-based materials was proposed by Bhattacharjee, Fishman and Coqblin (1983) and explained the behavior of α and β. The model is based on the virtual-bound-state approach for the Mn-3d[5] electrons : two mechanisms contribute to the value of exchange constants.

Table I : Experimental sp-d exchange constants $N_o\alpha$ and $N_o\beta$ for wide-gap SMSC $A^{II} M B^{VI}$.

COMPOUND	COMPOSITION x	CRYSTAL STRUCTURE	T(K)	β/α	$- N_o(\alpha-\beta)$ (eV)	$N_o\alpha$ (eV)	$N_o\beta$ (eV)	EXPERIMENTS	REFERENCES
Cd Mn Te	x ≤0.3	Z.B.	1.4	-4.0	-1.10+0.05	0.22+0.01	-0.88+0.04	magnetorefl.	Gaj et al. (1979)
Zn Mn Te	x ≤0.1	Z.B.	2.2	-5.7	-1.29+0.09	0.19+ 0.01	-1.09+0.08	magnetoabs.	Twardowski et al. (1984)
	x ≤0.25	Z.B.	1.8	-5.78+0.4	-1.23+0.07	0.18+0.01	-1.05+0.06	magnetorefl.	Barilero et al. (1985)
Cd Mn Se	0.106	W.	1.9 ≤ T ≤4.2			0.261+0.013		SFRS	Shapira et al. (1982)
	0.10	W.	1.5	-4.3+0.4	1.37+0.07		-1.11+0.07	magnetorefl.	Aggarwal et al. (1983)
	0.05	W.	1.7			0.257+0.009	-1.24+0.19	magnetoabs.	Arciszewska et al. (1986)
							-1.301+0.42		
Zn Mn S	x ≤0.10	Z.B.	2.2 ≤ T ≤10	-5.0	1.61+0.08	0.26+0.01	-1.31+0.07	magnetorefl.	Twardowski et al. (1984)
Cd Mn S	0.023	W.	1.9			0.22+0.01		SFRS	Heiman et al. (1983)
	0.13	W.	1.6				-1.80+0.08	magnetorefl.	Nawrocki et al. (1986)
	0.001	W.	1.7	-13			-3.3		Gubarev (1986)
Zn Fe Se	0.05	Z.B.	2.0	7.0+0.7	1.79+0.16	0.22+0.04	-1.57+0.12	magnetorefl.	Twardowski et al. (1987)
	x ≤ 0.043	Z.B.	4.5		1.560+0.024			magnetorefl.	Liu et al. (1988)
Zn Co Se	x ≤0.0104	Z.B.	4.2		2.42+0.04			magnetorefl.	Liu et al. (1989)
Cd Co Se	x ≤0.082	W.	1.8			0.320		SFRS	Bartholomew et al. (1989)

i) there always exists the direct "potential" exchange between band electrons s and holes p and the d electrons. This mecanism gives a positive (ferromagnetic) and relatively small contribution to the exchange constant.

ii) in addition, there exist the "kinetic" exchange arising from hybridization between Mn-3d^5 levels and valence band states. The "kinetic" exchange does not contribute for conduction band, s-d hybridation is forbidden, but its negative contribution (antiferromagnetic) is important in magnitude for valence band.

In these conditions, the exchange integral α describing the conduction bands arise from potential interaction and is positive. The exchange constant β describing the valence band contains both contributions and is dominated by "kinetic" exchange. The latter being larger, β is negative and larger than α. One must now examine the question of variation of exchange integrals on the very wide range of composition for the SMSC compounds. In the model of Bhattacharjee et al, β is expected to show only a small variation through the hybridization constant as the position of Mn-3d^5 levels is unchanged (Fujimori et al. 1985). On the other hand, Wei and Zunger, (1987) have estimated, in terms of a p-d repulsion mechanism, the exchange constants $N_o\alpha$ and $N_o\beta$ in $CdMnTe_2$. ($Cd_{1-x}Mn_xTe$ with x = 0.5) and MnTe (x = 1). Althrough an overstimation of these values, the model predicts a only small decrease by about 10% of both $N_o\alpha$ and $N_o\beta$ as x increase from x = 0.50 to 1.

2.5 Deviation from Mean Field Model for concentrated Mn-based SMSC compounds

The Zeeman splitting of exciton has been extensively investigated for low Mn concentration while such a study presents much more difficulties for high x values. In the concentrated regime (x > 0.4) due to alloy broadening, magnetoreflectivity spectra exhibit only one and broad feature in each σ polarization, assumed to represent the strong component. However the energy separation ΔE between σ^+ and σ^- structures remains resolvable up to very concentrated $Cd_{1-x}Mn_xTe$ and $Zn_{1-x}Mn_xTe$ compounds (Rebmann et al., 1983 ; Desjardins et al., 1986 ; Lascaray et al., 1987, 1988 ; Coquillat et al., 1988). The behavior of the energy difference ΔE plotted against $|x <S_z>|$ is illustrated in Fig. 6 for $Zn_{1-x}Mn_xTe$. Two points are worthy of note. First, in the dilute regime up to $x = 0.25$ and according to the mean field model, the Zeeman splitting ΔE versus $|x <S_z>|$ is a straight line of slope $N_o(\alpha-\beta)$ (Gaj et al., 1979). Second, a drastic decrease of slope is observed from $x = 0.45$ and the slope is reduced by about 65% as x passes from 0.25 to 0.71 . A similar decrease for large x has been observed in $Cd_{1-x}Mn_xTe$ (Lascaray et al., 1988). Taking into account, in one hand , the very weak decrease of $N_o\alpha$ and $N_o\beta$ calculate by Wei et al.,(1987) as x increases, and on the other hand the fact that the position of $Mn-3d^5$ level in the valence band appears to be at same energy for the whole range of composition for a given type of material (at about 3.4 eV for $Cd_{1-x}Mn_xTe$) (Taniguchi et al., 1986, 1987 ; Ley et al., 1987 ; Chab et al., 1988) such a large scale effect cannot be attributed to a variation of exchange constants. One is thus led to question the validity of the mean field formula, which actually corresponds to the first order perturbation treatment of the sp-d exchange hamiltonien. Bhattacharjee (1988) have reexamined the mean field model and studied the energy shift of each subband in second order of perturbation.

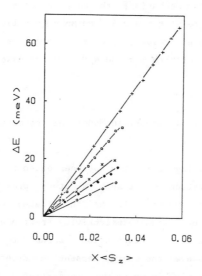

FIGURE 6 : Zeeman splitting between strong components in $Zn_{1-x}Mn_xTe$ versus magnetization expressed in $g\mu_B$ per unit cell. [After Lascaray et al.,1988].

Within the quasi-static approximation for the Mn spin system, the conduction band splitting is given by :

$$\Delta_c = - N_o \alpha \ x \ <S_z> - \delta_c$$

where δ_c is the second-order correction. Assuming that the Fourier transform of exchange coupling J_q for conduction band is constant and have a maximum at $q = 0$ ($J_q \simeq J_o = \alpha$) and neglecting the q-dependence of transverse spin-spin correlation function Γ_q at Γ point, correction for conduction band splitting, in paramagnetic system is :

$$\delta_c = \frac{3}{4} \ |N_o \alpha| \ r_c \ [-2x \ <S_z> + (\pi/3) \ x \ S(S+1) \ \overline{\sqrt{r_c} \ |x \ <S_z>|}] \qquad \text{Eq. 2}$$

with $x \ <S_z> \ <0$ and $r_c = |N_o \alpha/W_c|$, W_c being the width of the conduction band assumed parabolic. Analogous results are obtained for the valence band, leading to the energy difference between the strong components:

$$\Delta E \simeq N_o \beta \ (1 - \frac{3}{2} \ r_v) \ x \ <S_z> - \ N_o \alpha \ (1 - \frac{3}{2} \ r_c) \ x \ <S_z>$$

$$- \frac{\pi}{4} \ x \ S(S+1) \ [|N_o \alpha| \ r_c^{3/2} + |N_o \beta| \ r_v^{3/2}] \ |x \ <S_z>|^{1/2} \qquad \text{Eq. 3}$$

and the energy difference between the weak components :

$$\delta E \simeq N_o \beta/3 \ (1 - \frac{3}{2} \ r_v) \ x \ <S_z> + \ N_o \alpha \ (1 - \frac{3}{2} \ r_c) \ x \ <S_z>$$

$$+ \frac{\pi}{4} \ x \ S(S+1) \ [|N_o \alpha| \ r_c^{3/2} - \frac{1}{3} \ |N_o \beta| \ r_v^{3/2}] \ |x \ <S_z>|^{1/2} \qquad \text{Eq. 4}$$

The third term of Eq. 3 leads to a substantial decrease of band obtained in the first order. This correction is found to increase with the manganese concentration x and with the ratio of exchange parameter to the band width. Calculated curves ΔE versus $x \ <S_z>$ have an important x dependence even if $N_o \alpha$ and $N_o \beta$ are assumed constant, and for a given value of $|x \ <S_z>|$, ΔE decreases linearly with x. With second order correction, δE becomes very small. Finally, a rough numerical estimate shows that the anomalous behavior, i.e. the important decrease of the slope at high x values, is described with good agreement by a contribution of a second order.

3 - MAGNETIC CONTRIBUTION TO THE ENERGY GAP OF SMSC

The increase of energy gap of a semiconductor with decreasing temperature is well described by a phenomenological expression given by Varshni (1967) :

$$E_g(T) = E_g(0) - aT^2/(T+b) \qquad\qquad Eq. 5$$

where $E_g(T)$ and $E_g(0)$ are the band gap energies at T and 0 K respectively, a and b are constants of the material. The shift of $E_g(T)$ to the high energy , as T increases , is linear in T for high temperature (dE_g/dT = constant , < 0) and proportional to T^2 for low temperature ($dE_g/dT \rightarrow 0$ as $T \rightarrow 0K$). In semimagnetic semiconductors a review of the literature reveals anomalous behavior of the energy gap versus temperature and composition in absence of magnetic field, due to sp-d exchange interaction and related to magnetic degrees of freedom of SMSC. The first thing noted by several authors is the observed increase of $|dE_g/dT|$ with manganese concentration for ternary wide-gap alloys . A summary of values of temperature coefficient for SMSC has been presented by Becker (1988) . The second observation is that particularly stricking and curious behavior is observed at low temperature for the highest Mn concentration. The flat part observed in nonmagnetic semiconductor is replaced by a kind of a bump for concentrated alloys, as in the behavior of E_g shown in Fig. 7.

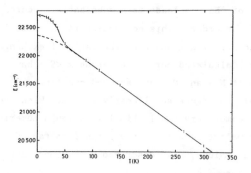

FIGURE 7 : Energy of absorption edge versus temperature for $Cd_{0.27}Mn_{0.73}Te$. The dotted curve is an extrapolation of $E_g(T)$ obtained linear high temperature part according to Varshni equation. [After Diouri et al.,1985].

Attempts to analyse the effect of magnetic order on energy gap in SMSC have been done by Diouri et al. (1985) in $Cd_{1-x}Mn_xTe$ and by Donofrio et al. (1985) in quaternary $Cd_{1-x-y}Zn_yMn_xTe$. Diouri et al. have observed variation of band edge absorption on samples thin enough to obtain an absorption coefficient greater than 5000 cm^{-1}, in order to eliminate contribution of intramanganese bands. They have underscored the anomalous behavior at low temperature for middle x values and highest Mn concentration (apparent blue shift). Diouri et al. have proposed a simple analysis of this effect using second order of perturbation theory based on calculation of Rys et al. (1967) on magnetically ordered semiconductors .Diouri et al. have shown strong evidence of effect of magnetic order on energy gap . At temperature $T \ll T_N$, where T_N is the Néel temperature, in the antiferromagnetic phase of type III there is no significant effect due to the magnetic order and the correction is zero . At $T \gg T_N$, the correlation between ion-spin can be neglected and the exchange interaction give rise to a red shift. To estimate the amount of red shift of the band gap from experimental data, they have considered however that the slope dE_g/dT in the high temperature part of the experimental curve is described by the Varnshi expression where a et b are dependent of concentration. Donofrio et al. have determined the variation of E_g as function of T on a wide range of $Cd_{1-x}Zn_yMn_xTe$ alloys by wavelength modulated reflectance measurements and observed the anomalous behavior at low temperature with composition. The authors have considered that all the magnetic effects are negligible in the higher temperature region of curves. In this way, they have calculated magnetic contribution at low temperature , and analyzed it in terms of the theoretical relation of Alexander et al. (1976) in the vinicity of the spin-glass temperature.

Suceeded quantitative analyses of the magnetic effect on energy gap has been proposed by Bylsma et al. (1986) and independently by Gaj et al. (1987a) ,both based on idea of Rys et al., as a simple model in which magnetic contribution to the energy gap is found to be proportional to the product of temperature and magnetic susceptibility. The first aspect mentionned above, i.e. the variation of the slope dE_g/dT of the linear high temperature part with x and T is underscored and described with these two models. The second anomalous aspect of the behavior, at low temperature from x = 0 to high concentration where appears the kind

62

of bump, is reasonnably fit by Gaj et al. (1987b) (see $Zn_{1-x}Mn_xTe$
alloys illustrated Fig. 8). A third aspect visible in the dependence of
E_g on x (Fig. 9) for low Mn concentration in $Zn_{1-x}Mn_xSe$ and showing a
minimum at x $\neq$ 0 is observed and explained by Bylsma et al. Such
behavior take place also in $Cd_{1-x}Mn_xS$ (Ideka et al., 1968). Each of
these two theoretical studies considers the magnetic contribution in
terms of second order perturbation theory in s-p and p-d interaction,
and indicates how E_g will quantitatively change as function of
temperature and crystal composition. The exchange interaction between
band carrier and magnetic ion is described by the Heisenberg
Hamiltonian with several approximation. A common approximation for the
two models is that, J_q the Fourier transform of the exchange coupling
J(x) for conduction and valence band are constant and have a maximum at
q = 0:

$$J_q = \begin{array}{l} J_o \text{ for } |q| < q_c \\ 0 \text{ for } |q| > q_c \end{array}$$

where q_c correspond to a cutoff wavevector value.

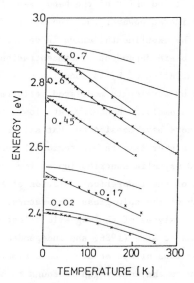

FIGURE 8 : Energy gap of $Zn_{1-x}Mn_xTe$ alloys versus
temperature (crosses are experimental data). Lower lines
represent model of Gaj et al. described in text. Upper
lines were calculated without magnetic contribution.
[After Gaj et al.,1986].

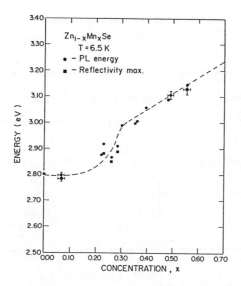

FIGURE 9 : Energy gap of $Zn_{1-x}Mn_xSe$ alloys versus composition at $T = 6.5K$, taken from photoluminescence and reflectivity maxima. [After Bylsma et al.,1986].

Gaj et al. have calculated theoretical $E_g(T)$ dependencies including second order magnetic correction and also nonmagnetic correction in paramagnetic, spin glass and antiferromagnetic phase, assuming flat magnetization fluctuations spectrum in the vinicity of Γ point in this case, $(\Gamma_q = \Gamma_0)$ and neglecting the contribution of conduction band. Variation of energy gap is given by :

$$E_g(x,T) = A(x) + (1+\alpha x) [E_g(0,T) - E_g(0,0)] + b\chi T \qquad \text{Eq. 6}$$

where parameters α and b are common for all the compositions, the parameter $A(x)$ describes the composition variation of the energy gap extrapoled to $T = 0$ and is fitted independly for each composition , χ is the magnetic susceptibility . This equation was fitted to the experimental data for $Cd_{1-x}Mn_xTe$ (Gaj et al., 1987a) and for $Zn_{1-x}Mn_xTe$ (Fig. 8, Gaj et al., 1987b) alloys , using parameter of Table II. The parameter b give corresponding cutoff wavector values and shows that the exchange coupling constant for the valence band decreases with the wawevector (see Table II).

Byslma et al. have explained the dependence of $E_g(T)$ coherently for

Table II : Parameters obtained to fit the temperature and composition dependence of energy gap.

COMPOUND	PARAMETERS	COMPOSITION x	VALUES	REFERENCES
$Cd_{1-x}Mn_xTe$	α	$0.2 \leq x \leq 0.4$	0.86	
	$b(eV\ Gs^2 erg^{-2}K^{-1})$	$0.2 \leq x \leq 0.4$	-0.002	
		0.2	1.919	Gaj et al.
	$A(x)(eV)$	0.3	2.088	
		0.4	2.254	(1987a)
	$qc(cm^{-1})$	$0.2 \leq x \leq 0.4$	$6.6\ 10^{-7}$	
$Zn_{1-x}Mn_xTe$	α	$0.02 \leq x \leq 0.703$	0.2	
	$b(eV\ Gs^2\ erg^{-2}K^{-1})$	$0.02 \leq x \leq 0.703$	-0.123	
		0.02	2.408	Gaj et al.
		0.17	2.536	
	$A(x)(eV)$	0.45	2.755	(1987b)
		0.60	2.856	
		0.703	2.915	
	$qc(cm^{-1})$	$0.02 \leq x \leq 0.703$	$9.6\ 10^{-7}$	
$Zn_{1-x}Mn_xSe$	$b(eV\ cm^3/emu\ K)$ obtained to fit temperature dependence	0.066	2.0	Bylsma et al.
		0.360	2.0	
		0.554	1.6	(1986)
	$b(eV\ cm^3/emu\ K)$	$0.066 \leq x \leq 0.554$	7.8	
	$qc(cm^{-1})$ obtained to fit concentration dependence	$0.066 \leq x \leq 0.554$	$3.3\ 10^{-7}$	

differentes values of crystal composition of $Zn_{1-x}Mn_xSe$ ($0 < x < 0.35$ limit for zinc blende phase) and the dip in E_g (x) seen at small values of x and low temperature in these alloys. They have taken into account the four fold degeneracy of the valence band and have expressed the results in paramagnetic phase in terms of empirically obtained magnetic susceptibility assumed independent of q. The expression of the band gap is : (Eq . 7)

$$E_g(x,T) = E_{og}(x,T) - (\frac{V}{2})^2 \frac{kT}{g\mu B^2} \frac{q_c}{\pi^2 h^2} \chi [3m_c\alpha^2 + (m_{hh} + \frac{2}{3}m_{lh})\beta^2]$$

where V is the unit-cell volume, m_c, m_{hh} and m_{lh} are effective masses for conduction electrons , heavy holes, and light holes respectively. α and β are exchange constants. $E_{og}(x,T)$ contains all the temperature and concentration dependence of the energy gap except for that due to the exchange interaction. Their model predicts a splitting of heavy hole and light hole valence bands. The theoretical expression (Eq. 7) fit reasonnably experimental data $E_g(T)$ with value of b given in table II. The linear relationship between magnetic contribution and χT predicted by the model is observed on Fig. 10 as function of Mn concentration and explained the minimum in the E_g versus x curve. The value of parameter b (table II) fitting E_g versus T is four times smaller than that needed to reproduce E_g versus x behavior. This example shows the difficulties to obtain some information on wavevector dependence of carrier-ion exchange coupling constant.

The variation of the energy gap with temperature was recently study in a new SMSC such $(CuIn)_{1-x}Mn_{2x}Te_2$ alloys with x = 0.4 and 0.5 (Quintero et al.,1989). The dominant phase in these alloys was the ordered zinc blende. Behavior at low temperature for the two concentrations is similar that in $Cd_{1-x}Mn_xTe$.

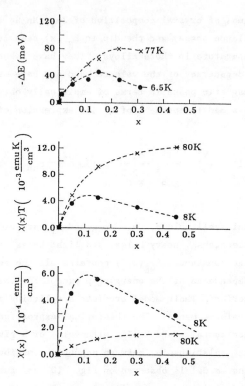

FIGURE 10 : Magnetic susceptibility , product χT , and non linear part of $E_g(x)$ as function of composition. The lines are added for clarity. [After Bylsma et al.,1985].

4. OPTICAL TRANSITIONS AND ZEEMAN SPLITTING IN THE VINICITY OF THE L POINT OF THE BRILLOUIN ZONE

4.1. Optical transitions in the vinicity of E_1 and $E_1 + \Delta_1$

Optical properties far from the Γ point are of interest because direct informations about the band structure of SMSC compounds can be obtained. First experiments of reflectivity in a SMSC were performed by Kendelewicz et al. (1978) on $Hg_{1-x}Mn_xTe$ alloys and transitions E_1 and $E_1 + \Delta_1$ (according to the notations used by Cardona and Greenaway,

1963) were observed for x < 0.30 at room temperature .The E_1 and $E_1+\Delta_1$ transitions correspond to a M_1-type critical point in the vicinity of L point in the Λ direction (Chelikowsky et al., 1976). The width of the E_1 and $E_1+\Delta_1$ structures is greatly influenced by the presence of Mn in the crystal. In contrast to the behavior of the $A^{II}_{1-x}B^{II}_{x}C^{VI}$ nonmagnetic ternary compounds, a deterioration of these structures appears in manganese rich samples. The effect of blurring show in Fig. 11 in $Cd_{1-x}Mn_xTe$ has been observed by reflectivity experiments (Kendelewicz,1980; Kendelewicz et al.,1981; Zimnal-Starnawska et al., 1984; Kisiel et al.,1987; Oleszkiewicz et al.,1987) in $Zn_{1-x}Mn_xSe$ and $Zn_{1-x}Mn_xS$ (Zimnal-Starnawska et al.,1984) . This blurring tendency is attributed not only to a modification of the shape and energy position

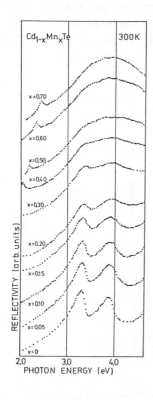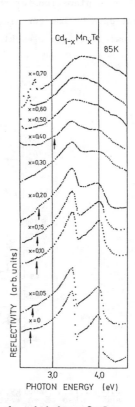

FIGURE 11 : Reflectivity spectra in the vicinity of L point (E_1 and $E_1+\Delta_1$ transitions) in $Cd_{1-x}Mn_xTe$ alloys with $0 < x < 0.7$ at room temperature and 85 K. [After Kendelewicz et al.,1981].

of conduction and valence bands, or to the modification of the
critical point in the k-space, but it is caused by new transitions due
to the presence on Mn-3d states in valence and conduction band. The
occupied Mn-3d states are split by the tetrahedral field and give
sharp peak of density of state 3.4 eV below the valence band maximum.
The location of the unoccupied Mn-3d states is not determined
precisely . For some authors they lie about 3.5-3.6 eV (Ehrenreich et
al.,1986; Furdyna,1988) above the top of the valence band. Taking into
account that photoemission experiments give a shift of the valence band
at the L point compare to valence band maximum of about 0.9 eV for x =0
(Taniguchi et al.,1986), unoccupied states lie at about 4.5 eV above
the region of the upper valence band of L point. This provide a
possible explanation for the anomalous structure seen near 4.5 eV in
optical data by Lautenschlager et al. (1985) and between 4 and 5 eV by
Kendelewicz (1981). Oleszkiewicz et al. (1987) (see also Kisiel et al.

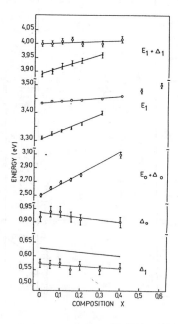

FIGURE 12 : Energy of reflectivity maxima of E_1, $E_1 + \Delta_1$
and also $E_o + \Delta_o$ structures (where Δ_o in spin-orbit
splitting at Γ^o point) in $Cd_{1-x}Mn_xTe$ alloys versus
composition at room temperature (full circles) and 85K
(open circles). Solid lines represent linear fit of the
experimental data. [After Kendelewicz et al.,1981].

1987) found presence of Mn-3d transitions by soustracting from reflectivity spectra of the $Cd_{1-x}Mn_xTe$ compounds the spectrum of CdTe . They have extracted contribution of Mn-3d states and observed three maxima of reflectivity at about 4.0, 6.0 and 8.5 eV with growing of the manganese content. The width of E_1 and $E_1+\Delta_1$ reflectivity features are more larger and more rapidly broad when x increases in $Zn_{1-x}Mn_xTe$ who have high energy transitions at L point (about 3.80 eV at 4.2 K for E_1) than in $Cd_{1-x}Mn_xTe$ or in $Hg_{1-x}Mn_xTe$, whereas at Γ point for equivalent Mn content E_o structure is narrower for $Zn_{1-x}Mn_xTe$ than for $Cd_{1-x}Mn_xTe$ (Coquillat et al.1986,1989; Moulin et al.,1990) .This could be an other manifestation of additional Mn-3d transitions .

The dependence of energy of E_1 and $E_1+\Delta_1$ transitions as function of Mn composition for $Cd_{1-x}Mn_xTe$ alloys is a subject of some controversy. This is often due to the fact that structures are large, broad with Mn content and consequently the choice of the point (inflexion point, middle of the decreasing part, minima, maxima of the structure) assumed to represent the energy position of the transition, influences the

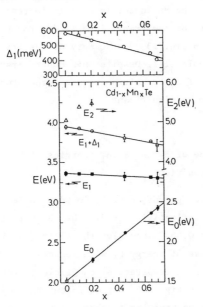

FIGURE 13 : Energy of critical point of E_1, $E_1+\Delta_1$ and E_o transitions in $Cd_{1-x}Mn_xTe$ alloys versus composition. [After Lautenschlager et al.,1985].

results. In $Cd_{1-x}Mn_xTe$ alloys, for example, reflectivity measurements and wawemodulated reflectivity experiments show a weak increase of energy of E_1 transition when x increases at room and low temperature (Fig. 12, Kendelewicz,1980; Bücker et al.,1985; Montegu et al.,1984) . Unlike these experiments, Ellipsometric measurements of dielectric function performed by Lautenschlager et al. (1985), at room temperature for $Cd_{1-x}Mn_xTe$ alloys with (0 < x < 0.7), through a critical-point analysis exhibit a clear linear decrease of both E_1 and $E_1+\Delta_1$ when x increases (Fig. 13).Despite the difference of sign of E_1 variation with x, the spectra of ellipsometric measurements and reflectivity remain in good agreement : Lautenschlager et al. have calculated the normal incidence reflectivity spectra from the dielectric function and have found that the positions of reflectivity maxima agree with the direct reflectivity measurements . They have related the observed decrease of E_1 and $E_1+\Delta_1$ in elliptometric measurements to an increasing amixture of Mn-3d states in the valence band. The valence band which participate in the E_1 and $E_1+\Delta_1$ transitions will be repelled to higher energies by the interaction with the Mn-3d levels as x is increasing. This should lead to a decrease of the transitions larger for $E_1+\Delta_1$ than for E_1, as it is observed in Fig. 13 . A summary of results obtained by these authors is given table III . Despite this controversy , the observed variation of energy of E_1 transition versus temperature and composition are strong enough to be significant .

4.2. Dependence of E_1 and $E_1+\Delta_1$ energies with temperature and Manganese composition

The temperature dependence of E_1 in $Cd_{1-x}Mn_xTe$ has been evaluated experimentally from reflectivity studies by Bücher et al. (1985) for composition 0 < x < 0.45. They give the first quantitative indication of the linear decrease of energy E_1 with temperature. The transition E_1 has a negative linear temperature coefficient ($dE_1/dT < 0$) ; its dependence is given in Fig. 14 for different Mn composition. In that, E_1 shows an opposite behavior to those of E_g at the Γ point upon composition. The study of variation of E_1 and $E_1+\Delta_1$ with temperature reported by Moulin et al. (1990) has been stimulated by the need to

Table III : Variation of E_1 and $E_1 + \Delta_1$ energy transition with composition for $Cd_{1-x}Mn_xTe$.

COMPOSITION	T(K)	VARIATION OF E1 (ev)	VARIATION OF E1+ Δ1 (ev)	EXPERIMENTS	REFERENCES
0 ≤ x ≤ 0.30	300	3.307 + 0.289x	3.889 + 0.229x	maximum of reflectivity structure	Kendelewicz (1980)
0 ≤ x ≤ 0.40	85	3.432 + 0.068x	3.997 + 0.032x		
0 ≤ x ≤ 0.45	300	superlinearly increase	...	maximum of reflectivity structure	Bucker et al. (1985)
	20	3.51	...		
0 ≤ x ≤ 0.31	300	3.390 + 0.240x	3.995 + 0.350x	negative peak of wawelength modulated reflectivity	Montegu et al. (1984)
0 ≤ x ≤ 0.28	82	3.513 + 0.185x	4.092 + 0.360x		
0.10 ≤ x ≤ 0.65	300	3.36 – 0.088x	3.95 – 0.33x	critical point by ellipsometry of dielectric function	Lautenschlager et al. (1985)
0.01 ≤ x ≤ 0.39	200	3.43 + 0.34x	...	inflexion point of reflectivity structure	Moulin et al. (1990)
	100	3.49 + 0.25x	...		
	4.2	3.52 + 0.20x	...		

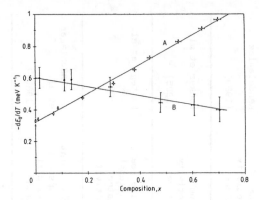

FIGURE 14 : Dependence of temperature coefficient dE_g/dT and dE_1/dT versus composition in $Cd_{1-x}Mn_xTe$ alloys. [After Bücher et al.,1985].

know if there is magnetic effect due to sp-d exchange interaction, as it can be seen at Γ point in absence of magnetic field. The energy values of transition E_1 are measured at the inflexion point of reflectivity feature corresponding to magneto-circular-dichroïsm extrema as it will be describe it in the next subsequent section .The shape of energy E_1 versus T is flat for x<0.4 in low temperature region and passes to decrase at high temperature (T > 50 K) . The linear temperature coefficient are given Table IV. As pointed out previously, the average value of the coefficient $|dE_1/dT|$ in the linear high temperature part decreases strongly when composition is growing, contrarly to the Γ point. The ionic radii between Mn^{2+} and Zn^{2+} being very similar in the host lattice CdTe, one can compare the evolution of the slope of linear high temperature part at the Γ point for ,on one hand $Cd_{1-x}Mn_xTe$ calculated without the magnetic contribution (only contribution of the lattice, Gaj et al., 1987a) and, on the other hand $Cd_{1-x}Zn_xTe$. In $Cd_{1-x}Zn_xTe$ alloys, the slope increase from x = 0 to x = 0.5 by about 13% as can be seen in table IV. The slope calculated without magnetic contribution for $Cd_{0.60} Mn_{0.40}Te$ is augmented by about 28% with respect to its value at x = 0, i.e. in the same way and with comparable magnitude that for $Cd_{1-x}Zn_xTe$.Now, at the L point, the absolute value of the linear high temperature part of $Cd_{1-x}Zn_xTe$ alloys increases very weakly (same sign that at Γ point) by about 12% from x = 0.13 to x = 0.40 (Moulin et al., 1990) . Consequently, one might

Table IV : Temperature coefficients dE_g/dT and dE_1/dT versus composition x for $Cd_{1-x}Zn_xTe$ and $Cd_{1-x}Mn_xTe$ compounds.

COMPOUND	COMPOSITION x	dE_g/dT in 10^{-4} eV/K	dE_g/dT without magnetic contribution in 10^{-4} eV/K	dE_1/dT in 10^{-4} eV/K	REFERENCES
$Cd_{1-x}Zn_xTe$	0	-3.53			Donofrio et al. (1985)
	0.5	-3.99			
	1	-4.10			
	0.13			-4.47	Moulin et al. (1990)
	0.23			-4.57	
	0.40			-4.99	
$Cd_{1-x}Mn_xTe$	0	-3.53			Donofrio et al. (1985)
	0.2		-3.97		Gaj et al. (1986)
	0.3		-4.47		
	0.4		-4.53		
	0.01			-5.54	Moulin et al. (1990)
	0.09			-5.06	
	0.14			-4.41	
	0.25			-3.03	
	0.30			-2.58	
	0.39			-1.04	

automaticly assume that at L point, the slope of $Cd_{1-x}Mn_xTe$ alloys
without magnetic effect increases itself very weakly with about
comparable magnitude that $Cd_{1-x}Zn_xTe$. In reality ,the variation of
slope is of opposite sign and is considerably greater as it is shown in
table IV , suggesting that an effect other than those due to lattice,
contribute to variation of energy of E_1 structure with temperature.
Such an effect could be magnetic contribution, but no quantitative
information is available on q dependence of Fourier transform of
exchange coupling constants J_q and of correlation function Γ_q to
conclude in this way (see section 3). The origin of this deviation
remains still an open question.

4.3.Zeeman splitting at L point of the Brillouin zone

Magneto-optical results at L point was initiated by Dudziak et
al. (1982) and investigated on the two wide-gap compounds $Cd_{1-x}Mn_xTe$
and $Zn_{1-x}Mn_xTe$ and on the narrow-gap material $Hg_{1-x}Mn_xTe$ (Ginter et al.
1983; Coquillat et al. 1986,1989) . First experimental observation by
Dudziak et al. on $Cd_{1-x}Mn_xTe$ revealed no significant difference in the
maxima reflectivity position measured with σ^+ and σ^- polarization,
since Zeeman splitting of E_1 was found so small and width of E_1
structure is large. Using a method of modulated polarization, the
authors were finally able to show a certain effect of magnetic field
and have finally evaluate the splitting of E_1. It is much smaller than
that at Γ point.
The first quantitative results on splitting of E_1 structure were
published by Ginter et al. (1983) . Taking into account the effect of
blurring on reflectivity feature ,magneto-optical results have been
investigated by several authors only for x up to 0.39. An example of
two spectra obtained, degree of polarization $P=(I^+-I^-)/(I^++I^-)$ and
reflectivity for the doublet E_1, $E_1+\Delta_1$ are shown in Fig 15. The
splitting for each structure between the two components σ^+ and σ^- may
be determined from the equation given by Ginter et al., relating the
measured degree of polarization to the logarithmic derivative of
reflectivity spectrum. Figure 16 represented splitting of structure E_1
of $Cd_{1-x}Mn_xTe$ as function of magnetic field . The observed behavior is

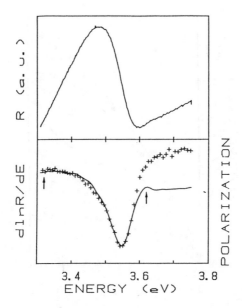

FIGURE 15 : Reflectivity structure of E_1 transition and its logarithmic derivative (continuous line below) compared with polarization spectra (+++) for $Cd_{0.91}Mn_{0.09}Te$ at T = 4.5K and H=5.5 T. Scales of dlnR/dE and of polarization have been chosen to obtain the best fit. [After Coquillat et al.,1986].

found resemble the behavior of splitting of the strong σ component at Γ point. The proportionality of splittings at L and Γ points is visible in Figure 17 for the compounds $Cd_{1-x}Mn_xTe$ and $Zn_{1-x}Mn_xTe$. Variation of temperature does not influence the proportionality of these two splittings. The comparison of splitting at Γ and L is not possible for $Hg_{1-x}Mn_xTe$ alloys of zero or small energy gap. The magneto-optical effects at Γ point in $Hg_{1-x}Mn_xTe$ alloys are more complex, because of splitting pattern of energy band is a combination of strong Landau quantification and band mixing effects, and there is some controversy in the literature about the exchange parameters $N_o\alpha$ and $N_o\beta$ in the $Hg_{1-x}Mn_xTe$ alloys . Figure 18 shows the plot of splitting of E_1 transition as function of magnetization expressed in $g\mu_B$ per unit cell $|x <S_z>|$ in dilute $Cd_{1-x}Mn_xTe$, $Zn_{1-x}Mn_xTe$ and $Hg_{1-x}Mn_xTe$. The plot shows a roughly linear behavior independant of x. Observations can be summarized as follows :

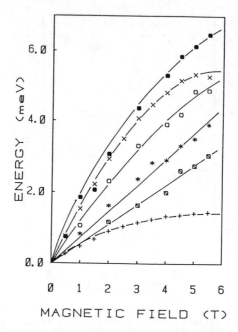

FIGURE 16 : Values of Zeeman splitting of E_1 structure at T = 4.5K as a function of magnetic field for $Cd_{1-x}Mn_xTe$ and values of crystal composition x : 0.01 (+), 0.09 (x), 0.14 (●), 0.25 (O), 0.30 (*) , 0.39 (∅). Continuous lines are dded for clarity. [After Coquillat et al.,1986].

1) At low temperature the Zeeman splitting of E_1 and $E_1+\Delta_1$ was systematically equal but opposite in sign.

2) The splitting of E_1 is of same sign and is reduced in magnitude in comparison with the splitting between the two strong σ components observed at the Γ point by a factor 15 for $Cd_{1-x}Mn_xTe$ and 20 for $Zn_{1-x}Mn_xTe$ in the wide range of composition (0.01 < x < 0.39) and temperature (4.5 < T < 100 K).

3) Splitting of E_1 structure is easier observable in $Cd_{1-x}Mn_xTe$ than in $Zn_{1-x}Mn_xTe$ for similar concentration. For $Zn_{1-x}Mn_xTe$ alloys, the degree of polarization can be correctly fit by the logarithmic derivative of reflectivity only up to x ≃ 0.17 (up to 0.39 for $Cd_{1-x}Mn_xTe$), due to presence of additional Mn transitions in the vicinity of L point.

4) Zeeman splitting in $Hg_{1-x}Mn_xTe$ shows the same slope in splitting versus $|x<S_z>|$ for alloys with x = 0.05 (zero gap) and x = 0.20 (narrow gap). No difference is observed even if a discontinuitie in the

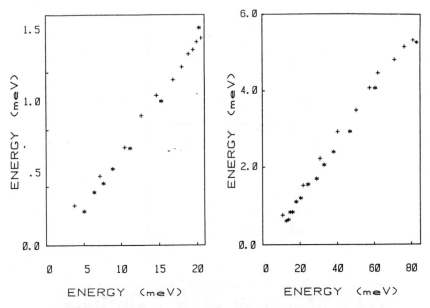

FIGURE 17 : Zeeman splitting of E_1 structure versus Zeeman splitting between strong components ΔE at Γ points for $Cd_{1-x}Mn_xTe$ x = 0.01 (a) and x = 0.09 (b). (*) H = 5.5T and temperature varies from 4.5K to 40K (a) or 100K (b). (+) T = 4.5K and H varies from 0.5T to 5.5T. [After Coquillat et al.,1986].

concentration dependence of position of E_1 transition is obtained at x = 0.05 (Amirtharaj et al., 1981).

5) The splitting of E_1 in $Hg_{1-x}Mn_xTe$ is much larger than in both $Cd_{1-x}Mn_xTe$ and $Zn_{1-x}Mn_xTe$ alloys. The slope of splitting of E_1 versus $|x<S_z>|$ is about 0.06 for $Cd_{1-x}Mn_xTe$ and $Zn_{1-x}Mn_xTe$,and about 0.18 for $Hg_{1-x}Mn_xTe$.

The first theoretical description of the effect of magnetic field at the L point of Brillouin zone is presented by Ginter et al. (1981) (see also Gaj, 1988) who used a simple four-orbital tight binding calculation . In this model employed to explain the observed difference in Zeeman splitting between the Γ and L points,the assomption that exchange constants α and β are wawevector independant is made. Thus this calculation predicts a reduction of four times of the splitting at L so that at Γ, and for $E_1+\Delta_1$ transition a Zeeman

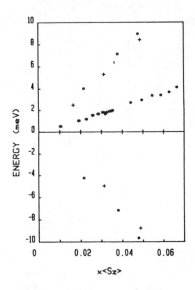

FIGURE 18 : Zeeman splitting of E_1 (positive values) and $E_1 + \Delta_1$ (negative values) structure in $Hg_{1-x}Mn_xTe$ $x = 0.05$ for crosses and $x = 0.20$ for empty circles , $Zn_{1-x}Mn_xTe$ (solid circle) and $Cd_{1-x}Mn_xTe$ (asterisks) as function of magnetization expressed in $g\mu_B$ per unit cell. [After Coquillat et al. 1989].

splitting equal in magnitude and opposite in sign to that of the E_1 transitions. This last prediction confirmed by experimental observation appears more general than the simple tight binding model used. Comparison between theoretical analysis and experimental results shows a difference in the factor of reduction between L and Γ points : 4 for theoritical calculation in contrast to 15 or 20 for experimental values in $Cd_{1-x}Mn_xTe$ and $Zn_{1-x}Mn_xTe$ respectively. Finally, in spite of the fact that certain physical aspect are not described by the model (mixing of the s-like and p-like states in the conduction band ; the component of magnetic field perpendicular to the k vector produces no splitting of the heavy hole states; selection rules are not strict as at Γ point) , theorical and experimental results suggest that the exchange integrals are significantly reduced at the L point compared to the Γ point . This idea remains in good agreement with the obtained decreasing of sp-d exchange interaction for valence band with the wawevector found in investigation of gap dependence with composition and temperature described in section 3 (values of cut off wawevector are given in Table II) . These both results (energy gap variation and Zeeman splitting at L point) show the necessity of obtain more information on variation of exchange coupling constant and on correlation function with wawevector.

Recently , Bhattacharjee (1990) has expressed the Zeeman splitting at the L point and the k-dependence of the conduction and valence band exchange parameters within the framework of the model of Bhattacharjee, Fishman and Coqblin (1983) . Bhattacharjee has estimated the Zeeman splitting ΔE_1 of E_1 as proportional to $N_o(\alpha-\beta/4)|x <S_z>|$. Reporting the known values of exchange parameters , the comparison of calculated values of $\Delta E_1/|x <S_z>|$ with experimental results in $Cd_{1-x}Mn_xTe$, $Zn_{1-x}Mn_xTe$ and $Hg_{1-x}Mn_xTe$ alloys gives calculated values about 3 times larger than experimental results and relative values in good agreement for the three compounds .

REFERENCES:

Aggarwal,R.L.,Japerson,S.N.,Stankiewicz,J.,Shapira,Y.,Foner,S.,Khazai,B. and Wold,A. (1983) Phys.Rev.B28,6907

Abramishvili,A.G.,Gubarev,S.I.,Komarov,A.V.,and Ryabchenko,S.M. (1984) Fiz. Tverd. Tela 26,1095 (Sov. Phys. Solid State 26,666).

Alexander,S.,Helman,J.S.,and Balberg,I. (1976) Phys. Rev. B13,304.

Amirtharaj,P.M.,Pollak,F.H.,and Furdyna,J.K. (1981) Solid State Commun.39,35.

Antoszewski,J.,and Kierzek-Pecold,E. (1980) Solid State Commun. 34,733.

Arciszewska,M.,and Nawrocki,M. (1982) Proceedings of the 10th Conf. on Physics of Semiconducting Compounds, Jaszowiec,1981. Warsaw, Poland; Polish Acad. Sci.,225.

Arciszewska,M.,and Nawrocki,M. (1986) J. Phys. Chem. Solids 47,309.

Barilero,G.,Rigaux,C.,Menant,M.,Nguyen Hy Hau,and Giriat,W. (1985) Phys. Rev. B32,5144.

Bartholomew,D.U.,Furdyna,J.K.,and Ramdas,A.K. (1986) Phys. Rev. B34,6943.

Bartholomew,D.U.,Suh,E-K.,Ramdas,A.K.,Rodriguez,S.,Debska,U.,and Furdyna,J.K. (1989) Phys. Rev. B399,5865.

Bastard,G.,Rigaux,C.,Gulner,Y.,Mycielski,J.,and Mycielski,A. (1978) J. Phys. 39,87.

Becker,W.M. (1988) Semiconductors and Semimetals edited by J.K. Furdyna and J. Kossut (Academic,Boston,1988) Vol 25,35.

Beckett,D.J.S.,Chehab,S.F.,Lamarche,G.,and Woolley,J.C. (1987) J. Magn. Magn. Mat. 69,311.

Bhattacharjee,A.K.,Fishman,G.,and Coqblin,B. (1983) Physica 117B&118B ,449.

Bhattacharjee,A.K. (1988) Solid State Commun. 65,275.

Bhattacharjee,A.K. (1990) unpublished .

Brandt,N.B.,and Moshchalkov,V.V. (1984) Adv. Phys. 33,193.

Bücker,R.,Gumlich,H-E.,and Krause,M. (1985) J. Phys.C:Solid State Phys.,18,661.

Bylsma,R.B.,Becker,W.M.,Kossut,J.,and Debska,U. (1986) Phys. Rev. B33,8207.

Cardona,M.,and Greenaway,D.L. (1963) Phys. Rev. 131,98.

Chab,V.,Paolucci,G.,Prince,K.C.,Surman,M.,and Bradshaw,A.M. (1988)
Phys. Rev. B38,12353.

Chelikowski,J.R.,and Cohen,M.L. (1976) Phys. Rev. B14,556.

Coquillat,D.,Lascaray,J-P.,Deruelle,M.C.,Gaj,J.A.,and Triboulet,R. (1986)
Solid State Commun. 59,25.

Coquillat,D.,Desjardins-Deruelle,M.C.,Lascaray,J-P.,Deportes,J.,and
Bhattacharjee,A.K. (1988) J. Phys. Suppl.n°12 ,tome 49,877.

Coquillat,D.,Lascaray,J-P.,Gaj,J.A.,Deportes,J.,and Furdyna,J.K. (1989)
Phys. Rev. B39,10088.

Desjardins-Deruelle,M.C.,Lascaray,J-P.,Coquillat,D.,and Triboulet,R. (1986)
Phys. stat. sol.(b) 135,227.

Diouri,J.,Lascaray,J-P.,and El Amrani,M. (1985) Phys. Rev. B31,7995.

Donofrio,T.,Lamarche,G.,and Woolley,J.C. (1985)
J. Appl. Phys. 57(6),1932.

Dudziak,E.,Brzezinski,J.,and Jedral,L. (1981)
Proceedings of 14th International Conference on the Physics of Semicond.
Compounds ,Jaszowiec,Poland (Polish Academy of Science,Warsaw),166.

Ehrenreich,H.,Hass,K.C.,Johnson,N.F.,Larson,B.E.,and Lempert,R.J. (1987)
Proceedings of 18th International Conference on Physics of Semiconductors
Stockholm 1986,edited by O. Engsstrom (Word Scientific,Singapore),1727.

El Amrani,M.,Lascaray,J-P.,and Diouri,J. (1983) Solid State Commun. 45,351.

Franciosi,A.,Shu Chang,Reifenberger,R.,Debska,U.,and Riedel,R. (1985)
Phys. Rev. B32,6682.

Furdyna,J.K. (1982) J. Appl. Phys. 53(1),7637.

Furdyna,J.K. (1988) J. Appl. Phys. 64(4),R29.

Gaj,J.A.,Galazka,R.R.,and Nawrocki,M. (1978a) Solid State Commun. 25,193.

Gaj,J.A.,Byszewski,P.,Cieplak,M.Z.,Fishman,G.,Galazka,R.R.,Ginter,J.,
Nawrocki,M.,Nguyen The Khoi,Planel,R.,Ranvaud,R.,and Twardowski,A.(1978b)
Proceedings of 14th International Conference of Semiconductors ,Edinburg,
edited by B.L.H. Wilson,Inst. of Phys.,Conf. Series 43 (Inst. of Phys.,
London,1978).

Gaj,J.A.,Ginter,J.,and Galazka,R.R. (1978c) Phys. stat. sol.(b) 89,655.

Gaj,J.A.,Planel,R.,and Fishman,G. (1979) Solid State Commun. 29,435.

Gaj,J.A. (1980) J. Phys. Soc. Japan , Suppl.. A49,747.

Gaj,J.A.,and Golnik,A. (1987a) Acta Phys. Pol. A71,197.

Gaj,J.A.,Golnik,A.,Lascaray,J-P.,Coquillat,D.,and Desjardins-Deruelle,M.C.
    (1987b) Proceedings of Material Research Society,Fall Meeting, edited by
    R.L. Aggarwal,J.K. Furdyna and S. von Molnar (MRS,Pittsburg,PA)Vol 89,59.

Gaj,J.A. (1988) Semiconductors and Semimetals edited by J.K. Furdyna
    and J. Kossut (Academic,Boston) Vol 25,276.

Galazka,R.R. (1985) J. Crystal Growth 72,364.

Ginter,A.,Gaj,J.A.,and Le Si Dang (1983) Solid State Commun. 48,849.

Goede,O.,and Heimbrodt,W. (1988) Phys. stat. sol.(b) 146,11.

Gubarev,S.I. (1981) Zh. Eksp. Teor. Fiz. 80,1174 (Sov. Phys. JETP 53,601).

Gubarev,S.I.,and Tyazholv,M.G. (1986)
    Pis'ma Zh. Eksp. Teor. Fiz. 44,385 (JETP Lett. 44,494).

Gubarev,S.I.,and Tyazholv,M.G. (1988) Pis'ma Zh. Eksp. Teor. Fiz. 48,437.

Heiman,D.,Shapira,Y.,Foner,S. (1983) Solid State Commun. 45,899.

Heiman,D.,Petrou,A.,Isaacs,E.D.,Bloom,S.H.,Shapira,Y.,and Giriat,W.(1988)
    Phys.Rev. Lett. 60,1876.

Ideka,M.,Itoh,K.,and Sato,H. (1968) J. Phys. Soc. Japan 25,455.

Jonker,B.T.,Krebs,J.J.,and Prinz,G.A. (1988) Appl. Phys. Lett. 53(5),450.

Kendelewicz,T.,and Kierzek-Pecold,E. (1978) Solid State Commun. 25,579.

Kendelewicz T. (1980) Solid State Commun. 36,127.

Kendelewicz,T. (1981) J. Phys. C: Solid State Phys. 14,L407.

Kett,H.,Gebhardt,W.,Krey,U.,and Furdyna,J.K. (1981)
    J. Magn. Magn. Mat. 25,215.

Kierzek-Pecold,E.,Szymanska,W.,and Galazka,R.R. (1984)
    Solid State Commun. 50,685.

Kisiel,A.,Oleszkiewicz,J.,Rodzik,A.,Antonangeli,F.,Placentini,M.,Zema,M,
    Balzarotti,A.,and Mycielsky,A. (1987) Acta Phys. Pol. A71,231.

Komarov,A.V.,Ryabchencko,S.M.,Terletskii,O.V.,Zheru,I.I.,Ivanchuk,R.D.
    (1977) Zh. Eksp. Teor. Fiz. 73,608 (Sov. Phys. JETP 46,318)

Komarov,A.V.,Ryabchenko,S.M.,Terletskii,O.V.,Ivanchuk,R.D.,Savitskii,A.V.
    (1980a) Sov. Phys. Semicond. 14(1),9.

Komarov,A.V.,Ryabchenko,S.M.,and Terletskii,O.V. (1980b)
    Phys. stat. sol.(b) 102,603.

Kossut,J. (1976) Phys. stat. sol.(b) 78,537.

Kullendorf,N.,and Hok,B. (1985) Appl.Phys. Lett. 46(11),1016.

Lascaray,J-P,Deruelle,M.C.D.,and Coquillat,D. (1987)
    Phys. Rev. B35,675.

Lascaray,J-P.,Coquillat,D.,Deportes,J.,and Bhattacharjee,A.K. (1988)
   Phys. Rev. B38,7602.
Lautenschlager,P.,Logothetidis,S.,Vina,L.,and Cardona,M. (1985)
   Phys. Rev. B32,3811.
Lee,Y.R.,Ramdas,A.K.,and Aggarwal,R.L. (1988) Phys. Rev. B387,10600.
Lewicki,A.,Schlinder,A.I.,Furdyna,J.K.,and Giriat,W. (1989)
   Phys. Rev. B40,2379.
Ley,L.,Taniguchi,M.,Ghijsen,J.,Johnson,R.L.,and Fujimori,A (1987)
   Phys. Rev. B35,2839.
Liu,X.,Petrou,A.,Jonker,B.T.,Prinz,G.A.,Krebs,J.J.,and Warnock,J. (1988)
   J. Vac. Sci. Technol. A6(3),1508.
Liu,X;,Petrou,A.,Jonker,B.T.,Prinz,G.A.,Krebs,J.J.,and Warnock,J. (1989)
   Appl. Phys. Lett. 55(10),1023.
Montegu,B.,Laugier,A.,and Triboulet,R. (1984)
   J. Appl. Phys. 56(11),3061.
Moulin,N.,Coquillat,D.,Lascaray,J-P. (1990) unpublished.
Nawrocki,M.,Lascaray,J-P.,Coquillat,D.,and Demanuiuk,M. (1987)
   Proceedings of Material Rechearch Society , Fall meeting , edited by
   R.L. Aggarwal,J.K. Furdyna,and S. von Molnar(MRS,Pittsburg,PA)Vol 89,65.
Nguyen The Khoi,Ginter,J.,and Twardowski,A. (1983)
   Phys. stat. sol.(b) 117,67.
Oleszkiewicz,J.,Kisiel,A.,and Rodzik,A. (1987)
   Solid State Commnun. 63,77.
Petrou,A,Liu,X.,Waytena,G.,Warnock,J.,and Giriat,W. (1987)
   Solid State Commun. 61,767.
Quintero,M.,Marks,B.D.,and Woolley,J.P. (1980)
   J. Appl. Phys.
Rebmann,G.,Rigaux,C.,Bastard,G.,Menant,M.,Triboulet,R.,Giriat,W. (1983)
   Phys. 117B&118B,452.
Rys,F.,Helman,J.S.,and Baltensperger,W. (1967)
   Phys. Kondens. Materie. 6,105.
Shapira,Y.,Heiman,D.,and Foner,S. (1982) Solid State Commun. 44,1248.
Sundersheshu,B.S.,and Kendelewicz,T. (1982)
   Phys. stat. sol.(a) 69,4467.
Swiderski,P.,and Twardowski,A. (1984) Phys. stat. sol.(b) 122,K147.
Taniguchi,M.,Ley,L.,Johnson,R.L.,Ghijsen,J.,and Cardona,M. (1986)
   Phys. Rev. B33,1206.

84

Taniguchi,M.,Fujimori,M.,Fujisawa,M.,Mori,T.,Souma,I.,and Oka,Y. (1987)
  Solid State Commun. 62,431.

Twardowski,A.,Nawrocki,M.,and Ginter,J. (1979)
  Phys. stat. sol.(b) 96,497.

Twardowski,A. (1981) unpublished .

Twardowski,A.,and Ginter,J. (1982) Phys. stat. sol.(b) 114,331.

Twardowski,A. (1983a) Physics Lett. 94A,103.

Twardowski,A.,Dietl,T.,and Demianiuk,M. (1983b)
  Solid State Commun. 48,845.

Twardowski,A.,von Ortenberg,M.,Demianiuk,M.,and Pauthenet,R. (1984)
  Solid State Commun. 51,849.

Twardowski,A.,von Ortenberg,M.,and Demianiuk,M. (1985)
  J. Cryst. Growth 401.

Twardowski,A.,Glod,P.,de Jonge,W.J.M.,and Demianuik,M. (1987)
  Solid State Commun. 64,63.

Wei,S.H.,Zunger,A. (1987) Phys. Rev. B35,2340.

THE LUMINESCENCE OF WIDE BAND GAP II-Mn-VI SEMIMAGNETIC SEMICONDUCTORS

by

Carsten Benecke[+] and Hans-Eckhart Gumlich[*]

Institut für Festkörperphysik der Technischen Universität Berlin, Germany

+*Central Research Units, F.Hoffmann-La Roche & Co. Ltd. Basel, Switzerland*

CONTENT

1 INTRODUCTION

Until now only special aspects of the luminescence of highly doped II-Mn-VI semimagnetic semiconductors (SMSCs) with a large forbidden band gap have been reviewed in a systematic way.[1-15] Therefore, it is the aim of the present paper to give a survey of the Mn related luminescence properties of the II-VI group with large band gap, as far as bulk properties are concerned. The luminescence of quantum wells of these materials is reviewed by other authors in this volume.[16]

It is worthy to note that there are striking similarities in the emission of II-Mn-VI compounds when the concentration of Mn exceeds about one percent. On the other hand it seems to be an important feature that the appearance of some luminescence phenomena of these materials are correlated to the semimagnetic properties. In most of these compounds a very characteristic emission due to the internal transition $^4T_1(G) \to {}^6A_1(S)$ within the 3d shell of tetrahedrally coordinated Mn^{2+} is observed even at very low Mn concentrations. However, if the Mn concentration exceeds a threshold, a typical high doping luminescence is observed. In the following chapters we describe some general features of the optical properties of II-Mn-VI compounds and we report on specific details of the luminescence of $Zn_{1-x}Mn_xS$, $Zn_{1-x}Mn_xSe$, $Cd_{1-x}Mn_xS$, $Cd_{1-x}Mn_xSe$, $Zn_{1-x}Mn_xTe$ and $Cd_{1-x}Mn_xTe$. Some ideas on the chemical trend of the luminescence are presented and some models of the infrared emission are discussed. Finally, we give arguments and data of a model which involves a cubic coordination of Mn^{2+} ions.[17]

2 BASIC CONSIDERATIONS

2.1 The Forbidden Gap As A Function Of The Mn Concentration

The energies of the band gap E_g of the II–Mn–VI compounds depend considerably upon the Mn concentration x. The values of E_g have been determined either by absorption measurements or by reflection spectroscopy taking into account the shift of the exitons, see e. g. ref. 18.

Also the temperature has a pronounced influence upon the band gap. In a useful approximation one can describe both the concentration and the temperature dependence as linear functions of x and T.[4] Experimental data of E_g can be given only within the limits of perfect alloys, which vary considerably for different II-VIs. With the exception of $Zn_{1-x}Mn_xS$ the gap energies of all compounds which we consider increase with increasing amount of Mn.

2.2 The Ground State Of Mn^{2+} In The II-VI Compounds

One of the matters discussed intensively in the case of Mn in the II-VIs is the question of the relation of the Mn^{2+} ground state with respect to the band structure of the parent compounds. The experimental situation seems to be quite confusing. On the one hand the experimental data suggest a strong localization of the wavefunctions of Mn^{2+}. EPR measurements and the multiplett structure of the optical spectra point in this direction.[19,20]

However, on the other hand we definitely know results obtained by ENDOR proving the delocalization of the Mn wavefunction.[21] The most important insight in this discussion comes from photoelectron spectroscopy and reflection measurements. ARUPS experiments of Orlowski et al. showed for instance a structure 3.5eV below the top of the valence band, which is assigned to the ground state of the 3d shell of Mn^{2+} in e-symmetry.[22] Beside of this the blurring of another peak at 1.3eV is explained by a p-d hybridization of p-electrons in t_2-symmetry of Se with the d-electrons of Mn^{2+} in t_2-symmetry.[23-25] By now it seems to be generally accepted that the e-ground state of Mn^{2+} in II-VI compounds lies about 3.5eV below the top of the valence band and gives rises to the multiplett structures. This energy does not depend upon the concentration of Mn. A detailed model of the ground state of Mn in II-VI compounds has been given by Wei and Zunger.[26]

2.3 The Dependence Of The Luminescence Upon The Mn Concentration

As already mentioned the luminescence properties of the II-Mn—VIs depend in a characteristic way upon the Mn concentration.

At low concentrations ($x \leq 0.001$) only the emission due to the internal transition $^4T_1(G) \rightarrow ^6A_1(S)$ within the 3d shell of tetrahedrally coordinated Mn^{2+} can be observed, provided the energy gap is large enough to permit the transmission of this emission band. The width of the band gap determines also the possibility of observation of the excitation bands due to internal transitions within the 3d shell of Mn^{2+}. For instance, $Zn_{1-x}Mn_xS$ shows five excitation bands with a pronounced fine structure of zero phonon lines and phonon satellites. These bands and their structure have been studied intensively during the last decade so that even the influence of the crystal structure and of neighbouring stacking faults on the zero phonon lines could be explained.[27-29]

In an intermediate concentration range ($0.001 \leq x \leq 0.01$) zero phonon lines due to Mn-Mn pairs are observed.[30] Very recently, pair lines are analysed in $Zn_{1-x}Mn_xS$ due to Mn-Mn pairs interacting from the first to the sixth nearest neighbours. [31] Roughly speaking at Mn-concentrations above $x \geq 0.01$ the fine structure both in emission and excitation disappears and, depending on the anion, one, two or three additional emission bands are observed. These emission bands, which are likely to be connected to the other SMSCs properties, are the main topic of the present paper.

2.4 The Selfactivated Luminescence

When considering the luminescence of II-VI compounds, the selfactivated emission (SA) has to be taken into account. The main questions in this context are: What is the model of the SA luminescence? Is the SA emission observed in all II-Mn—VIs? Is there any connection between the SA and the Mn related centers and does the SA emission depend upon the Mn concentration? And finally, are there any differences of the SA emission of different II-Mn—VIs?

There are models of the SA luminescence, which differ in detail, but all of them explain the emission by a donor acceptor recombination. The donors and the acceptors respectiviley can consist of vacancies, interstitials or impurities. The strong tendency toward selfcompensation of the II-VIs favours the creation of vacancies.[32]

The SA emission and the luminescence due to deep level impurities behave in some respect quite differently. First, the peak of the SA emission shifts as a function of time after the end of the excitation towards lower energies. In the beginning of the decay predominantly the carriers of next nearest neighbours recombine, whereas lateron the more distant DA pairs follow. Second, the same shift is observed, when the intensity of emission is diminished. When the intensity of the excitation increases the decay of emission becomes relatively faster. Thirdly, by raising the temperature the shallow donors and/or acceptors are ionized sometimes leading to a drastic change of the spectral distribution of the emission. At low temperatures a bound-to-bound transition is prevailing, whereas at higher temperatures a free-to-bound transition takes places.

It should be mentioned that in some cases double ionized vacancies neighbouring to alkalihalid ions play an important role in the SA luminescence.

As for the influence of the concentration of Mn upon the SA emission there are obviously differences between different host compounds. In case of $Zn_{1-x}Mn_xS$ the SA emission disappears at Mn concentration $x \geq 0.01$.[33] $Zn_{1-x}Mn_xSe$ and $Cd_{1-x}Mn_xS$ however show SA emission over the whole possible concentration range of Mn.

2.5 The Time Dependence Of Luminescence Emission

The registration of the time dependence of emission is a powerfull tool to discriminate different emission centers and to analyze energy transport processes. One important feature of the time dependence is the built-up phenomenon. If we assume a direct excitation of a luminescence center, there should be no time delay between the excitation by laser pulses and the begin of the emission, as far as we consider the problem within the framework of ns-spectroscopy. However if there is any energy transfer from one system to some other type of luminescence center, a time delay might be involved which can be detected by the built-up phenomenon in emission. This is an important question as far as the assignments of the emission bands to the sometimes complicated excitation structures are concerned. The second point of interest is the law of decay of emission. As it is wellknown since many years, the decay of emission due to the isolated Mn^{2+} in the II-VIs can be described by one simple exponential function. Whenever there is any interaction among the Mn ions or if a transport of free carriers within the bands takes place the law of decay becomes more complicated.[34] The build-up and the decay behaviour of the different emission bands and compounds will be discussed in the following chapters.

3 LUMINESCENCE PROPERTIES OF DIFFERENT WIDE BAND GAP II-Mn-VI SMSCs

3.1 $Zn_{1-x}Mn_xS$

3.1.1 Emission

As already mentioned four emission bands can be detected if the Mn concentration is sufficiently high ($x>0.01$). The emission energies of these bands are found in the visible and in the near infrared region. For low temperatures the so-called yellow emission dominates. If the temperature increases an orange, red and IR emission band occures. The former two bands have been already observed by Gumlich et al.[35] and Neumann[36] whereas the IR band was first described by Busse et al.[9] and were confirmed by Benecke et al.[8] and Anastassiadou et al..[37] Also Benoit et al.[11] and Dang Dinh Thong et al.[10] report on different emission bands but the peak positions described by these authors differ from the energies given by others.

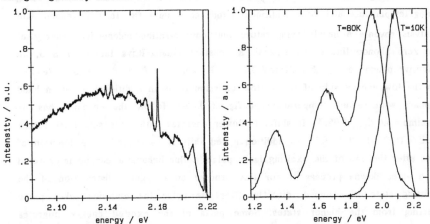

Fig.1 Emission band: $^4T_1(G) \rightarrow {}^6A_1(S)$ of $Zn_{0.999}Mn_{0.001}S$; T=10K, E_{ex}=2.6eV (ref. 17)

Fig.2 Emission spectra of $Zn_{0.5}Mn_{0.5}S$ at different temperatures; E_{ex}=2.68eV (ref. 17)

For very low Mn concentrations ($x<10^{-3}$) one observes a well structured emission band which is due to the energetically lowest internal transition of the $3d^5$ electrons of a Mn^{2+} ion on a metal site of the ZnS crystal: $^4T_1(G) \rightarrow {}^6A_1(S)$[27,28] (fig. 1). Besides the zero phonon line ($h\nu$=2.2184eV) and its thermalised sublevel (2.2196eV) different phonon replica can be clearly discriminated (TA, LA, TO, LO). With increasing temperatures (T>40K) these structures blur out and only a smooth broad band with a nearly poissonian shape remains. At lower energies no emission occurs, but at higher energies an additional emission band appears. This emission will be described in the next section.

At low Mn concentrations and low temperatures the SA luminescence at $E_{em} \approx 2.6 eV$ can be observed if the excitation energy reaches the band gap energy E_g. This emission is discussed as a donor acceptor pair recombination and is experimentally confirmed by pressure dependent emission and excitation measurements[38] and by ODMR measurements.[39,40] If the Mn concentration exceeds $x=10^{-2}$ the SA luminescence disappears.[33]

For higher concentrations ($x \geq 10^{-2}$) only one broad emission band is detected as far as the temperature does not exceed critical values. The shape of the emission band is a poissonian one. The lack of structure within the band is due to imperfections and strain of the crystals. If the temperature increases the emission spectra change drastically. The yellow emission intensity decreases fast with rising temperature and three additional bands grow up (fig. 1b). Extended data analysis reveals a nearly temperature and concentration independent energy of the zero phonon line of $h\nu \approx 2.22 eV$, a constant Huang-Rhys factor of $S \approx 3$, an effective phonon energy $\hbar\omega \approx 32 meV$ for T=4K, and a shift of $dE/dT \approx 6 \cdot 10^{-5} eV/K$. On the low energy side of the yellow emission band an additional emission band appears with raising temperatures (fig. 3). At T~30K the emission has its maximum at $E_{em}=1.95 eV$. It shifts to lower energies with increasing temperature ($dE/dT=-2.8 \cdot 10^{-4} eV/K$). In addition a red shift is registered as a function of time after the end of the exciting laser pulse.[8] This behaviour can be interpreted as a »step down« process.[41] For this process an energetic distribution of the emitting state is assumed: directly after the pulse the emission takes place starting from the higher states. Some parts of the excited energy migrates radiationless into emission centers with lower excited states where the luminescence, however shifted to lower energies, is emitted. In this way the migration of excitation energy leads to a time correlated red shift of the orange emission.

A third band peaking at $E_{em}=1.67 eV$ is observed in the red spectral range (fig. 3). This emission band shifts with a shift of $dE/dT=+1.25 \cdot 10^{-4} eV/K$ but can only be registered for T>40K. It should be noted that no genuine excitation structure or band of the red emission could be found even at very low temperatures T<4K.

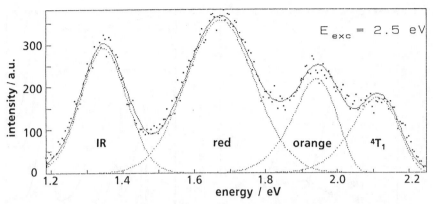

Fig.3 Emission spectrum of $Zn_{0.594}Mn_{0.406}S$ at T=100K, E_{ex}=2.5eV. The fitted bandshape are represented by lines. (ref. 17)

Another striking point is the fact that for one highly Mn doped crystal (x=0.062) no red luminescence could be detected whereas for other crystals even with the same Mn concentration this emission occurs. How far this is due to some possible variation of the growing conditions could not be clarified.[17] Finally, it should be noted that the shape of this emission band is nearly a gaussian one.

The observed IR emission band, peaking at E_{em}=1.34eV, is a common feature in all investigated II-VI compounds which have Mn concentrations above x≥0.005. Below this concentration threshold no IR emission occurs. It is very important to note that the IR emission peak energy does not depend neither on temperature nor on Mn concentration. When the Mn^{2+} ions on metal sites are excited within the $^4T_2(G)$ band (i.e. $^6A_1(S) \rightarrow {}^4T_2(G)$), the emission intensity increases with increasing temperature and reaches its maximum at T≈100K. When the temperature increases to T=200K a slight reduction of the integral intensity of emissi on is observed (fig. 4). This behaviour reflects an effective energy transfer from the ordinary Mn^{2+} yellow emitting luminescence center to the IR emitting center which will be discussed in a following section. If only the IR centers are excited (E_{ex}≤2.1eV) the integral luminescence intensity decreases monotonously; fig. 4. No energy shift as a function of time after the end of the laser pulse has been detected. The IR emission band can be fitted by a gaussian intensity distribution. From the temperature dependence of the half width FWHM of the intensity distribu- tion the effective phonon energy can be deduced from $FWHM = H_0 \sqrt{\tanh(\hbar\omega_{eff}/kT)}$. The value of $\hbar\omega_{eff}$≈32meV obtained by this procedure agrees quite well with the energy $\hbar\omega_{TO}$≈37meV of the binary compound ZnS.[42]

92

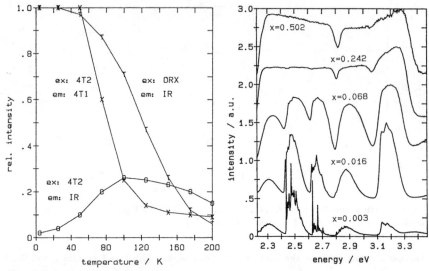

Fig. 4 Relative intensities of the yellow and of the IR emission bands as a function of temperature at different excitation energies of a ZnMnS crystal. ORX: see fig.6.

Fig. 5 Excitation spectra of ZnMnS for different Mn concentrations; T=10K, E_{ex}=2.2eV. (ref. 17)

3.1.2 Excitation

Fig. 5 shows the excitation bands of $Zn_{1-x}Mn_xS$ crystals with different amounts x of Mn (0.003<x<0.5). As it is demonstrated the structure of the bands blurs out if the Mn concentration increases. This loss of intensity modulation is due to the strong absorption A of the crystals [$A \sim \exp(-\alpha_{d-d} \cdot x_{Mn} \cdot d)$ with $\alpha_{d-d} \approx 10^3 cm^{-1}$].[2] For thinner crystals (d of the order of 20μm) the expected modulation is detectable.[43,8] The individual bands can be assigned to the crystal field transitions (in energetic order and for x≥0.1): $^6A_1(S) \to {}^4T_1(G)$ (E_{ex}=2.34eV); $^6A_1(S) \to {}^4T_2(G)$ (2.52); $^6A_1(S) \to {}^4E(G)$ (2.68); $^6A_1(S) \to {}^4T_2(D)$ (2.92eV)[44] and a 5th band at 3.16eV. One striking feature is the intensity ratio of the $^4T_1(G)/^4T_2(G)$ bands as a function of the Mn concentration x (fig. 5). The change of this ratio is also observed for all other wide band gap II–Mn–VI compounds. Until now no reasonable explaination of this phenomenon is known. It should be mentioned, however, that this feature is not confirmed by absorption measurements done by Mayrhofer et al.[43] and Giriat.[45]

As already mentioned in chapter 2 the band gap decreases with increasing Mn concentration. Detailed studies of the excitation structures reveal a 6th excitation band. The peak of this band at $E_{ex} \approx 3.66eV$ coincides with the peak found by excited state absorption measurements (ESA).[46,47] If only quartet states are taken into account this excitation band can be assigned to the $^6A_1(S) \rightarrow ^4T_1(P)$ transition within the Mn^{2+}. This assignment still has to be confirmed.

Recently, Mayrhofer et al.[43] studied the luminescence behaviour of highly doped $Zn_{1-x}Mn_xS$ crystals under hydrostatic pressure. The results agree with crystal field theory and are qualitatively similar to those of $Zn_{1-x}Mn_xSe$ and $Zn_{1-x}Mn_xTe$. For the sake of clarity we divide the spectral range of excitation into three parts:

(1) a low enery range of $1.6eV \leq E_{ex} \leq 2.2eV$, this part is just below the
   $^6A_1(S) \rightarrow ^4T_1(G)$ excitation band;

(2) the spectral region of the Mn^{2+} excitation bands leading to the yellow
   emission: $2.2eV \leq E_{ex} \leq 3.3eV$; and

(3) a high energy range of $3.3eV \leq E_{ex} \leq 3.84eV$.

(1) As a consequence of the Mn concentration threshold mentioned above the excitation structures in this range can only be observed for higher Mn concentrations ($x > 0.01$). Fig. 6 gives as an example a low energy excitation spectrum. Five bands are observed the peak energies of which are partly listed in table 2. The temperature dependence of emission when exciting in this range is the same for all low energy excitation bands: the intensity decreases with increasing temperatures whereas the excitation peak energies do not shift. For low Mn concentrations ($x \approx 0.06$) fine structures are found within the bands YX1 and YX2. The latter band is not shown in fig. 6. The difference of the energies of these fine structures are of the same order as it was found for the phonon energies of pure ZnS crystals. Therefore it seems to be safe to assign these fine structures to phonon replica.

The excitation bands YX2, YX1, ORX can only be detected when the IR emission band is registered, but not when the red emission band is recorded. It turns out that the red and the IR emission centers are independend of each other.

(2) In this region one observes an overlapping of the excitation structure of the regular incorporated Mn^{2+} ions (yellow emission) and the excitation structures of the IR emitting centers. Therefore a discrimination of the different influences of each luminescence center is difficult. Nevertheless, the influence of temperature on the excitation spectra gives the possibility to discriminate two ways of excitation.

Fig. 7 shows three excitation spectra: spectra (1) and (2) belongs to the IR emission whereas spectrum (3) represents the excitation spectra of the yellow emission. At low temperatures (T≈10K) spectrum (1) and (3) are anticoincident to each other, which means that a maximum in spectrum (3) coincides with a minimum in (1) and vice versa. With increasing temperature up to T≈100K the spectral distribution of the excitation of the IR emission changes. The spectra (2) and (3) coincide. The excitation spectrum of the IR emission reflects the excitation behaviour of the yellow emission center. This can be explained by an effective energy transport from the yellow center to the IR emitting center in a way that the specific excitation structure of the IR emitting center is masked by that of the yellow luminescence center.

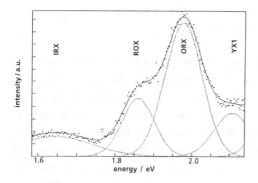

Fig. 6 Excitation spectrum of $Zn_{1-x}Mn_xS$ (x=0.406) at T=10K; E_{em}=1.34eV. (ref. 17)

Fig. 7 Excitation spectra of $Zn_{1-x}Mn_xS$ for different emission energies and temperatures. For details see text. (ref. 17)

(3) This anticoincidental behaviour is also observed in the high energy region. The maxima at E_{ex}≈3.16eV and E_{ex}≈3.7eV of spectrum (1) are attributed to the 5th and 6th excitation band of the yellow emitting center, which implies an internal transition within the 3d shell. Registration of the IR excitation energy (2) shows two additional bands labeled UV1 and UV2. The band peak maximum of UV1 follows the shift of the band gap as a function of Mn concentration x (dE/dx~-0.12eV). However the peak of the UV2 band shifts into the opposite direction (dE/dx~+0.18eV).[17]

3.1.3 Transient measurements

As already mentioned transient measurements are used to study the interdependence of different emission centers. Especially a build-up phenomenon of the emission after the end of a laser pulse can be interpreted as a result of an energy transport from the excited to the emitting center. A decision whether a relaxation process is accompanied by an energy transport or not can be based on the occurence of such a build-up phenomenon. Of course this decision is at present limited by the specific time resolution, which is reported to be $\Delta t = 30 ns$.[17]

For very low Mn concentration ($x \leq 0.001$) the decay of the transition $^4T_1(G) \to {}^6A_1(S)$ is a purely exponential one with a temperature independent characteristic decay constant of $\tau \approx 1800 \mu s$.[48] For higher concentrations this decay is not an exponential one. The decay becomes faster with both increasing temperature and concentration. This change in decay is obviously due to the opening of relaxation channels to other centers emitting in the orange, red and IR spectral range or to take radiationless channels. It should be mentioned that also the energy transfer to higher Mn clusters (pairs, triples etc.)[30,31] reduces the decay time.

When exciting the low energy excitation bands (i.e. ORX, IRX) of the IR emission no build-up phenomenon was registered. Nevertheless in all cases the decay is a non-exponential one. Assuming an additional radiationless energy transfer to other channels the decay curves can be fitted by Förster's law. As a result of this fit procedure one obtains a decay constant of about $\tau \approx 100 \mu s$. This decay constant does neither depend on the Mn concentration x nor on the temperature nor on the excitation energy ($E_{ex} \leq 2.2 eV$).[17] When assuming the diffusion model of Yokota-Tanimoto one obtains a nearly vanishing diffusion constant. This implies a very small energy migration between the IR centers, and a small abundance of these centers.

In all cases an energy transfer from the Mn^{2+} ions on metal sites to the orange, red and infrared centers is observed. A temperature dependent build-up process is registered as it is shown in fig. 8. The fact that we are probably dealing with *one* single energy donor (Mn^{2+} ions) and at least *four* different energy acceptors (orange, red, IR centers and radiationless transitions) renders the analysis of the energy transfer problem difficult. However, in order to get some information from the experimental data one can extract the modal value t_m (time for which the emisison decay curve reaches the intensity maximum) and plot this value as a function of temperature for different Mn concentration x. As for the orange and the red centers the t_m value decreases with increasing temperature which reflects an increasing energy transfer rate.

In the case of the IR centers the situation is more complicated. Obviously two processes overlap. One part of the IR emitting centers can be excited within the spectral range of internal transitions of Mn^{2+} on isolated metal sites directly without any energy transport and without phonon assistance. This part is temperature independent. As for the other part the light is absorbed first by internal transitions of the Mn^{2+}. Then the energy migrates phonon assisted to the IR emitting center.

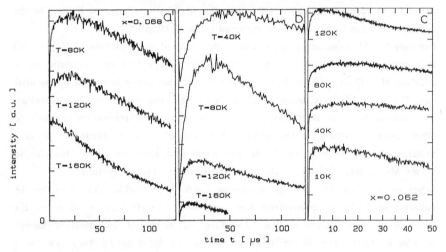

Fig. 8 Transient measurements of $Zn_{1-x}Mn_xS$ at different temperatures;
E_{ex}=2.55eV (ref. 17)
a) E_{em}=1.92eV, x=0.068; b) E_{em}=1.64eV, x=0.068; c) E_{em}=1.34eV, x=0.062

3.2 $Cd_{1-x}Mn_xS$

3.2.1 Emission

With respect to the luminescence emission the differences between $Cd_{1-x}Mn_xS$ and $Zn_{1-x}Mn_xS$ are small. As in $Zn_{1-x}Mn_xS$ four emision bands occure at high Mn concentrations (x≥0.01).[49] The peak energy of the yellow emission band is registered at E_{em}≈2.1eV and at low Mn concentrations a zero phonon line is observed (2.308eV). This emission is assigned to the $^4T_1(G) \rightarrow {}^6A_1(S)$ transition which is confirmed by Zeeman measurements. However it should be mentioned that the notation $^4T_1$ refers to a cubic crystal field, whereas CdS has hexagonal structure. For simplicity we keep the notation $^4T_1$ and consider the hexagonality as an additional perturbation of the cubic structure.

In contrast to what has been observed with $Zn_{1-x}Mn_xS$ crystals a red emission is registered even at low temperatures (E_{em}≈1.68eV; x=0.4; T≈10K). Also in contrast to $Zn_{1-x}Mn_xS$ the halfwidth of this emission increases with increasing temperatures. This broadening is accompanied by a red shift of the peak energy. An interpretation of the origin of this band is difficult because a red band can also be observed at crystals with no Mn content.[50-52] However, according to Tschierse[53] the most likely assignment of this band is to a self-activated luminescence center.

The IR emission of highly doped $Cd_{1-x}Mn_xS$ crystals is very similar to that of the $Zn_{1-x}Mn_xS$. The peak energy $E_{em} \approx 1.36eV$ does not depend on the Mn concentration and shows a temperature shift of $dE_{em}/dT \approx 2 \cdot 10^{-4}eV/K$.

3.2.2 Excitation

The band gap energy E_g of $Cd_{1-x}Mn_xS$ depends significantly on the Mn concentration x: $dE_g/dx \approx 1.04eV$ with $E_g \approx 2.56eV$ for CdS at T=4.2K. This implies that the excitation bands can be observed only at crystals with high Mn concentration. For x=0.4 three excitation bands are registered with the assignments: $^6A_1(S) \rightarrow {}^4T_1(G)$ (E_{ex}=2.42eV); $^4T_2(G)$ (2.55) and $^4E(G)$ (2.72) at T=10K. This is in accordance to the absorption measurements of Ikeda et al.[54] and Goede et al.[2]; fig. 9. Also in the case of $Cd_{1-x}Mn_xS$ the intensity ratio $^4T_1/^4T_2$ changes both in absorption and excitation. With increasing temperature these structures blur out.

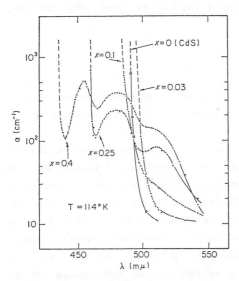

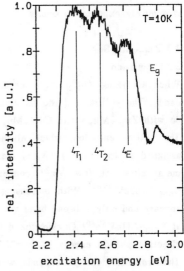

Fig. 9a Absorption spectra of $Cd_{1-x}Mn_xS$ crystals at different Mn concentration at T=114K (ref. 54)

Fig. 9b Excitation spectra of $Cd_{1-x}Mn_xS$ (x=0.4) at T=10K; E_{em}=2.06eV (ref. 17)

In order to describe the excitation of the IR emission we divide the spectral range into two regions: A low energy region ($1.5eV \leq E_{ex} \leq 2.3eV$) and a higher one ($2.3eV \leq E_{ex} \leq E_g$). In the low energy range at least three excitation bands are detectable. The peak positions of these bands do not depend on temperature but the intensity of these bands decreases slowly with increasing temperature. For the high energy region $E_{ex} \geq 2.3eV$ one finds the same IR excitation behaviour as it was found for $Zn_{1-x}Mn_xS$ crystals. At low temperatures the excitation spectra of the IR and the yellow emission are anticoincident. With increasing temperatures the shape of the IR excitation spectrum changes to that of the yellow one with the coincidence of the maxima of both spectra. As it was discussed before this effect can be explained by an effective temperature dependent energy transport from the yellow to the IR emitting centers.

3.2.3 Transient measurements

Following the work of Ehrlich et al.[49] and Tschierse[53] the decay of $Cd_{1-x}Mn_xS$ with $x \approx 0.01$ is a non-exponential one. The authors try to describe their result in the framework of percolation theory.[55] In the case of $x \geq 0.01$ they found a decay law of $I(t) = I_0 exp[- \sqrt{(t/\tau)}]$ which fits the experimental results quite well over 5 decades.

However, if the Mn concentration is small and does not exeed $x = 0.01$ the decay is a purely exponential one.

3.3 $Zn_{1-x}Mn_xSe$

3.3.1 Emission

When analyzing $Zn_{1-x}Mn_xSe$ crystals one observes three luminescence emission bands. The yellow and the IR emission bands show the same behaviour as they do with $Zn_{1-x}Mn_xS$ and $Cd_{1-x}Mn_xS$.

The yellow emission, which already has been studied by Langer et al.[56] is assigned to the $^4T_1(G) \rightarrow ^6A_1(S)$ transition of $3d^5$ electrons of isolated Mn^{2+} on metal sites. At low Mn^{2+} concentrations a zero phonon line is observed at $E_{em} = 2.235eV$.[57,58] With increasing Mn concentration the zero phonon line disappears and only a broad band with a poisson like shape can be detected peaking at $E_{em} = 2.11eV$.[59,60] Oczkiewicz determines in a crystal field theory approach an effective phonon energy of $\hbar\omega = 21.3meV$ and a Huang Rhys factor of $S = 4.79$.[59] The transition energy depends weakly on both the Mn concentration and the temperature. The luminescence intensity decreases if the temperature increases.

It is important to note that it depends on the crystal growing conditions wether a selfactivated luminescence emission is detectable[40] peaking at E_{em}=1.92eV or not. When raising the temperature from T=30K to 60K the typical change from bound-to-bound to free-to-bound transition is observed. Therefore the assignment of this band to a SA center seems to be confirmed. The intensity of the SA band increases with increasing temperature. No shift of the SA luminescence with increasing Mn concentration x is observed although the band gap increases with increasing x.

As all other II-Mn-VIs $Zn_{1-x}Mn_xSe$ shows an IR emission peaking at E_{em}=1.30eV with a gradient of $dE/dx \approx 70 meV$. This emission occures only if the Mn concentration exeeds x=0.001.[60] The temperature dependence is very similar to that of the IR emission of $Zn_{1-x}Mn_xS$: with increasing temperature the intensity ratio of IR to the yellow band increases, whereas the intensity of the yellow emission decreases monotonously.

3.3.2 Excitation

Depending on the Mn concentration and the related opening of the band gap three or four excitation bands are observed to: $^6A_1(S) \rightarrow {}^4T_1(G)$ (E_{ex}=2.34eV); $^4T_2(G)$ (2.50eV); $^4E(G)$ (2.72eV) and probably $^4T_2(D)$ (2.9eV).[60] More recently, Dreyhsig et al. confirmed these results by excited state absorption measurements.[61] As we might have expected the excitation of the SA luminescence shows some significant structure only for E_{ex}>2.7eV. This structure is related to donor-acceptor pairs (DAP) recombination, consisting of flat donors (i.e. Cl) and singly charged deep acceptors (i.e. V_{Zn}-Cl).

As for the low energy range ($E_{ex} \leq 2.25 eV$) five excitation bands of the IR emission can be detected. The striking similarity of these structures in both $Zn_{1-x}Mn_xS$ and $Zn_{1-x}Mn_xSe$ should be noted. The peak energies are concentration and temperature independent. The intensity of these structures decreases slowly with increasing temperature, the absolute intensities of the IR low energy excitations bands are some 100 times weaker as the excitation bands of the yellow emission. We take this as an indication of the ratio of yellow to IR emitting centers. In the energy range of the yellow Mn^{2+} excitation bands one observes the same strong temperature dependent excitation behaviour of the IR centers as it was discussed for $Zn_{1-x}Mn_xS$. At low temperatures the excitation maxima of the yellow and the IR emission respectively are anticoincident, whereas at higher temperatures the bands coincide, which is also an expression for an effective temperature dependent energy transfer.

3.3.3 Transient measurements

As for Mn concentrations below $x \leq 10^{-3}$ the decay of the yellow emission can be described by a single exponential function with a decay constant of $\tau \approx 240 \mu s$.[62] With increasing concentration x the decay becomes faster and the behaviour can not any more described by one single exponential function. This is obviously related to an energy transfer from the Mn^{2+} to the DAPs and to the IR centers. The decay of the IR emission shows no build up effect in the limits of the time resolution $\Delta t = 10 \mu s$. Generally speaking the decay becomes faster if the Mn concentration and/or the temperature increases.

3.4 $Cd_{1-x}Mn_xSe$

3.4.1 Emission

The literature with respect to the luminescene of $Cd_{1-x}Mn_xSe$ is quite modest especially as far as the Mn related luminescence is concerned. The small energy gap (CdSe: $E_g=1.83eV$ at $T=4.2K$ and $dE_{gap}/dx=1.47eV$) and the expected yellow emission energy near 2eV are conflicting. Therefore a yellow Mn^{2+} related emission can be observed only for $x \geq 0.3$. As a consequence no zero phonon line can be detected due to the inhomogenous broadening of the emission band. Moriwaki et al.[63] reported on five emission bands, although only two are specified: a yellow one ($E_{em}=2.13eV$) and an IR one (1.35eV). In contrast to Moriwaki et al. Benecke[17] registered only three emission bands. As fig. 10 shows the yellow emission, peaking at $E_{em}=2.14eV$, decreases with increasing temperature. In the same range of temperatures a second emission band occurs at $E_{em}=1.96eV$. By analogy to the luminescence of the $Zn_{1-x}Mn_xSe$ this emission is tentativily assigned to a DAP recombination.

3.4.2 Excitation

Due to the small energy gap only one excitation band of the yellow Mn^{2+} emission can be observed without doubt. This band is attributed to the transition $^6A_1(S) \rightarrow ^4T_1(G)$ ($E_{ex}=2.35eV$). A second excitation band seems to be hidden partly by the band-band transitions.

The excitation spectra for the DAP emission show two broad bands just below E_g at $T=10K$. With increasing temperature the intensity of the high energy band increases with respect to the low energy band. There seems to be a correlation to the band gap shift $E_g(T)$. A broad excitation band peaking at $E_{ex}=2.68eV$ ($T=10K$; notice $E_g(x=0.45)=2.5eV$) is clearly registered but not yet explained.

Finally, the excitation spectra for the IR emission band show the low energy bands as expected. However, these bands are not stuctured to the same extend as it is the case of $Zn_{1-x}Mn_xS$ and therefore only the peak energy of the lowest band could be given, as $E_{ex}=1.61eV$ (T=10K). Also Morales et al.[64] showed in this energy range a small absorption structure ($\alpha \approx 1\text{-}10cm^{-1}$); fig. 11. With increasing temperatures a small shift to lower energies occurs. This absorption is nearly independent of the polarisation (E ∥ c; E ⊥ c) and the absorption coefficient α does not change when the temperature increases to T=77K.

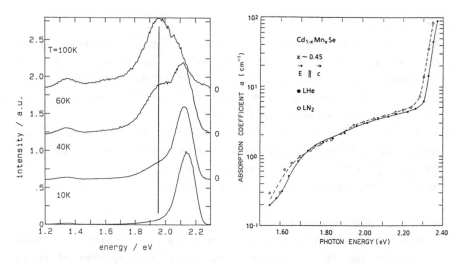

Fig. 10 Emission spectra of $Cd_{1-x}Mn_xSe$ (x=0.45) at different temperatures; $E_{ex}=2.7eV$. (ref. 17)

Fig. 11 Absorption spectra of $Cd_{1-x}Mn_xSe$ (x=0.45) at two temperatures. (ref. 64)

3.5 $Zn_{1-x}Mn_xTe$

3.5.1 Emission

$Zn_{1-x}Mn_xTe$ shows two emission bands: At low temperatures an intensive yellow emission at $E_{em}=1.95eV$ and a weaker one at $E_{em} \approx 1.1eV$. The high energy band is due to the internal transition of the $3d^5$ electrons of Mn^{2+} ions.[65] When the Mn concentration decreases the yellow emission shifts to lower energies. In contrast to this behaviour Lee et al.[66] found by piezomodulated measurements a concentration independent emission energy of the yellow band. According to these authors this energy decreases monotonously with increasing temperature in the same way as E_g does. On the other hand Müller et al.[67] observed a minimum at $T \approx 50K$ for the energy as a function of T in absorption spectra.

It should be noted that for experimental reasons the peak energy of the IR emission near $E_{em} \approx 1.1eV$ is not yet well established. However, from band shape analysis a shift to higher energies with increasing temperature is a save finding. The intensity of the IR band decreases with increasing temperature slowly and monotonously. The drop of the intensity of the IR emission is less pronounced compared to that of the yellow one, when the excitation energy is $E_{ex}=2.2eV$. This supportes the idea of an energy transfer from the yellow to the IR emitting centers.

3.5.2 Excitation

The number of excitation peaks of the yellow Mn emission depends on the Mn concentration, whereas the two low energy bands are not well resolved. At $x \approx 0.6$ four bands can be detected and assigned to: $^6A_1(S) \rightarrow {}^4T_1(G)$ ($E_{ex}=2.25eV$); $^4T_2(G)$ (2.40eV); $^4E(G)$ (2.65eV) and a 4th band at 2.74eV. These energies do not depend on temperature and concentration.[68,17]

As it was found with the other II-Mn-VI compounds one observes in the low energy range excitation bands of the IR emission, which behave quite similar to the other II-Mn-VIs. Also the same coincidence-anticoincidence feature as a function of temperature as described above was found.

3 5.3 Transient measurements

Müller et al.[67] studied the lifetime of the yellow emission band for different temperatures and Mn concentrations. They report a decay time of $\tau=43\mu s$ (x=0.1; T=2K) which agrees well with the given data given by Krause et al.[69] ($\tau=52\mu s$ at T=10K and x=0.008). Also in this case the character of the decay curve changes from an exponential to a non-exponential one if the Mn concentration increases.

3.6 $Cd_{1-x}Mn_x Te$

$Cd_{1-x}Mn_x Te$ is probably the most prominent SMSC. Compared to other II-Mn-VI SMSCs CdTe has the smallest energy gap of $E_g=1.6eV$ (T=4.2K; $dE_g/dx \approx 1.6eV$). Therefore the yellow emission appears only for $x \geq 0.4$. It should be mentioned that care should be taken when using the term yellow because sometimes in literature the term red is used for the same emission.

3.6.1 Emission

When the Mn concentration is appropriately high ($x \geq 0.4$) a yellow emission band peaking at $E_{em} \approx 2eV$ is observed. This emission is attributed to the $^4T_1(G) \to {}^6A_1(S)$ transition[13] which is confirmed by experiments with hydrostatic pressure and more recently by Zeeman measurements.[65,15] With increasing temperature the peak shifts in a complicated way as fig. 12 shows. As it is the case of all the other considered substances the intensity of yellow emission decreases with increasing temperature.[63]

The IR emission has been first observed by Vecchi et al.[12] who found that this emission is only detectable if the concentration exceeds $x \geq 0.05$. The temperature and concentration dependence of the peak energy is shown in fig. 13. The effect of hydrostatic pressure on the peak energy is manifested in a red shift with a gradient of $dE/dp = -4 \cdot 10^{-5} eV/MPa$.[70] This value is to be compared to the corresponding coefficient of the yellow emission: $dE/dp = -8.1 \cdot 10^{-5} eV/MPa$.[15]

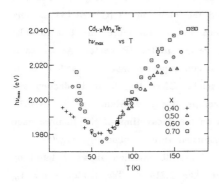

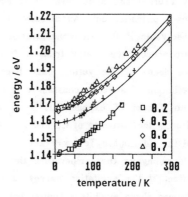

Fig. 12 Peak position of the emission band $^4T_1(G) \to {}^6A_1(S)$ of $Cd_{1-x}Mn_x Te$ of different Mn concentration as a function of temperature. (ref. 13)

Fig. 13 Peak position of the IR emission band at different Mn concentration and temperatures. (after ref. 70)

3.6.2 Excitation

In spite of a few attempts undoubtly clear assignments of the excitation bands of the yellow emission of Mn^{2+} could not established.[71] Tentatively we give the following assignments: $^6A_1(S) \to {}^4T_1(G)$ ($E_{ex} = 2.23eV$); $^4T_2(G)$ (2.46) and $^4E(G)$ (2.60). Fig. 14 demonstrates the uncertainty of the energy values. Due to the low energy gap even for Mn concentrations at the solubility limit a fourth excitation band was not detected. A temperature shift of the transition energies could not be found. The intensity of the unstructured bands decreases as the temperature increases whereas the band gap energy shows a distinct red shift.

It is interesting to note that, very similar to the low energy excitation bands of the $Zn_{1-x}Mn_xTe$, one observes also in $Cd_{1-x}Mn_xTe$ six bands. The most striking feature is the strong excitation peak at $E_{ex}=1.98eV$. The extraordinary large intensity can be explained by a resonant energy transfer from the yellow center ($E_{em}=2.0eV$) to the IR centers. With increasing temperature the intensity of these bands decreases and blurs out for $T \geq 60K$. As a consequence of the energy transfer already discussed the coincidence vs. anticoincidence behaviour is also observed in $Cd_{1-x}Mn_xTe$.[14)]

3.6.3 Transient measurements

As in the case of $Zn_{1-x}Mn_xTe$ the decay of the yellow emission band of $Cd_{1-x}Mn_xTe$ is relatively fast: $\tau(Cd_{0.35}Mn_{0.65}Te; T=2K)=20\mu s$.[67)] Similar results are obtained by Prüll[72)] and Schliwinski.[73)] It should be noted that the decay is always a non-exponential one. As already mentioned the emission $^4T_1(G) \rightarrow {}^6A_1(S)$ is only observable if the Mn concentration exceeds $x \approx 0.5$. Therefore one has to take into account the energy transfer from the yellow to the IR center. As a result the decay of the yellow band becomes a non-exponentional one and a clear definition of a decay constant τ is not possible. Often the decay is decomposed into a sum of exponentials[48)] or an average decay constant $<\tau>$ is used.[67)] Studing the influence of Cd and Zn cations on the decay time of the luminescence of the $Cd_xZn_yMn_zTe$ ($x+y+z=1$) as an empirical rule is found: $\tau_{Cd}>\tau_{Zn}$.[74)]

Depending on the temperature the decay of the IR emission shows a build up effect if the excitation energy is greater than 2.214eV. The modal value t_m defined above exhibits a similar behaviour: an increasing temperature leads to an increasing of the modal values. This reflects an increasing energy transfer from the Mn^{2+} to the IR center. A quantitative description of this decay behaviour is difficult because of the different kind and partly unknow relaxation channels.

4 ON THE NATURE OF THE Mn CORRELATED LUMINESCENCE CENTERS AND CHEMICAL TRENDS

The results reported in the foregoing chapters are leading to the conclusion that there are at least two independent luminescence centers in the wide band gap II–Mn–VI semimagnetic semiconductors: The yellow and the IR emission are present in all II–VIs, as fare as the Mn–concentration x exceeds a critical value. An orange and a red emission band are only observed in some compounds. As for the yellow emission, which is observed in all II–Mn–VIs with high Mn–concentration, the assignement to the internal transition $^4T_1(G) \rightarrow {}^6A_1(S)$ is generally accepted.

As for the orange and red emission bands, which are registered only in some crystals, selfactivation (SA) or Mn–clustering respectively are discussed.

The most interesting feature however is the appearence of the infrared emission band in all wide band gap II–Mn–VIs, as far as the Mn–concentration exceeds a value of about one percent. It should be mentioned that the origin of this IR emission and the nature of the IR emitting center is still controversial.

The first proposal came from Vecchi et al. in 1981, who tried to explain the 1.2eV emission band in $Cd_{1-x}Mn_xTe$.[12] In this model the Mn^{2+} ion is excited via the $^6A_1 \rightarrow {}^4T_1$ transition, where the excited state interacts with the valence band. The resulting recombination of an excited electron in the $^4T_1$–state with a free hole in the valence band gives rise to the 1.2eV IR emission band. To conserve the isoelectronic state of the Mn^{2+} ion the authors assume the transfer of a valence electron to the Mn^{2+} ground state $^6A_1$. The weak point of this model seems to be that the ground state of the Mn^{2+} ion is supposed to be 0.8eV below the top of the valence band. This is in contradiction to a number of findings, which lead to values near –3.5 eV. Taniguchi et al. gave a value of –3.8eV.[23] In the concept of Vecchi et al. transitions leading to the yellow and the IR emissions have the same starting point, namely the $^4T_1$ state of the Mn^{2+} ion. This assumption is in contrast to the experimental finding of a built-up-effect in the temporal evolution of the IR emission after the end of the exciting laser pulse, as Benecke et al.[8] and Busse et al. pointed out.[9] There is obviously some temperature dependent energy transfer between different emission centers involved.

In order to explain the two emission bands Becker et al. discuss a different model in $Cd_{1-x}Mn_xTe$, which is based on a static Jahn–Teller-effect (JTE).[80] Due to the coupling of vibronic states to the electronic triplet states of the

Mn^{2+} two different energetic minima (1,2) in the $^4T_1$-APES (excited states adiabatic potential energy surface) occur. Depending on temperature some part of the excitation energy is transfered from the $^4T_1$ minimum (1) into the $^4T_1$-APES minimum (2). The energy relaxation from minimum (1) leads to the yellow 2.0eV emission, whereas the 1.2eV emission is due to energy relaxation from minimum (2).

The inconvenience of this model seems to be the value of the energetic difference of 0.8eV between the two minima as the result of a static JTE. This value is obviously exceedingly high. Besides of this inconvenience according to this model one would expect an IR emission for (Zn,Mn)- or (Cd,Mn)-chalcogenides with similiar electron-phonon-coupling of the $^4T_1(G)$ state for lower Mn concentrations ($x \leq 10^{-3}$). This never has been reported.

The theoretical considerations of Wei and Zunger on this subject should be mentioned.[26] They assume that the 2.2eV emission of $Cd_{1-x}Mn_xTe$ is a result of a p-d interaction. After the excitation of the crystal a valence band electron is added to the lowest unoccupied Mn orbital. This leads to a Mn^{1+} ion plus a hole in the valence band. The recombination of this loosely bound electron with a delocalized valence band hole is supposed to give rise to the 2.0eV emission band.

In contradiction to Wei and Zunger's assignments the 2.2eV emission belongs to the group of concentration dependent phenomena as already reported.[26] The concentration dependence reflects the energetic shift of the fundamental gap. In accordance to Vecchi et al. we assume that this emission is due to an excitonic recombination. Following this line we assign the concentration independent 2.0eV emission band to the internal $3d^5$ Mn^{2+} transition $^4T_1(G) \rightarrow {}^6A_1(S)$, as already reported by other authors.[35]

Ambrazevicius et al. used the p-d interaction in order to explain the IR emission peaking at 1.2 eV in $Cd_{1-x}Mn_xTe$.[14] These authors observed a blue shift of the IR emission with increasing Mn concentration. The increasing Mn concentration leads to a reduction of the lattice constant of $Cd_{1-x}Mn_xTe$ which in turn should amplify the hybridization.[2] From this enhancement of the hybridization the authors deduce a blue shift of the corresponding transition from the excited level into the valence band.

As for $Zn_{1-x}Mn_xSe$ the IR emission shows a blue shift with increasing Mn concentration too.[60] It is important to note, however, that the lattice constant of this material increases with x. This is clearly in contrast to the arguments used by Ambrazevicius et al..[14]

To summarize the deficiencies of these models discussed so far all previous attempts lead to some discrepancies in the light of the experimental findings:

a) The existence of a concentration threshold of the IR emission is not included in these concepts.

b) The temperature dependent excitation spectra, in particular the anticoincidence phenomenon as a qualitative change cannot be explained by the proposed assumptions.

c) The existence of the low energy excitation bands of the IR emission does not fit into these models.

d) The built-up phenomenon depends upon temperature: at low temperatures the intensity ratio $I_{IR}/I_{4_{T_1}}$ is small. This leads to the assumption of a small energy transfer probability to the IR center, which is equivalent to a long built-up-effect. The experimental data however show a vanishing built-up-effect with decreasing temperatures.[8]

These discrepancies of phenomena and models, backed up by the measurements and arguments of Goede et al. lead us to the assumption of an independent IR center, which appears in addition to the yellow emitting Mn^{2+} center (Mn^{2+} on a metal site) at Mn concentrations $x » 0.01$.[2] The strongest experimental support for this assumption is given by the time dependence of the emission.

While it is evident that the existence of the IR centers are closely connected to the presence of higher Mn concentrations, the precise nature of them is not yet definitly established. Therefore for the time being we use for the IR center the notation Mn_x. However the shape of both the emission and the excitation spectra of the IR luminescence lead us to the idea of dealing with Mn^{2+} in another highly symmetric coordination, for instance Mn^{2+} in octrahedral coordination.

Two possibilities of different coordination are discussed.[17]

First the Mn^{2+} ion may go on interstitial sites, which are available in the compounds. Site one is near to the anion in a position $(1/2,1/2,1/2)a$, where a is the cubic lattice constant. Site two is nearest to the cation $(3/4,3/4,3/4)a$.[75] The interstitial ion in a zincblende lattice has four tetrahedrally coordinated next nearest neighbours and six octahedrally coordinated next nearest neighbours. Also in a wurtzite lattice octahedrally coordinated interstitial sites are available, as has been demonstrated for instance by Schallenberger and Hausmann who deduced from EPR measurements the existence of Cu ions on octahedral interstitial sites in ZnO:Cu.[76]

As a matter of fact Goede et al. have demonstrated the shift of the emission and excitation peaks as a result of a phase transition from tetrahedrally to octrahedrally coordinated Mn^{2+} in MnS. The authors report a drastic change of the energetic positions of peaks both in emission and excitation of MnS thin films induced by thermal treatement. The energy of the $^4T_1(G) \rightarrow {}^6A_1(S)$ transition shifts from 1.95 eV in the tetrahedral environment to 1.45 eV in the octrahedral one. Corresponding red shifts of the emission bands have been reported by other authors who studied the luminescence emission of Mn in crystals with different site symmetries, among them Lanver and Lehmann.[77] A corresponding possibility, which has to be discussed, is the formation of domains of octahedral MnS, MnSe and MnTe, respectively. Obviously the probability of formation of such domains increases with increasing amount x of Mn. The appearence of the IR emission only at concentrations x»0.01 could be reasonably explained in that way. However the energies of the IR emission and their excitation bands of the II-VI-crystals with high amounts of Mn should be consistent with the energies of pure octahedral MnS, given by Goede et al..[79] As table 2 shows the difference of values in both cases cannot be neglected. One may argue that the formation of MnS, MnSe and MnTe domains within II-VI crystals induces some strain leading to an energy shift of the bands with respect to the strainless structure of the MnS-films. This argument is certainly a convincing one but at this very moment the existence of IR emitting MnS, MnSe or MnTe domains in the II-VIs has not yet been demonstrated unambiguously.

Tab. 1   Energies of the excitation bands for the substitutional Mn^{2+} in T_d symmetry, the calculated Racah parameters B, C, Dq, and the ratio C/B

a) Band position values taken from Goede et al.[79]

symmetry	substance	band position (eV)			Racah parameter (meV)			
		$^4T_1$	$^4T_2$	^4E	B	C	$|Dq|$	C/B
T_d	CdMnTe	2.23	2.46 ?	2.60	60	400	63	6.8
	ZnMnTe	2.25	2.40	2.65	30	469	72	15.4
	CdMnSe	2.35	2.45 ?					
	ZnMnSe	2.34	2.50	2.67	38	459	63	12.2
	CdMnS	2.42	2.55	2.72	30	483	61	15.9
	ZnMnS	2.34	2.52	2.68	44	448	64	10.2
	β-MnS[a]	2.38	2.53	2.67	38	457	58	11.9
O_h	α-MnS[a]	2.03	2.44	2.75	80	390	100	4.9

Following this line we are restricted to the statement that the IR emission and their low energy excitation bands are very likely to be due to Mn in a cubic environment, which differs from the wellknown tetrahedrally coordinated Mn on a metal site, giving rise to the yellow emission band. This cubic environment may either be realized by Mn on interstitial sites or by Mn in domains of different phases, for instance rocksalt phases as MnS, within the II-VI-compounds.

It is tempting to use the tools of basic crystal field theory in order to describe the IR emission and their excitation spectra in the terms of the Racah parameters B and C and by the crystal field parameter Dq. For the octahedral site O_h and for the tetrahedral site T_d the energy matrices have the same form, only the appropriate crystal field parameters Dq have to be choosen: $Dq_{octa}/Dq_{tetra}=-9/4$.[78] The Racah parameters B and C for cubic symmetry, which are treated independently, are listed in table 1. The spin-orbit-interaction is neglected and only quartet terms are taken into account. The energetic input data are received from the band position of $^4T_1(G)$, $^4T_2(G)$ and $^4E(G)$.

Tab. 2   Energies of the low excitation bands for the Mn_x^{2+} center in cubic symmetry, the calculated Racah parameters B, C, Dq, and the ratio C/B

substance	band position (eV)			Racah parameter (meV)			
	$^4T_1$	$^4T_2$	^4E	B	C	$|Dq|$	C/B
CdMnTe	1.43	1.65	1.75	55	240	48	4.3
ZnMnTe	1.40	1.60	1.72	46	251	49	5.4
CdMnSe	1.61		2.10				
ZnMnSe	1.60	1.81	1.92	53	279	51	5.1
CdMnS	1.68	1.88	2.05	43	323	60	7.4
ZnMnS	1.64	1.85	1.98	50	296	54	5.9

One problem of this consideration should be mentioned. As a consequence of crystal field theory one should expect for the ratio C/B a value of approximatly 4 to 5.[29] Therefore, if the parameters B, C and Dq are treated in a linearely independent way, the ratio C/B can be used to check the validity of the model. As table 1 shows, there is a drastic difference between the theoretically required value and the value obtained by fitting the experimental data of the tetrahedral coordination. It should be noted, however, that similiar differences have already been reported by Pohl.[29] Obviously the large ratios C/B belong to the well-known unsolved problems of the crystal field theory. In contrast to the tetrahedral case the values calculated for the cubic Mn_X center, which may be realized as interstitials or in a domain of different phase, agree remarkedly well with the theoretical requested values of 4 or 5.

It should be also mentioned that the analysis of the spectra within the framework of the crystal field theory agrees well with other usual approaches, e.g. the configuration coordination model of luminescence. Fitting the shape of the low energy excitation bands in $Zn_{1-x}Mn_xS$ lead to a Huang-Rhys-factor of about $S=3$ and an effective phonon energy near 32meV. This is in good agreement with the corresponding values of the tetrahedrally coordinated Mn^{2+} ions on metal sites.

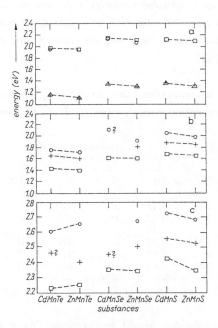

Fig. 14 Energies of the observed transitions of different wide band gap SMSCs. The dashed lines are to guide the eyes.

(?) means that the spectral position is uncertain.

a) emission bands ($\square$ $^4T_1$(T=10K), + IR (T=10K),

$\qquad$ O $^4T_1$(T=90K), $\triangle$ IR (T=90K));

b) excitation bands of the Mn_X center (O ^4E, + $^4T_2$, $\square$ $^4T_1$);

c) excitation bands of substitutional Mn^{2+} (O ^4E, + $^4T_2$, $\square$ $^4T_1$).

Having in mind some discrepancies of the treatment of the IR emission and their excitation bands by the crystal field theory, this treatment is backed up by a striking feature of the spectra, which indicates a chemical shift: both emission and excitation peaks are shifted with the same energy, if one compares different II-VIs going from one compound to the other. The values of the shift follow the line: Te→Se→S.

Fig. 14 visualize the observed trends. The dashed lines are guiding the eyes. In fig. 14a the peak positions of the emission bands are given. It is obviously the anion which essentially determines the energy of the emission, whereas a small shift to lower energies is observed if the Cd is replaced by Zn cations. It should also be mentioned that a small shift of the peak position as a function of the concentration x has been registered in some spectra.

Also in the case of the low energy excitation bands the trend is obvious (fig. 14b). The peak energies of the Te- and S-compounds are nearly constant, neglecting some uncertainty with the Se-crystals. Also the trend with respect to the cations still holds: Te→Se→S.

For the excitation energies of substitutional, tetrahedrally coordinated Mn^{2+} ions on metal sites the situation seems to be more complicated (fig. 14c). In the case of Te the excitation peak energy shifts to higher values, if the cations are interchanged. The opposite is observed for the S-compounds. However also in this case the general trend is valid again: Te→Se→ S. This may reflect the trend of the ionic radii:

$$Te^{2-} \ (0.221nm) \to Se^{2-} (0.198nm) \to S^{2-} \ (0.184nm).$$

It is interesting to note that also the observed Stoke's shift of the IR luminescence of about 300meV is in good agreement with the published data of the considered compounds.[59]

However one problem remains unsolved, if the basic crystal field approach for the IR emission of II-Mn-VIs is accepted. This is the behaviour of the ^4E state of Mn^{2+} in variing crystal fields. Within the framework of the crystal field theory the state ^4E should not depend on the crystal field strength Dq. As the experimental results show the energetic difference $^4E_{tetra} - {}^4E_{Mn_x}$ of $Zn_{1-x}Mn_xS$ is 0.7eV. This again seems to indicate the limits of basic crystal field theory, when one does not take into account other effects, for instance the effect of covalency.

Generally speaking it seems to be reasonable to explain the IR emission of wide band gap $II_{1-x}Mn_x VI$-compounds by the model of Mn_x on cubic sites either as interstitials or in octahedrally coordinated domains. It should be the goal of the future work to find out wether this specific cubic environment is realized on

112

interstitial sites or in microscopic domains of MnS, MnSe or MnTe. Nevertheless the luminescence phenomena of the bulk semimagnetic semiconductors give already by now evidence of the inhomogeniety of the crystals. All semimagnetic semiconductors with Mn contents of more than one percent are far from the ideal statistical distribution of the Mn and perhaps include more than only one crystallographic phase.

Acknowledgement.

The authors wish to thank W.Busse and H.Hoffmann for helpfull discussions.

5 REFERENCES

1 Brandt, N.B. and Moshchalkov, V.V., Adv. Phys. $\underline{33}$, 193 (1984)
2 Goede, O. and Heimbrodt, W., phys. stat. sol. (b) $\underline{146}$, 11 (1988)
3 Furdyna, J.K. and Kossut, J., (Ed) "Diluted Magnetic Semiconductors"
     in Semiconductors and Semimetals $\underline{25}$, (1988)
4 Furdyna, J.K., J. Appl. Phys. $\underline{62}$(4), R29 (1988)
5 Galazka, R.R., J. Cryst. Growth $\underline{72}$, 364 (1985)
6 Gaj, J.A., J. Phys. Soc. Japan Suppl. $\underline{A49}$, 747 (1980)
7 Gebhardt, W., "Excited-State Spectroscopy in Solids", Bologna, 111 (1987)
8 Benecke, C., Busse, W., Gumlich, H.-E. and Moros, H.-J.,
     phys. stat. sol. (b) $\underline{142}$, 301 (1987)
9 Busse, W., Gumlich, H.-E., Krause, M., Moros, H.-J., Schliwinski, J., and
     Tschierse, D., J. Lum. $\underline{31\&32}$, 421 (1984)
10 Dang Dinh Thong, Goede, O., phys. stat. sol. (b) $\underline{120}$, K145 (1983)
11 Benoit, J., Benalloul, P., Geoffroy, A., Balbo, N., Bartou, C., Denis, J.P. and
     Blanzat, B., phys. stat. sol. (a) $\underline{83}$, 709 (1984)
12 Vecchi, M.P., Giriat, W. and Videla, L., Appl. Phys. Lett. $\underline{38}$ 99 (1981)
13 Moriwaki, M.M., Becker, W.M., Gebhardt, W. and Galazka, R.R.,
     Phys. Rev. B $\underline{26}$, 3165 (1982)
14 Ambrazevicius, G., Babonas, G. and Rud, Yu. V.,
     phys. stat. sol. (b) $\underline{125}$, 759 (1984)
15 Müller, E., Thesis, Regensburg (1986)
16 Goede, O. and Heimbrodt, W., this volume
17 Benecke, C., Thesis, TU Berlin D83 (1989)

18 Bylsma, R.B., Becker, W.M., Kossut, J., Debska, U. and Yoder-Short, D.
   Phys. Rev. B $\underline{33}$, 8207 (1986)

19 Röhrig, , R. Thesis, Freiburg (1973)

20 Tanabe, S., Tanabe, Y. and Kamimura, H.,
   "Multiplets of Transition-metal Ions in Crystals", New York (1970)

21 Vogl, P. and Baranowski, J.M., Acta Phys. Pol. $\underline{A67}$, 133 (1985)

22 Orlowski, B.A., Kopalko, K. and Chap, W., Solid State Commun. $\underline{50}$(8),
   749 (1984)

23 Taniguchi, M., Ley, L., Johnson, R.L., Ghijsen, J. and Cardona, M.,
   Phys. Rev. $\underline{B33}$, 1206 (1986)

24 Chab, W., Paolucci, G., Prince, K.C. and Bradshaw, A.M.,
   Phys. Rev. B $\underline{38}$, 12353 (1988)

25 Burmester, B., Gumlich, H.-E., Jung, Ch., Krost, A., Middelmann, H.-U.,
   Ricken, D., Weidemann, R., Becker, U. and Kupsch, M., 19th Int. Conf.
   Phys. Semicond., Warsaw, Polish Academy of Science, 1583 (1988)

26 Wei, S.-H. and Zunger, A., Phys. Rev. $\underline{B35}$, 2340 (1987)

27 Langer, D. and Ibuki, S., Phys. Rev. $\underline{138}$, A809 (1965)

28 Parrot, R., Geoffroy, A., Naud, C., Busse, W. and Gumlich, H.-E.,
   Phys. Rev. B $\underline{23}$, 5288 (1981)

29 Pohl, U.W., Thesis, TU Berlin D83 (1988)

30 Benecke, C., Busse, W., Gumlich, H.-E. and Pohl, U.-W.,
   phys. stat. sol. (b) $\underline{128}$, 701 (1985)

31 Pohl, U.W. and Gumlich, H.-E., Phys. Rev. B $\underline{40}$, 1194 (1989)

32 Hartmann, H., Mach, R. and Selle, B.,
   in Current Topics in Materials Sciences (Ed. E. Kaldis) $\underline{9}$, 1-414 (1982)

33 Gobrecht, H. and Gumlich, H.-E., Z.Physik $\underline{156}$, 436 (1958)

34 Di Bartolo, B., "Energy Transfer Processes in Condensed Matter"
   New York (1985)

35 Gumlich, H.-E., Moser, R. and Neumann, E., phys. stat. sol. $\underline{24}$, K13 (1967)

36 Neumann, E., Thesis, TU Berlin D83 (1971)

37 Anastassiadou, A., Liarokapis, E., Anastassakis, E. and Stoyanov, S.,
   Physica Scripta $\underline{38}$, 444 (1988)

38 Jaszcyn-Kopec, P., Pinceaux, J.P., Zigone, M., Kennedy, J.M., Stadtmuller, A.,
   Solid State Commun. $\underline{32}$, 473 (1979)

39 Zink, K., private communication

40 Davies, J.J., J. Cryst. Growth $\underline{72}$, 317 (1985)

41 Müller, E., Gebhardt, W., Gerhardt, V., phys. stat. sol. (b) $\underline{113}$, 209 (1982)

114

42 Landoldt-Börnstein, Physik der II–VI und der I–VII Verbindungen 17(b)
       Ed. Madelung, O., Berlin (1982)

43 Mayrhofer, K., Hochberger, K. and Gebhardt, W.,
       J. Phys. C: Solid State Phys. 21, 4393 (1988)

44 Pohl, U.W., Gumlich, H.-E., Busse, W., phys. stat. sol. (b) 125, 773 (1984)

45 Giriat, W., Prog. Cryst. Growth Charact. (Ed. Pamplin, B.R.) 10 (1985)

46 Dreyhsig, J. and Allen, J.W. J.Phys.: Condensed Matter 1, 1087 (1989)

47 Kushida, T., Tanaka, Y. and Oka, Y., J. Phys. Soc. Japan, 37, 1341 (1974)

48 Busse, W., Gumlich, H.-E., Meissner, B. and Theis, D.,
       J. Lum. 12 & 13, 693 (1976)

49 Ehrlich, C., Busse, W., Gumlich, H.-E. and Tschierse D.,
       J. Cryst. Growth 23C, 371 (1985)

50 Shiraki, Y., Shimada, T., Komatsubara, K.F., J. Appl. Phys. 45, 3554 (1974)

51 Ermolovich, I.B., Gorbunov V.V. and Matvievskaja, G.I.,
       phys. stat. sol. (b) 106, 45 (1981)

52 Ermolovich, I.B., Matvievskaja, G.I., Pekar, G.S. and Sheinkman, M.K.,
       Ukr. fiz. Zh. 18, 732 (1973)

53 Tschierse, D., Thesis, TU Berlin D83 (1986)

54 Ikeda, M., Itho, K. and Sato, H. J. Phys. Soc. Japan 25, 455 (1968)

55 Evesque, P., J. Physique 44, 1217 (1983)

56 Langer, D.W. and Richter, H.-J., Phys. Rev. 146, 554 (1966)

57 Allen, J.W. and Leslie, T.C., phys. stat. sol. (a) 65, 545 (1981)

58 Fournier, D., Boccara, A.C., Rivoal, J.C., Parrot, R., Naud, C. and Porte, C.,
       Phys. Rev. 17, 3 (1978)

59 Oczkiewicz, B., Twardowski, A. and Demianiuk, M., Solid State Commun.
       46, 107 (1987)

60 Waldmann, H., Benecke, C., Busse, W., Gumlich, H.-E. and Krost, A.,
       Semicond. Sci. Technol. 4, 71 (1989)

61 Dreyhsig, J., Stutenbäumer, U., Gumlich, H.-E. and Allen, J,W.
       J. Cryst. Growth 101, 443 (1990)

62 Pradella, H. and Pohl, U.W., phys. stat. sol. (b) 141, K143 (1987)

63 Moriwaki, M.M., Tao, R.Y., Galazka, R.R., Becker, W.M., Richardson, J.W.,
       Physica 117B & 118B, 467 (1983)

64 Morales, J.E., Becker, W.M. and Debska, U., Phys. Rev. B32, 5202 (1985)

65 Lee, Y.R, Ramdas, A.R. and Aggarwal, R.L., Phys. Rev. B38, 10600 (1988)

66 Lee, Y.R., Ramdas, A.R. and Aggarwal, R.L., 18th Int. Conf. Physics of
       Semiconductors, Stockholm 1759 (1986)

67 Müller, E. and Gebhardt, W., phys. stat. sol. (b) 137, 259 (1986)

68 Morales Toro, J.E., Becker, W.M., Wang, B. Debska, U. and Richardson, J.W.
   Solid State Commun. 52, 41 (1984)

69 Krause, M., Gumlich, H.-E. and Krost, A.,
   EPS Conf. Berlin We-02-034 (1985)

70 Ambrazevicius, G., Babonas, G. and Buinevicius, A., Sov. Phys. Collect.
   27(1), 61 (1987)

71 Lascaray, J.P., Diouri, J., El Amrani, M. and Coquillat, D.,
   Solid State Commun. 47(9), 709 (1983)

72 Prüll, C. private communication (1985)

73 Schliwinski, J., private communication (1986)

74 Benecke, C., Busse, W., Gumlich, H.-E. and Jung, Ch.,
   Semicond. Sci. Techn. 5, (1990) in press

75 Zunger, A., Solid State Physics 39, 275 (1986)

76 Schallenberger, B. and Hausmann, A., Z.Phys. B 40 183, (1980)

77 Lanver, U. and Lehmann, G., J.Lum. 17, 225 (1978)

78 McClure, D.S. "Electronic Spectra of Molecules and Ions in Crystals"
   Academic Press New York p.175 (1959)

79 Goede, O., Heimbrodt, W., Weinhold, V., Schnürer, E. and Eberle, H.G.
   phys.stat.sol. (b) 143 511 (1987)

80 Becker, W.M., Bylsma, R., Moriwaki, M.M., Tao, R.Y. and Richardson, J.W.
   Solid State Comm. 49 245 (1984)

PHOTOEMISSION SPECTROSCOPY AND THE ELECTRONIC STRUCTURE OF DILUTED MAGNETIC SEMICONDUCTORS

A. Fujimori

Department of Physics, University of Tokyo, Hongo 7-3-1
Bunkyo-ku, Tokyo 113
JAPAN

1.   INTRODUCTION

Manganese-doped II-VI compounds such as $Cd_{1-x}Mn_xTe$[1,2] exhibit interesting magnetic, transport and optical properties including bound magnetic polarons,[3] spin glasses, cluster antiferromagnetism,[4] and pronounced magneto-optical effects.[5] These properties are thought to arise from the combination of localized Mn $3d$ and band-like host electronic states and the random distribution of Mn atoms substituting the cation sites.

However, the degree of localization of the Mn $3d$ states or their interaction with the host band states, which are primarily of anion p character, have been quite controversial.   Namely, there has been no general consensus on the strength of hybridization between the Mn $3d$ and the anion p states and on the importance of electron correlation among the Mn $3d$ electrons. One expects that if the p-d hybridization is sufficiently weak, the Mn d electrons form the high-spin state of the localized d^5 configuration of the Mn^{2+} ion [corresponding to the ^6S (S = 5/2 and L = 0) atomic term] in the ground state. The observed magnitude of the magnetic moment of Mn ($\sim$5 μ_B)[5] is consistent with this high-spin state. The observation of intra-atomic d-d optical transitions characteristic of the Mn^{2+} ion[6-10] also support this localized picture.

If the hybridization is strong, on the other hand, the Mn $3d$ states are expected to have tendency toward delocalization and the one-electron picture, which does not take explicit account of electron correlation, would become more appropriate. Band theories have been developed for the Mn-doped compounds by several groups[11-13] and have successfully explained the magnetic interactions between the Mn $3d$ and host band electrons and those between the Mn ions.[11,12] These band-structure calculations have also explained the unusual structural behavior,[12] and the *apparent* position of the occupied Mn $3d$ levels and their hybridization with the anion-p valence-band states as observed by photoemission spectroscopy.[14,15] According to band-structure calculations, e.g., on the antiferromagnetic state of the hypothetical $x = 1$ compound MnTe[14] or of CdMnTe$_2$ supercells,[13] the Mn $3d$ band is split into the fully occupied majority-spin band and the empty minority-spin band, leading to a Mn magnetic moment of ~$5\mu_B$, which is as large as that of the localized ^6S state.

The purpose of the present article is to give a unified picture on the electronic structure of the Mn-doped II-VI compounds in which the extended and localized states coexist. We will present configuration-interaction theories based on the impurity Anderson model and a cluster model, in which electron correlation for the Mn $3d$ electrons and their hybridization with the uncorrelated band electrons are explicitly treated. It will be clarified that the seemingly conflicting localized and itinerant behaviors of the Mn $3d$ electrons are, in fact, different aspects of the same physical models.

Photoemission spectroscopy has provided crucial experimental information about the behaviors of the Mn $3d$ electrons and their hybridization with the valence band states. Indeed, the configuration-interaction description became necessary in order to interpret the Mn d-derived photoemission spectra, including the so-called "multielectron satellites" appearing at higher binding energies, of Cd$_{1-x}$Mn$_x$Te and related compounds.[16,17]

Results of photoemission studies are collected in Sec. 2, focusing on the experimental effort to obtain reliable Mn $3d$-derived photoemission spectra. In Sec. 3, we present two types of existing theories. One is the one-electron model, which is known to be a good approximation in the limit of weak electron correlation. The other is the multiplet theory or ligand-field theory, corresponding to the strong electron correlation limit in the Mn 3d shell. In Sec. 4, the Anderson-model and cluster-model descriptions of the electronic structure are presented for the substitutional Mn^{2+} impurity in the CdTe host. Derivation of the ground-state and spectroscopic properties of the Mn-doped materials from the latter models are described in Sec. 5.

2.   RESULTS OF PHOTOEMISSION SPECTROSCOPY

Photoemission spectroscopy is a useful tool to directly probe the occupied electronic states of solids. In the angle-integrated mode, photoelectrons emitted in all directions are collected, and the photoemission spectra give occupied electronic density of states modulated by the photoionization cross sections of relevant atomic orbitals. In the angle-resolved mode, the momenta of photoelectrons are analyzed, and thus photoemission spectra of a single crystal give k-dependent dispersions of the energy bands of the crystal. Many photoemission studies have been performed so far in order to determine the positions of the occupied Mn 3d levels and their hybridization with the host band states.[14-24] Conclusions of these studies, however, were not necessarily consistent with each other as described below.

Olehafen et al.[20] measured angle-resolved photoemission spectra on cleaved single-crystal $Cd_{1-x}Mn_xTe$ and CdTe samples and found no clear evidence for the presence of localized Mn 3d states. Instead, they found essentially the same k-dependent dispersions both for the pure CdTe and doped $Cd_{1-x}Mn_xTe$, based on which they concluded that the Mn 3d states are completely delocalized due to strong hybridization with the host p-band states. However, they concentrated on the energy positions of various spectral features and have overlooked differences in the line shapes and intensities of spectral features between the pure and Mn-doped samples.

In the other photoemission studies on $Cd_{1-x}Mn_xTe$,[15,17,19,24]

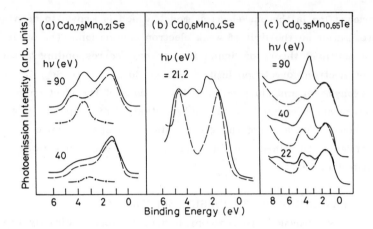

Fig. 1. Photoemission spectra of $Cd_{1-x}Mn_xTe$ ($Cd_{1-x}Mn_xSe$) compared to those of CdTe (CdSe). The spectra for the Mn-doped and pure samples are given by solid and dashed curves, respectively. Their differences represent Mn $3d$-derived photoemission spectra. (a) Franciosi et al.[23] The $Cd_{1-x}Mn_xSe$-CdSe difference curves are shown by dash-dotted lines. (b) Orlowski et al.[18]. (c) Taniguchi et al.[15]

$Cd_{1-x}Mn_xSe$,[14,23] and $Hg_{1-x}Mn_xSe$,[22] the intensity of emission centered ~3.4 eV below the top of the valence band was found to increase with Mn concentration and was assigned to localized Mn $3d$ levels. Franciosi et al.[22,23] made subtractions between the angle-integrated photoemission spectra of pure and Mn-doped compounds in order to deduce Mn $3d$ contributions as shown in Fig. 1(a) and found Mn $3d$ contributions only for the 3.4 eV peak. Based on this, they concluded that the Mn d electrons are localized and that there exists no sizable amount of p-d hybridization between the Mn d and the band electrons. On the other hand, in their angle-resolved photoemission study of $Cd_{1-x}Mn_xSe$ Orlowski et al.[14] observed, in addition to the Mn d-derived emission at ~3.4 eV, some changes within ~3 eV of the top of the valence band upon Mn substitution [Fig. 1(b)], which they attributed to the effect of p-d hybridization. Webb et al.[19] identified additional Mn $3d$ contributions around 6.5 eV below the top of the valence band in their photoemission spectra of $Cd_{1-x}Mn_xTe$, which they attributed to a multielectron satellite[14] as observed for many $3d$ transition-metal compounds.[25,26]

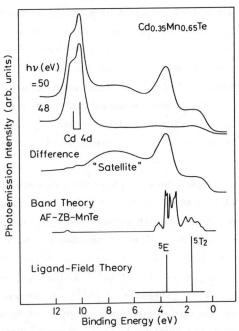

Fig. 2. Photoemission spectra of $Cd_{1-x}Mn_xTe$ in the Mn 3p → $3d$ core absorption region by Ley et al.[16] The Mn $3d$-derived photoemission spectrum is obtained by subtracting the off-resonance spectrum ($h\nu$ = 48 eV) from the on-resonance spectrum ($h\nu$ = 50 eV). The Mn d-derived spectrum is compared with the Mn $3d$ partial density of states of antiferromagnetic MnTe having a hypothetical zinc-blende structure as calculated by a local-density band-structure calculation.[12] Mn d-derived peaks according to the ligand-field theory are also given.

Taniguchi et al.[15] also made subtractions between the angle-integrated photoemission spectra of $Cd_{1-x}Mn_xTe$ and CdTe using various photon energies and confirmed the existence of Mn $3d$ contributions at 3.4 eV and around 6.5 eV [Fig. 1(c)]. The absence of Mn $3d$ contributions at around 6.5 eV in the Franciosi et al.'s data may be attributed to their data analysis procedure, in which the background due to secondary electrons has been subtracted in the energy range from the top of the valence band to about 6 eV below it prior to subtraction under the assumption that the signals at around 6 eV are mostly due to secondary electrons. Taniguchi *et al.* identified significant Mn $3d$ contributions also within ~3 eV of the top of the valence band using a resonant photoemission technique.[26] In this technique, one utilizes the enhancement of Mn $3d$-derived photoemission for photon energies in the Mn $3p \rightarrow 3d$ core-absorption region: The enhancement of Mn 3d photoemission occurs through interference between the $3p$-core absorption $3p^6 3d^5 + h\nu \rightarrow 3p^5 3d^6$ followed by an Auger-type decay $3d^5 3d^6 \rightarrow 3p^6 3d^4 + e$ and the direct photoemission $3d^5 + h\nu \rightarrow 3d^4 + e$, where e denotes an emitted photoelectron.

The resonant photoemission phenomena can also be utilized to obtain a more reliable measure of the Mn $3d$-derived photoemission by subtracting a spectrum taken with photon energy below the Mn $3p \rightarrow 3d$ threshold (off-resonance; $h\nu$ ~ 48 eV) from that taken at the threshold (on-resonance; $h\nu$ ~ 50 eV). Such measurements were performed for $Cd_{1-x}Mn_xTe$ by Ley *et al.*[16] and subsequently for $Cd_{1-x}Mn_xSe$ and $Cd_{1-x}Mn_xS$ by Taniguchi *et al.*[17] As shown in Fig. 2, subtraction between the on- and off-resonance spectra, which have been normalized to photon flux, leads to a near cancellation of the host valence-band emission as well as of the Cd $4d$ shallow core level located about 10 eV below the valence-band top, resulting in the Mn $3d$-derived photoemission spectrum. Now one can clearly see that, in addition to the Mn d-derived feature at 3.4 eV and around 6.5 eV, Mn $3d$ character is distributed over the whole valence-band region from the top of the valence band to below it, implying strong hybridization between the Mn $3d$ states and the host valence-band states.

3. PREVIOUS THEORETICAL MODELS
3.1. One-Electron Models

One-electron energy levels of $3d$ ions doped in semiconductors have been studied by band-structure, molecular-orbital and impurity Green's function calculations.[28,29] In the first-principles calculations, the local-density approximation has usually been used for the exchange-correlation potential. In the one-electron energy-level scheme, as shown in Fig. 3, the largest interaction is the intra-atomic exchange at the Mn site. This interaction splits the ten-fold Mn $3d$ levels into the majority-spin (up-spin) and minority-spin (down-spin) levels separated by ~5 eV according to the band-structure calculations.[12,13] The up-spin levels are fully occupied and the down-spin levels are empty, giving rise to a nearly full magnetic moment (~5 μ_B) of the high-spin ($S = 5/2$) ground state of the free Mn^{2+} ion. Each of the up-spin an down-spin d states are further split into t_2 (yz, zx, xy) and e ($3z^2-r^2$, x^2-y^2) sublevels

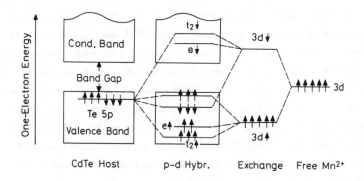

Fig. 3. One-electron energy-level scheme for the Mn^{2+} impurity substituting a Cd atom in CdTe.

under the tetrahedral (T_d) point-group symmetry at the Mn site since this site is tetrahedrally coordinated by four nearest-neighbor Te ions. The t_2 orbitals are directed toward the nearest-neighbor Te ions and are therefore strongly hybridized with the Te p orbitals whereas the e orbitals are directed toward a between two nearest-neighbor Te ions and are only weakly hybridized with the Te p states. Thus the e states are relatively localized, which could explain the sharp Mn $3d$ peak observed at 3.4-eV in the photoemission spectra.

As shown in Fig. 2, there is good correspondence between the

band-theoretical Mn d partial density of states of the
antiferromagnetic MnTe (with a hypothetical zinc-blende structure)[12)]
and the Mn $3d$-derived photoemission spectrum of $Cd_{1-x}Mn_xTe$ within ~4
eV of the top of the valence band. As we will see below, however,
the *apparent* good correspondence between photoemission and band
theory is rather fortuitous. Band theory cannot explain the
"multielectron satellite" at ~6.5 eV. Nor can it explain the intra-
atomic $d-d$ optical absorption observed at $h\nu$ ~ 2.2 eV for $Cd_{1-x}Mn_xTe$
with $x > 0.4$.[6-10)] (For $x < 0.4$, the absorption edge is due to a
transition from the top of the host valence band to the bottom of the
host conduction band, whose energy increases linearly with x.) Some
authors attributed the 2.2-eV absorption to transitions from the top
of the valence band to the unoccupied Mn $3d$ states, which may be
consistent with the one-electron picture,[13,27)] but this possibility
has been excluded by a recent magneto-optical study.[8)] Within the
same one-electron energy-level scheme, the multielectron satellite at
6.5 eV could be explained as due to a *shake-up* transition[14)] in which
an electron is excited from the Te $5p$ to Mn $3d$ orbitals accompanying
the photoemission of a Mn $3d$ electron. However, we will show below
that the Te-to-Mn charge transfer in the photoemission final state is
actually of *shake-down* type, by which the positive charge of the Mn d
hole is screened and the energy of the system is lowered.

3.2. Ligand-Field Theory

Although the results of photoemission studies (Sec. 2) seem
to suggest strong p-d hybridization and are not consistent with
completely localized Mn d electrons, it should be noted that there is
also correspondence between the one-electron energy levels described
above and the d^4 photoemission final states as given by the ligand-
field theory[30)] if we allow for hybridization of the d orbitals with
the ligand p orbitals.

Possible d^4 final states that can be reached from the $^6A_1$
ground state by d-electron emission are ^5E and $^5T_2$. The latter state
is lower in energy (i.e., located at lower binding energies in the
photoemission spectra) than the former by the crystal-field splitting

$10Dq$. The ^5E final state is more localized and could give rise to the sharp peak at 3.4 eV since the e hole in this final state is only weakly hybridized with the neighboring Te p orbitals. The $^5T_2$ final state, on the other hand, could be more spread in energy due to its stronger hybridization with the valence p orbitals, and could be assigned to the broad Mn d-derived emission within ~3 eV of the valence-band top. Such correspondence between the ligand-field theory and the one-electron theory is owing to the closed-shell structure (with respect to the up-spin orbitals) of the Mn $5d^5$ configuration, and is not generally guaranteed for the other $3d$ transition-metal ions. The satellite at ~6.5 eV cannot be explained within the ligand-field theory, too, and has to be considered as a shake-up satellite as in the case of the one-electron model.

4.   CONFIGURATION-INTERACTION THEORIES
4.1.   Anderson Model and its Relation to Other Models

In order to give a consistent explanation to the photoemission results of $Cd_{1-x}Mn_xTe$ including the multielectron satellite at ~6.5 eV, Ley et $al.$[16] presented a configuration-interaction theoretical analysis in which both the p-d hybridization and the Mn d electron correlation are explicitly taken into account within a $MnTe_4$ cluster model. Although this cluster is very small, the results reproduce surprisingly well the characteristic features of the Mn $3d$-derived photoemission spectra. Before handling the $MnTe_4$ cluster specifically, we start with a more general model, i.e., the Anderson model, to describe a system consisting of localized, correlated electrons (such as Mn $3d$) and uncorrelated, band-like electrons (such as CdTe host) hybridizing with each other. The model is represented by a Hamiltonian:

$$H = \sum_{im} \epsilon_d n_{im} + \sum_{mm'nn'} U_{mm'nn'} d^+_{im} d_{im'} d^+_{in} d_{in'} + \sum_{kv} \epsilon_{kv} n_{kv}$$
$$+ \sum_{kvim} (V_{kvim} d^+_{im} c_{kv} + V^*_{kvim} c^+_{kv} d_{im}), \tag{1}$$

with $n_{im} = d^+_{im} d_{im}$ and $n_{kv} = c^+_{kv} c_{kv}$, where the first two terms represent d electrons at the i-th Mn site with orbital energy ϵ_d and on-site electron-electron interaction $U_{mm'nn'}$ ($= \langle mm'|(e^2/r_{12})|nn'\rangle$), the third term host (valence- and conduction-)band electrons with wave vector k and band index v, and the last term hybridization between the Mn $3d$ and the band electrons. The band electrons of our interest are mainly consist of Te $5p$ and Cd $5s,p$ atomic orbitals. Of these orbitals, the Te $5p$ component is most significantly hybridized with the Mn $3d$ states since the Te atoms are coordinated to the Mn atoms. The Mn sites i's are randomly distributed on the Cd sublattice of the zinc-blende-type CdTe.

The Anderson model (1) includes most of relevant interactions in the Mn-doped CdTe: The correlation term gives rise to the high-spin ground state and the multiplet structure as observed by optical absorption; The hybridization term leads to the Mn $3d$-conduction/valence band exchange interaction[5] as well as the intersite Mn-Mn exchange interaction as a second-order process of the p-d hybridization. These different types of interactions in (1) lead to the seemingly conflicting observed physical properties.

In (1), we may consider a single Mn site and reduce the model into a single-impurity Anderson model, if the intersite Mn-Mn interaction can be ignored. Indeed, the magnitude of the intersite Mn-Mn exchange interaction ($JS_i\cdot S_j$) is smaller than 0.1 eV,[4,11] and the single-impurity model can be used as a good approximation as far as large energy-scale properties are concerned as in the case of photoemission, optical absorption, intra-atomic exchange and exchange interaction between the Mn $3d$ and band electrons.

In the cluster approximation, the valence band is replaced by molecular orbitals consisting of sp orbitals of the neighboring Te atoms (ligands). The cluster model can be derived by neglecting the finite width of the valence band in the impurity Anderson model, so that the calculated energy levels have to be suitably broadened when one compares the calculated photoemission spectra with experiment. We note that although we consider a very small cluster, it includes all relevant interactions and is as useful as the impurity Anderson model in many cases.

Since the Anderson model (1) is a quite general model for semiconductors doped with $3d$ transition-metal ions, the other models can be derived from (1) under certain approximations. The one-electron models including band theory[11,12,13] (for which Mn atoms have to be periodically arranged in the Cd sublattice) and the impurity Green's function theory[28,29] can be derived from the original Anderson Hamiltonian by applying a mean-field approximation to the Mn $3d$ correlation term. Multiplet theory or ligand-filed theory[30-32] can also be viewed as a special case of the single-impurity Anderson model in which the number of the localized d electrons is fixed, i.e., charge-transfer degree of freedom between the Mn ion and the host is suppressed, while the electron correlation at the Mn site is explicitly treated. The p-d hybridization effects are incorporated in the ligand-filed theory through the use of Mn $3d$ orbitals hybridizing with ligand molecular orbitals (i.e., Mn $3d$-ligand antibonding orbitals) in stead of using the unhybridized Mn $3d$ orbitals, but the weights of the Mn $3d$ and ligand components in the hybridized Mn d-ligand orbitals are fixed. This treatment of the hybridization is found to be insufficient to explain the photoemission spectra where charge transfer between the Mn and the host occurs during the photoemission process.

4.2   Configuration Interaction Theory

In order to take into account charge fluctuations between the Mn atom and the host, we must resort to a theory which allows this charge-transfer degree of freedom. We consider interactions between various electronic configurations of the Mn and the host mediated by p-d hybridization. For the formally divalent single Mn impurity, we start from the purely ionic configuration of Mn^{2+} ($3d^5$) and the filled valence band, which we denote by $|d^5\rangle$. Into this state are mixed other configurations produced by p-to-d charge transfer such as $3d^6\underline{L}$ and $3d^7\underline{L}^2$ via the p-d hybridization, V's in Eq. (1), where $\underline{L}$ denotes a hole in the host valence band. If we denote the energy required for the charge transfer $d^5 \rightarrow d^6\underline{L}$ by Δ, the $d^6\underline{L} \rightarrow d^7\underline{L}^2$ energy is given as $\Delta + U$ within the model (1), where U is the intra-atomic Coulomb energy at the Mn site. Here Δ is defined as $\Delta = \epsilon_d -$ $\epsilon_L + 5U$ using the parameters in the original Hamiltonian (1), where ϵ_L is an appropriate average of ϵ_{kv} [In the Anderson model (1), interatomic Coulomb interaction is neglected. If it is included, there will be interatomic contributions to the charge-transfer energies.[32)] The $d^7\underline{L}^2$ and higher terms can be neglected if U is large enough. The hybridization of the $d^6\underline{L}$ configuration into the d^5-dominated ground state yields a covalent contribution to the Mn-Te bonding.

As shown in Fig. 4, the ground state of the free Mn^{2+} (d^5) ion ^6S, becomes a $^6A_1$ irreducible representation of the T_d point group in the crystal, and is mixed with $d^6\underline{L}$ states with the same $^6A_1$ symmetry as

$$\Phi_g^{(N)}(^6A_1) = a\,|\,e^2t_2^3\ ^6A_1> + b_1\,|\,e^2t_2^4\underline{L}_\sigma{}^6A_1> + b_2\,|\,e^3t_2^3\underline{L}_\pi{}^6A_1>, \quad (2)$$

where $\underline{L}_\pi$ and $\underline{L}_\sigma$ represent ligand holes having E and T_2 symmetries, respectively. (The $\underline{L}_\pi$ and $\underline{L}_\sigma$ form π and σ bonds with the d orbitals, respectively. Note that for the octahedral coordination, E_g- and T_{2g}-symmetry ligand holes are denoted as $\underline{L}_\sigma$ and $\underline{L}_\pi$, respectively.[33)]) State (2) itself is the ground state of the $(MnTe_4)^{-6}$ model cluster, whereas the $\underline{L}$'s in (2) are actually continua in the original impurity

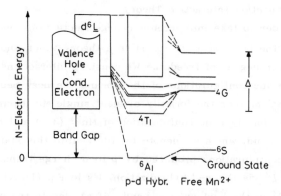

Fig. 4. Schematic energy levels for the ground state and electron-hole pair excited states (N-electron states) of the Mn^{2+} impurity in the configuration-interaction picture.

Anderson model, and each $b|d^6\underline{L}>$ term has to be replaced by an integral of the form $\int d\epsilon\, b(\epsilon)|d^6\underline{L}(\epsilon)>$. The diagonal matrix elements of the Hamiltonian are such that $\Delta = <d^6\underline{L}_m|H|d^6\underline{L}_m> - <d^5|H|d^5>$ ($m = \sigma$ and π) when averaged over the multiplet terms. The off-diagonal elements are $<d^5|H|d^6\underline{L}_\sigma> = 2(pd\sigma)$ and $<d^5|H|d^6\underline{L}_\pi> = (8/3)\sqrt{(2/3)}(pd\pi)$.

The first excited state of the free Mn^{2+} ion, ^4G, is split into $^4A_1$, ^4E, $^4T_1$ and $^4T_2$ representations of the T_d point group when hybridized with the $d^6\underline{L}$ configuration, as shown in Fig. 4. Of these states, $^4T_1$ has the lowest energy. If the $^6A_1 \rightarrow {}^4T_1$ excitation energy is less than the energy of the host band gap, this d-d intra-atomic transition is observable in optical absorption spectra. The wavefunction of the $^4T_1$ state is also given by the same form as Eq. (2).

Ionized states (photoemission final states; referred to as N-1-electron states as opposed to the N-electron ground state and the N-electron electron-hole pair excited states discussed above) of the Mn^{2+} impurity are expanded using basis functions $|d^4>$, $|d^5\underline{L}>$ and $|d^6\underline{L}^2>$ as

$$\Phi_f(N-1)(^{2S+1}\Gamma) = c|d^4\ {}^{2S+1}\Gamma> + e|d^5\underline{L}\ {}^{2S+1}\Gamma> + f|d^6\underline{L}^2\ {}^{2S+1}\Gamma>, \quad (3)$$

where $^{2S+1}\Gamma = {}^5E$ or $^5T_2$ for the emission of a Mn d electron. The diagonal matrix elements of the Hamiltonian in the N-1-electron state averaged over the multiplet terms are such that $<d^5\underline{L}|H|d^5\underline{L}> - <d^4|H|d^4> = \Delta - U$ and $<d^6\underline{L}^2|H|d^6\underline{L}^2> - <d^5\underline{L}|H|d^5\underline{L}> = 2\Delta$. The matrix elements including the multiplet terms are listed in Ref. 34. As will be shown below (Sec. 5.1), U (~ 7.5 eV) is larger than Δ (~ 3.5 eV) for the Mn impurity in $Cd_{1-x}Mn_xTe$, i.e., $\Delta - U < 0$, which means that the lowest ionization state is more $d^5\underline{L}$-like than d^4-like or that the lowest energy holes are distributed mainly on the Te atoms and not on the Mn atoms.

Although the lowest energy holes are $d^5\underline{L}$-like, the energy difference between the d^4 and $d^5\underline{L}$, $\Delta - U \sim -4$ eV, is not very large compared to the magnitudes of the off-diagonal matrix elements of the final-state Hamiltonian, which are of the order of a few eV.[16,34)

130

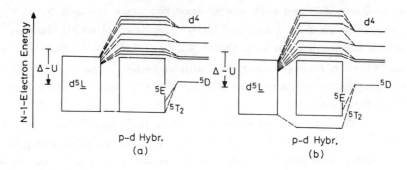

Fig. 5. Schematic energy levels for the ionized states (N-1-electron states) of the Mn^{2+} impurity in the configuration-interaction picture. The vertical scale is the *total* energy of the N-1-electron system or the *hole* energy. (a) Weak p-d hybridization. (b) Strong p-d hybridization.

This causes a strong hybridization between the d^4 and $d^5\underline{L}$ final-state configurations. The $d^5\underline{L}$ configuration is a continuum in which there is a hole in the host valence band. This situation is schematically described in Fig. 5. Thus the strong d^4-$d^5\underline{L}$ hybridization leads to the distribution of the Mn $3d$-derived photoemission intensity over the whole anion-p valence band.

The d^4 final states lie *higher* in energy than the $d^5\underline{L}$ by $U - \Delta$ ~ 4 eV. Because of this, the d^4 final-state signals appear around 6.5 eV *below* the top of the valence band as the satellite in the photoemission spectra. The satellite has become a broad feature due to the lifetime broadening although the d^4 states are originally discrete multiplet states. We expect that if the d^4-$d^5\underline{L}$ hybridization is strong enough, it may push some bound states of $d^5\underline{L}$ character out of the $d^5\underline{L}$ continuum as depicted in Fig. 5(b), i.e., as discrete states just *above* the top of the host valence band.[35] It is not clear, however, within the present energy resolution of photoemission spectroscopy whether such bound states are indeed formed within the band gap or not. [Here it should be noticed that the sign of energy is inverted between the one-electron and N-1-electron energies.]

As described above, experiments involving excited states of the Mn^{2+} ion such as intra-atomic d-d transition and photoemission cannot be accounted for within the framework of the one-electron picture. On the other hand, the ground state properties such as charge and spin distribution may be well described by the one-electron theories using the local-density approximation. In the one-electron energy-level scheme, net charge transfer occurs only through hybridization between occupied and unoccupied states: Hybridization within occupied states or within unoccupied states does not contribute to net charge transfer. Therefore, hybridization between the filled down-spin p band and the empty down-spin Mn $3d$ states results in p-to-d charge transfer, i.e., covalent contribution to the Mn-Te bonding. The hybridization between the occupied up-spin Mn d and up-spin Te p bands does not lead to net charge transfer. The latter hybridization can be observed only when a hole is created by photoemission or by hole doping. However, the one-electron energy-level scheme of the ground state as shown in Fig. 3 becomes no longer valid once a hole is created because of the strong electron correlation, so that explicit treatment of the electron correlation is necessary to consider photoemission or hole doping.

The hybridization of the $d^6\underline{L}$ configuration into the dominant d^5 configuration in the ground state as in Eq. (2) results in charge transfer from the occupied down-spin p band to the unoccupied up-spin Mn $3d$ states as in the case of the one-electron picture. If the charge-transfer energy Δ is similar to the difference between the one-electron energies, $\epsilon(Mn\ 3d\!\downarrow) - \epsilon(Te\ 5p)$ in the one-electron energy levels, the configuration-interaction theory gives essentially the same charge and spin distribution for the ground state as those given by the one-electron theories.

5. COMPARISON WITH EXPERIMENTS

5.1. Photoemission Spectra

The Mn $3d$-derived photoemission spectrum $I(E_B)$ of $Cd_{1-x}Mn_xTe$ within the impurity model can be calculated from the ground-state and final-state wave functions, (2) and (3), as

$$I(E_B) \sim \sum_f |\langle\Phi_f(N-1)|T|\Phi_g(N)\rangle|^2 \delta(E_f(N-1)-E_g(N)-E_B), \qquad (4)$$

where E_B is the electron binding energy measured from the top of the valence band, and $E_g(N)$ and $E_f(N-1)$ are the energies of the N-electron ground state and the $N-1$-electron final state, respectively. In the sudden approximation, in which interaction between the photoelectron and the remaining electrons in the final state is neglected, the transition matrix elements from the N-electron to the $N-1$-electron states are those of annihilation operators for the d electron.[33,34] Therefore, $\langle d^4|T|d^5\rangle$ and $\langle d^5\underline{L}|T|d^6\underline{L}\rangle$ are nonzero, and the intensity is given essentially by $\sim|ac + be|^2$, where a, b, c and e are the coefficients in Eqs. (2) and (3).

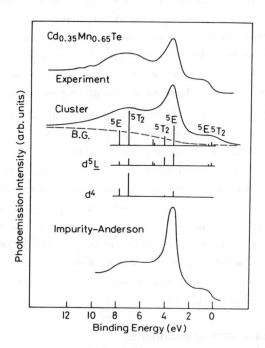

Fig. 6. Mn $3d$-derived photoemission spectra of $Cd_{1-x}Mn_xTe$ calculated using the configuration-interaction cluster model[16] and the impurity Anderson model.[44] The result of the cluster-model calculation is decomposed into d^4 and $d^5\underline{L}$ final-state components.

Fitting of the calculated spectrum to the experimental Mn $3d$-derived photoemission spectrum is performed by adjusting the model parameters, Δ, U and $(pd\sigma)$. [$(pd\sigma)/(pd\pi)$ = -2.2 has been assumed according to Harrison[36] in order to reduce the number of independent parameters.] The best fit to the Mn d-derived spectrum of $Cd_{1-x}Mn_xTe$ using the $MnTe_4$ cluster model is shown in Fig. 6, where the calculated spectrum is compared with the experimental Mn d-derived photoemission spectrum. The resulting parameter values are $U \sim 7.5$ eV, $\Delta \sim 3.5$ eV and $(pd\sigma)$ = $-2.2(pd\pi) \sim -1$ eV.[16] In Fig. 6, the calculated spectrum originally consists of sharp lines because of the finite size of the cluster, so that the lines have been convoluted with Gaussian and Lorentzian functions in order to faciliate comparison with experiment. Figure 6 also shows a decomposition of the spectrum into d^4 and $d^5\underline{L}$ final-state components, indicating that the d^4 and $d^5\underline{L}$ configurations dominate the satellite (d^4) and main-

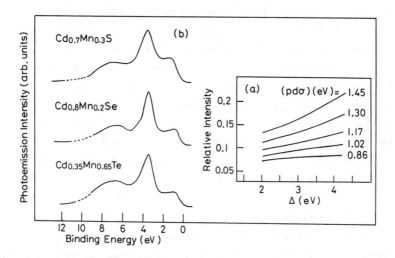

Fig. 7. (a) Relative intensities of the satellite to the 3.4-eV peak calculated using the configuration-interaction cluster model.[17] The intensity ratio is plotted for various p-d hybridization strengths $(pd\sigma)$ as a function of the charge-transfer energy Δ. U is fixed at 7.5 eV. (b) Mn $3d$-derived photoemission spectra of $Cd_{1-x}Mn_xS$, $Cd_{1-x}Mn_xSe$ and $Cd_{1-x}Mn_xTe$ obtained from the Mn $3p$-$3d$ on-resonance ($h\nu$ = 49.5 eV) and off-resonance ($h\nu$ = 47 eV) spectra.[17]

band ($d^5\underline{L}$) regions, respectively.

The present model [Eqs. (2)-(4)] predicts that the intensity of the satellite relative to the main band decreases with increasing charge-transfer energy Δ, or with increasing electronegativity of the ligand atoms, if U is fixed. In Fig. 7(a), the relative intensity of the satellite to the 3.4-eV peak is plotted as a function of Δ, with U = 7.5 eV.[17] This trend has indeed been observed for the comparative photoemission study of $Cd_{1-x}Mn_xS$, $Cd_{1-x}Mn_xSe$ and $Cd_{1-x}Mn_xTe$ [Fig. 7(b)],[17] giving support to the configuration-interaction theoretical treatment of the Mn^{2+} impurity.

Recently, the large intra-atomic Coulomb energy ($U \sim 7.5$ eV) was questioned by Balzarotti et al.[24] They compared Mn L_3VV ($2p_{3/2}$-valence-valence) Auger-electron spectra of $Cd_{1-x}Mn_xTe$ and the self-convolutions of Mn d-derived photoemission spectra (excluding the satellite region) and concluded that $U \sim 0$ eV. The large discrepancy between the photoemission and Auger results can be explained as due to the fact that they implicitly assumed in their analyses the Mn core-hole state to have an electronic configuration $\underline{2p}_{3/2}3d^5$ (where $\underline{2p}_{3/2}$ denotes a Mn $2p_{3/2}$ core hole) and the two-hole final states to have $3d^3$ configuration. In analogy with the final state of Mn d-electron photoemission, the energy of the core-hole state is expected to be lowered by ligand-to-metal charge transfer and becomes $\underline{2p}_{3/2}3d^6\underline{L}$.[34] Also, Auger final states may be extensively screened by the same type of charge transfer and become $3d^4\underline{L}$- and/or $3d^5\underline{L}^2$-like. If these effects are taken into account, the Auger results can be reconciled with the large U.[34]

5.2 d-d Optical Absorption

The intra-atomic d-d optical absorption of the Mn^{2+} ion observed at $\sim$2.2 eV[6-10] is due to a transition from the ground state (2) having $^6A_1$ symmetry to the lowest excited state having $^4T_1$ symmetry, as illustrated in Fig. 4. The energy difference between the $^6A_1$ and $^4T_1$ states, however, has not been calculated using the configuration-interaction theory, whereas this type of calculations have been performed for the Ni^{2+} ion in NiO[33,37] and good agreement with experiment and the ligand–field theory[38] has been obtained from the lowest $^3A_{2g} \rightarrow \,^3T_{2g}$ transition at ~1.1 eV to higher energies up to ~3.5 eV. In the case of the Mn^{2+} ions in II–VI compounds, unfortunately, detailed comparison between the theory and experiment may be hampered by the fact that only the lowest d–d peak is visible below the fundamental absorption edge of the host material.

5.3. Charge–transfer Optical Absorption and Inverse Photoemission

The photoemission studies and their analyses have concentrated on the occupied part of the Mn $3d$ and the host band states. Information about the unoccupied states, particularly on the unoccupied down–spin Mn $3d$ ($3d^6$) states, is also of great interest. In order to obtain such information, optical absorption of charge–transfer–type from the host valence band to the unoccupied Mn d states is expected to be useful. The optical absorption edge at ~2.2 eV was attributed by some authors to this type of transition,[12,13,27] but is now assigned to the intra-atomic d–d transition rather unambiguously based on the absence of Zeeman splitting that would be expected for the top of the valence band (Γ_8) in a magneto-optical study of $Zn_{1-x}Mn_xTe$.[8] An optical-reflectance study of $Cd_{1-x}Mn_xTe$ well above the fundamental absorption edge ($h\nu <$ 5.5 eV) revealed no structures that could be associated with transitions to from the valence band to the unoccupied Mn $3d$ states.[39]

Direct information on the unoccupied states can be obtained by inverse photoemission spectroscopy, in which one detects photons emitted from the sample when electrons are incident on it. Using the Anderson-model parameters for $Cd_{1-x}Mn_xTe$, the lowest unoccupied Mn$3d$ state (the ^5D component of the d^6 multiplet, which is further split into $^5T_2$ and ^5E crystal-field levels) is estimated to be lie as high as ~5 eV above the top of the valence band. Unfortunately, there have been no inverse photoemission experiments performed on the Mn-doped semiconductors so far.

5.4   Magnetic Interactions

The exchange constants between the Mn ion and the valence and conduction bands and those between the Mn ions have been calculated using band theory by Larson et al.[11] The calculated results are in good agreement with those obtained from the optical measurements[5,3] and neutron scattering experiments.

Within the configuration-interaction theory, the exchange coupling between the Mn ion and a valence hole is given by the energy difference between the lowest-energy N-1-electron state with $S = 3$ and $S = 2$. This procedure has been employed to evaluate the exchange coupling between the localized Cu d^9 spin and an extra hole in high-T_C superconductors using a CuO_5 cluster model.[40] Unfortunately, the N-1-electron energy levels of the $MnTe_4$ cluster calculated for the photoemission final state have been limited to $S = 2$ states because we have been interested in the emission of Mn d electrons.[16,17,34]

The Mn-Mn interatomic exchange interaction (superexchange interaction) would be calculated using a Mn-Te-Mn cluster, in analogy with the case of 3d transition-metal monoxides.[41] For the NaCl-type monoxides MO, a linear M-O-M cluster has been used to calculated the superexchange coupling constants between the next-nearest neighbor M-M pair. The Néel temperatures T_N's thus calculated are in much better agreement with experiment than those calculated using the Anderson's theory of superexchange interaction.[41]

6.   FIRST-PRINCIPLES ESTIMATION OF THE MODEL PARAMETERS

The fitting of photoemission spectra to configuration-interaction theory is a quite useful method to estimate the values of the parameters (Δ, U, ($pd\sigma$), etc.) in the Anderson Hamiltonian (1). Recently, this method has successfully been applied to estimate the model parameters of the Cu-oxide high-T_C superconductors.[40,42,43]

A quite different approach has been made by Gunnarosson et al.[44] to estimate the Anderson-model parameters of $Cd_{1-x}Mn_xTe$. They have made band-structure calculations on CdTe and MnTe within the local-density approximation using the linearized-muffin-tin-orbitals (LMTO) method. The latter method is able to directly give the energy-dependent p-d hybridization parameters $|V_m(\epsilon)|^2$ [$= \sum_k V^*_{km} V_{km} \delta(\epsilon - \epsilon_k)$], where m denotes an irreducible representation of the T_d point-group symmetry [not the band index v in Eq. (1)]. The intra-atomic Coulomb energy U can be estimated through $U = \partial\epsilon_{3d}/\partial n_{3d}$, where ϵ_{3d} is the one-electron energy of Mn 3d in the local-density approximation. With these parameters, Gunnarsson et al. have calculated the Mn 3d-derived photoemission spectra using the impurity Anderson model as shown in Fig. 6. Considering the fact that no empirical parameters were used in the calculation, the agreement between the theory and experiment is surprisingly good. The same method has been used to estimate the Anderson-model parameters for La$_2$CuO$_4$,[45] but the hybridization parameter are found to be considerably overestimated compared to those estimated from the fit to photoemission experiment.

The p-d hybridization parameters can also be estimated from band-structure calculations through fitting the k-dispersions of the calculated energy bands to tight-binding (LCAO) energy bands. This method has been applied to rare-earth compounds[46] and high-T_c superconductors,[47] but not to Mn compounds so far.

7. CONCLUSION

We have reviewed photoemission spectroscopic studies on the electronic structure of Mn-doped II-VI semiconductors. The photoemission spectra reveal strong hybridization between the Mn 3d and the host anion p-band states as well as the "multielectron" satellite features. In order to interpret consistently all these results, we have presented configuration-interaction theories, in which both the p-d hybridization and the electron correlation at the Mn site are exactly treated within the single Mn impurity models. Such a configuration-interaction approach can describe the localized Mn 3d^5 multiplet and the strongly hybridized photoemission spectra on the same ground, reconciling the localized behavior of the Mn 3d electrons with their strong hybridization with the host band states.

Although the basic physical picture and the model parameters for the Mn-doped compounds have been obtained through the photoemission studies, the applicability of the configuration-interaction theory to the d-d optical transitions and the exchange interactions in these compounds remain to be checked.

Finally, we remark that the physical picture given by the configuration-interaction treatment of the Anderson model, i.e., hybridization between the strongly correlated, localized electrons and uncorrelated band electrons is known to be an essential feature not only of the Mn-doped semiconductors but also of a wide range of $3d$ transition-metal compounds[48] including high-T_C superconductors and of valence-fluctuating, Kondo and heavy-fermion rare-earth ($4f$) and actinide ($5f$) compounds.[49]

ACKNOWLEDGEMENTS

The author would like to express his sincere thanks to Professor M. Taniguchi of Hiroshima University for drawing his attention to the present problem and for very fruitful collaboration.

REFERENCES

1. Furdyna, J. K., J. Appl. Phys. 53, 7637 (1982).
2. Galazka, R. R., in *Physics of Semiconductors*, edited by Wilson, B. L. H. (Institute of Physics, Bristol, 1979) p.133.
3. Heiman, D., Wolff, P. A. and Warnock, J., Phys. Rev. B28, 4848 (1983).
4. Galazka, R. R., Nagata, S. and Keesom, P. H., Phys. Rev. B22, 3344 (1980); Giebultowicz, T. M., Rhyne, J. J., Ching, W. Y. and Huber, D. L., J. Appl. Phys. 57, 3415 (1985).
5. Gaj, J. A., Planel, R. and Fishman, G., Solid State Commun., 29, 435 (1979).
6. Nguyen The Khoi and Gaj, J. A., Phys. Status Solidi B83, K133 (1977).
7. Müller, E., Gebbardt, W. and Rehwald, W., J. Phys. C16, L1141 (1983).
8. Lee, Y. R. and Ramdas, A. K., Solid State Commun. 51, 861 (1984).
9. Lascaray, J. P., Diouri, J., El Amrani, M. and Coquillat, D., Solid State Commun. 47, 709 (1983).
10. Lee, Y. R., Ramdas, A. K. and Aggarwal, R. L., Phys. Rev. B33, 7383 (1986).
11. Larson, B. E., Hass, K. C., Ehrenreich, H. and Carlsson, A. E., Solid State Commun. 56, 347 (1985).
12. Wei, S.-H. and Zunger, A., Phys. Rev. Lett. 56, 2391 (1986); Phys. Rev. B35, 2340 (1987).
13. Podgorny, M., Z. Phys. B69, 501 (1988).
14. Orlowski, B. A., Kopalko, K. and Chab, W., Solid State Commun. 50, 749 (1984).

15. Taniguchi, M., Ley, L., Johnson, R. L., Ghijsen, J. and Cardona, M., Phys. Rev. B33, 1206 (1986).
16. Ley, L., Taniguchi, M., Gijsen, J., Johnson, R. L. and Fujimori, A., Phys. Rev. B35, 2839 (1987).
17. Taniguchi, M., Fujimori, A., Fujisawa, M., Mori, T., Souma, I. and Oka, Y., Solid State Commun. 62, 431 (1987).
18. Orlowski, B. A., Phys. Status Solidi B95, K31 (1979).
19. Webb, C., Kaminska, M., Lichtensteiger, M. and Lagowski, J., Solid State Commun. 40, 609 (1981).
20. Olehafen, P., Vecchi, M. P., Freeouf, J. L. and Moruzzi, V. L., Solid State Commun. 44, 1547 (1982).
22. Franciosi, A., Caprile, C. and Reifenberger, Phys. Rev. B31, 8061 (1985).
23. Franciosi, A., Chang, S., Reifenberger, R., Debska, U. and Riedel, R., Phys. Rev. B32, 6682 (1985).
24. Balzarotti, A., De Crescenzi, M., Messi, R., Motta, N. and Patella, F., Phys. Rev. B36, 7428 (1987).
25. Kakizaki, A., Sugeno, K., Ishii, T., Sugawara, H., Nagakura, I. and Shin, S., Phys. Rev. B26, 4845 (1982).
26. Davis, L. C., J. Appl. Phys. 59, R25 (1986).
27. Abreu, R. A., Giriat, W. and Vecchi, M. P., Phys. Lett. 85A, 399 (1981); Lemasson, P., Wu, B.-L., Triboulet, R. and Gautron, J., Solid State Commun. 47, 669 (1983).
28. Zunger, A., in *Solid State Physics* Vol.39, eds. Seitz, D., Turnball, D. and Ehrenreich, H. (Academic, New York, 1986) p. 275.
29. Katayama-Yoshida, H. and Zunger, A., Phys. Rev. B33, 2961 (1986); Oshiyama, A., Hamada, N. and Katayama-Yoshida, Phys. Rev. B37, 1395 (1988).
30. Sugano, S., Tanabe, Y. and Kamimura, H., *Multiplets of Transition-Metal Ions in Crystals* (Academic, New York, 1970).
31. Fazzio, A. and Caldas, M. J. and Zunger, A., Phys. Rev. B30, 3430 (1984).
32. Fujimori, A. and Minami, F., Phys. Rev. B30 , 957 (1984).
33. Watanabe, S. and Kamimura, H., J. Phys. Soc. Jpn. 56, 1078 (1987); Mater. Sci. Eng. B3, 313 (1989).
34. Fujimori, A., Saeki, M., Kimizuka, N., Taniguchi, M. and Suga, S., Phys. Rev. B34, 7318 (1986).
35. Haldane, F. D. M. and Anderson, P. W., Phys. Rev. B13, 2553 (1976).
36. Harrison, W. A., *Electronic Structure and the Properties of Solids* (Freeman, San Francisco, 1980).
37. Zaanen, J., Ph. D. Thesis, University of Groningen (1986).
38. Newman, R. and Chrenko, R. M., Phys. Rev. 114, 1507 (1959).
39. Lautenschlager, P., Logothetidis, S., Vina, L. and Cardona, M., Phys. Rev. B32, 3811 (1985).
40. Fujimori, A., Takayama-Muromachi, E., Uchida, Y. and Okai, B., Phys. Rev. B35, 8814 (1987); Fujimori, A., Phys. Rev. B39, 793 (1989).
41. Zaanen, J. and Sawatzky, G. A., Can. J. Phys. 65, 1262 (1987).
42. Fujimori, A., Takayama-Muromachi, E., Uchida, Y. and Okai, B., Phys. Rev. B35, 8814 (1987).

43. Ghijsen, J., Tjen, L. H., van Elp, J., Eskes, H., Westerink, J., Sawatzky, G. A. and Czyzyk, M. T., Phys. Rev. B38, 11322 (1988).
44. Gunnarsson, O., Andersen, O. K., Jepsen, O. and Zaanen, J., Phys. Rev. B39, 1708 (1989).
45. McMahan, A. K., Martin, R. M. and Satpathy, S., Phys. Rev. B38, 6650 (1988).
46. Monnier, R., Degiorgi, L. and Koelling, D., Phys. Rev. Lett., 86, 2744 (1986).
47. Park, K.-T., Terakura, K., Oguchi, T., Yanase, A. and Ikeda, M., J. Phys. Soc. Jpn., 57, (1988).
48. Zaanen, J., Sawatzky, G. A. and Allen, J. W., Phys. Rev. Lett. 55, 418 (1985); Fujimori, A., in *Core-Level Spectroscopy in Condensed Systems*, eds. Kanamori, J. and Kotani, A. (Springer-Verlag, Berlin, 1988).
49. Kasuya, T. and Saso, T. eds., *Theory of Heavy Fermions and Valence Fluctuations* (Springer-Verlag, Berlin, 1985); Lee, P. A., Rice, T. M., Serene, J. W., Sham L. J. and Wilkins, J. W., Comments Cond. Mat. Phys. 12, 99 (1986);

TRANSPORT PROPERTIES OF DILUTED MAGNETIC SEMICONDUCTORS

Tomasz Dietl

Institute of Physics, Polish Academy of Sciences,
Al. Lotników 32/46, PL 02-668 Warszawa,
POLAND

ABSTRACT

A review is given of the numerous effects the sp–d exchange interaction may have upon electron transport phenomena in diluted magnetic semiconductors doped with hydrogenic–like impurities. The ranges of impurity concentrations that correspond to the metallic and to the insulating phases are considered. Particular attention is payed to the transitory region between the above two extreme cases where the metal–insulator transition occurs. In this region the presence of the exchange interaction manifests itself in a rather non–standard way, and constitutes a valuable tool to elucidate the driving mechanisms of the transition. The effects of the giant s–d spin–splitting of electronic levels, the fluctuations of magnetization, and the formation of magnetic polarons are described in detail, while the phenomena associated with long range magnetic ordering, such as electron–magnon interaction, are not discussed as diluted magnetic semiconductors exist almost exclusively in a paramagnetic or spin–glass phase. The theoretical survey is illustrated by experimental results for Mn–based II–VI compounds. The second part of the article concerns such diluted magnetic semiconductors, in which magnetic constituents act as resonant impurities. The remarkable properties of $Hg_{1-x}Fe_xSe$ in the mixed–valence region of Fe donors are presented. An interesting problem of electron localization in the case when the potential of the donating impurity has no bound states is also mentioned. Preceding the discussion of the above issues, basic information concerning the location of energy levels of transition metals in II–VI semiconductors is presented and the various models of the metal–insulator transition in electronic systems are recalled.

1.   INTRODUCTION

Early studies of magnetic semiconductors,[1,2] as well as the more recent investigations of diluted magnetic semiconductors (DMS)[3-5] have convincingly demonstrated that the presence of magnetic ions modifies dramatically the electron transport phenomena. The aim of the present contribution is to sum up the current view of the various aspects of electron transport in DMS.[6] We shall illustrate our considerations by experimental results obtain for II–VI DMS since this class of compounds is the most familiar to the author and, at the same time, it has been the most comprehensively studied one.

It has been know for a long time that the influence of magnetic atoms on transport properties depends crucially on the relative positions of the open d or f magnetic orbitals and the relevant band edges. Furthermore, this influence depends on whether the carriers reside in the localized or in the extended states. Therefore, in order to make our article self contained we begin by recalling the location of energy levels of transition metals in II–VI compounds, as well as by summarizing basic information about the proposed models of the metal–insulator transition and the electronic states in doped semiconductors.

1.1.   Transition Metals in II–VI Semiconductors

As already mentioned the influence of magnetic atoms on transport properties is ultimately related to the energetic position of d or f–like open shells in respect to the Fermi level and band edges. This position as well as the degree of electron localization on the magnetic shells are known to reflect the competition between two effects. The first is the on-site Coulomb repulsion between electrons (which tends to conserve the local character of d or f shells), while the second is the delocalizing influence of mixing (hybridization) between extended states and magnetic orbitals. Additionally, the nature of the magnetic ground state is determined by local distortions and spin–orbit coupling.

In the case of interest here, i.e., for transition metals in II–VI compounds, these effects result in a rich spectrum of possible situations. For instance, the uppermost occupied one-electron level created by Mn in the 2+ charge state is superimposed on the valence-band continuum–of–states, whereas that of Sc^{2+} lies probably above the bottom of the conduction band in, e.g., CdSe. A remarkable empirical observation is that the distances between the energy levels of different transition

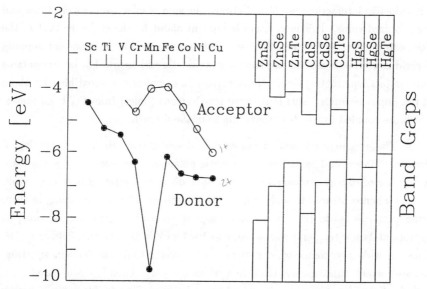

Figure 1: Approximate energy level positions of transition metals in the 2+ (solid symbols) and 1+ (open symbols) charge states relative to band edges of II–VI compounds (after Heiman et al.[7] and Refs. 8–10).

metals remain, to a good approximation, the same in various hosts. These distances are schematically shown in Fig. 1 together with distances to the band edges of various II–VI semiconductors.[8-10] This plot enables an approximate determination of the level position when no direct experimental result is available. Moreover, it points to the existence of a common "reference energy", from which band offsets between various semiconductors can be obtained.[9-11]

In the Mn-based compounds,[12] the Pauli exclusion principle allows such a distribution of electrons between states created by Mn^{2+} that the total spin attains its maximum possible value, $S = 5/2$. This configuration reduces strongly the Coulomb–repulsion energy, and hence the uppermost level of Mn^{2+} resides deeply in the valence band. At the same time, because an extra electron must have an opposite spin orientation, the correlation energy U is very large. As sketched in Fig. 1, the level of Mn^{1+} is located above the bottom of the conduction band. Therefore, Mn in II–VI compounds has rather unique properties: it acts neither as a donor, nor as an acceptor, and does not perturb directly the density–of–states in the vicinity of the Fermi energy. However, since there is a localized spin on Mn ions, there exists a Kondo–like interaction, $-J \boldsymbol{S} \cdot \boldsymbol{s}$, between the spin of effective mass electrons and the localized spins. This interaction is brought about by the exchange part of the Coulomb interaction between the two subsystems as well as by quantum hopping (hybridization) between Mn and band states (kinetic exchange). The latter operates (and, in fact, dominates) for electronic states of a symmetries compatible with those of the impurity levels. This is the case for electrons (holes) from Γ_{15} band states, which are coupled to t states originating from the d–shell.

The exchange interaction causes spin–disorder scattering, spin splitting of electronic states, and formation of magnetic polarons. Because of the small wave vector of electrons in semiconductors, the spin–disorder scattering is usually weak, at least when compared to ionized impurity scattering. The spin splitting, in turn, being proportional to the macroscopic magnetization of d–spins, can become, at appropriately low temperatures, as large as $1 - 3$ meV in the external field of 1 kOe. Thus, in wide–gap compounds, it is not only greater than the Zeeman splitting, but also several times greater than the cyclotron energy. Therefore Mn–based compounds offer an interesting opportunity to examine the influence of the spin–splitting on transport phenomena. Still another effect of the exchange interaction is the existence of ferromagnetic clouds around the singly occupied local electronic states. The formation of such complexes – known as bound magnetic polarons – enhances the binding energy and shrinks the localization radius of localized electrons.

In many respects, Fe–based compounds[13] exhibit entirely different properties than the materials containing Mn. In particular, since there is an even number of d–electrons involved, the Kramers theorem does not apply, and the ground state of the Fe^{2+} ion is a magnetic singlet. Furthermore, according to Fig. 1, the uppermost occupied level of Fe^{2+} resides at relatively high energies, in most cases above the top of the valence band. This state is of e symmetry, and therefore, in the limit of weak spin–orbit interaction is not coupled to the band states from the vicinity of the center of the Brillouin zone. This weak coupling is indirectly confirmed by magnetooptical studies of the exchange interactions between holes and d–spins in $Zn_{1-x}Fe_xSe$ and $Cd_{1-x}Fe_xSe$.[14] These studies imply a negative sign of the p–d exchange integral demonstrating that the kinetic exchange involves primarily t states, which are located below the top of the valence band. As shown by A. Mycielski et al.,[15] a particularly interesting situation occurs in the zero–gap $Hg_{1-x}Fe_xSe$, where the Fe–related level is superimposed on the conduction–band continuum–of–states. Thus, Fe acts as a resonant donor in HgSe. If the concentration of these donors is greater than the number of states available in the conduction band below the Fe states, the Fermi level is "pinned" by those states. Under such conditions various aspects of mixed valence systems can be studied.

Worth mentioning is also the case of CdSe:Sc and HgTe:Cu, where the transition metals appear to form resonant donors and acceptors, respectively. The problem of electron localization by impurities with no bound states can be studied in those systems, provided that the interaction between the impurity and band states will not result in band gap levels.

1.2. Semiconductors Doped with Hydrogenic–Like Impurities

It is known that electronic properties of semiconductors containing hydrogenic–like impurities depend crucially on the ratio of the mean distance between impurities $\bar{r} = (3/4\pi n)^{1/3}$ to their effective Bohr radius $a_B = \hbar^2 \kappa_0 / e^2 m^*$. In the dilute case, $\bar{r} \gg a_B$, electrons are bound to individual impurities, and low–temperature conduction proceeds by means of phonon–assisted tunneling between occupied and empty states (which are always present in real materials as a result of compensation). Since there are no phonons at zero temperature, the hopping conductivity vanishes in this limit, and the material can be regarded as an insulator. In this strongly localized regime, correlation effects are known to be of primary importance. For instance, it is the Coulomb repulsion between electrons localized on the same site that precludes the existence of doubly occupied states at low temperatures.

In the opposite limit, $\bar{r} \ll a_B$, electrons reside in the conduction band, and low–temperature mobility is determined by ionized impurity scattering. Here, the correlation effects are insignificant as, according to the Landau theory of Fermi liquids, they lead to the qualitatively and, for $\bar{r} \ll a_B$, quantitatively unimportant renormalization of the effective mass. Besides, provided that the temperature is smaller than the Debye and Fermi temperatures, no temperature dependence of conductivity is observed, because under these circumstances inelastic processes are inoperative and the electron gas is degenerate.

It has been experimentally established that the transition from the metallic phase [where $\sigma(T) > 0$ for $T \to 0$] to the insulating phase [where $\sigma(T) \to 0$ for $T \to 0$], the Anderson–Mott transition, occurs for roughly $\bar{r} = 2.5a_B$. This means that n–InSb remains metallic down to $n \simeq 10^{14} \mathrm{cm}^{-3}$ but n–CdSe, for instance, changes from being a metal to being an insulator at $n \simeq 3 \times 10^{17} \mathrm{cm}^{-3}$. In terms of the wave function of electrons at the Fermi level, the metal–insulator transition (MIT) marks a transformation from extended states into localized states. Thus, as the kinetic energy involves space derivatives of the wave function, the MIT is obviously associated with an increase of the kinetic energy of electrons. Therefore, quite generally, the MIT will occur provided that the localization results in such a drop of the potential energy which overcompensates the above mentioned increase in the kinetic energy.

A considerable experimental and theoretical effort has been devoted over the recent years to establish the dominant driving mechanisms of the MIT in bulk doped semiconductors [16,17]. In the subsequent section some of the proposed models are briefly described. The influence of the s–d exchange interaction upon n_c is discussed in Sec. 2. 3.

1.3. Models of the Metal–Insulator Transition

1.3.1. <u>Mott–Hubbard transition</u>. Numerical studies of electronic states in a crystal built up of hydrogen atoms suggest that the MIT occurs for $n_c^{1/3} a_B = 0.22$.[18] The driving force of this transition is the energy of the Coulomb interaction between electrons on the same site, U. For an appropriately large lattice constant the increase of the kinetic energy due to localization becomes smaller than the decrease of the potential energy related to the fact that electrons can no longer appear on the same site. In the insulating phase there is one electron localized around each atom, and thus the model explains how local magnetic moments can be formed. The MIT is, presumably, discontinuous in the sense that at n_c extended states undergo transformation to localized states with the localization radius $\sim a_B$. In contrast, in doped semiconductors, because of a random distribution of impurities and the presence of minority impurity charges, the localization could occur gradually. Within this scenario, the MIT marks the end–point of the transformation of the extended states with energies $\epsilon \leq \epsilon_F$ into the singly occupied local states characterized by the localization radius $\sim a_B$.

1.3.2. <u>Wigner crystallization</u>. In this model the MIT is also driven by the Coulomb interactions among electrons, but the impurities are assumed to act only as a uniformly charged neutralizing background. To some extend this situation is accomplished in the recently designed modulation doped heterostructures.[19] It has been also argued that such a model should apply if the conducting electrons originate from resonant donors with bound states above the Fermi energy ϵ_F.[20] Standard arguments demonstrate that the critical concentration of the Wigner condensation should scale with a_B^{-3}.[20,21] The proportionality factor is anticipated to be, however, at least one order of magnitude smaller than that observed in doped semiconductors.

1.3.3. <u>Percolation transition</u>. Imagine a degenerate liquid of electrons to be placed in a randomly fluctuating potential produced by impurities. If the rms amplitude of the potential fluctuations increases, the volume occupied by electrons shrinks. Once the electron droplets turn to occupy less than $\sim 16\%$ of the sample volume, they cease to form a percolation cluster, and the materials turns into an insulator.[22] Such an MIT is continuous. Experimental studies[23] of a mixture of conducting and nonconducting materials imply that the conductivity vanishes with the critical exponent $t = 1.8 \pm 0.2$. When approaching the MIT from the insulating phase, the dielectric susceptibility diverges with the critical exponent $s = 0.7 \pm 0.1$. Obviously, the percolation model disregards effects of electron–electron interactions as well as quantum phenomena, such as tunneling and interference, which tend to delocalize and localize electrons, respectively.

1.3.4. <u>Anderson localization</u>. Anderson's MIT is the quantum localization of the Fermi liquid induced by scattering. It occurs due to a collective action of many scattering centers and would take place even if the potential of a single impurity, or defect, had no bound states. In macroscopically homogeneous systems, the Anderson localization is thought to set–in before the percolation transition. An interesting exception is probably the localization in the two–dimensional systems under the quantum Hall effect conditions.[24]

It appears, by now, that there are two main mechanisms by which scattering may give rise to the localization.[17,25,26] One of them is the quantum interference of scattered waves. It is known that, when the positions of scattering centers are spatially correlated (as, e.g. , atoms in a crystal) the interference enhances or diminishes the transition probability between two points in space depending on the electron wavelength. If, on the other hand, the scattering centers are distributed randomly, the interference terms will vanish after averaging over all possible configurations of scattering centers. In other words, the mean transition probability between two points is well approximated by the sum of probabilities of all possible electron paths, in agreement with the classical Boltzmann description of diffusion. This is no longer true, however, if we consider the probability of returning to the starting point. This is because the "optical paths", and hence the phases of the transition probability amplitudes, are identical for the clockwise and counterclockwise trajectories of an electron wave along the same sequence of scattering centers. Since such interference is constructive, it increases the probability of return and thus, diminishes the conductivity.

Another consequence of scattering by impurities is a substantial modification of the interactions between electrons. A simplified picture behind it is that the diffusion motion allows two quasiparticles to meet and interact several times. If the two quasiparticles have similar energies they will acquire a small phase difference between the successive meetings. In such a case the successive interaction events cannot be regarded as independent. This invalidates several conclusions of the standard Fermi liquid theory and, in particular, leads to corrections to the Boltzmann conductivity.

The role played by the above quantum effects grows with the increase of $\hbar/\epsilon_F\tau$, where τ is the momentum relaxation time. Since, according to the Brooks–Herring formula (and its modification by Moore[27]) $\epsilon_F\tau$ for ionized impurity scattering depends only on $n^{1/3}a_B$, the model in question is compatible with the proportionality of n_c to a_B^{-3}. An interesting aspect of these quantum corrections is their unusual sensitivity to the temperature and to the so–called symmetry lowering perturbations. The corrections depend on the inelastic scattering processes (appearing at $T > 0$, $\tau_{in} \sim T^{-p}$, where $p \geq 1$) as well as on the magnetic field, spin–orbit– and spin–disorder scattering, because these change the phase of the wave functions. The contribution from the electron–electron interactions is affected additionally by thermal broadening of the Fermi–Dirac distribution function, and by the spin–splitting, as the latter influences the time–evolution of the phase difference if the two quasiparticles come from the different spin subbands. Depending on the symmetry lowering perturbations present (magnetic field, spin–splitting, spin–orbit– or spin–disorder scattering) various universality classes have been identified, each characterized by a different set of critical exponents controlling the behavior of relevant variables in the immediate vicinity of the MIT.[26]

The models of the MIT listed above by no means exhaust all the possibilities. In particular, an interesting question arises whether it is justified to treat, say, phonons or a subsystem of interacting localized spins only as a source of scattering. While for the inelastic processes the corresponding scattering rates vanishes at $T \to 0$, it seems plausible that in this limit the dynamic aspects of these couplings could show up, affecting the localization.

2. INFLUENCE OF THE sp–d EXCHANGE INTERACTION ON TRANSPORT PHENOMENA

2.1. Metallic Range

2.1.1. Spin–splitting. Perhaps the most studied property of DMS is the exchange–induced spin–splitting of effective mass states. According to unnumerable magnetooptical works,[28] this spin–splitting – in the case of extended states – is proportional to macroscopic magnetization of magnetic ions, $M_0(T, H)$. Thus, for an s–type conduction band the total spin–splitting assumes the form

$$\hbar\omega_s = g^*\mu_B H + \frac{\alpha}{g\mu_B}M_0, \tag{1}$$

where the first term describes the Zeeman splitting of the conduction band; α is the s–d exchange integral, and $g \simeq 2.0$ is the Mn–spin Landé factor.

One of effects of spin–splitting on transport phenomena is a *redistribution of carriers between spin–subbands*. In degenerate semiconductors the redistribution starts when $\hbar\omega_s$ becomes comparable to the Fermi energy ε_F, and leads to an increase in the Fermi wavevector of majority–spin carriers. Since the scattering rate depends usually on the Fermi wavevector, the redistribution may have an influence on transport phenomena. Such an approach is meaningful provided that neither the Landau quantization of the density of states, i. e, $\omega_c\tau \ll 1$, nor mixing of the spin states, i. e, $\omega_s\tau_s \gg 1$, are of significance (where τ and τ_s are the momentum and spin relaxation time, respectively). A detail quantitative discussion of the effect for various scattering mechanisms and degree of electron–gas degeneracy was given by Shapira and Kautz[29] in the context of their experimental results for EuS. The calculation was carried out within the first Born approximation and adopting the Thomas–Fermi dielectric function to describe the free–carrier screening of ionized impurity potentials. The above approximations should be valid deeply in the metallic phase, $n \gg n_c \approx (0.26/a_B)^3$. In this concentration range, Shapira and Kautz[29] predict up to twofold decrease of the low–temperature resistivity in the magnetic field, if mobility is limited by ionized impurity scattering. A similar calculation was performed for n-$Cd_{0.95}Mn_{0.05}Se$.[30] Since for this material in the metallic region $\hbar\omega_s < \varepsilon_F$, the field–induced redistribution of electrons between the spin–subbands is only partial, and thus its effect on the conductivity is rather small. In contrast, the exchange contribution to the spin–splitting modifies substantially the quantum (Shubnikov–de Haas) oscillations of the resistivity, visible best in narrow–and zero–

gap DMS. This issue was reviewed in detail previously,[4,5] and will not be explored here.

As shown in the pioneering works of Bastard et al.,[31] Jaczyński et al.,[32] Gaj et al.,[33] and Walukiewicz[34] particularly striking is the *influence of the p–d exchange interaction on the Γ_8 band*. According to Gaj et al.,[33] the exchange interaction splits the Γ_8 band into four components which become strongly non-parabolic and non–spherical. In the case of symmetry–induced zero–gap materials (such as $Hg_{1-x}Mn_xTe$ for $x \leq 0.07$) this splitting gives rise – for a certain range of x values and magnetic fields – to an overlap between the heavy hole and conduction bands.[31,32] Accordingly, the magnetoresistance of zero–gap $Hg_{1-x}Mn_xTe$ was found to contain a negative component,[35–39] in contrast to positive magnetoresistance of $Hg_{1-x}Cd_xTe$ with similar band structure parameters.[40] The negative magnetoresistance appears also in cases when no overlap seems to exist. It results presumably from delocalization of holes bound to resonant acceptor states.[37–39] Figure 2 shows the field dependence of the conductivity tensor components in p–$Hg_{0.935}Mn_{0.065}Te$, which was interpreted on this line.[38] In particular, a visible increase of σ_{xx} and σ_{zz} between 0.9 and 3 T, and the associated change of sign of the Hall coefficient, were ascribed to delocalization of holes originating from resonant acceptor states. Such kind of the field–induced insulator–to–metal transition is thought to result from the exchange–induced upward shift of the valence–band edge.

In the case of open-gap DMS, the uppermost hole band was predicted to acquire the light–hole character in the direction perpendicular to the macroscopic magnetization (magnetic field), while to retain the heavy–hole character for the motion along the magnetic field.[33] Thus, in p–type metallic samples, a strong negative magnetoresistance is expected, as the redistribution of holes between the four sub-bands causes the majority of holes to change their transverse mass from the heavy mass to light mass. Experimental studies[41–46] of p-type open–gap DMS have indeed revealed the presence of a giant negative magnetoresistance associated with unusual anisotropy; the conductivity was larger in the transverse configuration. Except for one sample of Germanenko et al.,[44] the other studied samples were on the insulating side of the metal–insulator transition and, therefore, the magnetoresistance reflected the influence of the p–d exchange interaction on acceptor states or on the transition. We shall comeback to these results in further sections.

An interesting case represents magnetoresistance of metal–insulator–semiconductor structures containing p–type narrow-gap $Hg_{1-x}Mn_xTe$.[47] The a.c. conduc-

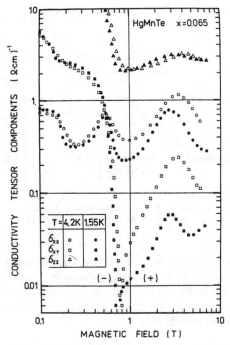

Figure 2: Conductivity tensor components as a function of the magnetic field in zero–gap p–Hg$_{0.935}$Mn$_{0.065}$Te at 4.2 and 1.55 K. The change of sign of the Hall conductivity σ_{xy} and an increase of σ_{xx} and σ_{zz} above 0.9 T are assigned to the exchange–induced delocalization of holes bound to resonant acceptor states (after Sawicki and Dietl[38]).

tivity of interfacial holes was probed by measuring absorption of far–infrared laser radiation in the accumulation regime. As shown in Fig. 3, a strong increase of conductivity was noted above a critical magnetic field, $B \simeq 2$ T. A quantitative calculation, depicted by a solid line in Fig. 3, has demonstrated that the critical field marks the onset of a spin–splitting–induced redistribution of holes between the heavy–hole and light–hole electric–subbands.

Another manifestation of the large spin–splitting could be the *extraordinary Hall effect*, known best from studies of ferromagnetic metals,[48] but also detected in non–magnetic semiconductors such as n–InSb and n–Ge.[49] The spin–dependent Hall effect appears under the presence of a spin–orbit interaction in the crystal, and its magnitude is proportional to the spin–polarization of the electron gas. Beside, the ratio of the extraordinary to ordinary Hall coefficient decreases with the electron

152

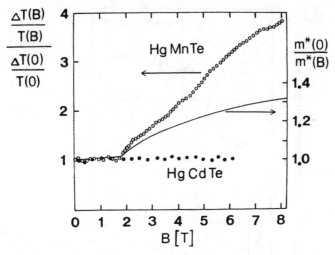

Figure 3: Transmission changes for the surface hole accumulation layer ($N_s =$ 10^{12}cm^{-2}) versus magnetic field, in p–Hg$_{0.87}$Mn$_{0.13}$Te (open symbols) and p–Hg$_{0.74}$Cd$_{0.26}$Te (solid symbols). The solid line presents the calculated inverse of the average effective mass at the Fermi level in an accumulation layer of p–Hg$_{0.87}$Mn$_{0.13}$Te (after Chmielowski et al.[47]).

mobility.[49] As far as DMS are concern, an anomaly of the Hall coefficient was noted by Brandt et al. and Gluzman et al.[50] near the spin–glass freezing temperature in Hg$_{1-x}$Mn$_x$Te. Though no detail microscopic mechanism has been proposed, it is probable that the effect is to be linked to a particularly large influence of the spin–orbit interaction, caused by the p–type character of the carrier wave functions in this narrow–gap semiconductor.

An interesting question arising in this context is whether the presence of magnetic ions would affect the Hall–resistivity quantization of the two dimensional electron gas. Appropriate experimental studies were undertaken by Grabecki et al.,[51] who performed quantum Hall effect measurements in bicrystals of the narrow–gap p–Hg$_{0.75}$Cd$_{0.23}$Mn$_{0.02}$Te. In such compounds, defects occurring in the grain boundary plane were found to have a donor character.[52] Thus, in the p–type material they lead to the appearance of the two–dimensional inversion–layer in the grain boundary vicinity. It was found that under hydrostatic pressure the concentration of inversion-layer electrons diminishes and at the same time their mobility increases,[51,53] so that the quantum Hall effect can be observed at a magnetic field as low as 60 kOe and at

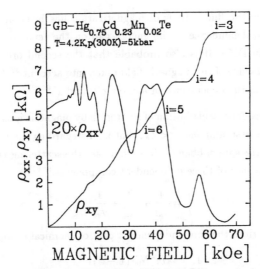

Figure 4: Diagonal and Hall resistivity at 4.2 K and under hydrostatic pressure of 5 kbar (applied at 300 K) versus magnetic field for two–dimensional electron gas adjusted to a grain boundary in p–Hg$_{0.75}$Cd$_{0.23}$Mn$_{0.02}$Te (after Grabecki et al.[51]).

temperature as high as 4.2 K, as shown in Fig. 4. No effect of magnetic ions upon the values of the Hall *plateaux* was detected.[51]

2.1.2. <u>Fluctuations of magnetization.</u> It is obvious that a replacement of non-magnetic cations by transition metals affects electron self–energy, as it changes crystalline potential, giving also rise to the appearance of a spin–dependent interaction. Within the first order perturbation theory, i.e., within the virtual–crystal and mean–field approximations, a modification of crystalline potential leads to a change of the energy gap, while the spin–dependent interaction produces a spin–splitting, which – according to Eq. 1 – is proportional to the macroscopic magnetization of d–spins. Proceeding further, the real part of the second–order term leads to an additional shift of electron energies,[54–59] while its imaginary part describes scattering between k-states.[29,60–64] Two distinct effects contribute to the second–order corrections to the electron self–energy in diluted magnetic systems: randomness of magnetic ion distribution (compositional fluctuations) and randomness of their spin orientations (fluctuations of magnetization). The latter depends on the spin–spin correlation function which, via the fluctuation–dissipation theorem and Kramers–Kronig relations, is determined by magnetic susceptibility of the system. The shift of electron energies induced by the fluctuations of magnetization has been shown to account for a non-standard dependence of the energy gap on the temperature T[56,65] and composition x[55] in DMS. It is also probable that the second order perturbation theory with the presence of the magnetic field taken into account [57,58] could explain some peculiarities in the dependence of $\hbar\omega_s$ on x[51] and T.[66,67]

We shall consider a contribution of scattering by magnetic ions to the momentum relaxation rate assuming $\hbar\omega_s \ll \varepsilon_F$, that is, disregarding the redistribution of electrons between the spin subbands. In such a case the scattering rate in question can be written as a sum of three independent components:[61,62]

$$\frac{1}{\tau} = \frac{1}{\tau_{al}} + \frac{2}{\tau_{sx}} + \frac{1}{\tau_{sz}}. \tag{2}$$

The first term describes the spin-independent part of chemical (alloy) scattering,

$$\frac{1}{\tau_{al}} = \frac{km^* N_0 x(1-x)V^2}{\pi\hbar^3}, \tag{3}$$

where k is the electron momentum, $N_0 x$ is the magnetic ion concentration, and VN_0 is the energy difference in the position of conduction band edges for the end-point materials, i.e., the heterostructure band offsets.[11] The remaining terms in Eq. 2 stands for electron scattering by fluctuations of magnetization. The corresponding scattering rate is determined by the time dependent spin-spin correlation function,

$$C\{S_i(t), S_j(0)\} = < (S_i(t) - x <S>), (S_j(0) - x <S>) >, \tag{4}$$

where indexes i and j run over all cation sites, and $-g\mu_B x <S>$ is the macroscopic magnetization. It is convenient to separate different mechanisms that contribute to the cross section. The first of them involves $< (S_i(t) - <S_i>), (S_j(0) - <S_j>) >$, and thus, via the fluctuation-dissipation theorem, the absorptive part of the dynamic susceptibility. This contribution corresponds to scattering by thermodynamic fluctuations of magnetization. Such scattering is inelastic and thus vanishes for $T \longrightarrow 0$. The remaining part of the correlation function, the static correlation function, $< (<S_i> - x <S>), (<S_j> - x <S>) >$, gives rise to elastic scattering. Two effects make the static correlation function to be non-zero in DMS: compositional fluctuations of magnetization specific to alloys, and spin-glass freezing.[62,63] Neglecting the latter we obtain

$$\frac{1}{\tau_{sx}} = \frac{km^* \alpha^2 k_B T}{4\pi\hbar^3 g^2 \mu_B^2} \tilde{\chi}_\perp(T, H); \tag{5}$$

$$\frac{1}{\tau_{sz}} = \frac{km^*}{4\pi\hbar^3} \left[\frac{\alpha^2 k_B T}{g^2 \mu_B^2} \tilde{\chi}_\|(T,H) + \frac{x(1-x)}{N_0} \left| \frac{d\hbar\omega_s(T,H)}{dx} \right|^2 \right], \tag{6}$$

where the second term in Eq. 6 arises from compositional fluctuations. The functions $\chi_{\perp,\|}(T,H)$ are given by

$$\tilde{\chi}_{\perp,\|} = \frac{\beta^2}{\pi} \int \frac{d\omega' \chi''_{xx,zz}(\omega')}{[\exp(\beta\omega') - 1][1 - \exp(-\beta\omega')]}, \tag{7}$$

where $\beta = \hbar/k_B T$; $\chi''_{xx}(\omega)$ and $\chi''_{zz}(\omega)$ denote, respectively, the transverse and longitudinal imaginary parts of the dynamic magnetic susceptibility of Mn spins for $q \leq 2k$.

If the energetic width of the Mn–spin excitation spectrum in the above q–range is smaller than $k_B T$, the denominator in Eq. 7 can be expanded to lowest order in $\beta\omega'$. In this high temperature case $\chi_{\perp,\|}(T,H)$ reduce, via the Kramers–Krönig relations, to the transverse and longitudinal static susceptibilities according to,

$$\tilde{\chi}_\perp(T,H) \simeq \chi_\perp(T,H) \equiv M_0(T,H)/H; \tag{8}$$

$$\tilde{\chi}_\|(T,H) \simeq \chi_\|(T,H) \equiv \partial M_0(T,H)/\partial H. \tag{9}$$

Inspection of Eq. 7 shows that the high temperature approximation (Eqs. 8,9) constitutes an upper limit estimate for actual values of $\chi_{\perp,\|}$. It is worth noting that while the efficiency of scattering by thermodynamic fluctuations decreases with the magnetic field, the efficiency of scattering by compositional fluctuations grows with the field. Thus, spin–disorder scattering in alloys may be responsible for both negative and positive magnetoresistance, depending on the field range, and character of the magnetization.

The above formalism can easily be adopted for *two–dimensional electronic systems*. If only one electric subbands is occupied, then k in Eqs. 4–6 is to be replaced by $\pi \int dz |f(z)|^4$, where $f(z)$ is the envelope function describing electron confinement, and the integration extends over the region where magnetic ions reside.[68,69]

In the *spin–glass phase*, in addition to inelastic scattering described by Eqs. 5 and 6, elastic scattering may operate. The corresponding scattering rate is proportional to the spin–glass order parameter $q = S^{-2} << S_i >^2>_i$.[62] Since at the spin freezing temperature T_f rather a slowing down of spin dynamics than a change in the spin configuration takes place, the total spin–disorder scattering rate should not exhibit any anomaly at T_f. This seems to be confirmed by experimental results for metallic spin glasses.[70] At the same time "after–effects" and a cusp in the Hall coefficient have been observed,[50,70] presumably because of the presence of remanent magnetization and the associated extraordinary Hall effect.

The formulae presented above describe the scattering rate within the lowest order Born approximation. After Kondo it is well known that higher order terms can modify the efficiency of scattering by the localized spin in a highly nontrivial way.[71] The sign of the *Kondo–effect corrections* depends on the sign of the s–d exchange integral α; for the ferromagnetic coupling the Born approximation overestimates the scattering rate. The corrections appears near and below the Kondo temperature, $T_k = \varepsilon_F \exp[-\rho(\varepsilon_F)|\alpha|/3]/k_B$. In semiconductors, T_k is very low because of the small density–of–states at ε_F, $\rho(\varepsilon_F) = m^* k_F/\pi^2 \hbar^2$. Furthermore, the Kondo effect may be weakness even more by the fact that the concentration of itinerant electrons in DMS is usually much smaller than that of localized spins.[72] It seems, therefore, that the Kondo effect may not appear in DMS.

So far we have considered electrons with the s–type Bloch wave functions. The case of zinc–blende *zero–and narrow–gap semiconductors*, in which the conduction band wave function contains p–type components, was treated by Kossut.[61,73] The momentum relaxation time for alloy scattering is still given by Eq. 3 with V^2 replaced as follows:

$$V^2 \rightarrow V^2 a^4 - \frac{2}{3} V W a^2 (b^2 + c^2) + \frac{W^2}{3} \left[(b^2 + c^2)^2 + 2b^2 \left(\frac{1}{2} b - \sqrt{2} c \right)^2 \right], \qquad (10)$$

where a represents a contribution of s–type component, while b and c p–type components in the Bloch function; W is the energy difference in the position of the Γ_8 valence–band edges for the end point materials, related to V and the energy gap $E_0 = E_{\Gamma_6} - E_{\Gamma_8}$ according to

$$\frac{\mathrm{d}E_o}{\mathrm{d}x} = N_o(V - W). \qquad (11)$$

The treatment of spin–disorder scattering is more complex because the scattering rate depends on the direction of the electron momentum $\boldsymbol{k}$. Averaging over $\boldsymbol{k}$ orientations we obtain Eqs. 5 and 6 with

$$\alpha^2 \rightarrow \alpha^2 a^4 - \frac{2}{3} \alpha \beta a^2 \left(c^2 - \frac{1}{3} b^2 \right) + \frac{\beta^2}{3} \left[c^4 + \frac{5}{6} b^4 + \frac{2 b^2 c^2}{3} + \frac{2 \sqrt{2}}{3} b^3 c \right], \qquad (12)$$

where β is the p–d exchange integral.

The formulae above give the momentum relaxation rate which contributes to the resistivity. Another quantity of interest is the *electron lifetime* in a given state k. The lifetime determines among other things the Dingle temperature,[67] that is, the damping of the quantum oscillations such as the Shubnikov – de Haas magnetoresistivity oscillations. For short range scattering potential and s–type electron wave functions, the momentum relaxation time and the lifetime coincide. If the electron wave function contains a p–type component, the lifetime is still determined by Eqs. 10 and 12, provided that the terms proportional to VW and $\alpha\beta$ are omitted.

Quantitative estimates show that spin–disorder scattering is usually much weaker than ionized impurity scattering in DMS. Its influence would therefore be hardly seen in standard resistivity or magnetoresistivity measurements. Nevertheless, the presence of spin–disorder scattering manifests itself in a number of situations. For instance, Wittlin et al.[75] have attributed a resistivity increase under EPR conditions of Mn spins to this scattering mode. As already noted, spin–dependent scattering play an important role near the metal–insulator transition, as it modifies quantum localization effects.[76] Furthermore, spin–disorder scattering, together with other spin relaxation mechanisms, determines the degree of luminescence polarization [77] and photomagnetization signal[78] under optical pumping conditions. As long as spin relaxation occurs in the conduction band the spin relaxation rate is $1/T_1 = 4/\tau_{sx}$, where τ_{sx} is given in Eqs. 5 and 8. The spin–disorder scattering contributes also to the width of the spin resonance line of conducting electrons. The corresponding transverse spin relaxation time assumes the form

$$\frac{1}{T_2} = \frac{2}{\tau_{sx}} + \frac{2}{\tau_{sz}}, \tag{13}$$

where τ_{sz} is given in Eqs. 6 and 9. Recently, a quantitative study of the linewidth of spin–flip Raman scattering in barely metallic $Cd_{0.95}Mn_{0.05}Se$:In has been carried out.[64] As shown in Fig. 5, the analysis of experimental results demonstrates that both thermodynamic and compositional fluctuations of magnetization are of relevance. Besides, the linewidth was found to be enhanced by electron–electron interactions and interference phenomena, both of paramount importance because of the proximity to the metal–insulator transition.

Within the molecular field approximation the fluctuations of magnetization can be regarded as a source of an additional spin–splitting, inversely proportional to the square root of the volume occupied by the electron.[79,80] This effect would result in a gaussian broadening and a shift of the spin–resonance line unless an ef-

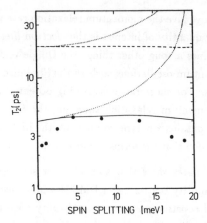

Figure 5: Spin–relaxation time versus spin–split energy for itinerant electrons, obtained by comparing linewidth of forward and backward spin–flip Raman scattering at 1.9 K in $Cd_{1-x}Mn_xSe$:In, $x = 0.05$, $n = 7 \times 10^{17} cm^{-3}$. The solid curve represents scattering by fluctuations of magnetization, including the enhancement due to electron–electron interactions and localization from nonmagnetic scattering. The dashed curve, Eqs. 5–9, 13, neglects these enhancements. The dotted curves show the effect when compositional fluctuations are ignored (after Dietl et al.[64]).

ficient averaging mechanism exists. For delocalized electrons a transformation of the gaussian into the lorenzian line occurs due to the motional narrowing. Our estimates show that the resulting linewidth assumes the same form as that given above (Eqs. 5–9, 13). This indicates the equivalence of the two approaches to the linewidth. This conclusion may call into question the interpretation of magnetooptical spectra of n-$Hg_{0.9}Mn_{0.1}Te$ in terms of the fluctuation–enhanced spin–splitting.[80] Actually, some anomalies were also detected in magnetooptical spectra of barely metallic non–magnetic n-$Hg_{0.8}Cd_{0.2}Te$.[81] They were interpreted as being due to the inter–impurity transitions, involving those donors which remained isolated.[81]

2.1.3. <u>RKKY interaction and free magnetic polaron.</u> So far we have neglected the influence of the electron liquid on the localized spins, and possible feedback effects on electron transport. It is well known that electrons mediate an exchange interaction between localized spins. Quite generally, the magnetization, $M_0(T, H)$ which determines the spin–splitting (Eq. 1) and spin–disorder scattering rate (Eqs. 5,6) has to take the presence of this electron–mediated exchange interaction into account. In other words, disregarding a small diamagnetic correction as well as the paramagnetic term involving the band Landé factor g^*, $M_0(T, H)$ represents the true macroscopic magnetization of a given sample. For itinerant electrons in a narrow band, the carrier–mediated exchange interaction is to be described within the double exchange formalism.[82] In the case of a wide conduction band the celebrated RKKY model applies. According to the latter, if – as in DMS – the electron concentration is smaller than that of localized spins the resulting spin–spin interaction is ferromagnetic. Its effect on $\chi(T)$ can be estimated considering the Landau free energy of the system with the magnetization of d spins, M, and the net concentration of spin–down electrons, $\Delta n = n_\downarrow - n_\uparrow$, treated as order parameters,

$$\mathcal{F}[M, \Delta n] = \frac{M^2}{2\chi_0} - M \cdot H + \frac{(\Delta n)^2}{2\rho(\epsilon_F)\left(1 + \frac{F}{2}\right)} - \frac{1}{2}\left(g^*\mu_B H + \frac{\alpha M}{g\mu_B}\right)\Delta n. \quad (14)$$

Here χ_0 is the magnetic susceptibility of localized spins in the absence of electrons, and F is the Fermi liquid parameter describing interactions between electrons. The individual terms in Eq. 13 represent respectively: (i) a change in the d–spin entropy and internal energy caused by the d–spin polarization; (ii) the Zeeman energy of d–spins; (iii) spin–polarization–induced change in the electron kinetic energy as well as in the potential energy of the electron–electron interaction; (iv) the Zeeman energy of electrons and the s–d energy in the molecular field approximation. By minimizing $\mathcal{F}[M, \Delta n]$ with respect to M and Δn we get their values corresponding to thermodynamic equilibrium, with the effects of electron–electron interactions taken into account. Inserting M and Δn to $\mathcal{F}$ we obtain the free energy F_{tot} of the system within the mean field approximation, which should be valid except for the proximity of the magnetic phase transition. The magnetic susceptibility $\chi_{tot} = -\partial^2 F_{tot}/\partial H^2$ becomes

$$\chi_{tot}(T) = \chi_0(T)\frac{\left[1 + \frac{1}{4}\left(1 + \frac{F}{2}\right)\rho(\epsilon_F)g^*\alpha/g\right]^2}{\left[1 - \frac{1}{4}\left(1 + \frac{1}{2}F\right)\rho(\epsilon_F)\frac{\alpha^2}{g^2\mu_B^2}\chi_0(T)\right]} + \frac{1}{4}\left(1 + \frac{F}{2}\right)\rho(\epsilon_F)g^{*2}\mu_B^2, \quad (15)$$

and is seen to diverge for T given by

$$\frac{1}{4}\left(1 + \frac{1}{2}F\right)\rho(\epsilon_F)\frac{\alpha^2}{g^2\mu_B^2}\chi_0(T) = 1. \quad (16)$$

Story et al.[83] discovered a ferromagnetic phase transition induced by free carriers in p–$Pb_{1-x-y}Sn_yMn_xTe$. The transition occurred once holes started to occupied a side band Σ, characterized by the large density of states $\rho(\epsilon_F)$.[84] The transition was accompanied by the appearance of a giant extraordinary Hall effect.[85]

A great deal of interest has attracted the question of magnetic polarons in DMS. At least three kinds of magnetic polarons can be distinguished: (i) bound magnetic polaron, i.e., electron localized by a defect or impurity potential, the s–d interaction supplying an additional binding energy; (ii) self–trapped magnetic polaron, i.e., electron in a localized, symmetry–broken state, the s–d coupling constituting the main binding force; (iii) free magnetic polaron, i.e., delocalized electron surrounded in its way through the crystal by a ferromagnetic cloud of d–spins.

The issue of bound magnetic polaron will be discussed in subsequent sections. As far as *self-trapped magnetic polaron* is concerned, its stability was calculated[86] to require a rather large sp–d coupling constant, effective mass, and magnetic susceptibility. Hence, such a complex can probably exist only in the form of the trapped heavy hole, particularly in systems of reduced dimensionality such as quantum wells.[88]

The case of the *free magnetic polaron* has proven to be very involved and, therefore, we limit ourselves to qualitative comments. Quantitative approaches[89] to this problem in DMS seem to suffer from improper treatment of the kinetic energy, and have turned out to be unsuccessful vis a vis experimental data.[64,67,90] The effects of the s–d interaction which we have discussed above were described within the adiabatic approximation, i.e., electronic energies were determined for frozen configurations of d–spins. From previous studies of the electron–phonon system we know that in order to describe the free polaron this approximation has to be relaxed. In the magnetic systems the adiabatic approximation is valid as long as electron–spin–density fluctuations decay faster than fluctuations of d–spin magnetization. This is not the case for quasiparticles with energies (calculated from the Fermi energy) smaller than the energetic width of the excitation spectrum of the d–spin subsystem. If, therefore, $k_B T$ is smaller than this width polaronic (dynamic) corrections to, say, transport coefficients may become important. Since there is no long range magnetic order in DMS, the magnetic excitations are, presumably, localized in space even at $T \to 0$. Thus, theoretical treatment of the problem cannot follow methods developed for magnetic semiconductors[91] or copper oxides.[92] Recently, Sachdev[93] has examined effects of the exchange coupling between the Fermi liquid electrons and a disordered system of localized spins. The analysis implies that the dynamic effects do affect the spin–spin correlation function of itinerant electrons and, in particular, the Fermi liquid parameter F. Whether these effects will perturb the conductivity is to be elucidated.

2.2. Insulating Region

2.2.1. Spin–splitting. It is easy to show that if the spin–splitting of relevant donor levels and of the conduction band is the same, it does not affect the concentration of thermally activated electrons. Furthermore, within a standard Miller–Abrahams[94] model of the hopping conduction, no influence of the spin–splitting on the low–temperature conductivity is expected. A number of materials exhibits, however, a magnetoresistance in the hopping region. In order to account for the experimental findings, several modifications of the Miller–Abrahams model has been put forward. For instance, Goldman and Drew[95] in an attempt to explain the *negative magnetoresistance* of n-$Cd_{0.5}Mn_{0.1}Se$ have taken into consideration the influence of an exchange interaction between neighboring donor electrons upon the shape of the impurity band. As this inter–site exchange interaction is predominately antiferromagnetic, the spin–splitting may convert pairs of coupled electrons from singlets into triplets. According to Goldman and Drew,[95] this field–induced spin polarization diminishes the width of the impurity band, thereby increasing the density of states at ϵ_F, and thus the hopping conductivity.

A model for the spin–splitting–induced *positive magnetoresistance* in the hopping region has been proposed by Kurobe and Kamimura.[96] Their model assumes the width of the impurity band to be greater than the on–site correlation energy, so that singly and doubly occupied impurity states can coexist near ϵ_F. In such a case, in addition to hops between occupied and empty states, transitions between the two singly occupied neighboring impurities are also possible. Since the ground state of doubly occupied donors is singlet, these additional transitions are suppressed when electrons become spin–polarized. While the above models assumes the hopping process to be spin–conserving, Movaghar and Schweitzer[97] have argued that the hopping conduction is actually limited by the efficiency of transitions involving spin–flip of the electron. Although Movaghar and Schweitzer[97] have developed their model under assumption $\hbar\omega_s \ll k_B T$, which is usually not fulfilled in DMS, von Ortenberg et al.[98] have applied this model to describe the low–temperature magnetoresistance of n-$Zn_{1-x}Mn_xSe$.

Because of a complex influence of the exchange interaction on the fourfold degenerate *acceptor states*, the case of p–type materials requires a separate discussion. Delves[41] already in 1966 discovered a giant negative magnetoresistance of p-$Hg_{1-x}Mn_xTe$ and interpreted it correctly as being caused by a different spin–splitting of the acceptor levels and the valence band. It has been theoretically

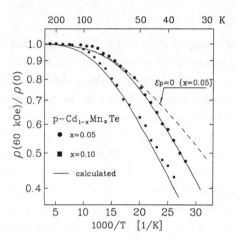

Figure 6: Temperature dependence of magnetoresistance in p–$Cd_{1-x}Mn_xTe$ at $H = 60$ kOe in the temperature range, where conduction is due to free holes thermally activated to the valence band. Solid lines were calculated with no adjustable parameters taking orbital quenching of the acceptor spin–splitting and polaronic effects into account; the dashed line neglects the polaronic effects (after Jaroszyński and Dietl[45]).

demonstrated by Mycielski and Rigaux[99] that the p–d exchange spin–splitting of the acceptor state in the zinc–blende DMS is indeed about 20% smaller than that of the valence band at $k = 0$. Accordingly, the concentration of holes thermally activated to the valence band increases with the magnetic field. As shown in Fig. 6, this "boil off" effect, together with a destructive influence of the magnetic field on a polaronic contribution to the acceptor binding energy, explains quantitatively the negative magnetoresistance of p–$Cd_{1-x}Mn_xTe$ at temperatures, where conduction is due to free holes in the valence band.[45] When the spin–splitting becomes comparable to the acceptor binding energy – a situation occurring in narrow–gap DMS at moderately strong magnetic fields – the spin–splitting is accompanied by a substantial increase in the transverse Bohr radius. This is because the uppermost valence subband acquires the light hole character. As noted by Mycielski and Mycielski,[100] and then elaborated by their successors,[101] this increasing trend turns into decreasing one when the magnetic length $l_H = (e\hbar/cH)^{1/2}$ becomes smaller than the Bohr radius. A number of groups[42–44,46,102] has undertaken magnetoresistivity studies of p–$Hg_{1-x}Mn_xTe$ and p–$Hg_{1-x-y}Mn_xCd_yTe$ in the hopping region, where the conductivity is particularly sensitive to the acceptor wave function. These studies confirm

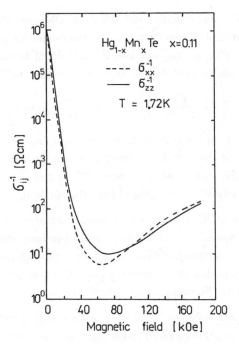

Figure 7: Inverse of the conductivity tensor components, σ_{xx}^{-1} and σ_{zz}^{-1}, versus magnetic field at 1.72 K for p–Hg$_{0.89}$Mn$_{0.11}$Te in the hopping region (after Wojtowicz et al.[102]).

the existence of both giant magnetoresistance and unusual anisotropy. As shown in Fig. 7, the magnetoresistance and its anisotropy change their sign in appropriately strong magnetic fields.[102] Such behavior supports theoretical expectations referred to above. A meaningful quantitative description of experimental results is, however, rather difficult since the change in the Bohr radius alters the hopping conductivity through at least three factors: (i) matrix element of transition probability;[101] (ii) width of the impurity band being determined by fluctuations of composition;[103] (iii) energy of bound magnetic polarons.[42] Furthermore, it is unclear whether acceptor concentrations in the studied samples were low enough to neglect precursory effects of the insulator–to–metal transition.

2.2.2. Fluctuations of magnetization. It is clear that the hopping conductivity drops when the width of the impurity band increases, that is, the density–of–states near ϵ_F decreases. In ordinary semiconductors, the band width is primarily determined by electric fields of minority charges and defects. In DMS, an additional broadening of the impurity band arises from fluctuations of the composition and magnetization. For random distribution of magnetic ions this additional broadening of the donor band is gaussian with the field–dependent variance σ given by[104]

$$\sigma^2 = \left[\left(V - \frac{1}{2} \frac{d|\hbar\omega_s|}{dx} \right)^2 x(1-x) + \frac{\alpha^2}{2g^2\mu_B^2} \chi_\| k_B T \right] \frac{1}{8\pi a_B^3}. \tag{17}$$

If V and $d|\omega_s|/dx$ are of the same sign, the variance σ decreases with the magnetic field. A magnitude of the resulting negative magnetoresistance has been estimated for n–$Cd_{0.95}Mn_{0.05}Se$, and found to be small.[104] To the same conclusion leads a detail numerical study of the far–infrared light absorption in n–$Cd_{0.9}Mn_{0.1}Se$.[105]

It is worth mentioning that thermodynamic fluctuations may prompt electron hopping between impurities. Such hopping mechanism could dominate at very low temperatures, where the magnetic excitations appear to determine specific heat of the system. It seems that this interesting possibility has not yet been explicitly considered in the DMS literature.

2.2.3. <u>Bound magnetic polarons.</u> DMS offer an appealing opportunity to elucidate how bound magnetic polarons (BMP) affect transport phenomena. Such favorable situation is due to detail spectroscopic measurements and their successful theoretical interpretation,[106] indicating that the basic properties of BMP in DMS are well understood. Those studies have demonstrated that the local magnetization which gives rise to zero–field spin–splitting is induced by a molecular field of the donor electron as well as by thermodynamic fluctuations of magnetization.[79] When these two factors are taken into account the free energy of the donor electron assumes the form:[79]

$$F_p = \frac{\hbar^2}{2m^*a_B^2} - \frac{e^2}{\kappa_o a_B} - \frac{\varepsilon_p}{2} - k_B T \ln\left[2\cosh\left(\frac{\hbar\omega_s}{2k_B T} \right) + \frac{4\varepsilon_p}{\hbar\omega_s}\sinh\left(\frac{\hbar\omega_s}{2k_B T} \right) \right], \tag{18}$$

where ω_s is given in Eq. 1, and ε_p is a characteristic energy of the BMP,

$$\varepsilon_p = \frac{\alpha^2 \chi_\|(T,H)}{32\pi g^2 \mu_B^2 a_B^3}. \tag{19}$$

By minimizing F_p with respect to a_B we get an estimate for the value of the effective Bohr radius renormalized by the s–d coupling. In terms of F_p the donor binding energy E_p becomes

$$E_p = -F_p - k_B T \ln\left[2\cosh\left(\frac{\hbar\omega_s}{2k_B T} \right) \right], \tag{20}$$

where the second term describes the free energy of the electron at the bottom of the conduction band. In the temperature range, where the conduction is due the thermally activated electrons in the conduction band but their concentration is smaller than that of compensating acceptors, $E_p(T)$ is related in a simple way to the activation energy of the conductivity, $E_a(T)$,

$$E_a(T) \equiv -k_B \frac{d \ln \sigma(T)}{d1/T} \simeq E_p(T) - T\frac{dE_p(T)}{dT}, \tag{21}$$

where the temperature dependence of the preexponential factor in $\sigma(T)$ has been disregarded.

The above model has been developed describing the subsystem of localized d spins within the continuous medium and Gaussian approximations. The former can be partly relaxed by excluding from the partition function those shapes of magnetization which correspond to strong variations of magnetization over distances smaller than the average distance between magnetic ions.[108] This procedure reduces ε_p according to $\varepsilon_p \to \varepsilon_p f(\Lambda a_B/2)$, where $\Lambda = (6/\pi)^{1/3}(\pi/a_o)x^{1/3}$, a_o is the lattice constant, and

$$f(y) = \frac{2}{\pi}\left[\arctan(y) + \frac{y(3y^4 + 8y^2 - 3)}{3(1 + y^2)^3}\right]. \tag{22}$$

The Gaussian approximation breaks down when the molecular field produced by the donor electron is so large that the response of the localized spins to it becomes nonlinear. The form of F_p at $H = 0$, derived going beyond the Gaussian approximation, is given by[87]

$$F_p = <\phi \mid \frac{p^2}{2m^*} + V(r) \mid \phi> \; - \int dr \int_0^{h(r)} dH\, M_0(T, H)$$

$$- k_B T \ln\left[2 + \frac{2}{k_B T}\int dr\, h(r) M_0(T, h(r))\right], \tag{23}$$

where $h(r) = \mid \phi(r) \mid^2 \alpha/2g\mu_B$ is the molecular field of the electron that resides in a state described by a variational wavefunction $\phi(r)$ to be chosen in the form which minimizes F_p; $M_0(T, H)$ is the experimentally available macroscopic magnetization of d spins. Obviously, Eq. 23 in the absence of the impurity potential, i. e., for $V(r) = 0$, represents the free energy of the self–trapped magnetic polaron.

The polaronic corrections to the impurity binding energy turned out to explain a strong increase of the acceptor ionization energy with the Mn concentration, estimated from conductivity measurements as a function of temperature in p–Cd_{1-x} Mn_xTe.[108] As already mentioned these corrections are also partly responsible for the negative magnetoresistance of p–$Cd_{1-x}Mn_xTe$, as depicted in Fig. 6.[45]

It has been realized by Kasuya and Yanase,[107] in the context of europium chalcogenides, that the formation of BMP may severely diminish the hopping conductivity. The influence of the BMP on this type of conduction in DMS has been considered by the present author and his co–workers[109] as well as by Ioselevich,[110] and Belyaev et al.[46] It has been suggested in the former work that the activation energy of the hopping conductivity in DMS is augmented because of a difference in the time–averaged spin–splitting of the occupied and empty impurity. If the thermodynamic fluctuations are taken into account, this model predicts a positive and negative magnetoresistance in small, $\hbar\omega_s \geq k_BT$, and strong, $g\mu_BH \geq k_BT$, magnetic fields, respectively. Such shape of the magnetoresistance is indeed observed experimentally.[98,109,111-114] Its calculated magnitude turns out, however, to be usually smaller than that observed.[109] This discrepancy was taken[109] as indicative of conduction proceeding through a simultaneous hopping of many electrons,[94] binding energy of each being altered by the polaronic corrections. However, it seems, by now, that other precursory effects of the insulator–to–metal transition may also be involved.

Ioselevich[110] analyzed theoretically polaronic effects under assumption that resistances forming the Miller–Abrahams network are determined by the inverse of the time–averaged hopping probabilities. Because of thermodynamic fluctuations, the energy difference between the ground states of the two impurities fluctuates around its time–averaged value. It follows that the time–averaged probability, and thus the hopping conductivity, is even less affected by the BMP than that calculated from the previous model,[109], by using the time–averaged spin–splittings. Finally, Belyaev et al.[46] have detected two activation energies of the hopping conductivity in p–$Hg_{1-x}Mn_xTe$. The data have been analyzed supplementing Ioselevich's model by taking into account the influence of the exchange interaction on the localization radius. Obviously, the polaronic effects leads to a decrease of the ground–state localization radius, and to its increase if it happens that the carrier appears in the spin–reverse state. Thus, at high temperatures the hopping proceeds mainly through the spin excited states and is, therefore, characterized by a large activation energy. At low temperatures only the ground state is involved.

2.3. Vicinity of the Metal–Insulator Transition

2.3.1. A view on experimental results. In order to visualize a strong and complex influence of the s–d exchange interaction on electron transport near the metal–insulator transition (MIT) it is instructive to contrast dependencies of the resistivity on the magnetic field and temperature in non–magnetic and magnetic materials. The magnetoresistances of n–CdSe and n–$Cd_{0.95}Mn_{0.05}Se$ at $T \simeq 2$ K are shown in Figs. 8 and 9.[104] The data span both sides of the MIT, as the Mott critical concentration n_c is about $3 \cdot 10^{17} cm^{-3}$ in n–CdSe.[115] In the case of CdSe, the magnetoresistance contains both negative and positive components, the latter taking over at appropriately strong magnetic fields and low donor concentrations. The magnetoresistance of n–$Cd_{0.95}Mn_{0.05}Se$ has a completely different character and much greater magnitude. As shown in Fig. 9, it is positive in low fields and negative in high fields. Qualitatively similar dependencies have also been observed for n–$Cd_{1-x}Mn_xSe$ with other Mn concentrations,[111,112] as well as for n–$Cd_{0.9765}Mn_{0.0235}S$,[113] n–$Zn_{1-x}Mn_xSe$,[98] and n–$Cd_{0.95}Mn_{0.05}Te$.[114] An interesting aspect of the data is that the magnetoresistance exhibits a similar character on the both sides of the MIT, its magnitude

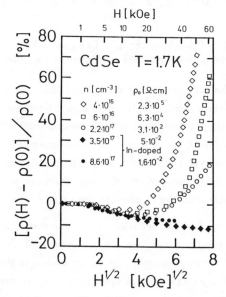

Figure 8: Magnetoresistance of n–CdSe as a function of the square root of the magnetic field at 1.7 K for selected electron concentrations n. The metal–insulator transition occurs at about $3 \times 10^{17} cm^{-3}$ (after Dietl et al.[104]).

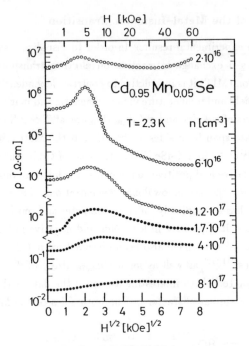

Figure 9: Resistivity of n–Cd$_{0.95}$Mn$_{0.05}$Se as a function of the square root of the magnetic field at 2.3 K for selected electron concentrations n (after Dietl et al.[104]).

decreasing away from the MIT. This suggests strongly that the magnetoresistance is primarily caused by the influence of the exchange interaction on the MIT. In light of this conclusion those descriptions of experimental results,[98,109] which have ignored the proximity of the MIT may require a reconsideration. Furthermore, since – according to Fig. 9 – n–type DMS exhibit a sizable negative magnetoresistance, its occurrence in p–type DMS may not entirely result from the peculiar influence of the p–d exchange interaction on the Γ_8 effective mass states.

DMS show also an unusual behavior in the absence of an external magnetic field. Figure 10 depicts a temperature dependence of the resistivity ρ in Cd$_{0.95}$Mn$_{0.05}$Se:In with the donor concentration corresponding to the close proximity to the MIT.[104] As shown, ρ increases with decreasing temperature T according to $\rho(T) \sim \exp(T_0/T)^\alpha$, where $\alpha > 1$. Such a dependence points out that this sample is on the insulating side of the transition. In similar samples of non–magnetic materials, particularly of n–CdSe,[115] the temperature dependencies of the resistivity are,

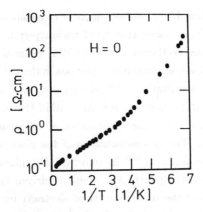

Figure 10: Resistivity of n–$Cd_{0.95}Mn_{0.05}Se$ as a function of the inverse temperature for $n = 4 \times 10^{17} cm^{-3}$ (after Dietl et al.[104]).

however, much weaker: $1/4 \leq \alpha \leq 1/2$. The strong temperature dependence of ρ observed in magnetic materials is referred to as temperature dependent localization. A precursory of the effect was discovered by Sawicki et al.[116] for n–$Cd_{0.95}Mn_{0.05}Se$ on the metallic side of the MIT in the form of a sudden and unexpected increase of ρ below 0.5 K. The enhanced value of ρ disappeared under the presence of the magnetic field.[116] Similar anomalies were also detected by Germanenko at al.[39] in p–$Hg_{1-x}Mn_xTe$.

Thus, theoretical description of the experimental results ought to provide explanation for the low–field positive and high–field negative magnetoresistances as well as for the temperature dependent localization and the associated negative magnetoresistance. At the same time, the data on DMS may constitute a valuable tool to tell the dominant driving mechanisms of the MIT in doped semiconductors.

2.3.2. Influence of the s–d exchange interaction on the metal–insulator transition. Early experiments on magnetic semiconductors already demonstrated that the exchange interaction exerted a strong influence on the MIT.[1,2,117] Model descriptions of experimental findings have emphasized the role played by the redistribution of electrons between spin subbands,[118] spin–disorder scattering,[119] and formation of bound magnetic polarons.[118,120] Obviously, in what direction the perturbations in question will shift the critical carrier concentration n_c or how they will make n_c to vary with the magnetic field and temperature depends on the adopted model of the MIT.

As already mentioned, *the redistribution of electrons induced by the spin–splitting* increases the Fermi wavevector k_F of majority–spin electrons. From the previous discussion devoted to the models of the MIT we know that the increase of k_F will shift n_c towards lower electron concentrations in the case of both percolation transition and Anderson localization.[96,121] Making a natural assumption that the smaller n_c, the greater conductivity we see that within the mentioned models the spin–splitting leads to a negative magnetoresistance, appearing in relatively high magnetic fields $\hbar\omega_s \gtrsim \epsilon_F$. The spin–polarization of the electron–gas diminishes n_c also for the Wigner crystallization.[122] In contrast, the spin–polarization will increase n_c, and thus the resistivity if the MIT would be driven by the Mott–Hubbard mechanism. Indeed, since the redistribution of electrons between the spin sub–bands increases the screening length, it enforces the binding ability of the screened Coulomb potential.[118] A similar conclusion emerges when considering the insulator side of the MIT; the spin–polarization precludes quantum hopping between occupied states, increasing in this way an effective on–site correlation energy U, and thus n_c.

An interesting influence of the spin–splitting and spin–disorder scattering upon n_c shows up in the case of Anderson's localization. If electron–electron interactions are taken into consideration, these *symmetry lowering perturbations* shift n_c towards higher electron concentrations.[26] Accordingly, once $\hbar\omega_s$ is greater than $k_B T$ and $\hbar/\tau_{si}$ – the spin–disorder scattering rate – the resulting magnetoresistance is positive. A contribution of these quantum effects can be analyzed quantitatively on the metallic side of the MIT, i.e., in the so–called weakly localized regime, where the perturbation approach should be valid.[25]

In principle, the mean free path, and thus n_c for the Anderson transition is directly altered by *alloy and spin–disorder scattering*. In most cases, however, the effect is of minor importance as these scattering modes are usually of a small efficiency in comparison to the ionized impurity scattering. Actually, the action of spin–disorder scattering upon n_c trough the quantum effects refereed to above appears to be stronger that through its direct influence on the mean free path. It is worth noting that near the MIT there may exist an additional source of spin–disorder scattering, that induced by those fluctuations of d–spin magnetization which are produced, via s–d coupling, by the inhomogeneous distribution of electron spin density, inherent to disordered systems.[104,118,123,124] If all electrons would be delocalized, a non–zero time–averaged local inhomogeneity in the s–spin density, and thus the fluctuation of d–spin magnetization may appear only under the presence of an external magnetic field.[118] The effect will operate as long as the magnetic field will not saturate all Mn spins in the crystal. A quantitative study[64] of the spin-flip Raman scattering spectra of conducting electrons in $Cd_{0.95}Mn_{0.05}Se$:In near the MIT does not point out to the significance of this scattering mechanism. Another source of additional fluctuations of magnetization could be the molecular field produced by localized electrons, i.e., *bound magnetic polarons*.[104,118,123] Obviously, localization associated with the polaron formation will lead to an increase of n_c independently of the model of the MIT. Since, in DMS, the polaron binding energy and the magnitude of the local magnetization drop with the temperature and magnetic field, the magnetic polarons may account for the temperature dependent localization, and the related negative magnetoresistance. A quantitative approach to the problem turns out, however, to be difficult because of our poor knowledge concerning mechanisms of the formation of the local electron moments near the MIT. We shall return to this topic when discussing experimental results.

As already mentioned, the s–d interaction, when treated going *beyond the adiabatic approximation*, mediates an interaction between the Fermi liquid electrons. Because of the crucial role played by electron–electron interactions in the localization phenomena, a contribution of the s–d coupling to the interaction amplitude may be of particular importance near the MIT. This interesting problem deserves further experimental and theoretical studies.

2.3.3. Discussion of selected experimental results. Results of one of the experimental attempts to detect the presence of quantum corrections to the conductivity of disordered metals are shown in Figs. 11 and 12, where conductivities of the barely metallic CdSe:In and $Cd_{0.95}Mn_{0.05}Se$:In are plotted as a function of the magnetic field.[116]

Though there is a qualitative difference between the results for CdSe and CdMnSe, in both materials the magnetoresistance (MR) appears at very weak magnetic fields, at which the cyclotron and spin splittings are much smaller than both the inverse momentum relaxation time and the Fermi energy. The Boltzmann–type theories of magnetoconductivity do not predict any MR in such a field range. In contrast, the theory[25] which takes the quantum corrections into account (solid lines in Figs. 11 and 12) is shown to describe correctly the experimental data. The MR of CdSe is negative reflecting the destructive effect of the magnetic field (vector potential) on the interference of scattered waves. A small positive MR visible below 1 K turns out to be due to the influence of spin–orbit scattering that operates in

172

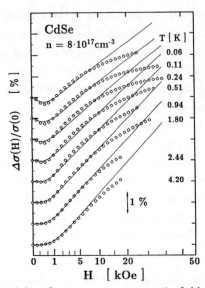

Figure 11: Conductivity changes versus magnetic field in metallic n–CdSe:In at various temperatures (dots). Solid lines were calculated taking quantum corrections to the conductivity of disordered systems[25] into account (after Sawicki et al.[116]).

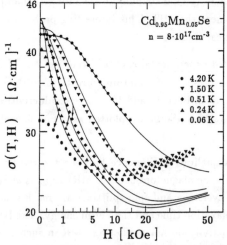

Figure 12: Dependence of the conductivity on the magnetic field in metallic n–$Cd_{0.95}Mn_{0.05}$Se:In at various temperatures. Solid lines were calculated taking the quantum corrections[25] into account but neglecting possible effects of the spin-splitting–induced redistribution of electrons between spin–subbands, and of the formation of magnetic polarons (after Sawicki et al.[116]).

this acentric crystal upon the interference. In the case of $Cd_{0.95}Mn_{0.05}Se$, the MR stemming from the interference is found to be totally masked by the positive MR resulting from the influence of the giant s–d spin splitting upon electron–electron interactions. Since the calculation of $\sigma(H)$ in CdMnSe involves only one adjustable parameter – the effective amplitude of the Coulomb interaction – a good agreement between experimental and theoretical results has to be regarded as significant.

Although the present theory[25] is successful in weak magnetic fields, it fails to describe the high–field negative component of the MR, visible in Fig. 9 and 12. As argued by Shapira et al.,[111] this negative MR is to be assigned to the spin–splitting–induced redistribution of electrons between the spin subbands; the redistribution increases the kinetic energy of carriers at the Fermi level diminishing their tendency towards Anderson localization.[121] The effect is particularly strong in p-type DMS, as the redistribution of holes between the Γ_8 subbands is connected with a change of the transverse hole mass from a heavy mass to a light one. The redistribution, together with the destructive effect of the magnetic field on the binding energy of bound magnetic polarons, is most probably responsible for the field–induced insulator-to-metal transition observed in magnetic semiconductors[117] as well as in n–$Cd_{0.95}Mn_{0.05}Se$[125] and p–$Hg_{0.92}Mn_{0.08}Te$.[125,126]

Figure 13 presents zero–temperature transverse and longitudinal conductivity, as well as the inverse Hall coefficient and dielectric constant, as functions of the magnetic field for p–$Hg_{0.92}Mn_{0.08}Te$.[126] Similar data for n–$Hg_{0.85}Mn_{0.15}Te$[127] are shown in Fig. 14. The data points were obtain from resistivity and capacitance measurements in the temperature range 30 mK$\leq T \leq$ 0.5 K. It is seen that p–HgMnTe undergoes an insulator–to–metal transition when the magnetic field gets stronger. The narrow–gap n–HgMnTe undergoes, in turn, a traditional field–induced metal–to–insulator transition. Such a transition takes place because of the diamagnetic effects, which dominate when the electrons are confined at the lowest Landau level, causing the kinetic energy to decrease with increasing magnetic field.

Several important conclusions can be drawn from the data displayed in Figs. 13 and 14. Firstly, in both cases the transition is continuous. This underlines the leading role of randomness, as MITs occurring in ordered systems (Mott–Hubbard transition, Wigner crystallization) would be, presumably, discontinuous in the 3d case. Secondly, the Hall coefficient R_H is seen to remain finite in the magnetic fields at which the conductivity vanishes. This contradicts the simple freeze–out picture in the framework of which the carrier concentration should vanish at the same field as the conductivity. The third conclusion comes from the fact that the

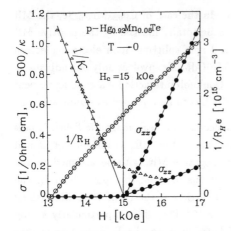

Figure 13: Conductivity tensor components σ_{xx} and σ_{zz}, the inverse of the Hall coefficient R_H and the dielectric constant κ_{xx}, extrapolated to zero-temperature for p-$Hg_{0.92}Mn_{0.08}$Te. The solid lines indicate critical behavior. The finite values of κ in the metallic phase result from the schottky barrier capacitance (after Jaroszyński et al.[126]).

Figure 14: Conductivity tensor components and the inverse Hall coefficient for n-$Hg_{0.85}Mn_{0.15}$Te near the field-induced localization. Solid lines indicate critical behavior. Above H_c the conductivity seems to be perturb by accumulation or inversion layers (after Wróbel et al.[127]).

diagonal components of the conductivity tensor tend to exhibit critical behavior of the form $\sigma_{ii} \sim |H - H_c|^t$, where $t = 1.0 \pm 0.2$. Such a value of t, observed commonly in doped semiconductors,[17] is in apparent disagreement with $t = 1.8 \pm 0.2$ expected from the percolation theory. This indicates that the transition is caused neither by long–range fluctuations of the impurity potentials nor by macroscopic inhomogeneities in the impurity distribution. At the same time $t = 1$ is in accord with theoretical predictions for Anderson's MIT, provided that effects of scattering-modified electron–electron interactions are taken into account.[26] This conclusion is supported by the temperature dependence of the conductivity near the MIT,[125] as well as by the character of the divergence of the localization length ξ at H_c, evaluated from the hopping conductivity[128] and dielectric constant[126] measurements on the insulating side of the MIT. The large values of ξ indicate also that the localization of electrons at the Fermi energy is caused by many scattering centers and must *not* be regarded as a freeze–out of carriers on individual impurities. Finally, in spite of the large field–induced anisotropy, visible in Figs. 13 and 14, the same values of

H_c and t are observed for both σ_{xx} and σ_{zz}. This finding confirms the theoretical expectations concerning the quantum localization in anisotropic systems.[129]

The experimental results discussed above substantiate the picture of the MIT as the quantum localization of the Fermi liquid. According to this model, a Fermi–liquid–type description of electronic states remains valid as long as the size of the system is smaller than the localization length ξ. On the other hand, it is well known that deep in the insulating phase the electrons are bound to individual impurities forming, owing to the Hubbard energy U, localized magnetic moments. Obviously, in the latter case the electronic states cannot be described within the Fermi–liquid formalism at any length scale. A question then arises when does a transformation of the Fermi liquid electrons into local moments occurs. There is a growing number of evidences that this transformation begins already on the metallic side of the MIT, leading to the existence of local moments in metallic samples.[130] At the same time, it has been speculated[127] that the divergence of the Hall coefficient, visible in Figs. 13 and 14, marks the point where the Fermi–liquid electrons disappear totally.

The presence of local moments in metallic samples has been inferred, in particular, from the unusual temperature dependence of the conductivity discovered in $Cd_{0.95}Mn_{0.05}Se$:In below 0.5 K,[116] as shown in Fig. 15.[131] In magnetic materials the local moments can polarize, via s–d interaction, the neighboring Mn spins (there are about 200 Mn ions within the Bohr orbit in $Cd_{0.95}Mn_{0.05}Se$). The ferromagnetic bubbles (bound magnetic polarons) formed in this way constitute centers of spin–dependent scattering for the itinerant electrons. The efficiency of this scattering increases rather steeply with decreasing temperature, as the degree of the bubble polarization is proportional to the magnetic susceptibility of Mn spins, $\chi(T)$. It has been demonstrated that the drop of conductivity below 0.5 K can be semiquantitatively explained by taking into account the influence of this scattering mode on the quantum corrections to the conductivity.[131] It is also probable that the drop of σ is partly caused by the temperature dependent transformation of the Fermi–liquid electrons into local moments, as the binding energy of bound magnetic polarons becomes greater when the temperature is lowered.

To summarized, results of extensive studies discussed above indicate that the low–field positive magnetoresistance is primarily caused by the influence of the spin–splitting on those quantum corrections to the conductivity which originate from the scattering–modified electron–electron interactions. This interpretation is strongly supported by a quantitative agreement between the measured and calculated conductivities in the weakly localized regime,[30,114,116,132] where the perturbation approach

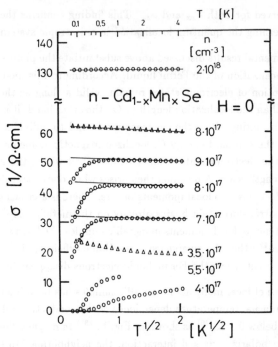

Figure 15: Zero–field conductivity of n–CdSe ($\triangle$) and n–Cd$_{0.95}$Mn$_{0.05}$Se (o). Solid lines were calculated taking quantum corrections into account but ignoring the presence of bound magnetic polarons (after Sawicki et al.[116] and Dietl et al.[131]).

should be valid.[25] The current theory turns out, however, to be less successful in the case of the Hall coefficient[132] as well as at very low temperatures, where effects of the bound magnetic polarons seem to show up.[104,116,131] The high–field negative magnetoresistance results presumably from the spin–splitting–induced repopulation of spin–subbands, and the associated increase in the kinetic energy of majority–spin electrons.[125] Such an increase, in terms of the Anderson model, reduces the relative magnitude of the quantum corrections to the conductivity, leading in this way to the negative magnetoresistance.[121] The relevance of Anderson's localization mechanism is further supported by the critical behavior of the conductivity[125–128] and the dielectric susceptibility[126] near the MIT. At the same time some experimental results, such as the temperature dependent localization[104] and the critical behavior of the Hall coefficient,[126,127] seem to indicate that certain elements of the Mott–Hubbard mechanism may also be of relevance in doped semiconductors.

3. MAGNETIC ATOMS AS RESONANT IMPURITIES

3.1. Charge Correlation in Mixed–Valence Region

The experimental evidence that substitutional Fe impurity acts as a resonant donor in HgSe is presented in Fig. 16, where the low–temperature electron concentration n, as deduced from Hall data, is plotted as a function of the Fe concentration N_{Fe}.[133–136] It is seen that for small values of N_{Fe}, n increases with N_{Fe} reflecting the donor character of Fe. However, for N_{Fe} greater than a critical value $N_{Fe}^* \approx 4 \times 10^{18} \mathrm{cm}^{-3}$, n tends to saturate indicating a pinning of the Fermi level to the resonant d–states. Surprisingly, the low temperature electron mobility μ increases sharply at the onset of the mixed valence phase, as shown in Fig. 16. For $N_{Fe} \approx 2 \times 10^{19} \mathrm{cm}^{-3}$, μ attains a value which is about four times larger than that theoretically expected when considering scattering by a system of randomly distributed

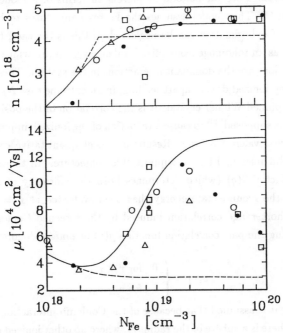

Figure 16: Electron concentration and mobility at 4.2 K for HgSe:Fe versus Fe concentration. Experimental points are taken from (o) Ref. 133; (•) Ref. 134; (□) Ref. 135; (△) Ref. 136. Solid lines were calculated within the short range correlation model; dashed lines ignoring inter–site Coulomb interactions (after Wilamowski et al.[141]).

ionized donors, and neglecting resonant enhancement of the cross section. The high mobility values remain unchanged down to 40 mK[137] indicating a minor influence of the resonant scattering and the Kondo effect even for quasi–particle energies as low as $\sim 4~\mu$eV. The above results were substantiated by studies of the Shubnikov–de Haas and de Haas–van Alfen effects.[133,138,139] The quantum oscillations yielded the cross section of the Fermi sphere (electron concentration) in an accord with Hall data demonstrating the one–band character of the electron transport. Furthermore, the cyclotron effective masses were found to indicate a negligible modification of the conduction–band shape by the resonant states. In turn, the Dingle temperature in HgSe:Fe,[133,138] as well as in related compounds such as $Hg_{1-x}Mn_xSe$:Fe,[139] is diminished in comparison to that of HgSe:Ga[74] proving independently that there is a significant reduction in the scattering rate of conduction–band electrons in HgSe:Fe.

These findings are understood at present in terms of a model put forward by J. Mycielski.[140] This model assumes the Fe resonant states to be long living, so that effects of hybridization can be neglected. If this is the case, the material is to be regarded as an inhomogeneous mixed–valence system, in which the inter–site Coulomb repulsions are the dominant interactions in the system. These interactions will tend to keep ionized donors apart leading, in an extreme case $N_{Fe} \gg N_{Fe}^*$, to a formation of a pinned Wigner crystal. It is the correlation of the positions of donor charges, which is supposed[140] to cause a reduction of the ionized–impurity scattering rate in the mixed–valence regime. Results of recent quantitative calculations,[141] depicted by solid lines in Fig. 16, support this conjecture. In those calculations, the structure factor $S(q)$ (which determines both the efficiency of ionized–donor scattering and the ground–state energy) has been evaluated at $T = 0$ taking into account only short–range correlation induced by the screened Coulomb repulsion, that is, by taking the pair correlation function $g(r)$ of *ionized* donors in the simple form,

$$g(r) = \begin{cases} 0 & \text{for } r < r_c \\ 1 & \text{for } r \geq r_c. \end{cases} \tag{24}$$

In other words, it is assumed that because of the Coulomb interaction, around each ionized donor there is a sphere of the radius r_c, where no other ionized donors reside. At the same time correlations in spatial position of ionized donors more distant than r_c are totally neglected. Let us stress that the correlation just mentioned emerge because of the redistribution of donor electrons among available donor sites and not because of a displacement of the impurities themselves. In order to determine r_c we note that the average number of the *neutral* Fe atoms in the spheres of the volume $V_c = 4\pi r_c^3/3$, which are centred at the *ionized* Fe donors, is given by

$$(N_{Fe} - n)V_c = \sum_{k=2}^{\infty}(k-1)P_k(V_c). \tag{25}$$

Here $P_k(V_c)$ is Poisson's probability of finding k Fe donors in the volume V_c. Making use of the identities $\sum_{k=0}^{\infty} P_k(V_c) = 1$, $\sum_{k=0}^{\infty} kP_k(V_c) = N_{Fe}V_c$, and $P_o(V_c) = \exp(-N_{Fe}V_c)$ we readily get V_c as a solution of the equation

$$nV_c = 1 - \exp(-N_{Fe}V_c). \tag{26}$$

Having $g(r)$ we can calculate the gain in the Coulomb energy $E_c(n)$ due to redistribution of charges among Fe donors,

$$E_c(n) = \frac{2\pi e^2 n^2}{\kappa_o}\int_0^{\infty} dr\, r[g(r) - 1]\exp(-r/\lambda), \tag{27}$$

where $\lambda = \lambda(n)$ is the screening length of the band electrons. Neglecting the exchange–correlation corrections to the energy of the conduction electrons the total energy per unit volume is

$$E(n) = E_c(n) + (N_{Fe} - n)\epsilon_d + E_{kin}, \tag{28}$$

where ϵ_d is the bare energy of Fe impurity, and E_{kin} represents the kinetic energy of the conduction electrons. By demanding that $\partial E(n)/\partial n = 0$ a relation between n and N_{Fe} is obtained with the interactions taken into account.

From the pair correlation function we obtain also the structure factor $S(q)$:

$$S(q) = 1 + 4\pi n \int_0^{\infty} dr\, r[g(r) - 1]\sin(qr)/q. \tag{29}$$

This in turn can be used to calculate the momentum relaxation time τ and the lifetime τ_D for scattering by ionized donors:

$$\frac{1}{\tau} = \frac{m^* n}{4\pi\hbar^3 k^3}\int_0^{2k} dq\, q^3 S(q)|V(q)|^2; \tag{30}$$

$$\frac{1}{\tau_D} = \frac{m^* n}{2\pi\hbar^3 k}\int_0^{2k} dq\, q S(q)|V(q)|^2, \tag{31}$$

where $V(q)$ is an appropriate matrix element of the scattering potential, which for zinc–blende narrow–gap semiconductors assumes the form

$$|V(q)|^2 = \left[\frac{4\pi e^2}{\kappa_o q^2(1 + 1/\lambda^2 q^2)}\right]^2\left(1 + \frac{q^2}{k^2}A + \frac{q^4}{4k^4}B\right), \tag{32}$$

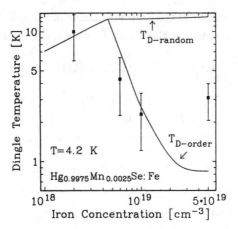

Figure 17: Dingle temperature determined from the field dependence of the Shubnikov–de Haas oscillation amplitude in $Hg_{1-x}Mn_xSe$:Fe as a function of the iron concentration. The line labelled "T_D random" was calculated without and that labelled "T_D order" with the inter–donor Coulomb interactions taken into account (after Dobrowolski et al.[139]).

with

$$A = b^2(b/2 - \sqrt{2}c)^2 - b^2 - c^2; \tag{33}$$

$$B = (b^2 + c^2)^2 - b^2(b/2 - \sqrt{2}c), \tag{34}$$

where b and c denote a contribution of the p-type component in the Bloch wavefunction, defined explicitly in, for example, Ref. 73.

The results of the calculation of the mobility and of the Dingle temperature $T_D = \hbar/2\pi k_B \tau_D$, which were carried out[139,141] making use of the above model, are depicted in Figs. 16 and 17 in comparison to the experimental results.[133–136,139]

In addition to the calculation based on the short–range correlation model, numerical simulations of the system have been carried out.[141] The simulations confirmed the correctness of the short–range correlation model and yielded, among other things, the value of the one–particle density of impurity states. As shown in Fig .18, the interactions were found to lead to a considerable broadening of this density–of–states as well as to the appearance of the so–called Coulomb gap at the Fermi energy ϵ_F.[142,143] It is the lack of impurity states at ϵ_F, which causes the resonant scattering to be suppressed. It is possible, however, that a residual contribution of

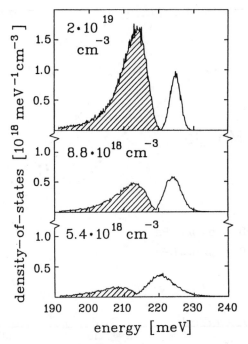

Figure 18: One particle density of Fe impurity states showing the Coulomb gap, obtained from numerical simulations for three values of the iron concentration. The bare energy of the Fe impurity is at 210 meV. The occupied states are shown by the shaded area (after Wilamowski et al.[141]).

this scattering is responsible for the mobility drop observed at high doping levels, $N_{Fe} > 2 \times 10^{19} \mathrm{cm}^{-3}$.

Extremely high electron mobilities were also found to occur in zero–gap p–type $Hg_{1-x}Mn_xTe$ at low temperatures with the Fermi level pinned to resonant acceptor states.[37,144] Shown in Fig. 19 is the value of $\mu = 2 \times 10^7 \mathrm{cm^2/Vs}$ observed under such conditions,[37] which seems to be the highest ever reported for a semiconductor. This reduction of scattering efficiency is also thought to be due to correlation induces by the inter–impurity Coulomb interactions.[4,145] In this material the ionized donors are likely to be in the vicinity of ionized acceptors owing to their mutual Coulomb attraction. Thus a system of dipoles develops which scatters the band carriers less efficiently then isolated donors do.

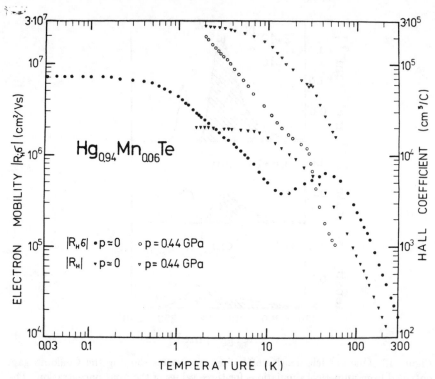

Figure 19: Temperature dependence of the Hall coefficient and mobility of electrons in zero–gap p–$Hg_{0.94}Mn_{0.06}Te$ with the Fermi level pinned to the acceptor state. The data are for two values of the hydrogenic pressure p. Note the peak value of the Hall mobility $|R_H|\sigma \simeq 2 \times 10^7$ cm^2/Vs, the highest value ever reported for a semiconductor (after Sawicki et al.[37]).

In summary, there is a growing amount of evidences that there exist systems in which the Fermi–liquid and local states coexist. This coexistence, while contradicting the one–electron approach, can occur due to the on–site repulsion and the decoupling effect of the Coulomb gap that appears at ϵ_F in any system of localized charges interacting by means of the Coulomb forces. The same inter–site interactions impose spatial correlations in positions of impurity charges resulting in a dramatic reduction of the scattering rate of conduction–band electrons. This leads to the conclusion, which is of considerable practical significance, that the use of resonant donors or acceptors may result in semiconducting materials having much greater mobility values than those characteristic of systems incorporating hydrogenic–like impurities.[141]

3.2. Localization by Scattering

Semiconductors in which carriers originate from resonant impurities offer a novel opportunity to study electron localization. This is because in the case when potential of individual impurities has no bound states, the localization by the Mott–Hubbard mechanism should be inoperative. At the same time Anderson's metal–to–insulator transition may occur as a result of the collective action of many scattering centers on the electron motion.

Recently, a study of $Cd_{1-x}Mn_xSe$ doped with Sc, which according to Fig. 1 acts as a resonant donor, has been undertaken.[146] Preliminary results indicate that both $Cd_{1-x}Mn_xSe$:Sc and $Cd_{1-x}Mn_xSe$ doped with hydrogenic–like impurities, In and Ga, exhibit a similar dependence of the conductivity on the electron concentration, temperature, and magnetic field in the vicinity of the metal–to–insulator transition. This may suggest that localization in doped semiconductors is caused primarily by Anderson's mechanism. There is some work in progress to substantiate this conclusion and, in particular, to find out whether Sc creates resonant states only.

4.   SUMMARIZING REMARKS

The above presented results demonstrate clearly that the extensive transport studies of DMS, carried out over the last ten years or so, have considerably enlarged our knowledge about the influence of magnetic atoms on semiconducting properties. We understand better, for instance, the relation between the band structure and the exchange spin–splittings of effective mass states, the role of the spin–splitting and spin–disorder scattering in the localization phenomena, the nature of the peculiarities in the mixed–valence region. Other works, particularly those devoted to the polaronic effects and RKKY interaction, show us how the effective mass electrons, by affecting the magnetic subsystem, can indirectly alter the transport properties. The accumulated results have also helped us to elucidate the origin of the giant magnetoresistance effects known for a long time to exist in magnetic semiconductors.[1,2] Furthermore, they have made it possible to verify a number of general ideas about electron transport in solids. To give just one example: the studies we have discussed in this article have convincingly demonstrated the crucial role of the electron–electron interaction in the disordered as well as in the inhomogeneous mixed–valence systems. It is clear that the materials examined so–far by no means exhaust all the interesting possibilities that are offered by DMS incorporating transition metals, or possibly, rare earth elements. Futures studies of novel systems, their hetero– and nanostructures will certainly lead to surprising discoveries, which will shed new light on various fundamental aspects of electron transport in solids.

ACKNOWLEDGEMENTS

The author is very indebted to W. D. Dobrowolski, G. Grabecki, J. Jaroszyński, J. Kossut, A. Lenard, W. Plesiewicz, M. Sawicki, W. Suski, L. Świerkowski, Z. Wilamowski, T. Wojtowicz, and J. Wróbel for collaboration concerning various aspects of electron transport in DMS, and to R. R. Gałązka, A. Mycielski, and T. Skośkiewicz for continuous support and encouragement.

REFERENCES

1. S. Methfessel, and D. C. Mattis, *Magnetic Semiconductors* in: *Handbuch der Physik*, Vol. XVIII (Springer, Berlin 1968); P. Wachter, in: *Handbook on the Physics and Chemistry of Rare Earths*, Vol. 2; Eds. K. A. Gschneidner and LeRoy Eyring (North Holland, Amsterdam, 1979) p. 507.

2. E. L. Nagaev, *Physics of Magnetic Semiconductors* (MIR, Moscow, 1983).

3. see, e.g., R. R. Gałązka and J. Kossut, *Landolt–Borstein, New Series Group III*, Eds. O. Madelung et al. (Springer, Berlin, 1983) Vol. 17b, p. 245; J. K. Furdyna, *J. Appl. Phys.* **64** (1988) R 29.

4. N. B. Brandt and V. V. Moshchalkov, *Adv. Phys.* **33** (1984) 193; I. I. Lyapilin and I. M. Tsidilkovskii, *Usp. Fiz. Nauk* **146** (1985) 35.

5. J. Kossut, in: *Diluted Magnetic Semiconductors*, Eds. J. K. Furdyna and J. Kossut, in: *Semiconductors and Semimetals*, Vol. 25, Eds. R. K. Willardson and A. C. Beer (Academic Press, New York, 1988) p. 183.

6. some of the topics described here are also discussed in lectures notes for the 21° Course of the International School of Material Science, Erice 1990, T. Dietl, in: *Semimagnetic Semiconductors and Diluted Magnetic Semiconductors*, Eds. M. Averous and M. Balkanski (Plenum Press) in print; T. Dietl, *J. Crystal Growth* **101** (1990) 808; T. Dietl, M. Sawicki, J. Jaroszyński, J. Wróbel, T. Wojtowicz, and A. Lenard, in: *Localization and Confinement of Electrons in Semiconductors*, Eds. F. Kuchar et al. (Springer, Berlin, 1990) p.127.

7. D. Heiman, M. Dahl, X. Wang, P. A. Wolff, P. Becla, A. Petrou, and A. Mycielski, *Mat. Res. Soc. Sym. Proc.* **161** (1990).

8. see, e.g., P. Vogl and J. M. Baranowski, *Acta Phys. Polon.* **A67** (1985) 133; A. Zunger, in: *Solid State Physics*, Vol. **39**, Eds. F. Seitz and D. Turnbull (Academic Press, New York, 1986) p. 275; V. I. Sokolov, *Fiz. Tverd. Tela* **29** (1987) 1848 [*Soviet Phys.–Solid State* **29** (1987) 1061].

9. J. M. Langer, C. Delerue, M. Lannoo, and H. Heinrich, *Phys. Rev.* **B38** (1988) 7723; A. Zunger, in Ref. 8.

10. W. Dobrowolski, K. Dybko, C. Skierbiszewski, T. Suski, E. Litwin–Staszewska J. Kossut and A. Mycielski, in: *Proc. 19th Intern. Conf. on Physics of Semiconductors*, Ed. W. Zawadzki (Institute of Physics, Warszawa, 1988) p. 1247.

11. T. Dietl and J. Kossut, *Phys. Rev.* **B38** (1988) 10941.

12. For a review of Mn–based semimagnetics see, e. g., *Diluted Magnetic Semiconductors*, Eds. J. K. Furdyna and J. Kossut, in: *Semiconductors and Semimetals*, Vol. 25, Eds. R. K. Willardson and A. C. Beer (Academic Press, New York, 1988).

13. For review of Fe–based semimagnetics see, e. g., A. Mycielski, *J. Appl. Phys.* **63** (1988) 3279; A. Twardowski, this volume.

14. A. Twardowski, P. Głód, W. J. M. de Jonge and M. Demianiuk, *Solid State Commun.* **64** (1987) 63; A. Twardowski, K. Pakuła M. Arciszewska, and A. Mycielski, *Solid State Commun.* **73** (1990) 601.

15. A. Mycielski, P. Dzwonkowski, B. Kowalski, B. A. Orłowski, M. Dobrowolska, M. Arciszewska, W. Dobrowolski, and J. M. Baranowski, *J. Phys.* **C19** (1986) 3605.

16. for review see, e.g., *The Metal Non–Metal Transition in Disordered Systems*, Eds. L. R. Friedman and D. P. Tunstall (SUSSP Publ., Edinburgh, 1978).

17. see, e.g., *Anderson Localization*, Eds. T. Ando and H. Fukuyama (Springer, Berlin, 1988).

18. see, e.g., T. M. Rice, *Phil. Mag.* **B42** (1985) 419.

19. M. Shayegan, T. Sayato, M. Santos, and C. Silvestre, *Appl. Phys. Lett.* **53** (1988) 791; M. Sundaran, A. C. Gossard, J. H. English, and R. M. Westervelt, *Superlattices and Microstructures* **4** (1988) 683.

20. G. Nimitz, in: *High Magnetic Fields in Semiconductor Physics*, Ed. G. Landwehr (Springer, Berlin, 1987) p. 491, and references therein; R. R. Gerhardts, *ibid.*, p. 482.

21. see, e.g., C. M. Care and N. H. March, *Adv. Phys.* **24** (1975) 101; I. M. Tsidilkovskii, *Usp. Fiz. Nauk* **152** (1987) 583 [*Sov. Phys. Usp.* **30** (1987) 676].

22. B. I. Shklovskii and A. L. Efros, *Electronic Properties of Doped Semiconductors*, (Springer, Berlin, 1984); R. Zallen, *The Physics of Amorphous Solids*

(John Willey & Sons, 1983); B. A. Aronzon and I. M. Tsidilkovskii, *Phys. Stat. Sol. (b)* **157** (1990) 17.

23. see, e.g., Y. Song, T. W. Noh, S.-I. Lee, and J. R. Gaines, *Phys. Rev.* **B33** (1986) 904.

24. see, e.g., S. Luryi, in Ref. 20, p. 16.

25. for review see, e.g., B. L. Altshuler, A. G. Aronov, D. E. Khmelnitskii, and A. I. Larkin, in: *Quantum Theory of Solids*, Ed. I. M. Lifshits (Mir Publishers, Moscow, 1982) p. 130; P. A. Lee and T. V. Ramakrishnan, *Rev. Mod. Phys.* **57** (1985) 287; B. L. Altshuler and A. G. Aronov, in: *Electron–Electron Interactions in Disordered Systems*, Eds. A. L. Efros and M. Pollak (North-Holland, Amsterdam, 1985) p. 1; H. Fukuyama, *ibid.*, p. 155.

26. A. M. Finkelstein, *Zh. Eksp. Teor. Fiz.* **86** (1984) 367 [*Sov. Phys. JETP* **59** (1984) 212]; C. Castellani, C. Di Castro, P.A. Lee, and M. Ma, *Phys. Rev.* **B30** (1984) 527; C. Di Castro, in Ref. 7, p. 96; G. Kotliar, in Ref. 7, p. 107; T. Dietl, M. Sawicki, J. Jaroszyński, T. Wojtowicz, W. Plesiewicz, and A. Lenard, in: *Proc. 19th Int. Conf. on Physics of Semiconductors, Warsaw 1988*, Ed. W. Zawadzki (Institute of Physics, Warszawa 1988) p. 1189.

27. E. J. Moore, *Phys. Rev.* **160** (1967) 607 and 618.

28. J. A. Gaj, in Ref. 22, p. 276.

29. Y. Shapira and R. L. Kautz, *Phys. Rev.* **B10** (1974) 4781; see also D. J. Kim and B. B. Schwartz, *Phys. Rev.* **B15** (1977) 377.

30. M. Sawicki, T. Dietl, and J. Kossut, *Acta Phys. Polon.* A **67** (1985) 399.

31. G. Bastard, C. Rigaux, Y. Guldner, J. Mycielski, and A. Mycielski, *J. de Phys. (Paris)* **39** (1978) 87.

32. M. Jaczyński, J. Kossut, and R. R. Gałązka, *Phys. Stat. Sol. (b)* **88** (1978) 73.

33. J. A. Gaj, J. Ginter, and R. R. Gałązka, *Phys. Stat. Sol. (b)* **89** (1978) 655.

34. W. Walukiewicz, *J. Magn, Magn., Mat.* **11** (1979) 157.

35. A. Sandauer and P. Byszewski, *Phys. Stat. Sol. (b)* **109** (1982) 167.

36. A. B. Davydov, B. B. Ponikarov, and I. M. Tsidilkovskii, *Phys. Stat. Sol. (b)* **101** (1981) 127; *Fiz. Tekh. Poluprov.* **15** (1981) 881; B. B. Ponikarov, I. M. Tsidilkovskii, and N. G. Shelushina, *Fiz. Tekh. Poluprov.* **15** (1981) 296.

37. M. Sawicki, T. Dietl, W. Plesiewicz, P. Sękowski, L. Śniadower, M. Baj, and L. Dmowski, in:*High Magnetic Fields in Semiconductor Physics*, ed G. Landwehr (Springer, Berlin 1983) p. 382.

38. M. Sawicki and T. Dietl, in: *Physics of Semiconducting Compounds*, Eds. R. R. Gałązka and J. Raułuszkiewicz (Ossolineum, Wrocław, 1983) p. 400.

39. A. B. Germanenko, L. P. Zverev, V. V. Kruzhaev, G. M. Minkov, and O. E. Rut, *Fiz. Tverd. Tela* **27** (1985) 1857; *Fiz. Tekh. Poluprov.* **20** (1986) 80.

40. see, e.g., I. M. Tsidilkovskii, W. Giriat, G. I. Kharus, and E. A. Neifeld, *Phys. Stat. Sol. (b)* **64** (1974) 717.

41. R. T. Delves, *Proc. Phys. Soc.* **87** (1966) 809.

42. T. Wojtowicz and A. Mycielski, *Physica* **117–118 B** (1983) 475.

43. W. B. Johnson, J. R. Anderson, and D. R. Stone, *Phys. Rev.* **B29** (1984) 6679; J. R. Anderson, *Physica B* **164** (1990) 67.

44. A. V. Germanenko, L. P. Zverev, V. V. Kruzhaev, G. M. Minkov, O. E. Rut, N. P. Gavaleshko, and W. M. Frasunyak, *Fiz. Tverd. Tela* **26** (1984) 1754; A. V. Germanenko, V. V. Kruzhaev, G. M. Minkov, and O. E. Rut, *Fiz. Tekh. Poluprov.* **20** (1986) 1662.

45. J. Jaroszyński and T. Dietl, *Solid State Commun.* **55** (1985) 491.

46. A. E. Belyaev, Yu. G. Semenov, and N. V. Schevchenko, *Pisma Zh. Eksp. Teor. Fiz.* **48** (1988) 623; see also, ibid. **51** (1990) 164 [*JETP Lett.* **48** (1988) 675; **51** (1990) 186].

47. M. Chmielowski, T. Dietl, P, Sobkowicz, and F. Koch, in: *Proc. 18th Intl. Conf. Physics of Semiconductors*, Ed. O. Engström (World Scientific, Singapore, 1987) p.1787.

48. see, e. g., F. E. Maranzana, *Phys. Rev.* **160** (1967) 421.

49. J. N. Chazalviel, *Phys. Rev.* **B11** (1975) 3918.

50. N. B. Brandt, V. V. Moshchalkov, A. O. Orlov, L. Skrbek, I. M. Tsidilkovskii, and S. M. Chudinov, *Zh. Eksp. Teor. Fiz.* **84** (1983) 1059; N. G. Gluzman, N. K. Lerinman, L. D. Sabirzyanova, I. M. Tsidilkovskii, and V. M. Frazunyak, *Pisma Zh. Eksp. Teor. Fiz.* **43** (1986) 600 [*Sov. Phys. JETP Lett.* **43** (1986) 777].

51. G. Grabecki, T. Suski, T. Dietl, T. Skośkiewicz, and M. Gliński, in: *High Magnetic Fields in Semiconductor Physics*, Ed. G. Landwehr (Springer, Berlin, 1987) p. 127.

52. G. Grabecki, T. Dietl, P. Sobkowicz, J. Kossut, and W. Zawadzki, *Appl. Phys. Letters* **45** (1984) 1214.

53. T. Suski, P. Wiśniewski, L. Dmowski, G. Grabecki, and T. Dietl, *J. Appl. Phys.* **65** (1989) 1203; P. Wiśniewski, T. Suski, G. Grabecki, P. Sobkowicz, and T. Dietl, in: *Polycrystalline Semiconductors*, Eds. H. J. Moller et al. (Springer, Berlin, 1989) p. 338.

54. F. Rys, J. S. Helman, and W. Baltensperger, *Phys. kondens. Materie* **6** (1967) 105.

55. R. B. Bylsma, J. Kossut, W. M. Becker, U. Dębska, and D. Yodershort, *Phys. Rev.* **B33** (1986) 8207.

56. J. A. Gaj and A. Golnik, *Acta Phys. Polon.* **A71** (1987) 197.

188

57. S. M. Ryabchenko, Yu. G. Semenov, and O. V. Terletskii, *Phys. Stat. Sol. (b)* **144** (1987) 661.

58. A. K. Bhattacharjee, *Solid State Commun.* **65** (1988) 275.

59. J. Blinowski and P. Kacman, *Acta Phys. Polon.* **A75** (1989) 215.

60. C. Haas, *Phys. Rev.* **168** (1968) 531; P. Leroux-Hugo, *J. Magn. Magn. Mat.* **3** (1976) 165.

61. J. Kossut, *Phys. Stat. Sol. (b)* **72** (1975) 359; **78** (1976) 537.

62. K. H. Fisher, *Z. Physik* **34** (1979) 45.

63. A. A. Abrikosov, *Adv. Phys.* **29** (1980) 869.

64. T. Dietl, M. Sawicki, E. D. Isaacs, M. Dahl, D. Heiman, M. J. Graf, S. I. Gubarev, and D. L. Alov, *Phys. Rev.* **B43** , in press.

65. J. A. Gaj, A. Golnik, J.-P. Lascaray, D. Coquillat, and M. C. Desjardins-Deruelle, in: *MRS Conf. Proceedings*, Vol. 89, Eds. R. L. Aggarwal et al. (MRS, Pittsburg, 1987) p. 59.

66. M. Dobrowolska, W. Dobrowolski, M. Otto, T. Dietl, and R. R. Gałązka, *J. Phys. Soc. Japan.* **49** (1980) 815, Suppl. A.

67. H. Pascher, P. Rothlein, G. Bauer, M. von Ortenberg, *Phys. Rev.* **B40** (1989) 10469.

68. G. Bastard, *Surf. Sci* **142** (1984) 284; *Appl. Phys. Lett.* **43** (1983) 591.

69. G. Bastard and L. L. Chang, *Phys. Rev.* **B41** (1990) 7899.

70. for review of spin-glasses, see, e.g., K. Binder and A. P. Young, *Rev. Mod. Phys.* **58** (1986) 801.

71. see, e. g., P. A. Wiegman, in: *Quantum Theory of Solids*, Ed. I. M. Lifshits (Mir, Moscow, 1982) p. 238.

72. P. Noziers, *Ann. Phys.* **10** (1985) 19.

73. J. Kossut, *Phys. Stat. Sol. (b)* **86** (1978) 593.

74. T. Dietl, *J. Physique (Paris)* **39** (1978) C6–1081.

75. A. Wittlin, M. Grynberg, W. Knap, J. Kossut, and Z. Wilamowski, *J. Phys. Soc. Japan* **49** (1980) 635, Suppl. A.

76. Y. Ono and J. Kossut, *J. Phys. Soc. Japan* **53** (1984) 1128.

77. J. A. Gaj, in: *Proc. 17th Intern. Conf. Physics of Semiconductors*, Eds. J. D. Chadi and W. A. Harrison (Springer, New York, 1984) p. 1423; J. Warnock, D. Heiman, P. A. Wolff, R. Kershaw, D. Ridgley, K. Dwight, A. Wold, and R. R. Gałązka, *ibid*, p. 1407.

78. H. Krenn, W. Zawadzki, and G. Bauer, *Phys. Rev. Lett.* **55** (1985) 1510; D. D. Awschalom, J. Warnock, and S. von Molnar, *ibid.* **58** (1987) 812; H. Krenn, K. Kaltenegger, T. Dietl, J. Spałek, and G. Bauer, *Phys. Rev.* **B39** (1989) 10918.

79. T. Dietl and J. Spałek, *Phys. Rev. Lett.* **48** (1982) 355; *Phys. Rev.* **B28** (1983) 1548.

80. R. Stępniewski, *Acta Phys. Polon.* **A69** (1986) 1001; *Solid State Commun.* **58** (1986) 19.

81. V. J. Goldman, H. D. Drew, M. Shayegan, and D. A. Nelson, *Phys. Rev. Lett.* **56** (1986) 968; J. B. Choi, L. S. Kim,, H. D. Drew, D. A. Nelson, *Solid state Commun.* **65** (1988) 547.

82. see, e.g., P. W. Anderson, in *Solid State Physics*, Eds. F. Seitz and D. Turnbull (Academic Press, New York, 1963) Vol. 14, p. 99.

83. T. Story, R. R. Gałązka, R. B. Frankel, and P. A. Wolff, *Phys. Rev. Lett.* **56** (1986) 777;

84. T. Story, L. Świerkowski, G. Karczewski, W. Staguhn, and R. R. Gałązka, in Ref. 20, p. 1567.

85. A. Sandauer, *Phys. Stat. Sol. (a)* **111** (1989) K 219.

86. S. M. Ryabchenko and Yu. G. Semenov, *Zh. Eksp. Teor. Fiz.* **84** (1983) 1419 [*Sov. Phys. JETP* **57** (1983) 825].

87. L. Świerkowski and T. Dietl, *Acta Phys. Polon.* **A73** (1988) 431; L. R. Ram–Mohan and P. A. Wolff, *Phys. Rev.* **B38** (1988) 1330.

88. C. E. T. Goncalves da Silva, *Phys. Rev.* **B32** (1985) 6962; **33** (1986) 2923; J.-W. Wu, A. V. Nurmikko, and J. J. Quinn, *Solid State Commun.* **57** (1986) 853; *Phys. Rev.* **B34** (1986) 1080; D. R. Yakovlev, W. Ossan, G. Landwehr, R. N. Bicknell–Tassius, A. Waag, and I. N. Uraltsev, *Solid State Commun.*, in press.

89. M. von Ortenberg, *Solid State Commun.* **52** (1984) 111; *Acta Phys. Polon.* **A69** (1986) 977; A. Golnik and J. Spałek, *J. Magn. Magn. Mat.* **54–57** (1986) 1207.

90. D. L. Alov, G. I. Gubarev, and V. B. Timofeev, *Zh. Eksp. Teor. Fiz.* **86** (1984) 1124.

91. see, e. g., A. L. Kuzemsky, D. I. Marvakov, and J. P. Vlahov, *Physica* **B138** (1986) 129.

92. see, e. g., J. R. Schrieffer, X. G. Wen, and S. C. Zhang, *Phys. Rev. Lett.* **60** (1988) 944.

93. S. Sachdev, *Phys. Rev.* **B39** (1989) 5297.

94. See, e. g., M. Pollak and M. Ortuno, in: *Electron–Electron Interactions in Disordered Systems*, Eds. A. L. Efros and M. Pollak (North–Holland, Amsterdam, 1985) p. 409; A. L. Efros and B. I. Shklovskii, *ibid.*, p. 483.

95. V. J. Goldman and H. D. Drew, *Phys. Rev.* **B30** (1984) 6221.

96. A. Kurobe and H. Kamimura, *J. Phys. Soc. Japan* **51** (1982) 1904.

97. B. Movaghar and L. Schweitzer, *J. Phys.* **C 11** (1978) 125; see also Y. Osaka, *J. Phys. Soc. Japan* **47** (1979) 729.

98. M. von Ortenberg, W. Erhardt, A. Twardowski, M. Demianiuk, in *High Magnetic Fields in Semiconductor Physics*, Ed. by G. Landwehr (Springer, Berlin, 1987) p. 446.

99. J. Mycielski and C. Rigaux, *J. de Phys. (Paris)* **44** (1985) 1041.

100. A. Mycielski and J. Mycielski, *J. Phys. Soc. Japan* **49** (1980) 807, Suppl. A.

101. T. R. Gawron, *J. Phys. C* **19** (1986) 21; **19** (1986) 29; A. D. Bykhovsky, E. M. Vakhabova, B. L. Gelmont, and A. L. Efros, *Fiz. Tekh. Poluprov.* **18** (1984) 2094; P. Janiszewski, in Ref. 10, p. 1285.

102. T. Wojtowicz, T. R. Gawron, J. L. Robert. A. Raymond, C. Bousquet, and A. Mycielski, *J. Cryst. Growth* **72** (1985) 385.

103. J. Mycielski, in Ref. 37, p. 431.

104. T. Dietl, L. Świerkowski, J. Jaroszyński, M. Sawicki, and T. Wojtowicz, *Physica Scripta* **T14** (1986) 29.

105. J. Mycielski, A. Witowski, M. Grynberg, and A. Wittlin, *Phys. Rev.* **B40** (1989) 8437; **B41** (1990) 5351.

106. see, e. g., P. A. Wolff, in Ref. 12, p. 413; U. Thibblin, *Intern. J. Mod. Phys.* **B3** (1989) 337.

107. T. Kasuya and A. Janase, *Rev. Mod. Phys.* **40** (1968) 684; A. Janase and T. Kasuya, *J. Phys. Soc. Japan* **25** (1968) 1025.

108. J. Jaroszyński, T. Dietl, M. Sawicki, and E. Janik, *Physica* **117–118 B** (1983) 473.

109. T. Dietl, J. Antoszewski, and L. Świerkowski, *Physica* **117–118 B** (1983) 491.

110. A. C. Ioselevich, *Pisma Zh. Eksp. Teor. Fiz.* **43** (1986) 148.

111. Y. Shapira, N. F. Oliveira, Jr., D. H. Ridgley, R. Kershaw, K. Dwight, and A. Wold, *Phys. Rev.* **B34** (1986) 4187.

112. T. Ichiguchi, H. D. Drew, and J. K. Furdyna, *Phys. Rev. Lett.* **50** (1983) 612; J. Stankiewicz, S. von Molnar, and W. Giriat, *Phys. Rev.* **B33** (1986) 3573; J. R. Anderson, W. B. Johnson, D. R. Stone, and J. K. Furdyna, *J. Phys. Chem. Solids* **48** (1987) 481.

113. D. Heiman, Y. Shapira, and S. Foner, *Solid State Commun.* **45** (1983) 899.

114. Y. Shapira, N. F. Oliveira Jr., P. Becla, and T. Q. Wu, *Phys. Rev.* **B41** (1990) 5931.

115. D. M. Finlayson, J. Irvine, and L. S. Peterkin, *Phil. Mag.* **B39** (1979) 253; Y. Zhang, P. Dai, M. Levy, and M. P. Sarachik, preprint.

116. M. Sawicki, T. Dietl, J. Kossut, J. Igalson, T. Wojtowicz, and W. Plesiewicz, *Phys. Rev. Letters* **56** (1986) 508; M. Sawicki, T. Wojtowicz, T. Dietl, J.

Jaroszyński, W. Plesiewicz, and J. Igalson, in: *Proc. 18th Intern. Conf. on Physics of Semiconductors*, Stockholm, 1986, Ed. O. Engström (World Scientific, Singapore, 1987) p. 1265; M. Sawicki, *Ph. D. Thesis*, Warszawa, 1990, unpublished.

117. see, e. g., Y. Shapira, S. Foner, and T. B. Reed, *Phys. Rev.* **B5** (1972) 4877; **B8** (1973) 2299; S. von Molnar, J. Flouquet, F. Holtzberg, and G. Remenyi, *Solid State Electron.* **28** (1985) 127; J. Stankiewicz, S. von Molnar, and F. Holtzberg, *J. Magn. Magn. Mat.* **54-57** (1986) 1217.

118. J. Kubler and D. F. Vigren, *Phys. Rev.* **B8** (1973) 2299.

119. M. I. Auslander, E. M. Kogan, and S. V. Tretyakov, *Phys. Stat. Sol. (b)* **148** (1988) 289.

120. P. Leroux–Hugon, *Phys. Rev. Lett* **29** (1972) 939; J. B. Torrance, M. W. Shafer, and T. R. McGuire, *ibid.* **29** (1972) 1168.

121. H. Fukuyama and K. Yosida, *Physica* **105 B+C** (1981) 132; *J. Phys. Soc. Japan* **46** (1979) 102.

122. D. Ceperly and B. Alder, *Phys. Rev. Lett.* **45** (1980) 566.

123. A. P. Grigin and E. L. Nagaev, *Pisma Zh. Eksp. Teor. Fiz.* **16** (1972) 438 [*Sov. Phys. JETP Lett.* **16** (1972) 312].

124. T. Dietl and L. Świerkowski, *Acta Phys. Polon.* **A71** (1987) 213.

125. T. Wojtowicz, T. Dietl, M. Sawicki, W. Plesiewicz, and J. Jaroszyński, *Phys. Rev. Lett.* **56** (1986) 2419; T. Wojtowicz, M. Sawicki, T. Dietl, W. Plesiewicz, and J. Jaroszyński, in: *High Magnetic Fields in Semiconductor Physics*, Ed. G. Landwehr (Springer, Berlin, 1987) p. 442;

126. J. Jaroszyński, T. Dietl, M. Sawicki, T. Wojtowicz, and W. Plesiewicz, in: *High Magnetic Fields in Semiconductor Physics II*, Ed. G. Landwehr (Springer, Berlin, 1989) p. 514.

127. J. Wróbel, T. Dietl, G. Karczewski, J. Jaroszyński, W. Plesiewicz A. Lenard, M. Dybiec, and M. Sawicki, *Semicond. Sci. Technol.* **5** (1990) S 299.

128. T. Wojtowicz, M. Sawicki, J. Jaroszyński, T. Dietl, and W. Plesiewicz, *Physica B* **153** (1989) 357;

129. R. N. Bhatt, P. Wölfle, and T. V. Ramakrishnan, *Phys. Rev.* **B32** (1985) 569.

130. R. N. Bhatt, M. A. Paalanen, and S. Sachdev, *J. Physique* **49–C8** (1988) 1179, and references therein; M. Milovanović, S. Sachdev, and R. N. Bhatt, *Phys. Rev. Lett.* **63** (1989) 82; M. Lakner and H. v. Lohneyson, *Phys. Rev. Lett.* **63** (1989) 648; D. Romero, M.-W. Lee, H. D. Drew, M. Shayegan, and B. S. Elman, in Ref. 17, p. 53.

131. T. Dietl, M. Sawicki, T. Wojtowicz, J. Jaroszyński, W. Plesiewicz, L Świerkowski, and J. Kossut, in Ref. 17, p. 58.

132. M. Sawicki and T. Dietl, in Ref. 10, p. 1217.

133. W. Dobrowolski, K. Dybko, A. Mycielski, J. Mycielski, J. Wróbel, S. Piechota, M. Palczewska, H. Szymczak, and Z. Wilamowski, in: *Proc. 18th Intern. Conf. on Physics of Semiconductors*, Stockholm 1986, Ed. O. Engström (World Scientific, Singapore, 1987) p. 1743.

134. F. Pool, J. Kossut, U. Debska, and R. Reifenberger, *Phys. Rev.* **B35** (1987) 3900.

135. N. G. Gluzman, L. D. Sabirzyanova, I. M. Tsidilkovskii, L. D. Paranchich, and S. Yu. Paranchich, *Fiz. Tekh. Poluprov.* **20** (1986) 94, 1996.

136. C. Skierbiszewski, T. Suski, E. Litwin–Staszewska, W. Dobrowolski, K. Dybko, and A. Mycielski, *Semicond. Sci. Technol.* **4** (1989) 293.

137. A. Lenard, T. Dietl, M. Sawicki, K. Dybko, W. Dobrowolski, T. Skośkiewicz W. Plesiewicz, and A. Mycielski, *Acta Phys. Polon.* **A75** (1989) 249; *J. Low. Temp. Phys.* **80** (1989) 15.

138. M. Miller and R. Reifenberger, *Phys. Rev.* **B38** (1988) 3423 and 4120.

139. W. Dobrowolski, J. Kossut, B. Witkowska, and R. R. Gałązka, *Acta phys. Polon.* **A77** (1990) 151.

140. J. Mycielski, *Solid State Commun.* **60** (1986) 165.

141. Z. Wilamowski, K. Świątek, T. Dietl, and J. Kossut, *Solid State Commun.* **78** (1990) 833; J. Kossut, W. Dobrowolski, Z. Wilamowski, T. Dietl, and K. Świątek, *Semicond. Sci. Technol.* **5** (1990) S260.

142. T. Dietl, *Japan. J. Appl. Phys.* **26**, Suppl. **26-3** (1987) 1907.

143. Z. Wilamowski, W. Jantsch, and G. Hendorfer, *Semicond. Sci. Technol.* **5** (1990) S266.

144. C. Morrissy, *Ph. D. Thesis*, Oxford 1973, unpublished; N. B. Brandt, I. M. Tsidilkovskii, V. V. Moshchalkov, B. B. Ponikarov, L. Skrbek, A. N. Taldenkov, and S. M. Chudinov,, *Phys. Tekh. Poluprov.* **17** (1983) 18; R. Mani and J. R. Anderson, *Phys. Rev.* **B38** (1988) 3354.

145. Y. G. Arapov, A. B. Davydov, and I. M. Tsidilkovskii, *Fiz. Tekh. Poluprov.* **17** (1983) 24.

146. P. Głód, M. Sawicki, A. Lenard, T. Dietl, and W. Plesiewicz, *Acta Phys. Polon.*, in press.

PHASE-TRANSITION AND ANHARMONIC PROPERTIES OF
DILUTED MAGNETIC SEMICONDUCTORS

R. K. SINGH

School of Physics, Barkatullah University,
Bhopal - 462 026, **India.**

I. INTRODUCTION

The diluted magnetic semiconductors are promising materials for application in electronic, laser and magneto-optic devices. The ternary alloys with chemical formula $Hg_{1-x}M_xTe$ (M=Cd, Mn or Zn) are known as the diluted magnetic semiconducting (DMS) materials[1-3] with unique combination of magnetic and semiconducting properties. These DMS materials are mixed crystals obtained by replacing a fraction of cations of tetrahedrally coordinated III-V and II-VI compound semiconductors by magnetic elements, like Mn, Cd or Zn. Thus, one component of these DMS mixed crystals is an ordinary semiconductor and the second one is a magnetic semiconductor. Due to this unique combination, these mixed crystals are also called as the semimagnetic semiconductors (SMS). At standard temperature and pressure, these DMS alloys crystallise in the tetrahedrally coordinated zincblende (or B3) structure.

A considerable attention has been focused on the high-pressure and alloying effects on crystal structure[2,5-10] and elastic properties[11] of these DMS mixed crystals. The energy dispersive X-ray measurements done by Quadri et.al.[5] and Raman Scattering measurements done by Arora et.al.[9,10] have revealed that the alloys of $Cd_{1-x}Mn_xTe$ ($0 \leq x \leq 0.7$), $Cd_{1-x}Mn_xSe$ and $Zn_{1-x}Mn_xSe$ undergo phase-transition from zincblende (B3) to rocksalt (B1) structure and shows a linear decrease in the phase-transition pressure (P_t) with increasing Mn concentration (x). Similar studies by Quadri et.al.[2,7,8] have shown that the ternary alloys of $Hg_{1-x}Mn_xTe$ (M = Cd, Mn or Zn) undergo phase-transitions from B3 - B9 (cinnabar) - B1 - A5 (β-Sn) at different P_t as listed in Table 1. The pressure dependence of the second order elastic constants (SOECs) has been investigated by Maheswaranathan et. al.[11] in semimagnetic semiconductors : $Cd_{1-x}Mn_xTe$ ($0 \le x \le 0.52$) and $Cd_{0.52}Zn_{0.48}Te$. The effect of pressure on optical absorption edge of DMS mixed crystals $Cd_{1-x}Mn_xTe$ ($0 \le x \le 0.71$) has been investigated by Shan et.al[12] using the diamond anvil cell (DAC) technique. Besides these investigations, Shapira and Oliveira[13] have observed the high field magnetisation steps and the nearest neighbour exchange constants in $Cd_{1-x}Mn_xS$, $Cd_{1-x}Mn_xTe$ and $Zn_{1-x}Mn_xSe$.

Recently, the high pressure phase-transition and elastic properties of II-VI and III-V compound semiconductors of zincblende structure (ZBS) have been investigated by Singh and coworkers[11-16] and others[17-20]. However, such detailed theoretical investigations were lacking for DMS materials until currently when Singh and his coworkers[21-23] applied their three-body interaction potential (TBP) model[24] to describe the observed phase-transition pressures[2,5-10] and pressure variations of elastic constants[11] in them.

This chapter intends to describe the experimental and theoretical investigations carried out on the phase-transition and high pressure elastic properties of DMS materials. These details have been traced in the following sections.

2. EXPERIMENTAL INVESTIGATIONS

The proposed experimental investigations will include the preparation of the DMS mixed crystal samples and the study of their phase-transition and anharmonic properties. Experimentally, the phase-transition pressures are determined by detecting the sudden changes in the compressibility, electrical resistivity, thermoelectric

Table 1. B3–B1 Transition Pressures (P_t) and fractional volumes
for MDS Materials.

x	P_t(GPa)	$\nabla V(P_t)/V(o)$	$\nabla V(P_t)/V(P_t)$	Ref.
0.00	3.4±0.1	0.175	0.185	5
0.10	3.0±0.1	0.171	0.182	5
0.13	3.2±0.1	0.164	0.175	5
0.30	2.8±0.1	0.169	0.181	5
0.45	2.7±0.1	0.173	0.193	5
0.60	2.5±0.0	0.180	0.187	5
0.70	2.4±0.0	0.178	0.174	5

power, dielectric constant, Hall resistivity, lattice parameter
or elastic constants. Some of these methods have been described
below starting from the sample preparation.

2.1 Sample Preparation

The samples of DMS mixed crystals were prepared via the Bridgman
technique at the Purdue Materials Research Laboratory. The analysis
of their chemical composition were done from the measurements of
their unit cell parameters and reference to the structural data
of Bottka et.al.[25]. The measurements of high pressure phase–transition and elastic constant variations have been made by various
workers as described below.

2.2 Phase–Transition Studies

There are the following two types of measurements used for
the study of composition dependence of the phase–transition pressures.

2.2.1 **X-ray Measurements:** The crushed fine powder of DMS mixed crystals are mixed with a similar powder of NaCl, which serves as a pressure calibrant. The powder mixtures are immersed in a 4:1:: Methanol: ethanol solution, which are contained in an inconel gasket between the two envils of a diamond-anvil pressure cell. The energy dispersive X-ray diffraction (EDXD) spectra are recorded at various fixed pressures for each compositions (x). The unit cell parameters are determined at each pressure from the (111), (220) and (311) diffraction peaks for B3 phase, and (200) and (220) peaks for B1 phase. These results in detail are reported elsewhere[2,5-8].

The phase-transition pressures have been determined by plotting the fractional volume changes (V/V_o), for both the B3 and B1 phases, obtained by fitting the data to the Birch equation:

$$V(P)/V(o) = 1 + a_o P + a_1 P^2 \qquad (1)$$

by the least square method for each composition (x). The coefficients a_o and a_1 are reported for various DMS materials elsewhere[5-8]. Using these plots and the expression for the phase-transition pressure:

$$P_t = 3.29 - 1.32x, \qquad (2)$$

earlier workers[5-8] have determined the composition dependence of P_t and the fractional change in volume associated with B3-to-B1 transition, as referenced both to the original volume (V(o)) and the volume $(V(P_t))$ of the B3 phase at the transition. These results will be compared with theoretical results in the next section.

Besides, Strossner et.al.[25] have made high pressure x-ray investigations of phase-transition in $Cd_{1-x}Mn_x Te$ and measured associated changes in volume bulk modulus (B_o) and its pressure derivatives (B'_o) by fitting the measured P(V) points with the Birch's equation of state[26]

$$P = \frac{3}{2} B_o \ (x^{7/3} - x^{5/3}) \ [1 - \frac{3}{4} \ (4 - B'_o)(x^{2/3} - 1)] \qquad (3)$$

with x = V(o)/V(P). These authors[25] have found two phase-transitions to occur in all DMS mixed crystal samples. The first transition between the ZB to RS is connected with a volume change of about 16%, whereas the second one, between RS and β -Sn structure, shows no significant volume change.

2.2.2 **Resistivity Measurements**: As described by Quadri et.al.[5], the values of resistivities (ρ) are calculated from the resistances directly measured during both the upstroke and downstroke, based on the starting dimension of the sample for different compositions (x). It has been found that the volume compressibilities are about 8% before the transition, 17-18% at the transition, and another 4% for the region above the transition to 6 GPa. This volume decrease would have the effect of increasing resistivity (ρ'), the slopes of the resistivity curves, and the size of the drop associated with phase-transitions. The composition dependence of the phase-transition pressures (P_t) measured from this method[5] and those obtained from X-ray measurements for $Cd_{1-x}Mn_xTe$ is depicted in Fig. 1. These P_t are found to decrease linearly with increasing Mn concentration. The absolute values of P_t, however, tend to be 0.20 ± 0.30 GPa higher when determined from the resistivity measurements as compared to the x-ray measurements.

2.3 **Pressure Variations of Elastic Constants**

Recently, Maheswaranathan and Sladek[11] have measured the variations of the second-order elastic constants (SOECs) with pressure at 296K and found that the longitudinal mode modulus C_L ($=(C_{11}+C_{12}+2C_{44})/2$) increases, while the shear mode modulus C_S($=(C_{11}-C_{12})/2$) decreases linearly with increasing pressure in all samples. The pressure dependence of the shear wave modulii C_S and C_{44}, the bulk modulus B ($=(C_{11}+2C_{12})/3$), and the lattice parameter a_L was also measured by them as is shown in Fig. 2. These investigations have revealed that the replacement of Cd, but not of Zn, in CdTe tends to destabilize the B3 (ZB) structure due to a weakend stability against shear strains. This phenomena is pressured to result from driminution in charge in the tetrahedral bonding orbitals as Cd

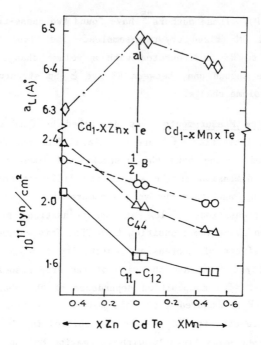

Fig. 1   Lattice Parameters a_L and elastic constants at 296K versus
concentration (x) of Zn or Mn. Taken from Ref. 11.

is replaced by Mn. This observation has motivated Singh and coworkers[21-23] to make theoretical investigations of the effects of charge
depletion (in terms of the three-body interactions[24,27,28]) on the
phase-transition and the pressure variations of elastic constants
as described below.

3.   THEORETICAL INVESTIGATIONS

A number of theoretical investigations on phase-transition
and high pressure elastic behaviour of II-VI and III-V semiconductors
have been carried out by means of valence force field (VFF)[29-31]
and pseudopotential total energy (PTE)[32-34] approaches. These
authors[29-34] have generally concluded that the rocksalt (RS) stru-

cture is preferred in these materials as the ionicity (or charge-transfer) is encreased under compression. In order to overcome some of the deficiences of these approaches[29-34], Singh and cowor-kers[14-16] have developed a three-body potential (TBP) model including the charge-transfer (or charge depletion) effects (via three-body interactions[24] (TBI)) and employed the same to predict the phase-transition and anharmonic properties of alkali halides[35], divalent metal oxides[38], and II-VI and III-V compound semiconductors[14-16].

Motivated from the versatility of the TBP approach and the need for inclusion of charger-transfer (or TBI) effects as remarked by Maheswaranathan and Sladek[11], this approach was considered appro-priate for the description of the phase-transition pressures and anharmonic properties of DMS mixed crystals. A review of these inves-tigations are given below.

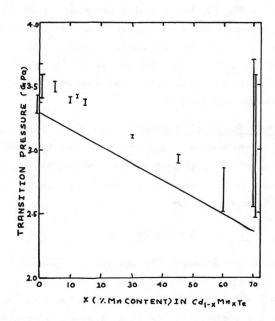

Fig. 2   Phase-Transition Pressures (P_t) determined from resistance
         (   ) and x-ray (——) measurements versus concentration
         (x). Taken from Ref. 5.

3.1 Pressure Induced Phase-Transition

In contrast to the alkali halides, oxides, and compound semi-conductors, the theoretical investigations for diluted magnetic semiconductors (DMS) have attracted only scant attention. Recently, Singh and coworkers[21-23,37] have made efforts to reveal the polymorphic phase-transitions in DMS materials as traced below.

3.1.1 Singh and Prabhakar Method :

During the course of investigation of the equation of state for $Cd_{1-x}Mn_xTe$ and $Cd_{0.52}Zn_{0.48}Te$, Singh and Prabhakar[27] have established an empirical relation

$$\log V = A + BP^{3/2} \qquad (4)$$

based on analogy to a relation between volume (V) and temperature (T) derived by Jain and Sinha[37]. A plot of $\log (V/V_o)$ against $P^{3/2}$, as shown in Fig. 3, has revealed deviations from their linear variations at pressures whose values are closer to the phase-transition pressures reported by the Maheswaranathan and Sladek[11] in $Cd_{1-x}Mn_xTe$ and $Cd_{0.52}Mn_{0.48}Te$. The values of the phase-transition pressures (P_t) obtained from Singh and Prabhakar method will be reported in the next sub-section.

3.1.2 Three-Body Potential Approach:

In recent years, Singh and coworkers[21-23] have adopted the three-body potential (TBP) approach[14-16] to describe the phase-transition pressures and associated volume collapses in DMS materials. For this purpose, they have expressed the Gibbs free energy

$$G = U + PV - TS \qquad (5)$$

as a function of the pressure (P) and the charge-transfer through three-body interaction potential (TBP) energy[14] at OK. Here, V is the unit cell volume and S is the entropy at absolute temperature T. At T = OK and pressure P, the Gibbs free energies for the real zincblende (B3) and hypothetical rocksalt (B1) structures are given by[14]

$$G_{B3}(r) = U_{B3}(r) + 3.08r^3P \qquad (6)$$

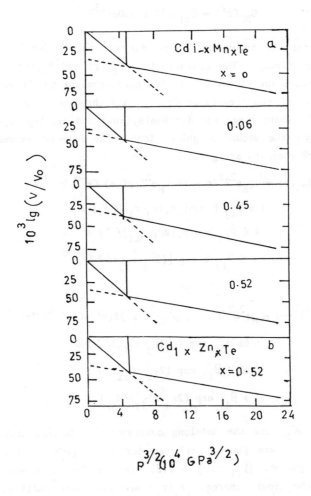

Fig. 3   Variation of log (V(P)/V(o)) with $P^{3/2}$ of $Cd_{1-x}Mn_xTe$ and $Cd_{0.52}Zn_{0.48}Te$. Taken from Ref. 21.

$$G_{B1}(r') = U_{B1}(r') + 3.00r'^3P \tag{7}$$

as the function of interatomic separations r and r' for ZB and RS phases, respectively. The abbreviations $U_{B3}(r)$ and $U_{B1}(r)$ are the lattice energies for ZB and RS studies corresponding to TBP energies which consist of the long-range Coulomb and three-body interactions (TBI) and the short-range van der Waals (vdW) and overlap repulsion effective upto the second neighbour ions. The relevant expressions for these TBP energies are given by

$$U_{B3}(r) = \alpha_M e^2 Z^2/r - (4\alpha_M e^2 Z/rf(r) - Cr^{-6} - Dr^{-8} +$$
$$+ 4\beta_{ij} b \exp[(r_i + r_j - r_{ij})/\rho] +$$
$$+ 6\beta_{ii} b \exp[(2r_i - r_{ii})/\rho] +$$
$$+ 6\beta_{jj} b \exp[(2r_j - r_{jj})/\rho] \tag{8}$$

$$U_{B1}(r') = -\alpha'_M e^2 Z^2/r' - (\alpha'_M e^2 Z/r')f(r') - C'r'^{-6} - D'r'^{-8}$$
$$+ 6b\beta_{ij} \exp(r_i + r_j - r'_{ij})/\rho)$$
$$+ 6b[\beta_{ii} \exp(2r_i - r'_{ii})/\rho)$$
$$+ \beta_{jj} \exp(2r_j - r'_{jj})/\rho)] \tag{9}$$

where α_M (α'_M) are the Madelung constants for ZB (RS) structures C(C') and D(D') are the overall van der Waals (vdW) coefficients of B3 (B1) phases. β_{ij} (i,j = 1,2) are the Pauling coefficients and Ze is the ionic charge. $r_i(r_j)$ are the ionic radii of ions i(j). r(r') and f(r) (f(r')) are the interionic separations and three-body interaction (TBI) parameters for ZB(RS) phases. b and ρ are the range and hardness parameters. $r_{ij} = r$, $r_{ii} = r\sqrt{3}/2 = r_{jj}$, and $r'_{ii} = r\sqrt{2} = r'_{jj}$.

The values of the overall vdW coefficients C and D have been evaluated from the knowledge of their dipole-dipole ($C_{kk'}$) and dipole quadrupole ($d_{kk'}$) interaction coefficients[38] derived from the Slater-Kirkwood variational (SKV) approach[39]. The values of these coeffi-

Table 2a. The values of vdW coefficients for the DMS Materials

Solids	C_{kk} (10^-60 erg.cm.6)	$c_{k'k}$ (10^-60 erg.cm.6)	$C_{kk'}$	d_{kk} (10^-76 erg. cm.2)	$d_{k'k'}$	$d_{kk'}$	$C(10^{-60}$ erg.cm.^{6}) ZnS	(NaCl)	$D(10^{-76}$ erg.cm.^{8}) ZnS	(NaCl)
CdTe	155	1911	503	58	1609	333	2878	(5184)	1568	(3918)
MnTe	75	1365	305	31	1027	187	1877	(3312)	897	(1573)
ZnTe	69	1687	314	25	1362	199	2077	(3657)	987	(1778)

Table 2b. Input Data for DMS Materials

Solids	Elastic Constants (10^{11} dyn. cm.$^{-2}$) C_{11}	C_{12}	C_{44}	Lattice Parameter a(A°)	Ref.
CdTe	5.3	3.7	1.99	3.2385	11
Cd$_{.9}$Mn$_{.1}$Te	5.3	3.7	1.96	3.2475	11
Cd$_{.7}$Mn$_{.3}$Te	5.2	3.6	1.89	3.2200	11
Cd$_{.4}$Mn$_{.6}$Te	5.0	3.4	1.79	3.2013	11
Cd$_{.52}$Zn$_{.48}$Te	6.17	3.8	2.53	3.1480	11

cients are listed in Table 2 a for the host crystals CdTe, MnTe and ZnTe. Their corresponding values for mixed DMS crystals (Cd-Mn-Te or Cd-Zn-Te) can be obtained from the Vegard's law.

The TBP energies cited above contain three model parameters (b, $\mathcal{f}$ and f(r)), which can be determined from the use of any two of the second order elastic constants.

$$C_{11} = (e^2/4a^4) \ [0.248Z(Z+8f(r_o)) + \frac{1}{3}\ (A_1+2B_1) +$$
$$+ (A_2+B_2) + 5.824Zaf'(r_o) \tag{10}$$

$$C_{12} = (e^2/4a^4) \ [-2.646Z(Z+8f(r_o)) + \frac{1}{3}\ (A_1-4B_1) -$$
$$- \frac{1}{4}\ (A_2-5B_2) + 5.824Zaf'(r_o)] \tag{11}$$

$$C_{44} = (e^2/4a^4) \ [-0.124Z(Z+8f(r_o)) + \frac{1}{3}\ (A_1+2B_1) +$$
$$+ \frac{1}{4}\ (A_2+2B_2) - \frac{\nabla}{3}\{-7.539Z(Z+8f(r_o)) +$$
$$(A_1 - B_1)\}\ ] \tag{12}$$

and the equilibrium condition :

$$B_1 + B_2 = -1.261e^2Z(Z+8f(r_o)) \tag{13}$$

The short-range parameters A_i and B_i (i = 1,2) have been defined in the next section. Here, $f'(r_o)$ is the derivative of the TBI parameter f(r) expressed as

$$f(r) = f_o\ \exp\ (-r/\mathcal{f}) \tag{14}$$

with f_o as the constant which can be evaluated from the knowledge of f(r). The expression for symbol $\triangle$ is the same as reported by Singh and Singh[14].

The values of the input data and model parameters are listed in Tables 2a and 3, respectively. The phase-transition pressures (P_t) for the DMS materials ($Cd_{1-x}Mn_xTe$ and $Cd_{1-x}Zn_xTe$) have been

Table 3. Model Parameters for DMS Materials

Solids	$\rho(A°)$	$b(10^{12}$ erg.)	$f(r_o)$
CdTe	0.372	0.303	-0.073
$Cd_{.9}Mn_{.1}Te$	0.360	0.258	-0.099
$Cd_{.7}Mn_{.3}Te$	0.374	0.233	-0.134
$Cd_{.4}Mn_{.6}Te$	0.398	0.210	-0.145
$Cd_{.52}Zn_{.48}Te$	0.378	0.228	-0.059

206

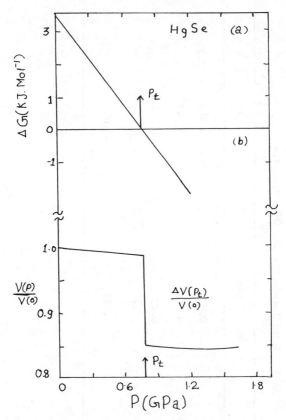

Fig. 4   Variation of Gibbs free energy differences G with pressures (P) for HgSe shown in (a). Phase-diagrams of HgSe
shown in (b). Taken from Ref. 23.

obtained by minimising the cohesive energies given by Eqs. (8)
and (9) for the equilibrium interionic separations r_1 and r_2
corresponding to ZBS (B_3) and RS(B_1) phases. The values of P_t have
been evaluated from the plot of the differences in Gibbs free energies, $\Delta G(=G_3 - G_1)$ against the pressures (P). The pressure at which
G approaches zero corresponds to the phase-transition pressure
(P_t) as indicated by the arrows in Fig. 4-7(a). The relative volume
changes, V(P)/V(o), have been plotted against the pressure to get
the phase diagrams as shown in Figs. 4-7 (b) for HgSe,   HgTe,

$Cd_{1-x}Mn_xTe$ and $Cd_{0.52}Zn_{0.48}Te$, respectively. The values of the phase-transition pressure (P_t) and the associated volume collapses $(-\Delta V(P_t)/V(o))$ along with their available experimental data are listed in Table 4.

A look at the Figs. 4-7(a) reveals that the values of G decrease with pressure and approach zero at the transition pressure (P_t). Below this pressure, ΔG is positive indicating that B3 phase is more stable and the negative values of ΔG beyond it implify the stability of B1 phase in that region. Further, it is seen from the Table 4 that the values of the phase-transition pressures obtained from three-body potential (TBP) approach are in closer agreement

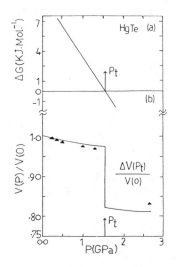

Fig. 5   Variation of Gibbs free energy differences G with pressures (P) for HgTe shown in (a). Phase-diagrams for HgTe shown in (b). Taken from Ref. 23.

with their available experimental data in all the DMS materials. The TBP approach has been found equally successful in revealing the P_t for both the host and mixed DMS crystals.

Table 4. Phase-Transition Pressure (in GPa) for DMS Materials.

Solids	Phase-Transition Pressures (P_t)		Volume Collapses ($\Delta V(P_t)/V(o)$)	
	Present[22]	Exptl.	Present[22]	Exptl.
CdTe	3.20	3.4 ± 0.1[a]	0.1448	0.175[a], 0.16[b]
$Cd_{.9}Mn_{.1}Te$	2.91	3.3 ± 0.5[b] 3.0[c], 3.5[d]		0.24[c], 0.165[b]
$Cd_{.9}Mn_{.1}Te$	2.91	3.0 ± 0.1[a]	0.1279	0.171[a]
$Cd_{.7}Mn_{.3}Te$	2.73	2.8 ± 0.1[a]	0.1343	0.169[a]
$Cd_{.4}Mn_{.6}Te$	2.58	2.5 ± 0.0[a]	0.150	0.180[a]
$Cd_{.52}Zn_{.48}Te$	8.38	6.7[e]		

a – Ref. 5 ; b – Ref. 6 ; c – Ref. 40 ; d – Ref. 41 ; e – Ref. 5.

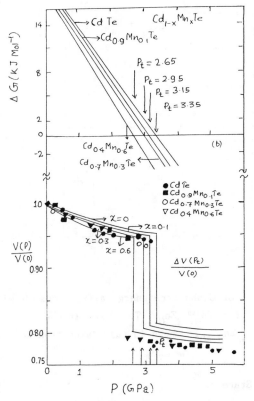

Fig. 6   Variations of Gibbs free energy differences G with pressu-
          res (P) for different concentrations (x) shown in (a).
          Phase-diagrams for different concentrations shown in (b).
          Taken from Ref. 22.

    An inspection of Table 4 also indicates that the values of
the volume collapses $(\Delta V(P_t)/V(o))$ at the phase-transition pressures obtained from TBP approach are in good agreement with their
available experimental data for the host and mixed DMS crystals.
Besides, the variations of the phase-transition pressures with
compositions (x), as shown in Fig. 8, are in reasonably good agreement with their measured values.

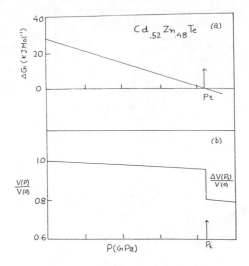

Fig. 7   Variation of Gibbs free energy differences G with pressures (P) for $Cd_{0.52}Zn_{0.48}Te$ shown in (a). Phase-diagrams
for $Cd_{0.52}Zn_{0.48}Te$ shown in (b). Taken from Ref. 22.

3.2 Equation of State

The values of the unit cell volumes V(P) at different pressures
(P) have been obtained as discussed in earlier section. The compression curves have been obtained by plotting the relative volumes
(V(P)/V(o)) against pressure (P) for $Cd_{1-x}Zn_xTe$ and $Cd_{1-x}Mn_xTe$ as
shown in Figs. 8(b) and 8(c), respectively. A comparison of the
compression curves with those obtained from experiments and plotted
in Fig. 8(a) shows that the trends followed by our theoretical
compression curves are similar to those revealed from the experimental measurements in the cases of CdTe, (circles), $Cd_{0.955}Zn_{0.045}$
Te (squares) and $Cd_{0.9}Mn_{0.1}$ Te (triangles). Thus, it is seen that
the TBP model approach[21] is capable to explain the equations of
state of DMS materials. The minor deviations can be eliminated
by including the effect of vdW attraction in the potential used
by Singh and Prabhakar[21].

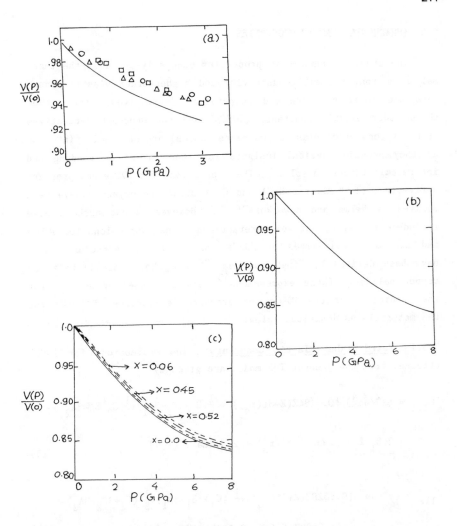

Fig. 8   Experimental compression curves (or equation of state) for $Cd_{1-x}Mn_xTe$ shown in (a) and taken from Ref. 6 (circles for CdTe, squares for $Cd_{0.045}Zn_{0.45}Te$ and traingles for $Cd_{0.9}Mn_{0.1}Te$, solid line is the data for CdTe produced by Clive and Stephens (J. Appl. Phys. 9 (1950) 2869). Theoretical compression curves for $Cd_{0.52}Zn_{0.48}Te$ shown in (b) and for $Cd_{1-x}Mn_xTe$ shown in (c) and taken from Ref. 21.

3.3 ANHARMONIC ELASTIC PROPERTIES

The study of anharmonic properties generally includes the thermal expansion, thermal conductivity and higher-order elastic constants and their pressure and temperature derivatives. Among them third order elastic constants (TOECs) and the pressure derivatives of the second order elastic constants (SOECs) are of special interest as they provide physical insights into the nature of bonding and interatomic forces in solids. The expressions for TOECs and pressure derivatives of SOECs for NaCl and CsCl structure crystals have been derived by Verma and coworkers[42-44]. However, these authors have excluded the effects of vdW interactions. The expressions for TOECs and pressure derivatives of SOECs for zincblende structure (ZBS) have been derived by Singh and Gupta[45] for their calculations for copper halides. These expressions[45] have been used by us[21-23] for the prediction of the TOECs and pressure derivatives of SOECs for DMS materials as described below.

3.3.1 Third Order Elastic Constants: The expressions for TOECs obtained from the present TBP model are given by[45]

$$C_{111} = (e^2/4a^4) \, [0.5184Z(Z+8f(r_o)) + \tfrac{1}{9}(C_1-6B_1-3A_1)+\tfrac{1}{4}(C_2-B_2-3A_2) - $$

$$- \, 2(B_1-B_2)-9.9326Zaf'(r_o) + 2.522Za^2f''(r_o)] \tag{15}$$

$$C_{112} = (e^2/4a^4)[0.3828Z(Z+8f(r_o))+\tfrac{1}{9} (C_1+3B_1-3A_1)+\tfrac{1}{8} (C_2+3B_2-3A_2)-$$

$$-11.6482Zaf'(r_o)+2.522Za^2f''(r_o)] \tag{16}$$

$$C_{123} = (e^2/4a^4)[6.1585Z(Z+8f(r_o))+\tfrac{1}{9} (C_1+3B_1-3A_1)-$$

$$-12.506Zaf'(r_o)+2.522Za^2f''(r_o)] \tag{17}$$

$$C_{144} = (e^2/4a^4)[6.1585Z(Z+8f(r_o)) + \frac{1}{9}(C_1+3B_1-3A_1)-4.1687Zaf'(r_o)+$$

$$+0.8407Za^2f''(r_o) +\nabla\{-3.3507\ Z(Z+8f(r_o))-\frac{2}{9}C_1+13.5486Zaf'(r_o)-$$

$$-1.6813Za^2f''(r_o)\} +\nabla^2\{-1.5637Z(Z+8f(r_o))+\frac{2}{3}(A_1-B_1)+\frac{1}{9}C_1-$$

$$-5.3138Zaf'(r_o)+2.935Za^2f''(r_o)\}\ ] \tag{18}$$

$$C_{166} = (e^2/4a^4)[-2.1392Z(Z+8f(r_o)) + \frac{1}{9}(C_1-6B_1-3A_1)+ \frac{1}{8}(C_2-5B_2-3A_2)-$$

$$-(B_1+B_2)-4.1687\,Za f'(r_o)+0.8407Za^2f''(r_o)+\nabla\{-8.3768Z(Z+8f(r_o))+$$

$$+ \frac{2}{3}(A_1-B_1) - \frac{2}{9}C_1+13.5486Zaf'(r_o)-1.6813Za^2f''(r_o)\} +$$

$$+ \nabla^2\{2.3527\ Z(Z+8f(r_o)) + \frac{1}{9}C_1-5.3138Zaf'(r_o)+2.935Za^2f''(r_o)\}\ ] \tag{19}$$

$$C_{456} = (e^2/4a^4)\ [4.8975Z(Z+8f(r_o))+ \frac{1}{9}(C_1-6B_1-3A_1)-B_2 +$$

$$+ \nabla\{-5.026Z(Z+8f(r_o))- \frac{1}{3}C_1\} +\nabla^2\{7.058Z(Z+8f(r_o))+ \frac{1}{3}C_1\} +$$

$$+ \nabla^3\{-4.8008Z(Z+8f(r_o)) + \frac{1}{3}(A_1-B_1) - \frac{1}{9}C_1\}\ ] \tag{20}$$

with

$$\nabla = \frac{-7.5391Z(Z+8f(r_o)) + A_1 - B_1}{-3.1416Z(Z+8f(r_o)) + A_1 + 2B + \underline{21.7656Zaf'(r_o)}} \tag{21}$$

Here, the underlined terms in Eqs. (15) and (21) the additional contributions to the expressions derived by Garg et.al.[46] for ZBS crystals. The parameters A_i, B_i and C_i ($i=1,2$), appearing in above equations, have been defined as

$$A_1 = \frac{4a^3}{e^2} [\frac{\partial^2 V_1(r)}{\partial r^2}]_{r=(\sqrt{3}/2)a} \quad ; \quad A_2= \frac{16a^3}{e^2} [\frac{\partial^2 V_2(r)}{\partial r^2}]_{r=\sqrt{2}a} \tag{22}$$

$$B_1 = \frac{4a^3}{e^2} \left[\frac{1}{r} \frac{\partial V_1(r)}{\partial r}\right]_{r=(\sqrt{3}/2)a} \quad ; \quad B_2 = \frac{16a^3}{e^2} \left[\frac{1}{r} \frac{\partial V_2(r)}{\partial r}\right]_{r=\sqrt{2}a}$$

$$\tag{23}$$

$$C_1 = \frac{4a^3}{e^2} \left[r \frac{\partial^3 V_1(r)}{\partial r^3}\right]_{r=(\sqrt{3}/2)a} \quad ; \quad C_2 = \frac{16a^3}{e^2} \left[r \frac{\partial^3 V_2(r)}{\partial r^3}\right]_{r=\sqrt{2}a}$$

$$\tag{24}$$

with $V_1(r)$ and $V_2(r)$ as the overlap repulsive potentials due to the nearest and next nearest neighbours and they can be expressed as

$$V_1(r) = \left\{b \left(1 + \sum_{i=1}^{2} \frac{Z_i}{n_i}\right) \exp\left[\sum_{i=1}^{2} (r_i - r)\right]/\rho\right\} - C_{12}r^{-6} - d_{12}r^{-8}$$

$$\tag{25}$$

$$V_2(r) = \sum_{i=1}^{2} \left\{(b(1 + \frac{2Z_i}{n_i}) \exp\left[(2r_i - r_{ii})/\rho\right]) - C_{ii}r_{ii}^{-6} - d_{ii}r_{ii}^{-8}\right\}$$

$$\tag{26}$$

Here, Z_i and n_i are the valence and number of electrons in the outermost shell of the ions.

The values of TOECs have been calculated from the Eqs. (15)–(20) using the values of the model parameters listed in Table 3. Their values have been depicted in Table 5 and found to follow a systematic trend of variation with the composition as is noted from the Fig. 9 for $Cd_{1-x}Mn_xTe$. A decisive comment on the reliability of these results will be deferred until the report of measured data on them. However, their magnitudes might serve as a guide to the experimental workers.

3.3.2 **Pressure Derivatives of SOECs** : Recently, Sharma and Verma[47] have derived the expressions for the pressure derivatives of SOECs by the direct differentiation of SOECs using the equilibrium condi-

Table 5. Values of TOECs ($\times 10^{11}$ dyn. cm.$^{-2}$) for DMS Materials

Solids	C_{111}	C_{112}	C_{123}	C_{166}	C_{144}	C_{456}
CdTe	-3.9976	-4.2006	-1.4802	-3.6048	-2.0255	2.1466
$Cd_{.9}Mn_{.1}Te$	-3.9754	-4.0100	-1.4087	-3.5409	-2.0453	0.1382
$Cd_{.7}Mn_{.3}Te$	-3.9469	-4.1945	-1.2669	-3.9381	-1.7740	0.0789
$Cd_{.4}Mn_{.6}Te$	-3.7696	-3.4241	-1.0909	-2.7501	-1.4165	0.0645
$Cd_{.52}Mn_{.48}Te$	-3.0509	-4.4181	-2.8988	-1.4645	-2.9200	2.0102

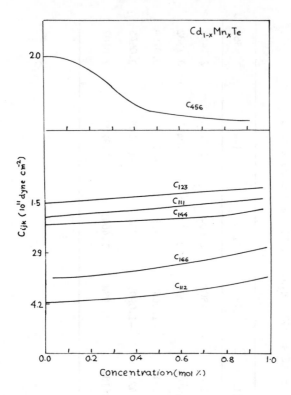

Fig. 9   Variation of TOECs with percentage concentration (x) for
         DMS materials as taken from Ref. 22.

tion. However, these expressions lack the inclusion of vdW interactions. Following their procedure[47], Singh and Gupta[45] have derived
the following expressions for the pressure derivatives of the bulk
modulus $B_T(=(C_{11}+2C_{12})/3)$ and the shear modulii $S(=(C_{11}-C_{12})/2)$
and C_{44} using the TBP model formalism:

$$\frac{dB_T}{dp} -(3\mathcal{N})^{-1}[20.1788Z(Z+8f(r_o))-3(A_1+A_2)+4(B_1+B_2) +$$

$$+C_1+C_2-104.8433Zaf'(r_o)+22.7008Za^2f''(r_o)] \qquad (27)$$

$$\frac{dS}{dp} = -(2\Omega)^{-1}[-11.5756Z(Z+8f(r_o))+2A_1-4B_1+\frac{3}{2}A_2-$$

$$-\frac{7}{2}B_2+\frac{1}{4}C_2+37.522Zaf'(r_o)] \tag{28}$$

$$\frac{dC_{44}}{dp} = -(\Omega)^{-1}[0.4952Z(Z+8f(r_o))+\frac{1}{3}(A_1-4B_1+C_1)+\frac{1}{4}(2A_2-6B_2+C_2)$$

$$+4.9667Zaf'(r_o)+2.5229Za^2f''(r_o)+\nabla\{-17.5913Z(Z+8f(r_o))+A_1-B_1-$$

$$-\frac{2}{3}C_1+40.6461Zaf'(r_o)-5.044Za^2f''(r_o)\}+$$

$$+\nabla^2\{\underline{3.1416Z(Z+8f(r_o))}+\frac{2}{3}(A_1-B_1)+\frac{1}{3}C_1-$$

$$-15.9412Zaf'(r_o)+8.8052Za^2f''(r_o)\}] \tag{29}$$

Here,
$$\Omega = 5.044Z(Z+8f(r_o))+A_1+A_2-2(B_1-B_2)+17.473Zaf'(r_o) \tag{30}$$

The underlined terms in the above equations are the additional contributions to the expressions derived by Garg et.al.[46].

The above expressions have been directly used to evaluate the pressure derivatives of the bulk and shear modulii (B_T, S, and C_{44}). Their values have been presented in Table 6 for DMS materials. The variations of the pressure derivatives of SOECs (C_{11}, C_{12} and C_{44}) with composition (x) have been depicted in Fig. 10 and found to follow a systematic trend similar to that observed in HgSe[19] and HgTe[20]. It is interesting to note from Table 6 that the pressure derivatives of bulk and shear modulii and SOECs are in almost excellent agreement with their experimental data[11] available only in the cases of CdTe and $Cd_{0.52}Zn_{0.48}Te$. Our decisive comments in other cases will be deferred until the report of measured data on them.

4. OVERVIEWS

This chapter presents a systematic and comprehensive description of the experimental and theoretical investigations on the pressure

Table 6. Values of the Pressure Derivatives of SOEC (Dimensionless Units)
of DMS Materials

Solids	dB_T/dp	dS/dp	dC_{11}/dp	dC_{12}/dp	dC_{44}/dp	Ref.
CdTe	4.8080	−0.3946	3.7581	4.5450	2.240	Theo.[22]
	(4.3)	(−0.42)	(3.7)	(4.5)	(2.05)	Expt.[11]
$Cd_{.9}Mn_{.1}Te$	4.4690	−0.3841	3.7011	4.7254	1.2615	Theo.[22]
$Cd_{.7}Mn_{1.3}Te$	3.4526	−0.4950	2.4626	3.7826	1.0844	
$Cd_{.4}Mn_{.6}Te$	3.3331	−0.5379	2.2731	3.6917	0.9779	
$Cd_{.52}Zn_{.48}Te$	3.7095	−0.1933	3.7936	4.1801	1.4700	Theo.[22]
	(4.2)	(−0.21)	(3.9)	(4.3)	(1.09)	Expt.[11]

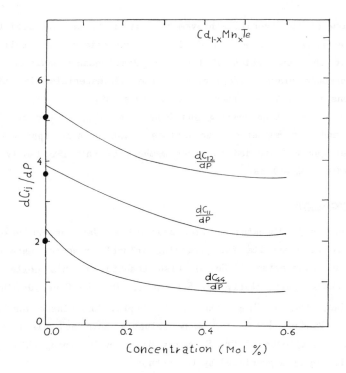

Fig. 10   Variation of Pressure Derivatives of SOECs with concentration for $Cd_{1-x}Mn_xTe$ as taken from Ref. 22.

induced phase–transition and the anharmonic elastic properties
of the diluted magnetic semiconducting (DMS) materials. The model
used for this purpose is called three-body force potential (TBP)
model, which consists of the long–range Coulomb and three–body
interactions (TBI) and the short-range van der Waals (vdW) attraction and overlap repulsion extended upto the second neighbour
ions. This TBP model has been found to be adequately successful
in describing the observed phase-transition pressures, volume collapses, equation of state, third order elastic constants, pressure
dependence of the second order elastic constants of the DMS
materials.

In view of the overall achievements, it might be concluded that the three-body potential model is an appropriate and realistic model for the description of the structural phase-transitions and the anharmonic elastic properties of the DMS materials. The minor deviations emerging in these descriptions might be due to the exclusion of the bond-bending and bond-stretching of the covalent bonds[19] and the magnetic interactions. The present approach of TBP model can be extended to other members of this DMS family and other systems of solids.

ACKNOWLEDGEMENTS

The author is thankful to Professor Mukesh Jain of the Western University of Australia for providing valuable research materials and useful discussions. He is also thankful to his colleagues and coworkers, particularly Dr. S.P.Sanyal, Dr. N.K.Gaur, Dr.(Mrs.) C. N. Rao, Mrs. S. Singh, Mr. D. C. Gupta, Mr. Dinesh Varshney, and Mrs. Vandana Batra for many valuable help and discussions. The skillful efforts of Mr. S. Surya Bhagawan in typing this manuscript is highly appreciated by the author.

REFERENCES

1. Shapira, Y., Foner, S., Ridgley, D.H., Dwight, K. and Wold, A., Phys. Rev. B30, 4021 (1984).

2. Quadri, S.B., Skelton, E.F., Webb, A.W., Colombo, L. and Furdyna, J.K., Phys. Rev. B40, 2432 (1989).

3. Spicer, W.E., Silberman, J.A., Landau, T., Chen, A.B., Sher, A. and Wilson, J.A., J. Vac. Sc. Technol. A1, 1735 (1983).

4. Galazka, A.R., in Physics of Semiconductors 1978, Edited by Wilson, W.H., IOP Conf. Proc. No. 43 (IOP, Bristol and London, 1978), p. 133.

5. Quadri, S.B., Skelton, E.F., Webb, A.W., Carpenter Jr. E.R., Schaefer, M.W. and Furdyna, J.F., Phys. Rev. B35, 6868 (1987).

6. Quadri, S.B., Skelton, E.F. and Webb, A.W., Physica 139 and 140B, 341 (1986).

7. Quadri, S.B., Skelton, E.F., Webb, A.W. and Dinam, J., J. Vac. Sc. Technol. A4, 1974 (1986).

8. Quadri, S.B., Skelton, E.F., Webb, A.W., Schaefer, M.W., Dinam, J.H., Chandra, D. and Colombo. L., J. Vac. Sc. Technol, A5, 3024 (1987).

9. Arora, A.K., Bartholomew, D.V., Peterson, D.L. and Ramdas, A.K., Phys. Rev. B35, 7966 (1987).

10. Arora, A.K. and Ramdas, A.K., Ind. J. Pure & Appl. Phys. 36, 182 (1988); Bull. Am. Phys. Soc. (U.S.A.). 32, 801, (1987); Phys. Rev. B35, 4345 (1987).

11. Maheswaranathan, P., Sladek, R.J. and Debska, U., Phys. Rev. B31, 5212 (1985).

12. Shan, W., Shen, S.C. and Zhu, H.R., Sol. Stat. Commun. 55, 475 (1985).

13. Shapira, Y. and Oliveira, Jr. N.F., Phys. Rev. B35, 6888 (1987).

14. Singh, R.K. and Singh, S., Phys. Rev. B39, 761 (1989).

15. Singh, R.K. and Singh, S., Phase Trans. 15, 127 (1989).

16. Singh, R.K. and Gupta, D.C., Phys. Rev. B 40, 11278 (1989) (Press)

17. Chelikowsky, J.R., Phys. Rev. B35, 1174 (1987).

18. Zhang, S.B. and Cohen, M.L., Phys. Rev. B35, 7604 (1987).

19. Miller, A.J., Saunders, G.A. and Yogurtcu, Y.K., Philos. Mag. A43, 1447 (1981); J. Phys. Chem. Solids 42, 49 (1981).

20. Ford, P.J., Miller, A.J., Saunders, G.A., Yogurtcu, Y.K., Furdyna, J.R. and Jaczynski, M., J. Phys. C15, 657 (1982).

21. Singh, R.K. and Prabhakar, N.V.K., Phys. Stat. Solidi (b) 146, 111 (1986).

22. Singh, R.K. and Batra, V., Phys. Rev. B (Communicated).

23. Singh, R.K., Sakalle, U.K. and Gupta, D.C., Phys. Rev. B. (Communicated).

24. Singh, R.K., Phys. Reports (Netherlands) 85, 259 (1982).

25. Strossner, K., Ves, S., Dieterich, W., Gebhardt, W., and Cardona, M., Sol. St. Commun. 56, 563 (1985).

26. Anderson, O.L., J. Phys. Chem., Solids 27, 547 (1966).

27. Lowdin, P.O., J. Chem. Phys. 8, 365 (1950).

28. Lundqvist, S.O., Ark. Fys. 6, 25 (1952); 9, 435 (1955).

29. Miller, A.J., Saunders, .A. and Yogurtcu, Y.K., Phil. Mag. A43, 1447 (1981); J. Phys. Chem. Solids 42, 49 (1981).

30. Yogurtcu, Y.K., Miller, A.J. and Saunders, G.A., J. Phys. Chem. Solids 38, 1355 (1977).

31. Ford, P., Miller, A.J., Saunders, G.A., Yogurteu, Y.K., Furdyna J.R. and Jaczynsky, M., J. Phys. C15, 657 (1982).

32. Froyen, S. and Cohen, M.L., Phys. Rev. B28, 3255 (1983).

33. Zhang, S.B. and Cohen M.L., Phys. Rev. B35, 7604 (1987).

34. Chelikowski, J.R., Phys. Rev. B35, 1174 (1987).

35. Prabhakar, N.V.K., Singh, R.K., Gaur, N.K. and Sharma, N.N., J. Phys. Codens. Matter 2,3345 (1990).

36. Jog, K.N., Singh, R.K. and Sanyal, S.P., Phys. Rev. B35, 5235 (1987).

37. Jain, R.K. and Sinha, R., J. Chem. Phys. 72, 1909 (1980).

38. Narayanan, R., J. Phys. Chem. Solids 38, 1097 (1977).

39. Slater, J.C. and Kirkwood, J.G., Phys. Rev. B37, 682 (1931).

40. Owen, N.B., Smith, P.L., Mastin, J.E., and Wright, A.J., J. Phys. Chem. Solids 24, 1519 (1963).

41. Brog, J.J. and Smith, D.K., J. Phys. Chem. Solids 28, 49 (1967)

42. Verma, M.P. and Puri, D.S., Sol. Stat. Commun. 18, 1295 (1976); Phys. Rev. B15, 2337 (1977).

43. Garg, V.K., Puri, D.S. and Verma, M.P., Phys. Stat. Sol.(b) 80, 63 (1977); Phys. Stat. Sol.(b) 78, 113 (1976).

44. Garg, V.K., Puri, D.S., and Verma, M.P., Phys. Stat. Sol.(b) 82, 325 (1977); 82, 481 (1977).

45. Singh, R.K. and Gupta, D.C., IL Nuovo Cimento 9D, 1258 (1987).

46. Garg, V.K., Puri, D.S. and Verma, M.P., Phys. Stat. Solidi (b) 87, 401 (1978).

47. Sharma, U.C. and Verma, M.P., Stat. Sol.(b) 102, 487 (1980).

LIGHT SCATTERING IN DILUTED MAGNETIC SEMICONDUCTORS

E. ANASTASSAKIS

National Technical University, Physics Department
Zografou Campus, Athens 157 73, Greece

CONTENTS

1.   INTRODUCTION

Diluted Magnetic Semiconductors (DMS) consist of $A^{II}B^{VI}$ non-magnetic semiconductors mixed with MnB magnetic semiconductors. The resulting compounds have the general form $A_{1-x}Mn_x B$, or (A,B)Mn, where x the atomic concentration. Although other members of the iron group (e.g. Fe, Co) are not excluded, the vast majority of DMS known to date are based on Mn, with A = Cd, Zn, Hg and B = Te, Se, S. Comprehensive reviews on growth, preparation and properties of DMS can be found in Refs. 1-6.

DMS are interesting materials in various ways. Being x-tunable, their non-magnetic properties supplement and extend those of the end compounds. Most importantly, the presence of magnetic ions Mn^{2+} leads to spectacular magnetic effects due to their strong interactions with free or impurity-bound carriers. Thus, phenomena like giant Zeeman splittings and Faraday rotations, magnetic phase transitions and bound magnetic polarons are commonplace in the literature of DMS. Successful preparation of multi-quantum-well or superlattice structures based on DMS components has also become possible in recent years. This extends further the unique properties of DMS, and creates more possibilities for applications in magneto-optic devices, sensors, solid state lasers, panel display technology, and IR detectors, to mention a few.

Substantial contributions to our understanding of the microscopic physical mechanisms underlying the properties of DMS have been made through their interactions with light, e.g., photoluminescence (PL), magneto-optics (MO), Far Infrared (FIR) transmission/reflection, and Raman Scattering (RS). The purpose of this article is to briefly focus on the type of knowledge that can be gained from RS experiments on DMS. This includes information on lattice dynamics (phonons), phase transitions, and a large variety of fundamental electronic interactions in the presence of a magnetic field H, or otherwise. Most of the work refers to wide-gap DMS, i.e., A=Cd, Zn.

After a brief summary of relevant properties of DMS and the basic principles of RS, the six wide-gap compounds will be treated separately. The review covers aspects of RS alone. Other types of spectroscopic studies (PL, MO, FIR) will not be touched, as falling beyont the scope of this article. Technical editorial limitations prohibit the author from referring in detail to every and each relevant source in the literature. This would require a different style of presentation. Omission of any important references in this regard is only accidental.

2. GENERAL PROPERTIES

2.1 Structure and Lattice Parameters

The Mn^{2+} ions in DMS occupy cation (A) sites in random. Some of the AB end compounds (x=0) crystallize in the cubic (zincblende, ZB) structure, some in the hexagonal (wurtzite, W) structure. The end compounds MnB (x=1) cannot be grown in the form of single crystals of ZB or W structure. MnS and MnSe occur in polycrystalline phases of ZB or W structure, but MnTe does not. The intermediate DMS crystals (0<x<1) occur in either ZB or W structure for certain range of x values each. Table I includes this information[4], while Fig. 1 shows the mean cation-cation distance d_c as a function of x[5,7]. These data are obtained from X-ray diffraction and obey Vegard's law; the corresponding lattice parameters are $a(x)=\sqrt{2}\ d_c(x)$ for ZB, and $a(x)=d_c(x)$ and $c(x)=\sqrt{8/3}\ d_c(x)$ for W. Dashed lines in Fig. 1 do not correspond to real ZB or

Table I. Structure of DMS (after Ref. 4).

Materials	Crystal structure	Range of composition
$Zn_{1-x}Mn_xS$	zinc blende	$0<x<0.10$
	wurtzite	$0.10<x<0.45$
$Zn_{1-x}Mn_xSe$	zinc blende	$0<x<0.30$
	wurtzite	$0.30<x<0.57$
$Zn_{1-x}Mn_xTe$	zinc blende	$0<x<0.86$
$Cd_{1-x}Mn_xS$	wurtzite	$0<x<0.45$
$Cd_{1-x}Mn_xSe$	wurtzite	$0<x<0.50$
$Cd_{1-x}Mn_xTe$	zinc blende	$0<x<0.77$
$Hg_{1-x}Mn_xS$	zinc blende	$0<x<0.37$
$Hg_{1-x}Mn_xSe$	zinc blende	$0<x<0.38$
$Hg_{1-x}Mn_xTe$	zinc blende	$0<x<0.75$

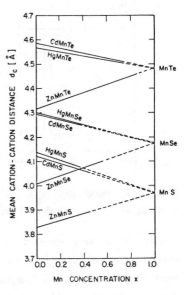

Fig. 1. Mean cation-cation distance of DMS as a function of x (after Ref. 7).

W structures, whereas solid lines do, consistently with Table I. Furthermore, MnTe may be regarded only as a "hypothetical" ZB structure. Fig. 1 provides a reliable method for determining x when d_c is available. The large range of a(x) values is of particular importance in the technology of strained superlattices; the built-in planar strain in these materials depends on the lattice constants of adjacent layers (lattice mismatch strains) and are known to influence critically the electronic and transport properties of the system.

228

2.2 Electronic Band Structure

The electronic band structure of the parent AB compounds is well known and understood. AB are direct-gap semiconductors with Γ-point extrema, and branches which originate from the s, p orbitals of A and B. Similar structures are anticipated for the MnB compounds, according to computational results based on virtual crystal approximation or other schemes. The intermediate DMS compounds behave alike, and in fact their energy gaps follow Vegard's law very closely. Fig. 2(a) presents the energy gap $E_g(x)$ for the DMS of our interest. The solid (dashed) lines correspond to linear fittings (extrapolations) based on data[5]. The continuous tuning of $E_g(x)$ from zero to

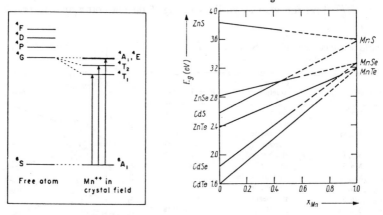

Fig. 2. (a):Energy gap $E_g(x)$ of DMS for $4 \leq T \leq 77K$ (after Ref. 5). (b):Energy levels of Mn^{2+} in a tetrahedral crystal field (after Ref. 4).

several eV is very essential in applications of band-gap engineering. These phenomenological findings have their microscopic origin on the electronic band structure of the AB compounds. In addition, there are properties due to the presence of the Mn^{2+} ions in the tetrahedral crystal field. A schematic diagram of Mn^{2+} energy levels before and after been placed in the crystal field is shown in Fig. 2. It is governed by the five electrons of the half-filled 3d shell of Mn^{2+}; the ground state of the latter is characterized by S=5/2, L=0, i.e., it is a six-fold degenerate state. Thus, in addition to the s,p-d exchange interactions which determine the gross electronic properties of DMS, one also observes, in a quite independent fashion, an entire set of properties associated with the electronic transition scheme of the Mn^{2+} ion. The lowest Mn^{2+} electronic transition $^6A_1 \rightarrow ^4T_1$ of Fig. 2(b) is the most prominent one; it occurs around 2.2 eV and determines the absorption edge of the crystal, provided $E_g(x) > 2.2$ eV, at least for large values of x. In some dilute $(x \sim 0.03)$ wide-band DMS, like (Zn,S)Mn or (Zn,Se)Mn, the $^6A_1 \rightarrow ^4T_1$ transition is responsible for intense electroluminescence phenomena.

Most of the information in this and the following subsection is based on Refs. 3-5, where the reader is referred to for more details and extensive literature.

2.3 Magnetic Properties

In this subsection we review only those magnetic properties of DMS which are relevant to RS. We start with the magnetic phase diagram. The main phase is the high-temperature paramagnetic phase which , for small x, can occur also at low tempratures. There is a low-temperature $[T < T_g(x)]$ frozen phase which is disordered and exhibits spin-glass behaviour, with short-range antiferromagnetic spin ordering. This phase requires $0.2 < x < 0.6$, in general, but can also occur for smaller x at very low temperatures. Finally, there is a moderately long-range antiferromagnetically-ordered phase which, at least for (Cd,Te)Mn, manifests itself for $x \geq 0.6$ and below a critical temperature $T_N(x)$. Above T_N the paramagnetic phase sets in.

Of fundamental importance for the magnetic properties of DMS are the exchange interactions of the Mn^{2+} ions with themselves and with carriers. The latter can be s-like conduction-band electrons, p-like valence-band electrons, impurity-bound electrons or even free or bound excitons. From the Mn^{2+} side, the interaction involves the outer-shell $3d^5$ electrons.

In the paramagnetic phase, the Mn^{2+} ions can be treated as independent, provided T_N is larger than the nearest-neighbor exchange interaction constant $|J_{NN}|$, to be discussed shortly. Usually this is the case. Therefore one may look at the six-fold degenerate ground level ^6A as an atomic level. A magnetic field H will lift the degeneracy of ^6A, producing Zeeman levels split by $g_{Mn}\mu_B H \equiv \hbar\omega_{PM}$, where g_{Mn} the g-factor of Mn^{2+} and μ_B the Bohr magneton. Transitions within this Zeeman multiplet are possible and are responsible for interesting RS processes (see later sections).   Pair and triangle exchange interactions among Mn^{2+} ions are known to be possible. These appear in the form of nearest-neighbor interactions and next-nearest-neighbor interactions, with exchange integrals J_{NN} and J_{NNN}, respectively. Both are antiferromagnetic and are largely governed by a super-exchange mechanism which requires two-hole processes and mediation of the anion B. Typical values of J_{NN} for all DMS studied, vary from -6K to -16K and can be deduced from a variety of experimental techniques, including Raman spectroscopy. The energy scheme of the Mn^{2+} pair system is more complicated than the single ion scheme; it involves a total spin $S = 0,..5$ with $M_S = -S,..+S$. With no field present, transitions between successive levels involve energy steps of $2|J_{NN}|$, and transitions from the ground $(S=0)$ to higher $(S=1,2,..)$ levels involve energy changes by $S(S+1)|J_{NN}|$, in general. Such energy can be measured either from magnetization experiments or from RS and yield values for $|J_{NN}|$. In the presence of a strong field each level produces a multiplet with energy steps equal to $M_S g_{Mn} \mu_B H = \hbar(M_S \omega_{PM})$. Since the total angular momentum is not conserved in these interactions, transitions with $\Delta M_S = 1,2,..$ are allowed; this leads to the appearance of RS at the overtones of ω_{PM}, i.e., at $2\omega_{PM}$, $3\omega_{PM}$,.. etc. (see later sections).

Next we turn to the exchange interactions between the $3d^5$ electrons and the s or p band electrons. These are commonly defined as the s,p-d interactions. They are based on two different mechanisms, (i) the $1/r$ potential exchange between s, p and d electrons, usually known as **positive ferromagnetic contribution** to the exchange constants, and (ii) the hybridization of the s, p and d electrons, known as **negative antiferromangetic contribution**. The latter mechanism is stronger than the former. It is because of the above magnetic interactions that DMS exhibit such impressive magneto-optical and magneto-transport phenomena, absent in the AB semiconductors.

The most profound effects involving s,p-d exchange interactions appear when H is present. Assuming for simplicity a parabolic conduction band Γ_6 in AB compounds the effect of H is to create Landau levels, each contributing to the energy of the s electrons a term $(l+1/2)\hbar\omega_c$, where ω_c the cyclotron frequency. Each level is further split into two spin sublevels $(m_j = \pm 1/2)$ separated by $g^*\mu_B H$ where g^* the g-factor of the band electron. The effect of Mn^{2+} is to increase the g-factor from g^* to

$$g_{eff} = g^* + \Delta g \tag{1a}$$

$$= g^* + \alpha M/(g_{Mn}\mu_B^2 H), \tag{1b}$$

where ($\alpha > 0$) is the exchange integral of s electrons. The positiveness of α is related to its exclussive origin from the potential exchange mechanism. The effective Zeeman splitting of each Landau level of the s electrons in this approximation now becomes $g_{eff}\mu_B H$. The entire energy diagram in this approximation is shown schematically in Fig. 3 for small (a), medium (b), and large (c) average spin $<S_z>$. This diagram describes a typical low- or

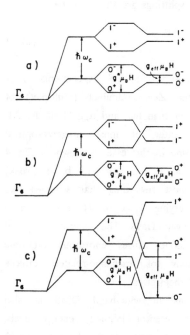

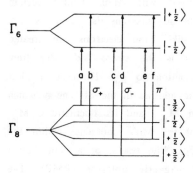

Fig. 3. Energy levels of the s-like (Γ_6) electrons in DMS of an arbitrary gap, with $H \neq 0$. Only the lowest two Landau levels are shown (ω_c) modified by spin splitting (g*), and d-exchange (g_{eff}). (a)→(c) indicate increasing average spin $<S_z>$ (after Ref. 4).

Fig. 4. Energy levels in wide-gap DMS as developed by the exchange interactions alone, with $H \neq 0$. $\Gamma_{6(8)}$ isthe bottom (top) of the c(v) band. $\sigma_+ (\sigma_-)$ dipole transitions require circular +(-) polarization of light. π transition requires linear ($\|H$) polarization (after Ref. 4).

medium-gap DMS (wide-gap DMS will be discussed shortly). Notice that depending on parameters, the energy levels may cross with each other, and also with the lower lying levels arising from the top of the covalence band, leading to semi-metallization. In the same approximation, the effect of Mn^{2+} on the four-fold split ($m_j = \pm 1/2$, $\pm 3/2$) p-like holes from the top of the valence band (Γ_8) is to modify their splitting by a correction which is proportional to the p-d exchange integral β. The latter includes contributions from potential (>0) mechanism and hybridization (<0) mechanism. Since the latter dominates, β is negative with $|\beta| > \alpha$.

A direct application of the above results is found in wide-gap DMS, where both, orbital (ω_c) and spin (g^*) shifts can be neglected in the presence of the much larger exchange term Δg. The two-fold (four-fold) split energy levels of the s-like (p-like) electrons (holes) are then governed entirely by the exchange interaction terms. The splittings are shown to be

$$
\begin{pmatrix} \Delta E_c \\ \Delta E_v \end{pmatrix} = \frac{M}{g_{Mn}\mu_B} \begin{pmatrix} \alpha \\ \beta/3 \end{pmatrix} \equiv \hbar \begin{pmatrix} \omega_{SF}^c \\ \omega_{SF}^v \end{pmatrix} . \tag{2}
$$

$\omega_{SF}^{c(v)}$ are the spin-flip (SF) Raman shifts observed in RS experiments. They involve electron (hole) transitions within the Zeeman-doublet (multiplet) of the conduction (valence) band. Being much larger than $\hbar\omega_c$ and $\mu_B g^* H$ of the AB compounds, these splittings are at the center of most magneto-optical phenomena in wide-gap DMS. The energy scheme in this case is shown in Fig. 4 together with the dipole-allowed transitions a, d (strong) and b, c (weak) and the light polarizations required. It is most rewarding that schemes like the one of Fig. 4 are easily tunable by varying T, x or H, and describe equally well the excitonic behaviour in wide-gap DMS. The clear-cut physical picture whereby the exchange interaction alone causes enormous excitonic Zeeman splittings, allows the direct study of exchange-related effects, including measurements of the parameters α, β themselves.

The strength of exchange interactions in wide-band DMS is also manifested by their influence on donor or acceptor impurity energy levels. Again, large Zeeman splittings are observed with $H \neq 0$. Under certain conditions the Zeeman splitting of donor-bound electrons is given by Eq. (2a), that is, donor-bound electrons and s-like band electrons are treated alike in this regard. The reason for this is the strong s-d interaction, which results in Δg being the main contribution to g_{eff}. However, at low temperatures and low fields an additional term appears in the SF energy which depends critically on the local magnetization $\mathbf{M}_{loc}$ and its fluctuations. $\mathbf{M}_{loc}$ is due to the electron-induced alingment of the Mn^{2+} moments within the electron orbit, yielding values as large as $25\mu_B$. The system of a donor-bound electron plus its $\mathbf{M}_{loc}$ is termed **bound magnetic polaron (BMP)**. The fluctuations of $\mathbf{M}_{loc}$ on the other hand can be, (i) time dependent, because of thermal fluctuations of the Mn^{2+} ions within the BMP volume, and (ii) space dependent, due to compositional fluctuations from donor to donor, of the Mn^{2+} ions comprising the BMP. For ions with non-zero magnetization in their ground state, such as Mn^{2+}, the mean value $<M_{loc}>$ is non-zero even with $H=0$, due to thermal fluctuations. Hence, the Zeeman splitting of the electron state is non-zero at $H=0$. On the other hand, if the ions exhibit no magnetization in their ground state, like Fe^{2+} ions, then $M_{loc}=0$ and there should be no Zeeman splitting at $H=0$. As H is applied, the usual linear-in-H Zeeman splitting should appear. The only known BMP, at least in connection with RS processes, involves donors; analogous phenomena due to acceptor-bound carriers are possible, in principle. RS phenomena arising from both types of ions, i.e., Mn^{2+}, Fe^{2+}, will be discussed in Sects. 4, 5 and 6, respectively.

In the antiferromagnetic phase, finally, the Mn^{2+}-Mn^{2+} interactions lead to quantized collective magnetic excitations, or **magnons**, with well-defined dispersion curves. The optical branches of such dispersion curves and their T-, x-, H-dependences have been very elegantly studied in DMS through RS, as we shall see later. It turns out, from Bloch's equation, that with $H \neq 0$ the long-wavelength magnon splits into two components with frequencies[6]

$$\omega_M^{\pm} = |\gamma|(H_A^2 + 2H_A H_E)^{1/2} \pm \gamma H, \tag{3}$$

where γ the gyromagnetic ratio of electrons, H_A the anisotropy field, and $H_E = \lambda M_S$ the isotropic exchange field; λ is an exchange constant and M_S the saturation value of the "sub-lattice magnetization". For $H=0$ Eq. (3) gives the bare frequency ω_M of the long-wavelength optical magnon. The parameters H_A, H_E are constants which can be measured independently from T_N, and the RS data for ω_M or $\omega_M^{\pm}$.

2.4 Lattice Dynamical Properties

The lattice dynamical properties of the end members AB in either ZB or W structure are well known from theoretical and experimental sources. The phonons of interest in first-order RS are those from the center of the Brillouin zone ($q \approx 0$) and obey the F_1 ($A_1 + E_1 + 2E_2$) symmetry of the point group T_d (C_{6v}) which is appropriate for ZB (W) structure. Three types of modes (F_1, A_1, E_1) are of polar character and exhibit different frequencies ω^T (ω^L) when the phonon polarization is transverse (longitudinal) relative to their wavevector **q**. Frequencies of the q≈0 optical phonons for the AB and MnB compounds are compiled in Table II.

When in small amounts (x<0.005), the Mn^{2+} ions can be treated as if they are isolated impurities occupying cation sites. Using Green's function techniques, it can then be shown that the impure AB system exhibits the so-called **AB-like modes** with frequencies $\omega_{AB}^{T,L}$ close to those of pure AB, and with an upward (downward) tendency, depending on whether $m_{Mn} < m_A$ ($> m_A$). In

Table II. Optical phonon frequencies (T,L) of AB and MnB, and Mn (A)-induced mode frequencies in AB (MnB) lattices (cm^{-1}). Type of mode behaviour 1 , 2, mixed (m). Structure:ZB, W. Values in parenthesis are averages of A_1, E_1 mode frequencies.

A	B	ω_M	ω_{AB}^T	ω_{AB}^L	ω_A	ω_{MnB}^T	ω_{MnB}^L	Type Struct.	T(K)	Ref.
	Te	195	147	173	147	185	216	2, ZB	80	9
Cd	Se	222.5	(171.5)	(213)	180	219.5	257	2, ZB	5	41
	S		(238)	(303)		293* / 286+	361* / 343+	W	300	49,68* / 69+
	Te	208	181	210	171	185	216	m, ZB	80	9
Zn	Se	233	206	253	197	217.5	255	m, ZB	300	41
								m, W	300	58
	S	222	271	349	197	293	361	1, ZB	300	68
						286	343	1, W	300	69
Hg	Te	191	119	137	125	185	216	2, ZB	77	73,74

addition, there appears an impurity-dependent localized mode at ω_{Mn} with the following characteristics[8]: (i) If $m_{Mn} < m_A$, the mode exhibits a frequency $\omega_{Mn} > \omega_{AB}^L(q)$ and is called a **local mode**. (ii) If $m_{Mn} > m_A$, it is called a **gap mode** with frequency $\omega_{Mn} < \omega_{AB}^T(q)$, but $\omega_{Mn} > \omega_{AB}^{ac}(q)$, where ω_{AB}^{ac} is the highest frequency of the acoustical branches of AB. (iii) It is possible, when m_A is not much different than m_{Mn}, to have **band modes** lying in the band of frequencies spanned by either the optical or the acoustical lattice modes. They are induced by impurities but otherwise are lattice modes off the center of the Brillouin zone, with non-local character. A **resonant mode** is a band mode with locally enhanced amplitude.

As the concentration of Mn^{2+} ions increases the frequencies $\omega_{AB}^{T,L}(q=0)$ and ω_{Mn} vary with x in a continuous but rather complicated way, since the lattice now is in a disordered state. From all relevant experimental and theoretical studies known thus far, it has been firmly established that as x varies from 0 to 1, the modes with frequencies $\omega_{AB}^{T,L}$, ω_{Mn} transform to analogous modes with frequencies $\omega_{MnB}^{T,L}$, ω_{A}, not necessarily one to one. The modes at $\omega_{MnB}^{T,L}$ are the so-called **MnB-like Modes** and correspond to the $q\approx0$ optical phonons of the MnB lattice (ZB or W); ω_{A} is the frequency of the localized mode of impurity A in the MnB lattice. The evolution from $\omega_{AB}^{T,L}$, ω_{Mn} to $\omega_{MnB}^{T,L}$, ω_{A} follows (at most) four branches which by tradition are labeled TO2, LO2, TO1, LO1 in order of increasing frequency[8]. Depending on which end frequencies are linked by which branch, there are three types of mode behaviour common to the majority of mixed crystals, DMS or otherwise[8]:

(i)   **One-mode behaviour**, where two of the four branches nearly coincide, and

$$\omega_{AB}^{T} \longrightarrow \omega_{MnB}^{T} \ , \quad \omega_{AB}^{L} \longrightarrow \omega_{MnB}^{L} \ , \quad \omega_{Mn} \longrightarrow \omega_{A} \ . \tag{4a}$$

(ii)   **Two-mode behaviour**,

$$\left. \begin{array}{c} \omega_{AB}^{L} \\ \omega_{AB}^{T} \end{array} \right\} \longrightarrow \omega_{A} \ , \quad \omega_{Mn} \longrightarrow \left\{ \begin{array}{c} \omega_{MnB}^{L} \\ \omega_{MnB}^{T} \end{array} \right. \tag{4b}$$

where the double arrows designate merging of the two branches into a single (impurity)   frequency.

(iii)   **Mixed-mode behaviour**, for which

$$\omega_{AB}^{L} \longrightarrow \omega_{MnB}^{L} \ , \quad \omega_{AB}^{T} \longrightarrow \omega_{A} \ , \quad \omega_{Mn} \longrightarrow \left\{ \begin{array}{c} \omega_{MnB}^{T} \\ \omega_{A} \end{array} \right. , \tag{4c}$$

with $\omega_{Mn} < \omega_{AB}^{T}$ and $\omega_{MnB}^{T} < \omega_{A}$, or vice versa $(>)$.

Various lattice dynamical models have been proposed in the literature, in connection with the three types of mode behaviour. The most recent and reasonably successful in its compatibility with the experimental data, is the **modified random element isodisplacement model** (MREI, Ref. 9). It is a modification of an earlier model by Genzel et. al.[10] which is based on macroscopic parameters of AB and MnB and makes use of local field arguments. In MREI it is assumed that for $q \approx 0$ all anions vibrate in phase and with the same amplitude (and likewise for cations) and that the net force experienced by each ion is a statistical average of its interaction with all its neighbors. Furthermore, the MREI model allows to combine data on the frequencies versus x for two DMS with a common anion, i.e., (A,B)Mn and (A′,B)Mn. In this way, it is possible to incorporate second-neighbor force constants and to limit the fitting parameters only to the four frequencies $\omega_{BMn}^{T, L}$ and ω_A, $\omega_{A'}$ for cubic MnB, even if the latter corresponds to a purely "hypothetical" structure like MnTe. Such information is included in Table II.

One final point of interest in relation to disordered lattices, is the reduction of translational symmetry, for large values of x; this, in turn, invalidates the momentum conservation principle in first-order Raman processes. As a result, normally forbidden first-order features are observed in the Raman spectra, due to **disorder-activated processes**.

3. BASIC CONCEPTS OF RAMAN SPECTROSCOPY

The usual optical techniques employed in pure and applied materials research are, photoluminescence, photoacoustic absorption, ellipsometry, non-linear spectroscopy, Raman spectroscopy, and IR spectroscopy. Raman spectroscopy, in particular, is a powerful experimental technique which can provide answers to questions as:

State, phase, and quality of materials, crystallinity, polycrystallinity, microstructures, etc.

Defect densities, and carrier characteristics, including concentration, mobility, life-time, surface depletion widths.

Composition of ternary and quaternary compounds, in bulk or hetero-structure systems.

Residual strains following material growth or treatment (implantation, thermal or laser annealing, polishing, etching) in bulk or in epilayers, and mismatched superlattices.

Collective excitations, and their interactions with lattice

and impurities.

Effects of external perturbations like temperature, pressure, electric and magnetic fields, etc.

In what follows, the necessary concepts and terminology from RS is presented in a brief and rather descriptive way, since the subject is exaustively covered in the literature[6,11,12].

3.1   Introduction to RS Processes

First-order RS is an inelastic process which involves the following three steps in arbitrary time sequence: (i) Through an electron-radiation interaction, an incident photon of frequency ω_1 and wavevector $\mathbf{k}_1$ raises the scattering medium (e.g., a crystalline material with lattice constant a) from the ground state to a higher virtual state $|\alpha>$ corresponding to an electron-hole pair, with excitation frequency ω_α. (ii) A second electron-radiation interaction causes the medium to emit (scatter) a photon ω_s, $\mathbf{k}_s$ and be brought to a new virtual pair state $|\beta>$ with excitation frequency ω_β, differing from $|\alpha>$ by the presence of a collective excitation within the medium (e.g., a phonon or magnon of frequency ω_j and wavevector $\mathbf{q}_j$). (iii) Through an electron-lattice interaction, the pair recombines, leaving the medium with one excitation, ω_j, $\mathbf{q}_j$. Conservation of energy requires that $\omega_s = \omega_1 \pm \omega_j$ where the $-(+)$ sign corresponds to production (destruction) of the phonon according to a Stokes (anti-Stokes), or S (A), process. **Conservation of momentum**, expressed by $\mathbf{k}_s = \mathbf{k}_1 \pm \mathbf{q}_j$, **is also valid provided the scattering medium exhibits translational symmetry.** The latter depends on the crystallographic quality of the lattice.

The scattering volume is determined by the region of the medium which is probed by the incident beam. If the medium is absorbent in the region of ω_1 and/or ω_s the scattering volume is limited to the light penetration depth δ. The scattering geometry in this case requires that $-\mathbf{k}_s \| \mathbf{k}_1$ and $q_j \simeq 2k_1$ and is known as **backward (or 180^0) scattering geometry**. For typical semiconducting materials the value of δ varies from 10^2 to 10^4 Å depending on ω. This fact introduces significant RS experimental possibilities, since detailed probing at variable depths under the surface now becomes possible by appropriate choice of the laser frequency alone.

In the case of RS by phonons, conservation of energy and momentum

238

require that only long-wavelength ($q_j \approx 0$) optical phonons may participate. Furthermore, crystal symmetry considerations impose additional restrictions. As a result, only some of these $q_j \approx 0$ optical phonons have the appropriate symmetry to allow RS in first order. Accordingly, these phonons are defined as **Raman active**. In more physical language, a mode is Raman active if its symmetry allows the medium polarizability α (a second-rank tensor) to be modulated by the **mode normal coordinate** u_j. In other words, the condition for Raman activity is $a_j = (\partial\alpha/\partial u_j) \neq 0$ and this depends entirely on the symmetry of the mode. The quantity a_j defines phenomenologically the so-called **Raman tensor** of the mode, and determines the scattering intensities through $|a_j|^2$. The matrix representation of a_j for the Raman active phonons of DMS in ZB (T_d) and W (C_{6v}) are given in Table III. The polarization selection rules for

Table III. Raman tensors for phonons from ZB and W structures. Symmetric and antisymmetric parts are shown separately, each accompanied by the appropriate basis function X, XY, R_x= YZ'-ZY', etc. (after Ref. 6).

A₁	E	F₂	F₁
$\begin{vmatrix} a & 0 & 0 \\ 0 & a & 0 \\ 0 & 0 & a \end{vmatrix}$	$\begin{vmatrix} -b & 0 & 0 \\ 0 & -b & 0 \\ 0 & 0 & 2b \end{vmatrix}, 2Z^2 - X^2 - Y^2$	$\begin{vmatrix} 0 & 0 & 0 \\ 0 & 0 & d \\ 0 & d & 0 \end{vmatrix}, X$	$\begin{vmatrix} 0 & 0 & 0 \\ 0 & 0 & c \\ 0 & -c & 0 \end{vmatrix}, R_x$
	$\begin{vmatrix} \sqrt{3}b & 0 & 0 \\ 0 & -\sqrt{3}b & 0 \\ 0 & 0 & 0 \end{vmatrix}, \sqrt{3}(X^2 - Y^2)$	$\begin{vmatrix} 0 & 0 & d \\ 0 & 0 & 0 \\ d & 0 & 0 \end{vmatrix}, Y$	$\begin{vmatrix} 0 & 0 & -c \\ 0 & 0 & 0 \\ c & 0 & 0 \end{vmatrix}, R_y$
		$\begin{vmatrix} 0 & d & 0 \\ d & 0 & 0 \\ 0 & 0 & 0 \end{vmatrix}, Z$	$\begin{vmatrix} 0 & c & 0 \\ -c & 0 & 0 \\ 0 & 0 & 0 \end{vmatrix}, R_z$

Symmetric			Antisymmetric	
A₁	E₁	E₂	A₂	E₁
$\begin{vmatrix} p & 0 & 0 \\ 0 & p & 0 \\ 0 & 0 & q \end{vmatrix}, Z$	$\begin{vmatrix} 0 & 0 & r \\ 0 & 0 & 0 \\ r & 0 & 0 \end{vmatrix}, X$	$\begin{vmatrix} s & 0 & 0 \\ 0 & -s & 0 \\ 0 & 0 & 0 \end{vmatrix}, X^2 - Y^2$	$\begin{vmatrix} 0 & a & 0 \\ -a & 0 & 0 \\ 0 & 0 & 0 \end{vmatrix}, R_z$	$\begin{vmatrix} 0 & 0 & -b \\ 0 & 0 & 0 \\ b & 0 & 0 \end{vmatrix}, X, R_x$
	$\begin{vmatrix} 0 & 0 & 0 \\ 0 & 0 & r \\ 0 & r & 0 \end{vmatrix}, Y$	$\begin{vmatrix} 0 & -s & 0 \\ -s & 0 & 0 \\ 0 & 0 & 0 \end{vmatrix}, -2XY$		$\begin{vmatrix} 0 & 0 & 0 \\ 0 & 0 & -b \\ 0 & b & 0 \end{vmatrix}, Y, R_y$

observing RS are based on these matrices, and are given in the standard coded form x(yz)x̄ ; this means that for incident (scattered) light along x(-x) with polarizations y(z) there will be RS by a particular phonon (e.g., E_1 of C_{6v}) only if its Raman matrix exhibits a non-zero component in the yz position (e.g., r, scattering intensity $|r|^2$, phonon polarization z). In the case of RS by magnetic excitations, such as SF transitions in Zeeman-split multiplets, the selection rules require circular polarizations for the incident and scattered light and antisymmetric Raman tensor components. The difference $\omega_1 - \omega_s$ in this case corresponds to the energy of the electronic transition.

Besides first-order processes it is also possible to have second- or higher-order RS, involving two or more excitations in each process. The excitations may have equal (**overtone RS**) or unequal (**combination RS**) frequencies, and may belong to any point of the Brillouin zone, provided some selection rules are obeyed.

3.2 Morphology of Raman Spectra

From spectroscopic point of view, the Stokes Raman band associated with a collective excitation at ω_j is detected in the region of the absolute scattered frequency ω_s or, more often, in the region of the relative frequency ω_j, provided the frequency scale is zeroed at ω_1. The scattering intensity at ω_s can be shown to be

$$I(\omega_s) \sim \sum_{\alpha,\beta} \frac{\omega_1^4}{(\omega_1-\omega_\alpha)^2(\omega_s-\omega_\beta)^2} , \qquad (5)$$

where ω_α and ω_β are the excitation frequencies of the electron-hole pairs in states $|\alpha>$ and $|\beta>$ involved in steps (i) and (ii) (Sect. 3.1) and the summation runs over all such states of the medium. The ratio I_S/I_A of the intensities of S/A is a measure of temperature of the scattering volume [$\sim \exp(\hbar\omega_j/kT)$, k = Boltzmann's constant]. In the case of perfect lattices and scattering by phonons the bandshape is a symmetric Lorentzian centered about ω_s (or ω_j). The corresponding full width at half maximum Γ_j is a measure of the phonon lifetime ($\sim 1/\Gamma_j$), that is, of lattice anharmonicity. If, on the other hand, the lattice is not perfect, the bandshape is a superposition of many Lorentzians, one for each (non-monochromatic) mode in the region $0 \leq q_j \leq 2\pi/L$, where L is the average range of lattice order. The net bandshape is asymmetrically broadened and its peak is shifted relative to the $q_j \simeq 0$ frequency of the perfect lattice. Such morphological characteristics are not interpreted in the same simple way as in the case of perfect crystals but, on the other hand, hold the key to characterization of the material quality.

3.3   Resonant Raman Scattering

The concept of **Resonant Raman Scattering** is of great theoretical and practical importance, as will become evident in subsequent chapters[11,12]. For a random choice of the excitation frequency ω_1, the two excited states of the medium which are involved in steps (i) and (ii) of the scattering process (Sect. 3.1) are virtual states in general. The corresponding effect is the **non-resonant Raman scattering** and its intensity is determined primarily by terms of the form of Eq. (5). However, ω_1 or ω_s may happen to be near one of the electronic transitions of the crystal (e.g., the fundamental energy gap, E_g). In that case one of the terms in Eq. (5) will have its first or second factor in the denominator diverging, thus leading to an increase of scattering intensity due to an **in-going** or **out-going single resonance,** respectively. Resonant scattering can become stronger than non-resonant by several orders of magnitude, thus allowing weak interactions to become observable.

Tuning ω_1 to the electronic gaps of the medium is nowadays an easy task due to the availability of lasers with a broad selection of single frequencies. Furthermore, it is possible to tune the electronic gap to the fixed frequency ω_1 by subjecting the crystal to variable temperature, hydrostatic pressure, uniaxial stress X, electric or magnetic fields. The case of **doubly-resonant RS** has also been demonstrated experimentally and theoretically in recent years in bulk and superlattice structures[13-16]. For instance, under a uniaxial stress of 6 Kb parallel to [111], the top of the valence band of GaAs develops a splitting equal to ω^L. Choosing ω_1 to match the larger $(E_g^{3/2})$ of the two gaps $E_g^{3/2}$ and $E_g^{1/2}$, leads essentially to double divergence of Eq. (5) and therefore to double resonance[13]. The effect concerns LO and TO phonons, in first- and higher-order scattering[14]. In superlattices, the conditions for doubly-resonant RS can be reached by either controlling the growth conditions[15] or by applying appropriate DC electric field to the superlattice[16]. It is also possible to reach doubly-resonant RS by magnetically tuning the Zeeman-split levels or the Landau levels.

3.4   Experimental Techniques

The main experimental approaches to Raman spectroscopy are outlined next. They have been developed in recent years to meet the ever increasing needs of RS users in all fields of interest, particularly in materials research.

3.4.1 Frequency scanning mode

The experimental setup shown in Fig. 5 has been developed and used for most routine applications in the last 25 years. The excitation frequency is provided by a cw laser. The beam is first filtered (F), polarized along the desired direction and then focused onto the sample. The scattered light is collected and focused on the entrance slit of the spectrometer, after been polarized and scrambled, to account for spectrometer anisotropies. A double or even triple spectrometer is necessary to achieve high rejection of stray light before the Raman signal reaches the detector (a cooled photomultiplier). The dispersive media of the spectrometer are high-quality holographically ruled gratings. The dispersed light is swept across the exit slit of the spectrometer and is recorded in the frequency domain (frequency scanning mode). Acquisition of the spectra may last anything from few minutes to tens of hours, depending on the scattering strength of the medium and the spectral quality desired. The main disadvantages of the frequency scanning mode are the long duration, and the fact that various regions of the spectrum are detected at different times. This may pose difficulties in case of photosensitive media or when accidental variation of experimental conditions occur during scanning. Both these problems can be remedied by use of the following technique.

3.4.2 Diode array detection systems

Suppose the exit slit of the spectrometer in Fig. 5 is kept wide open and the photomultiplier tube is replace by a one-dimensional array of photosensitive diodes. Each diode receives radiation from only an increment $\Delta\omega_s$ of the spectrum. The corresponding signal is stored in one channel of a multichannel analyser. This type of detector (**optical multichannel analyser or OMA**) allows the synchronous collection of the entire spectrum in a predetermined spectral range, much faster than in the frequency scanning mode, but with some loss of resolution. A state-of-art development of such devices involves two-dimensional diode arrays supported by microelectronic processors (**Charge Coupled Device** or **CCD**). The latter allow, (i) the synchronous mapping of the same spectral range originating from different scattering points at predetermined spatial resolution, and (ii) repetition of the spectrum from a fixed point of the scatterer at predetermined time

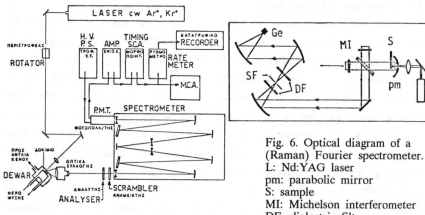

Fig. 5. Typical experimental setup for laser Raman spectroscopy.

Fig. 6. Optical diagram of a (Raman) Fourier spectrometer.
L: Nd:YAG laser
pm: parabolic mirror
S: sample
MI: Michelson interferometer
DF: dielectric filters
SP: spatial filter
Ge: Ge detector (after Ref. 17)

resolution. A comparative review of these and more recent devices operating on the same principle is presented by J. C. Tsang in Ref. 12, Vol. V, p. 233.

3.4.3 Fourier transform Raman spectroscopy (FTRS)

The use of laser excitation frequencies in the visible or UV often leads to one or both of the following difficulties, (i) the medium may strongly fluoresce in the Stokes region, and (ii) the incident photons may

cause degradation of the medium because of photochemical effects. Both these problems can be circumvented by using excitation lines in the near IR, such as the line at 1.06 μm of the Nd:YAG laser. Although the ω_1^4 factor of Eq. (5) is against such a choice, the possibility of using higher intensities without thermal or photochemical degradation effects overcomes the low scattering efficiency. More importantly, this particular photon energy is below the threshold of electronic excitations for most media and this results in fluorescence-free spectra.

From experimental point of view, it has been recognized in the last few years, that the most appropriate technique for obtaining Raman spectra in the IR is through **Fourier Transform (FT) spectroscopy** , with only minor modifications. Figure 6 shows the setup used by a number of research groups[17]. The scattered light is first collimated and then modulated by a Michelson interferometer. The intense background of semi-elastically scattered light is practically eliminated by the use of at least two dielectric filters, combined with spatial filtering. The Ge detector sees the Fourier spectrum which is then processed with standard FT methods to yield the direct Raman spectrum. It is emphasized that FTRS is most appropriate in cases where conventional Raman spectroscopy in the visible is hopeless, because of strong fluorescence. The technique has been developed for, and used in connection with RS studies in biological materials but could equally effectively applied to solid state materials like DMS where strong fluorescence often creates difficulties.

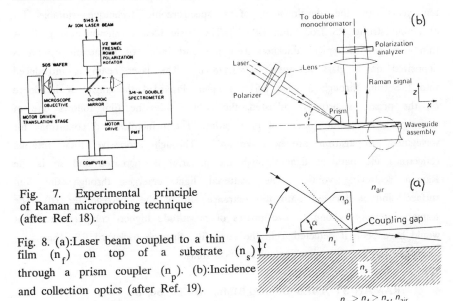

Fig. 7. Experimental principle of Raman microprobing technique (after Ref. 18).

Fig. 8. (a):Laser beam coupled to a thin film (n_f) on top of a substrate (n_s) through a prism coupler (n_p). (b):Incidence and collection optics (after Ref. 19).

3.4.4 Raman microprobing

Microprobing techniques have been used extensively in recent years to study

RS of minute media with physical dimensions approaching the laser wavelength. Thus, **Raman microprobing** offers a powerful non-desctructive microanalytical option. The principle of operation is shown in the simplified system of Fig. 7 (Ref. 18). The idea is to focus the laser beam onto a sample spot with diameter of the order of 1μm, using a microscope objective. The same objective collects the scattered light which, through a beam splitter, is focused on the entrance slit of the spectrometer. In this way it becomes possible to probe sample spots with spatial resolution of ≤1 μm and laser powers of a few mW only. The scattering spots are selected visually using the same microscope objective and may belong to large crystals, poly- or micro-crystals, individual small particles, single bacterial cells, macromolecules etc. The Raman signals are detected either in the frequency scanning mode or by an OMA system.

3.4.5 Waveguide Raman spectroscopy

A straightforward application of waveguide optics has led, in the last ten years, to the development of a spectroscopic technique suitable for observing strong RS from thin films[19]. The basic idea is shown in Fig. 8. A thin film (a) of typical thickness t~1 μm and index of refraction $n_f = 1.6$ is deposited on a glass substrate ($n_g = 1.478 < n_f$). The laser beam is introduced into the film through a glass prism coupler. Provided the angle of incidence on the prism is properly adjusted, the light inside the film follows one of the zig-zag waveguide modes of propagation. The principles and conditions for waveguide propagation are well known[20]. Through successive total internal reflections the path of light through the material is increased and so is the Raman scattering volume. The scattered light emerges through the film surface, and is focused onto the entrance slit of the spectrometer (b). The Raman intensity is one to two orders of magnitude higher, compared to what would have been if detected from the same film through backward RS. The technique has not yet been applied to DMS films.

4.   RAMAN STUDIES OF (Cd,B)Mn, B=Te, Se, S

Most of the Raman studies in DMS reported to date involve the wide-gap semiconductors (Cd,B)Mn, B=Te, Se, S. Raman scattering by phonons and magnetic excitations in these three compounds is reviewed separately.

4.1 (Cd,Te)Mn

4.1.1 Raman scattering by phonons

The first- and second-order phonon Raman spectra of this ZB structure have been studied by Venugopalan et. al.[31] and by Picquart et. al.[22] A typical spectrum is shown in Fig. 9. The upper part, $\omega > 130 \text{cm}^{-1}$, corresponds to the $q \approx 0$ optical modes; their x dependence, shown in Fig. 10, reveals two-mode behaviour. The solid lines are the results of MREI model and are in good agreement with the data. Their extrapolation to $x = 1$ specifies the phonon $(\omega_{MnTe}^{T,L})$ and impurity (ω_{Cd}) frequencies of MnTe at the "hypothetical" ZB (Td) structure. The fitting has been done jointly with the data of (Zn,Te)Mn which

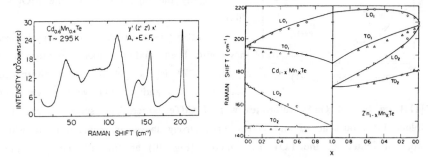

Fig. 9. Phonon Raman spectrum of (Cd,Te)Mn, x=0.4, at 1.833 eV (Ref. 21).
Fig. 10. The q≈0 optical phonon frequencies of (Cd,Te)Mn and (Zn,Te)Mn vs. x at 80K. Solid lines are the results of the MREI model (Ref. 9).

will be discussed in Sect. 5.1. In this way the only fitting parameters required are the frequencies $\omega_{MnTe}^{T,L}$, ω_{Cd}, ω_{Zn} (Table II).The lower part of the spectrum in Fig. 9 is attributed to disorder-induced first-order RS by $q \neq 0$ phonons. Comparison to neutron scattering data indicates that the peaks at ~ 41 and $\sim 110 \text{ cm}^{-1}$ correspond to transverse acoustical (TA) and longitudinal acoustical (LA) phonons, although other contributions to the latter are not excluded[21].

Resonant RS has also been seen to affect the intensity of the allowed TO and forbidden LO modes and of the various LO overtones and combinations, for $x \leq 0.3$, as ω_1 approaches Eg(x) from below (TO, LO) and above (LO) [21]. The resonances are mainly attributed to impurity effects for the LO, and to deformation potential interactions for the TO mode (Ref. 12, Vol. II).

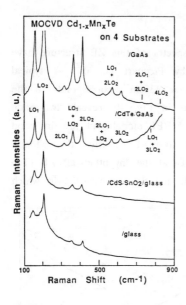

MOCVD $Cd_{1-x}Mn_xTe$ on 4 Substrates

Fig. 11. Raman spectra at 80K from (Cd,Te)Mn grown on various substrates (after Ref. 23).

Phonon RS was reported recently by Feng et. al.in high quality (Cd,Te)Mn films grown by MOCVD on various substrates. Data obtained by an OMA are displayed in Fig. 11 and show clearly RS by LO1, LO2 and their overtones and combinations up to fourth order. They were obtained without resonance conditions and signify the high quality of the MOCVD films.

MBE-grown (Cd,Te)Mn superlattices have also been studied under resonance[24]. A total of 270 layers (59 Å each) alternating between $x_1 = 0.5$ and $x_2 = 0.11$ were grown along [111] on top of GaAs substrates with a CdTe buffer. The spectra reveal two folded LA modes at 9.2 and 11 cm^{-1}, and strong above-gap resonant bands which were attributed to the LO1, LO2 modes (Fig. 10) at 167, 199 cm^{-1}, to their second and third overtones and their combinations.

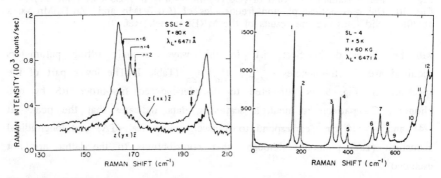

Fig. 12. (a):RS by optical phonons in the [001] CdTe/(Cd,Te)Mn, x=0.25, superlattice. Numbers designate optical phonons confined in the CdTe layer (after Ref. 25). (b): RS from interface optical modes (IF) in a [111] $(Cd,Te)Mn_{0.11}/(Cd,Te)Mn_{0.50}$ superlattice. 1, 2:Fundamental IF modes, 3-12: overtones and combinations of 1, 2 (after Refs. 25, 24).

RS studies from a variety of CdTe/(Cd,Te)Mn superlattices along [001],

[111] were reported by Suh et. al.[25], with $0.50 \leq x \leq 0.75$. Significant differences were found in the interpretation of the Raman bands observed from [001] and [111]. In the [001] CdTe/(Cd,Te)Mn, $x = 0.25$, superlattice the spectra consist of two pairs of folded LA phonons at 9.2, 11, 19, 20.7 cm^{-1}, and also of optical and interface (IF) modes. Fig. 12(a) summarizes these observations. The major peak at 165 cm^{-1} is the LO1 mode of the DMS layer which is a **propagating mode** (i.e., it can propagate through the CdTe layer too). The $n = 2$, 4, 6 bands correspond to the **confined modes** of the CdTe layer (they cannot penetrate into the DMS layer). The IF band corresponds to the **interface mode**. It characterizes both layers, propagates parallel to their interface and its frequency depends on ω_l. The peak at 200 cm^{-1} is the LO2 mode of the DMS layer. Selective tuning of ω_l to the gaps of either CdTe or (Cd,Te)Mn, $x = 0.25$, leads to independent resonance enchancement of the peaks of either layer. On the other hand, the [111] (Cd,Te)Mn$_{0.11}$/(Cd,Te)Mn$_{0.50}$ superlattice gave the Raman spectrum shown in Fig. 12(b); it consists of two fundamental modes (1, 2), the same ones identified as LO1, LO2 in Ref. 24. However, the way their frequencies depend on ω_l suggests that they actually correspond to allowed fundamental IF modes of this particular system. The remaining peaks (3-12) are resonance enhanced overtones and combinations up to fourth order. Out-going resonance is achieved by H tuning the split exciton levels, and explained by invoking a Fröhlich mechanism[24].

Before closing this subsection a brief reference should be made to the effects of hydrostatic pressure on the phonon RS spectra of (Cd,Te)Mn. Related Raman studies[26] confirm that these DMS switch from ZB to NaCl structure at pressures P_t between 25 and 35 Kb; it is also found that P_t shows little dependence on x (0-0.7). The spectra weaken and finally disappear as P approaches P_t, consistently with the NaCl structure. The optical phonon frequencies increase with P as usual but the defect-activated TA band appears to soften. Hysteresis phenomena are present and also indications that the increase (decrease) in the LO1-TO1 (LO1-TO2) splittings observed after cycling of the pressure, is due to redistribution of the mode effective charge. It is speculated that this is due to an irreversible change in the local atomic arrangement as a result of the phase transition.

4.1.2 Magnetic excitations in the paramagnetic phase

248

Both, paramagnetic and magnetically ordered phases of (Cd,Te)Mn exhibit a variety of first- and higher-order RS by magnetic excitations which can be probed very effectively by appropriate choice of all available tunable parameters, i.e., T, H, x_{Mn}, x_{donor}, ω_l and light polarizations. The necessary electronic transitions have been outlined in Sect. 2.3. We start here with RS in the paramagnetic phase.

The classical magnetic interaction leading to RS involves the Zeeman-split sublevels of the six-fold degenerate Mn^{2+} ground state (Fig. 2b). The localized excitation at ω_{PM} appears in the Raman spectrum in first order and overtones (Fig. 13). Details about the circular polarization selection rules and the intermediate electronic transitions involved can be found in Refs. 27, 28. The first-order effect (Fig. 13a) is equivalent to a Raman-EPR process, while the overtone effects (Fig. 13b) are attributed to $\Delta M_S = 1,2,..$ transitions within pairs of antiferromagnetically coupled Mn^{2+} ions (Sect. 2.3). The importance of in- or out-going resonances in these experiments cannot be overemphasized. Due to such resonance conditions (i.e., tuning of T, H, ω_l) it has become possible to observe even higher order

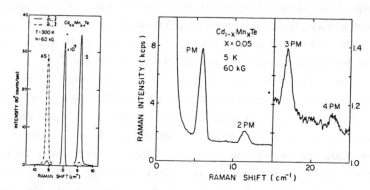

Fig. 13. Stokes, anti-Stokes RS by electronic transitions within the Zeeman-split multiplets. (a):First order at ω_{PM} ($\Delta m_j = \pm 1$, after Ref. 27). (b):Overtones at $n\omega_{PM}$ ($\Delta M_S = n$, after Ref. 28).

processess (Fig. 14) where the excitations involve combinations, in first[27,28] and second order[28], of ω_{PM} with the LO1,' LO2 (Fig. 10) mode frequencies. Fröhlich interactions provide the necessary coupling between the LO mode and the electronic transition. A variety of field-split exciton sublevels can serve as tunable intermediate levels to assure resonance conditions[28]. In physical terms, the Raman band at LO1-2PM of Fig. 14 for instance, designates a process whereby one LO1 mode is excited while a Mn^{2+}-Mn^{2+} pair in its $\Delta M_S = 2$ excited state, returns to its ground state releasing energy $2\hbar\omega_{PM}$. By invoking Zeeman-split excitonic levels as tunable intermediate states, Limmer et. al., have likewise studied recently the resonance profiles of LO1, TO1 modes of (Cd,Te)Mn, x=0.2, and described the results using Fröhlich interaction and deformation potential mechanisms[29].

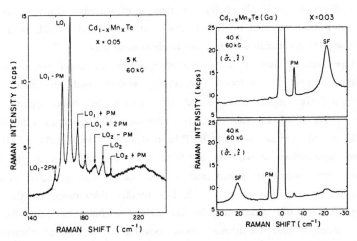

Fig. 14. RS combination processes involving LO phonons and electronic transitions within the Zeeman multiplet of Mn^{2+} (PM), with $\omega = 1.648$ eV (after Ref. 28).

Fig. 15. Raman processes for x=0.03 involving electronic transitions within Mn^{2+} (PM) as in Fig. 13, and SF of electrons bound to Ga donors (Ref. 31).

Equally impressive is the RS process which involves spin-flip ($\Delta m_j = \pm 1$) of either s-like band electrons[30] or donor-bound electrons[31]. The RS band appears at $\hbar\omega_{SF} = g_{eff}\mu_B H \simeq \Delta g\mu_B H$, and is shown in Fig. 15 for Ga donors in (Cd,Te)Mn, x=0.03 (Ref. 31). It should be distinguished from the PM band discussed above, i.e., the one due to electronic transitions in the Zeeman multiplet of the Mn^{2+} ground state. The large value of the frequency ω_{SF} is due to the large value of Δg which, according to Eq. (2), depends on the magnetization, i.e., on H, x and T. Detailed measurements of $\omega_{SF}(H,x,T)$ are in agreement with theory, for x≤0.15 at least[32]. Bound magnetic polarons also produce RS at zero field. Its frequency $\omega_{SF}(0,x,T)$ is observed around 4 cm^{-1} and exhibits weak dependence on x (≤ 0.1) and T ($\leq 20K$) in accordance with the molecular field approximation[32].

Spin-flip RS by s-like band electrons has been observed by Peterson et. al.[30] who studied its frequency ω_{SF}^c in (Cd,Te)Mn (no donors) as a function of T (5-50K), H (0-60 kG), x (0.01-0.18) and light polarization. Deviations from theory for x≤0.18 are attributed to clustering of Mn^{2+} ions, as a consequence of the increasing antiferromagnetic interaction among them. The value $N_o\alpha = 180$ meV (N_o being the density of cations) which is obtained from the saturation value of $\omega_{SF}^c(H) \cong 37$ cm^{-1} at T=5K and x=0.01, is in good agreement with that calculated from magneto-absorption data.

More recently Suh et. al.[33] reported spin-flip RS under resonance with the excitonic levels in In-doped (Cd,Te)Mn, x=0.16, and modulation-doped In:(Cd,Te)Mn/CdTe superlattices, successfully grown by photoassisted MBE techniques along [001], on top of CdTe substrates. The epilayers exhibited at 5K the PM and SF Raman bands, the BMP band at 9 cm^{-1} (H=0) and the LO1+PM band. The superlattices showed no SF band when the thickness of the CdTe layer was large enough to attract and accomodate nearly all donor electrons thus reducing to zero the effect of Δg. For smaller thicknesses, on the other hand, the SF band did appear and in fact yielded usefull information from its frequency, strength and bandshape, about possible influences of dimensional constraints on the magnetic phase transition. Additional support to this argument is provided by the absence of any Raman activity due to confined, propagating, or IF magnons, which are expected to exist in superlattices such as $(Cd,Te)Mn_{0.11}/(Cd,Te)Mn_{0.50}$. This particular superlattice, grown along [111], has been studied by Suh et. al.[25] under resonance conditions with H=0. In addition to the vibrational Raman bands, already discussed in Sect. 4.1.1, the spectra also include the PM Raman band and its overtones, up to $5\omega_{PM}$. The combinations of the IF1 and IF2 modes with the PM excitation via a Fröhlich interaction were also seen, but no evidence for confined, propagating, or IF magnons. A band around 9.5 cm^{-1} at 1.7K was attributed, to the S=0 →S=1 transition of the Mn^{2+}-Mn^{2+} pair, as evidenced by its T- and H-dependence, and by its polarization characteristics. The energy of this excitation is $\hbar\omega_{NN} = 2|J_{NN}|$; the value of the exchange constant J_{NN} determined in this way is equal to -6.8K, in agreement with independent determinations.

4.1.3 Magnetic excitations in the ordered phases

It is generally accepted that for $0.17 \leq x \leq 0.6$ and $T < T_g(x)$, the (Cd,Te)Mn compounds exhibit a spin-glass phase, while for $0.6 < x \leq 0.77$ and $T < T_N(x)$ the phase is antiferromagnetic (usually in the literature the distinction between T_g and T_N is lost and in fact is not yet fully understood). For $T > T_N(x)$ and $0 \leq x \leq 0.77$ the phase is paramagnetic. In both phases the ordering of neighbour Mn^{2+} ions is antiferromagnetic of short, and not so short range respectively, and justifies the presence of magnons. RS by the latter has been studied in Ref. 21. Spectra are shown in Fig. 16; in (a)

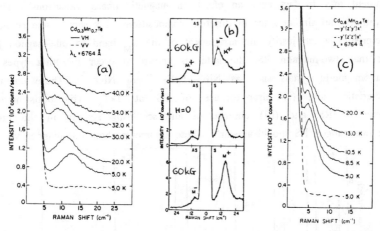

Fig. 16. Raman scattering by magnons (H=0), its T- dependence in the antiferromagnetic (a) and spin-glass (c) phase, and its H-induced splitting into two components $M^{\pm}$ (b) [after Refs. 21(a,c), 27(b)].

we have the T-dependence of the Raman line at ω_M (H=0) for x=0.7, i.e., in the antiferromagnetic phase. The line is seen only with crossed polarizations and disappears at $T \geq T_N = 40K$. These observations are consistent with the antisymmetric character of the Raman tensor of magnons which is of the form of F_1 in Table III. Its Stokes and anti-Stokes components for T=5K are shown in (b) (middle), always with H=0 [27]. The value of $T_N = 40K$ is obtained from the best fit of the theoretical model to the data of (a). The upper and lower parts of (b) show the splitting of the magnon in its two components with $H \neq 0$, according to Eq. (3) [27]. The polarization characteristics are as expected but the difference of their frequencies observed at 8.5 and 15.5 cm^{-1}, is less than the expected one (11 cm^{-1}). Finally, (c) shows the T-dependence of the magnon band for x=0.4, i.e., presumably in the spin-glass phase[21]. The behaviour is similar to that in (a) and the value suggested for $T_N \approx T_g$ is around 13K.

To conclude the discussion on the spin glass phase, it should be noted that (Cd,Te)Mn$_{0.5}$ layers in the superlattices studied in Ref. 24 with H=0 at 5K, produced no RS by magnons in spite of the fact that such layers should exhibit spin-glass behaviour at this temperature. On the contrary, with H=60 kG strong RS was observed at ω_{PM}, $2\omega_{PM}$ and $3\omega_{PM}$, implying that geometrical confinement may in fact bear an effect on magnetic phase transitions[24]. These observations and similar ones in different compositions of x show some common characteristics which justify the assignment of the ω_M band to one-magnon RS, rather than two-magnon RS; the latter is known to occur in other types of antiferromagnetic cystals such as NiF$_2$, FeF$_2$ [34] or NiO [35]. These common characteristics are, (i) disappearance of RS at $T \geq T_N$, (ii) antisymmetric character of RS, and (iii) absence of any scattering activity in the range where two-mangon RS should appear[21]. These arguments about one-magnon vs.

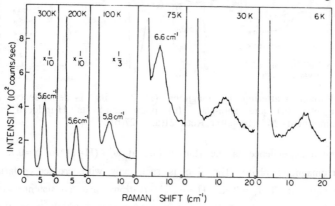

Fig. 17. Evolution of the PM band (T>T$_N$) to the M$^+$ band (T<T$_N$) with H=60 kG and x=0.7 (after Ref. 27).

two-magnon RS, oppose the conclusions reached in Ref. 36 whereby the peaks observed in (Cd,Te)Mn for x=0.50, 0.62, 0.64 and 0.68 at 2<T<40K, are due to two-magnon RS.

Finally, the evolution of $\omega_M^+(H)$ as a function of T from the antiferromagnetic to the paramagnetic phase, or vice-versa, reveals the elegant physical connection between magnon RS (or **Raman antiferromagnetic spin resonance**) on the one hand, and the **Raman paramagnetic spin resonance** (at ω_{PM}) on the other hand[27]. The collective character of the magnon excitation at $T < T_N$ is gradually lost as the thermal energy $k_B T$ prevails over J_{NN} when $T > T_N$. This means that the magnon peak gradually transforms to the PM peak for which we have assumed no Mn^{2+}-Mn^{2+} interactions. The sequence of such spectra is shown in Fig. 17 for $x = 0.7$, $H = 60$ kG, $\omega_1 = 1.833$ eV and (σ_+, z) polarizations.

4.2   (Cd,Se)Mn

This DMS crystallizes in hexagonal (W, C_{6v}) structure with $0 \leq x \leq 0.5$ (Table I). Stable cubic (ZB) structure has also been grown by MBE along [001], as epilayer on top of [001] GaAs substrates[37].

4.2.1   Raman scattering by phonons

The evolution of the phonon frequencies versus x for (Cd,Se)Mn of W structure follows two-mode behaviour and is shown in Fig. 18(a) (Raman and FIR measurements[38]). The Raman-active modes here are A_1, E_1 and $2E_2$. The A_1 and E_1 modes are also IR active and exhibit LO-TO splittings, i.e., $A_1(TO)$,

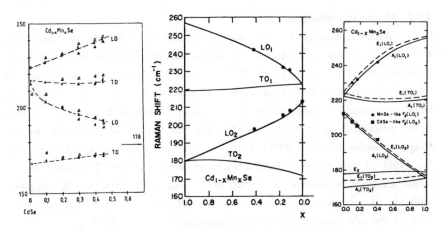

Fig. 18. (a):Phonon frequencies of (Cd,Se)Mn in the W structure from Raman (300K ▲, 77K △) and IR spectra (300K, x, •, after Ref. 38). (b):Same as (a) from Raman spectra of W structure (after Refs. 40, 41). (c):Same as (a) from Raman spectra of ZB epilayers; solid curves, MREI model (after Refs. 40, 41).

$A_1(LO)$, etc. The E_2 mode is IR inactive and its frequency is not expected to be much different from that of $A_1(TO)$, $E_1(TO)$ due to the small anisotropy of these W crystals. For this reason each of the branches in Fig. 18(a) includes A_1 and E_1 modes. One of the two E_2 modes is expected to coincide with the TO2 branch, the second E_2 mode being of too low frequency, ~ 34 cm^{-1}, to be observable[39]. More recent, yet unpublished[40], measurements on bulk samples of W structure yield the independent branches for the $A_1(LO,TO)$, $E_1(LO,TO)$ and E_2 modes shown in Fig. 18(b)[41]. The solid (dashed) curves follow the experimental data (not shown) for the A_1 (E_1) modes. The data designated by F_2 were taken from a ZB epilayer[41] and seem to match very closely the curves of the W structure. More complete data from the cubic epilayers are shown in Fig. 18(c) together with the MREI model results [41]. The latter was applied jointly to (Cd,Se)Mn and (Cd,Se)Zn. The fitting parameters are $\omega_{MnSe}^{T,L}$, $\omega_{Cd}^{T,L}$ and ω_{Mn} (Table II). Resonant RS by phonons in W structure (Cd,Se)Mn, x=0.15, has also been reported[42]. As ω_1 approaches from below the exciton frequency at 2 eV the intensity of the LO2 and LO1 modes increases monotonically (deformation potential and Frölich interactions); on the contrary, the intensity of the TO2 and TO1 modes first drops, (because of interferences[12]) and then increases (only deformation potential interaction).

4.2.2 Magnetic excitations

Raman scattering by the magnetic excitations described in subsections 4.1.2 and 4.1.3, has been observed in (Cd,Se)Mn by a number of investigators. Nawrocki et. al. have seen the SF band due to donor-bound electrons in n-type (Cd,Se)Mn, x=0.05, crystals of W structure, at 1.6K with H≠0 and ω_1 below resonance to avoid intense PL background[43]. The splitting $\hbar\omega_{SF}$ is shown to be practically equal to ΔE_c of Eq. (2). The "zero-field" band due to the "magnetic molecule" (i.e., BMP) is observed at ~ 8 cm^{-1}. It is calculated that the effect is due to an additional exchange field of ~ 0.05 T exerted on the bound electrons by as many as ~ 400 Mn^{2+} ions which constitute the magnetic molecule. It was these observations on the behaviour of BMP that stimulated most of the subsequent theoretical and experimental work on BMP.

Stimulated RS at ω_{SF} due to donor-bound electrons has been observed by Heiman[44] in (Cd,Se)Mn, x=0.012. Conversion coefficients larger than 5% were obtained with a pump intensity of 10 MW/cm^2, when ω_1 was tuned to the exciton energy (1.82 eV at 2K, H=2 kG). The threshold intensity was as low as 10^4 W/cm^2 and this renders the material suitable for stimulated laser action in the CW mode. Heiman et. al.[45,46] have reported extensive experimental and theoretical studies of spontaneous RS by SF of donor-bound electrons excited by a pulsed dye laser, as before, or by a He-Ne laser. Theoretically, the T- and H-dependence of ω_{SF} is accounted for by use of statistical mechanical models[32,45]. The thermal fluctuations δM of the local magnetization are shown to be of crucial importance in producing zero-field splitting on a time scale short compared to the spin-lattice relaxation of Mn^{2+}. The theoretical results, including bandshapes, are in good agreement with the observations of BMP from RS. The value of $\alpha N_o = 258$ meV obtained is insensitive to x for x≤0.2 and consistent with values in the literature.

The experimental work by Planel et. al. in (Cd,Se)Mn, x=0.05, also focuses, on the bandshape of BMP as a function of T [47]. A simple statistical model is used for δM, which successfully interprets the SF bandshapes observed in RS. Furthermore, it is estimated that the MBP carries an effective giant spin of ~81 which is sufficient to completely align the MBP along H. The saturation value of the latter is low (~1 kG) compared to that (20 kG) for the isolated Mn^{2+} ion. These results are supported by RS spectra of SF obtained with circular polarization selection rules.

Indium-doped (Cd,Se)Mn has been investigated by Peterson et. al[31]. The circular polarization selection rules predicted by theory for the W structure

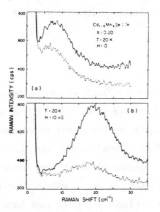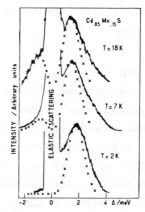

Fig. 19. (a,b):BMP and SF bands of In:(Cd,Se)Mn, x=0.2, for S (solid), A (dashed) (after Ref. 31). (c):BMP band for various T, in n-type (Cd,S)Mn, x=0.15 (x, theoretical points, after Ref. 47).

are confirmed. Extensive data of the form $\omega_{SF}(H,x,T)$ and $\omega_{BMP}(0,x,T)$ are fitted by the statistical mechanical model[32]. Fig. 19 shows the S, A bands due to RS by the BMP at H=0 and by SF at H=10 kG. In both cases the A frequency is slightly less than the S, their difference increasing as H is increased. It is argued that this is related to the fact that the probability of the excited state being occupied, is smaller than for the ground state, and becomes even smaller at higher H. The Mn^{2+}-Mn^{2+} pairs produce a band at $\hbar\omega_{NN} = 2|J_{NN}| = 11.3$ cm^{-1}, at 5K, for x=0.051, 0.104 and 0.203[48]. This is attributed to the S=0→S=1 transition of the pair. The value inferred for J_{NN} (-8.1K) is in excellent agreement with the results from direct magnetization measurements.

Cubic epilayers of (Cd,Se)Mn, x=0.16, have given distinct H-induced spectra of PM under resonance at 5K [41]. Likewise, a (Cd,Se)Mn$_{0.1}$/ZnSe superlattice has shown the fundamental PM band at ω_{PM} and all its overtones up to $7\omega_{PM}$ [41]. The multiple SF bands are explained in terms of excitations within the Mn^{2+}-Mn^{2+} pairs[48]. Furthermore, a (Cd,Se)Zn$_{0.1}$/(Cd,Se)Mn$_{0.1}$ superlattice has given under resonance at 5K,a RS band at ω_{SF} due to donor-bound electrons, with H=40 kG. From the slope of the $\omega_{SF}(H)$ straight line the value of g_{eff}=22 is obtained, for the electrons of the conduction band. Two peaks at 230, 210 cm^{-1} are assigned to the confined optical phonons ω_{MnSe}^L and the propagating optical phonons ω_{CdSe}^L, respectively. This particular superlattice is subject to sizable **bisotropic strains**, i.e., in-plane isotropic strains, due to the large lattice mismatch between its two layer components[41].

4.3 (Cd,S)Mn

This material grows in the W structure for 0≤x≤0.45. No systematic RS studies by phonons seem to exist in the literature. The CdS (W) frequencies are well known while the MnS (ZB, W) frequencies have been determined from independent studies in (Zn,S)Mn (Sect. 5.3). These values are included in Table II. Otherwise, the only known phonon spectra obtained from resonant RS in (Cd,S)Mn, x=0.0045, were reported by Zigone et. al.[49], with ω_1 approaching E_g from below. In addition to the $\omega_{CdS}^{T,L}$ modes (Table II), three new band modes were observed at 275, 283 and 293 cm^{-1} which were attributed to impurity-induced $q\neq0$ lattice modes (resonant modes). Because of lack of data, no further discussion on the nature of these modes and their x-dependence can be made at present. More recently, first- and second-order resonant RS by the ω_{CdS}^L mode of (Cd,S)Mn, $x=0.001$, was reported by Alov, with $H\neq0$ at $T=1.4K$ and ω_1 near the energy of the A-exciton ground state[50]. The exciton level is Zeeman split into two components which interact with opposite circular polarizations of light. The 90^0-scattered light at ω_{CdS}^L and $2\omega_{CdS}^L$ is enhanced by H tuning one of the exciton components to $\omega_1=2.497$ eV, thus producing in-going resonance conditions. The scattered light shows a degree of circular polarization with preference to right-handed polarization.

Most of the published RS work concerning this DMS refers to SF and BMP data. Alov et. al.[51] have examined in detail the frequency, bandshape and polarization properties of RS by spin flip of donor-bound electrons in n-type (Cd,S)Mn, $x=0.005$. As in the case of (Cd,Se)Mn[31], the anti-Stokes peak appears at slightly lower frequency than the Stokes peak. This is attributed to the sensitive variation of the ratio $I_S/I_{AS}\sim\exp(\hbar\omega_{SF}/kT)$ at these small frequencies (<10 cm^{-1}) and low T ($<10K$). It is estimated that $N_o\alpha=300$ meV, and that the BMP carries an effective moment of $\sim40\mu_B$ corresponding to eight Mn^{2+} ions. Depolarized RS by the BMP is observed at ~7 cm^{-1} with $T=18K$ and $H=0$. The theoretical approach of Ref. 32 is found sufficient to account for these data.

Contemporary to this work is the study of Heiman et. al.[52] who investigate the SF band in n-type (Cd,S)Mn, $x=0.023$, at 1.9K and fields near saturation; they find $N\alpha_o=217$ meV and a large effective g-value (~100). Likewise, Planel et. al.[47] have studied the SF and BMP characteristics in (Cd,S)Mn with $x=0.15$, 0.03, and analysed them using their simplified theoretical model (Fig. 19c, x). From the T-dependence of the BMP frequency for $x=0.15$ they obtain through fitting $N_o\alpha=220$ meV, and 25.5 Å for the effective Bohr radius. Douglas et. al.[53] have investigated an $x=0.1$ crystal in an extended range of H (≤120 kG). In addition to the low frequency PM band at 1.85K ($\omega_{PM}\leq12$ cm^{-1}) leading to an expected $g_{Mn}=2$, they observe donor-bound electron SF bands at 2K ($\omega_{SF}\leq200$ cm^{-1}) and their H-dependence; it is concluded that at high fields there are deviations from the Brillouin function dependence.

RS by SF of donor-bound electrons in crystals with $x=0.022$, 0.125 have been reported by Peterson et. al.[31] with 1.8≤T≤40 and H≤60 kG. The observed shifts $\omega_{BMP}(0,x,T)$ are more pronounced than in (Cd,Se)Mn. The data on $\omega_{SF}(x)$ indicate weak antiferromagnetism, and are well fitted by theory [32] which is particularly suitable for paramagnetic and weakly antiferromagnetic systems. The effective Bohr radius for x=0.125 turns out to be 22 Å, i.e., slightly less than that for CdS (23.5 Å), as predicted by theory[32]. It is found that the circular polarization selection rules predicted for the scattered light in W structures are well obeyed.

In a recent theoretical and experimental work on x=0.05 and 0.14 cystals, Alov has studied the limitations of existing theories[32]. A critical comparison is made between theory and the data, i.e., on $\omega_{SF}(H,x,T)$, $\omega_{BMP}(0,x,T)$, bandshapes and circular polarization characteristics of the bound-electron SF Raman bands[54]. It is concluded that the theory is adequate for good quality crystals, i.e., for low x, whereas for larger x additional interactions should be considered to account for the narrower bandshapes observed.

(Cd,S)Mn, x=0.125, turns out to be ideally suitable for studying RS by Mn^{2+}-Mn^{2+} pair excitations, as shown in Ref. 48. Such effects have also been observed in (Cd,Se)Mn (previous section) but in none of the ZB structures. The spectra, shown in Fig. 20, include one next-near-neighbor transition ($\hbar\omega_{NNN}=2J_{NNN}$) and three near-neighbor transitions at $\hbar\omega_{NN}=2J_{NN}$ ($\Delta S=1$-0), $4J_{NN}$ ($\Delta S=2$-1) and $6J_{NN}$ ($\Delta S=3$-2, not shown). The J_{NN} bands are all inherently split, they are not polarized and are not affected by H, though according to

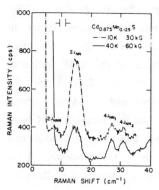

RAMAN SHIFT (cm^{-1})

Fig. 20. RS by Mn^{2+}-Mn^{2+} pair excitations ($2J_{NN}$, $2J_{NNN}$) for $\Delta S=1$-0 ($2J_{NN}$) and $\Delta S=2$-1 ($4J_{NN}$) (after Ref. 48).

theory they should exhibit Zeeman splittings. No satisfactory explanation is proposed, although the inherent splitting is probably associated with a slight anisotropy of J_{NN}, according to the relative positions of the Mn^{2+} ions within the W structure. The value of $J_{NN}=-10.6K$ obtained from these data is in excellent agreement with that obtained from magnetization experiments.

Finally, interesting observations have been made by Alov et. al.[55] in heavily doped In:(Cd,S)Mn, x=0.015, with a free electron density of $\sim 3.5 \times 10^{18}$ cm^{-3}. At 1.4K the system behaves like a degenerate conduction electron gas. In the presence of H a Zeeman splitting is developed which yields g≈90 from the RS band at ω_{SF} versus (small) H. Contrary to the case of RS by donor-bound electrons, there is no BMP effect here, i.e., no SF band at H=0. The SF banwidth for forward scattering (vanishingly small q) is found to be much narrower than in the bound-electron case, and to depend on scattering geometry through q^2, where q the momentum transfer to the electrons. As in the case of single particle RS, the mechanism of Doppler broadening is in effect, adding a contribution to the SF bandwidth which is linear in q. From the data on the bandwidth versus scattering angle the value of 10^{-14} sec is obtained for the momentum relaxation time.

5.   RAMAN STUDIES OF (Zn,B)Mn, B=Te, Se, S

These three types of DMS are also wide-gap semiconductors. Most of the RS studies, with phonons and magnetic excitations have been done on the Se compound. We examine them in the same order followed in chapter 4.

5.1   (Zn,Te)Mn

Spectroscopic evidence of impurity modes in this ZB crystal is found in the two Refs. 56 (RS and IR, respectively). This early work on RS reports an impurity-induced double peak at 197, 202 cm^{-1} for x=0.15. Both these frequencies lie in the band of optical phonons of ZnTe. One of them (197) appears in resonant RS (near E_g) in combination with the LO mode of ZnTe, and its overtone. This is a first hint that the impurity mode develops into two independent LO and TO branches with the frequencies of the former being lower than those of the latter. Such behavior precludes two-mode behaviour (notice, in addition that here $m_{Zn} \geq m_{Mn}$). According to the established code of sequencing the branches as TO2, LO2, TO1, LO1 in order of increasing frequencies, it is necessary to modify the notation of Ref. 56 and assign the two modes at 197, 202 cm^{-1} to the bands LO2 and TO1 respectively. The same comment applies to the notation used in Ref. 57, where various crystals with x≤0.15 are studied in first- and second-order RS, and IR spectroscopy. It is correctly concluded there that mixed-mode behavior is followed. Limitations however in the range of x and in the theoretical model used (REI, preceding to MREI) led to branches and values for $\omega_{MnTe}^{T,L}$ which now, in view of more recent and complete data[9], are in question.

The data obtained in Ref. 9 at 80K for $0.003 \leq x \leq 0.7$ are shown in Fig. 10, side-by-side with (Cd,Te)Mn. The two materials share the same end member, MnTe of "hypothetical" ZB structure, and this facilitates the application of the MREI model. The shape of the four branches and the frequencies $\omega_{MnTe}^{T,L}$ and ω_{Zn} are different than those in Ref. 57. A rich defect-activated first-order spectrum covers the entire region from 30 to 270 cm^{-1}; the assignments of the bands are in general agreement with those given in Ref. 57.

No Raman scattering by magnetic excitations in (Zn,Te)Mn has been reported to date.

5.2   (Zn,Se)Mn

This crystal grows in the ZB structure for $0 \leq x \leq 0.3$, then undergoes a phase transition and remains in the W structure for x up to 0.7.

5.2.1 Raman scattering by phonons

RS data from the cubic phase on phonon frequencies vs. x appeared only recently[58]. Based on four values of x, approximate full curves were ploted (Fig. 21) and revealed mixed-mode behaviour, similar to (Zn,Te)Mn and consistently with the criterion of $m_{Mn} \leq m_{Zn}$. Since no fitting by the MREI or any other model was made, the terminal frequency ω_{Zn} $(x=1)$ shown in Fig. 21 for the "hypothetical" cubic MnSe with Zn impurities should be regarded as speculative (Table II). The $\omega_{MnSe}^{T,L}$ frequencies agree with those reached via cubic (Cd,Se)Mn at $x=1$ (Fig. 18c). Recent FIR reflection and transmission spectra[59] from crystals with $x=0.18$, 0.31 reveal, after Kramers-Krönig and multi-oscillator analysis, distinct frequency shifts, which are in general agreement with Fig. 21 except for the TO1, LO1 points at $x=0.18$. The latter appear to be unreasonably high (294, 309 cm^{-1}) compared to the corresponding points of Fig. 21 (230, 254 cm^{-1}). [It is not clear whether this difference is of physical interest; more data are necessary in that region of x.]

Returning to Ref. 58, RS measurements under pressure have shown a phase transition at ~32±4 Kb from ZB to a new unidentified phase with three ands in the spectrum. This new phase cannot be of the NaCl structure for

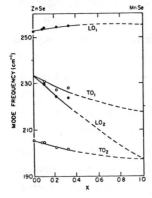

Fig. 21. Optical phonon frequencies of (Zn,Se)Mn at 300K. The lines are only to assist the eye (Ref.58).

which no RS is expected in first order. Hysteresis phenomena are observed and in fact one of the Raman bands remains even after complete release of the pressure; this indicates that after pressure cycling the crystal probably consists of a mixed phase.

Various ZnSe/(Zn,Se)Mn superlattices with $0.23 \leq x \leq 0.51$ have been grown along [001] and studied with RS by Suh et. al.[25] ZnSe or (Zn,Se)Mn buffers were inserted on purpose between the GaAs substrates and the superlattices, in order to regulate the lattice mismatch strains. Only the LO phonons were observed, below resonance, the TO phonons being forbidden in the [001] back-scattering geometry. Such LO bands were contributed by the GaAs substrate (~292 cm^{-1}), the ZnSe, (Zn,Se)Mn buffers, and the ZnSe, (Zn,Se)Mn layers of the superlattice, all in the range 250-256 cm^{-1}. The ZnSe-buffer LO frequency was used as a reference, to measure the shifts of the LO frequencies of the strained superlattice layers. Being under tensile (compressive) strain, the ZnSe (DMS) superlattice layers exhibited downward (upward) LO-frequency shifts equal to 3.5 (2.5) cm^{-1}. Using the lattice mismatch strain values, obtained from the superlattice parameters, and the theory of phonon frequencies under strain[60,61] it was possible to estimate the expected LO phonon shift for the ZnSe layer which was found 20% smaller than that observed. This may be due to the fact that the phonon-strain coefficients used in the calculation were those for the TO phonon[61]. In- and out-going resonant RS by the LO phonon at ω_{ZnSe}^{L} was observed by Leiderer et. al.[62] in crystals with x=0.03, in the region of the E_0 gap (~2.8 eV). Bound and continuum excitons were invoked as intermediate states, combined with scattering mechanisms via deformation potential and Fröhlich interaction. The out-going resonance was found stronger than the in-going resonance in both, allowed and forbidden scattering geometries.

5.2.2 Magnetic excitations

Relatively few studies of RS by magnetic excitations have been reported to date. Among the earlier ones is the combined SF and magnetization work in crystals with x=0.033 by Heiman et. al.[63] Measurements at 1.9K and 5≤H≤100 kOe led to the conclusion that the observed SF band concerned donor-bound electrons with Bohr radius 26 Å (slightly less than that, 29 Å, for ZnSe) and that $N_o \alpha$=243 meV. Scattering by BMP was also evident from the data.

The study of Douglas et. al.[53] refers to RS by SF in crystals with x=0.017, 0.041 at 2K and 0≤H≤120 kG. Giant SF shifts (≤50, 150 cm^{-1}, respectively) at 120 kG from donor-bound electrons follow a Brillouin-type function with adjusted effective temperature. Thermalized and non-thermalized spin states are concluded from the ratios I_S/I_A of the two samples, respectively. A PM band at ω_{PM}≤12 cm^{-1} appears to vary linearly with the field. It is concluded from the slope at low H for both samples that g=2, consistently with the intra-Mn^{2+} origin of the PM band.

The work of Alov[54] on SF scattering characteristics with crystals of ZB structure concludes, as in the case of (Cd,S)Mn, that the theoretical treatment[32] of the SF bandshapes and shifts is in good agreement with the data for low values of x (good crystal quality). The poor quality of the crystals is probably the reason for the observed broadening of the bandwidths.

5.3   (Zn,S)Mn

This material is the least studied DMS with RS, especially in the range of large values of x. Extensive Raman work has been published for small values of x, where the material is treated as impure ZnS[49,64-67] rather than as a mixed, disordered (Zn,S)Mn system. For x≤0.1 (Zn,S)Mn exhibits ZB structure and for 0.1<x≤0.45 to 0.6 it converts to W structure. As of today no RS work on magnetic excitations has been reported, while RS by phonons in the full range of x appeared only recently[68]. The following description of phonon data is based on the results of Ref. 68.

Raman spectra for x=0.03, 0.1, 0.2, 0.3 and ω_1=2.6 eV are shown in Fig. 22. For x=0.03 and ω_1=2.6 eV (a) the spectrum coincides with that of impure ZnS which has been studied extensively in the past[49]. In addition to the frequencies $\omega_{ZnS}^{T,L}$ at 271.5, 350 cm^{-1} (300K), three impurity-induced vibrational modes appear at $\omega_B = 297$ cm^{-1}, $\omega_1 = 313$ cm^{-1} and $\omega_2 = 333$ cm^{-1}. ω_B has been attributed to a disorder induced breathing mode of the S ions about the Mn impurity[49]. It is a localized mode and, according to lattice dynamical models, its frequency occurs in the spectral region where the normal mode density of states of the pure compound is zero. Similar breathing modes have been observed in (Cd,Te)Mn (Ref. 21) and (Zn,Te)Mn (Ref. 9). The modes at ω_1 and ω_2 on the other hand, have been identified as band modes (lattice modes) activated by the introduction of Mn (Ref. 49). Their frequencies correspond to high-density regions of the Brillouin zone. As such, they have been assigned to modes from the X and L points of the zone. All phonons shift to higher frequencies by about 3 cm^{-1} as the temperature is lowered from 300K to 25K. The Raman spectra for x=0.1, 0.2 and 0.5 taken at 300 K with $\omega_1 = 3.05$ eV (b) show that the frequencies of the impurity-induced vibrational modes

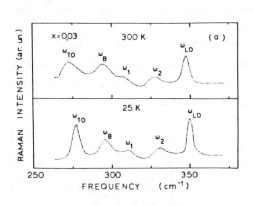

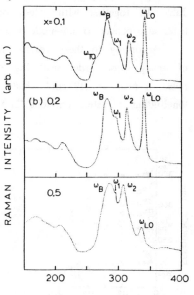

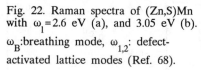

Fig. 22. Raman spectra of (Zn,S)Mn with $\omega_1 = 2.6$ eV (a), and 3.05 eV (b). ω_B:breathing mode, $\omega_{1,2}$: defect-activated lattice modes (Ref. 68).

appear rather insensitive to concentration. Therefore, it is assumed that they retain their nature as "breathing" (ω_B) and band modes (ω_1, ω_2) throughout the entire concentration range. The relative intensity of the impurity modes with respect to the TO and LO phonons increases with concentration. The TO phonon in particular, almost disappears for $x > 0.1$. In fact, the TO phonon is detectable for $x = 0.1$ and only with excitation wavelengths from the violet region of the spectrum.

Fig. 23 shows the observed optical phonon frequencies vs. x. The empty points correspond to the impurity-induced vibrational modes ω_B, ω_1, ω_2. The full points correspond to $\omega_{ZnS}^{T,L}$. The triangular points will be discussed separately. The dashed curves are the results of the MREI model, which predicts one-mode behaviour for the (Zn,S)Mn system. It also predicts splitting of the impurity-mode pair (designated by ω_i) with a maximum value of 0.2 cm^{-1} at $x = 0.5$, not shown in Fig. 23. The fitting parameters include $\omega_{MnS}^{T,L}$ and $\omega_{Zn} \equiv \omega_i$. We note that MnS normally exists in the rocksalt structure. The frequencies $\omega_{MnS}^{T,L}$ then correspond to an "hypothetical" ZB structure. Accordingly, a range of values for $\omega_{MnS}^{T,L}$ was used; the curves were rather insensitive to such choices. The end values of these curves are included in Table II, together with values obtained from recent IR experiments[69]. The latter also concur to one-mode behaviour.

According to these results and Fig. 22(b) the ω_i pair lies in the range of strong second-order scattering. Indeed measurements performed in pure ZnS reveal a peak at 219 cm^{-1} which is due to overtone scattering from zone-edge longitudinal acoustical phonons (2LA). The experimental points shown on the

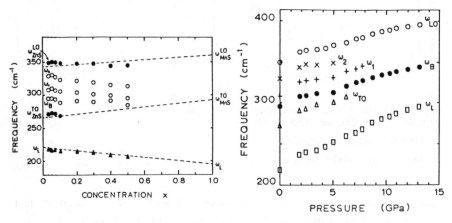

Fig. 23. Dependence on x of the frequencies of (Zn,S)Mn at 300K (Ref. 68).

Fig. 24. Pressure dependence of the optical phonon frequencies of (Zn,S)Mn for $x = 0.03$ and 300K (Ref. 68).

lowest branch of Fig. 23 actually correspond to the peak around 217 cm^{-1} in Fig. 22(b). The oscillator strength of the impurity mode pair in one-mode mixed crystals is, in general, weak[8]. Therefore, it is assumed that the peak at 217 cm^{-1} is largely due to overtone scattering with possible contributions from ω_i at ~222 cm^{-1}, the latter being too weak to be distinguished from the second-order background. Attempts to resolve the band at 217 cm^{-1} under pressure did not produce any additional information. Finally, the lower part of the spectrum of Fig. 22 is full of broad bands. They are similar to those of pure ZnS which are known to originate mainly from second-order RS processes.

Measurements under hydrostatic pressure have been reported in the past[70,71] and also more recently[68]. In Fig. 24 the pressure dependence of the mode frequencies for $x = 0.03$ are shown[68]. All frequencies shift to higher values as the pressure is increased. From the corresponding slopes the Gruneisen parameters are determined. The slopes of ω_B, ω_1 and ω_2 are comparable. The slope of the low frequency band at 217 cm^{-1} is twice that of ω_1 and ω_2 which is consistent with its overtone character.

Preliminary measurements of resonant RS have been performed in (Zn,S)Mn for the entire concentration range $0.03 \leq x \leq 0.5$ at 300K using excitation energies below 2.0 eV and above 2.4 eV. The strong PL band overcomes any Raman activity in the intermediate range, which hosts the following transitions of Fig. 2(b), $E(^6A_1 \rightarrow {}^4T_1) = 2.38$ eV, $E(^6A_1 \rightarrow {}^4T_2) = 2.52$ eV and $E(^6A_1 \rightarrow {}^4E, {}^4A_1) = 2.68$ eV. The characteristic PL emission is found to be Stokes-shifted and occurs at $E(^4T_1 \rightarrow {}^6A_1) = 2.13$ eV (Ref. 72). According to the results the ratio of the intensities of the impurity-induced modes (at ω_B, ω_1, ω_2) to that of the zone-center lattice modes (TO and LO) increases with increasing concentration x independently of ω_i. As the lattice disorder increases it is reasonable to expect more intense scattering by the impurity- (or disorder-) induced modes. In addition, a resonant behavior is observed for $\omega_i = 1.92$ eV. It is clearly seen for $x \geq 0.2$ and shows an enhancement of the impurity activated modes as the incident laser energy approaches the energy of the 2.13 eV transition. This enhancement suggests coupling of these modes to the electronic transition responsible for the PL emission. The strong coupling between lattice and the $3d^5$ electrons has been found responsible for the large Stockes shift of the PL band in this crystal. Zigone et. al.[49] have reported resonant RS measurements on Mn-doped ZnS. They suggest that the electron-phonon coupling is achieved predominantly through the breathing mode which, according to their data, is the only one showing an enhancement as ω_1 approaches the electronic transition energies. The measurements in Ref. 68 on the mixed system, on the other hand, suggest that all three impurity-activated modes are involved in the electronic transitions.

6.   OTHER POSSIBILITIES OF DMS

The majority of RS studies in DMS concerns the six wide-gap compounds reviewed in the previous two sections, i.e., (Cd,B)Mn and (Zn,B)Mn with B=Te, Se, S. Narrow-gap DMS based on II-VI materials, particularly (Hg,B)Mn, B=Te, Se, have attracted interest especially as regards their magneto-transport and magneto-optical properties (Ref. 2, Vol. 25, Chapters 5 by J. Kossut and 6 by C. Rigaux, respectively). Crystals of (Hg,B)Mn with ZB structure can be grown for $0 \leq x \leq 0.75$ (0.38) (Table I). Their energy gap is tunable through zero by varying x. With $H \neq 0$, their energy level schemes, modified by Landau and Zeeman splittings, render themselves suitable for novel RS processes. Nevertheless, the existing studies include only the (Hg,Te)Mn, x=0.11, crystal and concern phonon frequencies and SF observations. The spectrum at 77K reported by Amirtharaj et. al.[73] exhibits the $\omega_{HgTe}^{T,L}$ frequencies at 119, 137 cm^{-1}, a shoulder-like structure at 132 cm^{-1} (named "clustering mode") and a band at 191 cm^{-1} assigned to $\omega_{MnTe}^{T,L}$ (behaviour characterized as mixed mode). This assignment however is in conflict with the values $\omega_{MnTe}^{T,L} = 185$, 216 cm^{-1} established for the "hypothetical" ZB structure of MnTe, from the RS spectra of (Cd,Te)Mn and (Zn,Te)Mn [9]. An alternative interpretation of the data in Ref. 73 would be to assign the band at 191 cm^{-1} to ω_{Mn} and assume that the Hg impurity in MnTe induces a gap mode at $\omega_{Hg} \approx 125$ cm^{-1} (based on $m_{Hg} \gg m_{Mn}$ and the extrapolation of IR data in Ref. 74). The frequency evolution would then fit very well the two-mode behaviour, consistently with the $\omega_{MnTe}^{T,L}$ values and the IR data [74].

An interesting observation of stimulated RS by SF was demonstrated by Geyer and Fan[75], at $6 \leq T \leq 20K$ and $H > 25$ kG, using CO_2 laser radiation at 10.6 μm for excitation. The observed scattering is attributed to pure SF of conduction electrons (Γ band).

The latest newcomer to the wide-gap DMS studied by RS is (Cd,Se)Fe [76].

The ground state of $Fe^{+2}(3d^6)$ is a non-magnetic singlet (Van Vleck paramagnetic ion). Donor-bound electron SF scattering has already been reported[76] for x=0.006, 0.04 at T=1.9 K with $\omega_1 = 1.832$ eV (near the exciton energy). Figure 25(a) shows SF spectra for (Cd,Se)Mn and (Cd,Se)Fe for comparison, together with a schematic of the BMP for each case. The Mn^{2+} ion exhibits non-zero ground-state magnetic moments and large thermal fluctuations which at low T produce non-zero values for $<M_{loc}>$. It is this $<M_{loc}>$ that determines, at zero and low fields, the frequency ω_{SF} and the large full width at half maximum γ of the Gaussian-like bandshape, as shown in Figs. 25(b) and (c), respectively. At high fields the bandwidth is determined mainly by compositional fluctuations (Sect. 2.3). On the contrary, the ground state magnetic moment of Fe^{2+} is zero and so is the contribution of thermal fluctuations to the magnetic moment, i.e., there is no Zeeman-split ground state and no SF band at H=0. As H is applied, Zeeman splitting is developed and RS due to SF appears at a linear-in-H frequency ω_{SF} (b). The bandwidth γ is determined by the compositional fluctuations and varies linearly with ω_{SF} (c) or H. Furthermore, at low temperatures the SF band is found insensitive to T. This is expected for a Van Vleck ion, and is in sharp contrast to the T-dependent SF band of (Cd,Se)Mn. In short, all these observations are in good agreement with theory and show the strength of RS for probing magnetism on an atomic scale.

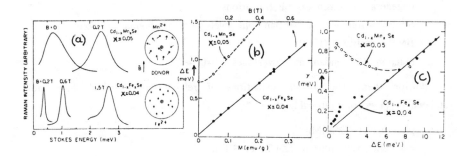

Fig. 25. (a):SF spectra of (Cd,Se)Mn and (Cd,Se)Fe at T=1.9 K for various fields B. (b):Frequency of the SF band vs. B. (c):Full width at half maximum vs. frquency (after Ref. 76).

7. CONCLUSION

The purpose of this work was to present an overview of the major contributions of RS to the field of DMS physics. A mere listing of excitations and/or transitions probed by RS in these systems suffices to demonstrate the range and wealth of physics behind these contributions:

lattice modes (q≈0 fundamental, overtones, combinations)

impurity modes (local, gap, breathing)

disorder-induced modes (q≠0 lattice modes)

superlattice, epilayer modes (confined, propagating, interface modes)

single Mn^{2+} ion electronic transitions (PM, combinations with lattice modes)

paired Mn^{2+} ion transitions at H=0 (ΔS=1 transitions within total-spin multiplets, J_{NN})

as before, with H≠0 (ΔM_S≠0 transitions within Zeeman-split multiplets, nPM, combinations with lattice modes)

triangle Mn^{2+} ion transitions (J_{NNN})

phonon replicas of free electron-to-acceptor transitions

pressure-induced structural phase transitions

temperature-induced magnetic transitions

single electron scattering

Landau levels, Zeeman split SF transitions (conduction band electrons)

degenerate electron gas SF transitions

SF transitions within Zeeman split levels of donor-bound electrons

BMP transitions (H=0)

magnon RS (fundamental, overtones, H=0, ≠0).

In addition there are obvious extentions of related effects which have not yet been observed or thoroughly studied:

phonons in (Cd,S)Mn

magnetic excitations in (Zn,Te or S)Mn

free magnetic polarons

acceptor-bound carrier SF

valence band related SF effects

doubly-resonant RS by H tuning Zeeman-split excitonic, or other, levels quantum mechanical treatment of the entire magnetic polaron subject.

The six compounds reviewed here are the ones that have attracted most of the experimental and theoretical interest so far, in connection with the above list of excitations. They are by no means the only ones and it is only a matter of time to see parallel and/or supplementary work been done on new types of DMS, chief example been the Fe^{2+} compounds. Furthermore, the field of superlattices and heterostructures, tightly connected to applications, is expected to introduce more material systems which will combine the strength of magneto-optical properties of II-VI bulk DMS with the novel possibilities of Solid State Physics in two dimensions. Raman spectroscopy undoubtly will continue to be present in all these exciting developments, by virtue of its power, directness, and the non-destructive and contactless character of it.

Acknowledgments

Useful exchange interactions with A. Anastassiadou and A. Petrou are acknowledged with appreciation. Partial support was provided by the General Secretariat for Research and Technology, Greece.

REFERENCES

1. Pajaczkowska, A., Prog. Cryst. Growth Charact. **1**, 289 (1978).

2. **Diluted Magnetic (Semimagnetic) Semiconductors,** Eds. R.L. Aggarwal, J.K. Furdyna, and S.Von Molnar, MRS Symposia Proceedings No.89 (Materials Research Society, Pittsburgh, 1987).

3. **Semiconductors and Semimetals,** R.K. Willardson and A.C. Beer, Treatise Editors; J.K.Furdyna and J.Kossut, Volume Editors (Academic, Boston, 1988), Vol. 25.

4. Furdyna, J.K., J. Appl. Phys. **64**, R29 (1988).

5. Goede, O. and Heimbrodt, W., phys. stat. sol. (b) **146**, 11 (1988).

6. Ramdas, A.K. and Rodriguez, S., in Ref. 3, Vol. 25, p. 345.

7. Joder-Short, D.R., Debska, M., and Furdyna, J.K., J. Appl. Phys. **58**, 4056 (1985).

8. Barker, A.S. and Sievers, A.J., Rev. Mod. Phys. **47**, Suppl. 2 (1975); Genzel, L. and Bauhofer, W., Zeitschrift fur Physik **B25**, 13 (1976).

9. Peterson, D.L., Petrou, A., Giriat, W., Ramdas, A.K., and Rodriguez, S., Phys. Rev. **B33**, 1160 (1986).

10. Genzel, L., Martin, T.P., and Perry, C.H., phys. stat. sol. (b) **62**, 83 (1974).

270

11. Hayes, W. and Loudon, R., in **Scattering of Light by Crystals** (J.Wiley, N.Y. 1978).

12. Cardona, M. and Guntherodt, G., Eds., **Light Scattering in Solids,** Vol. I-VI (Springer Verlag, Berlin, 1975, 1982, 1984, 1989, 1990).

13. Cerdeira, F., Anastassakis, E., Kauschke, W., and Cardona, M. Phys. Rev. Lett. **57**, 3209 (1986).

14. Alexandrou, A. and Cardona, M., Solid State Commun. **64**, 1029 (1987); Phys. Rev. **B38**, 2196 (1988).

15. Kleinman, D.A., Miller, R.C., and Gossard, A.C., Phys. Rev. **B35**, 664 (1987).

16. Argulo-Rueda, F., Mendez, E.E., and Hong, J.M., Phys. Rev. **B38**, 12720 (1988).

17. Hirschfeld, T. and Chase, B., Appl. Spectrosc. **40**, 133 (1986); Chase, D.B., J. Am. Chem. Soc. **108**, 7485 (1986); Chase, B., Anal. Chem. **59**, 881A (1987); Chase, B., in **International Conference on Raman Spectroscopy,** Eds. R.J.H. Clark and D.A. Long (J.Wiley, N.Y. 1988), p. 39; Rabolt, J.F., Zimba, C., and Hallmark, V., same conference, p. 987; Lewis, E.N., Kalasinsky, V.F., and Levin, I.W., Appl. Spectrosc. **42**, 1188 (1988).

18. Bruck, S.R.J., Tsaus, B-Y., Fan, J.C.C., Murphy, D.V., Deutsh,T.F., and Silversmith, J., Appl .Phys. Lett. **40**, 895 (1985).

19. Rabolt, J.F. and Swalen, J.D., in **Advances in Spectroscopy,** Vol. 16, Eds. R.J.H. Clark and R.E. Hester (Heyden, London 1988), p. 1.

20. Tien, P.K., Rev. Mod. Phys. **49**, 361 (1977).

21. Venugopalan, S., Petrou, A., Galaska, R.R., Ramdas, A.K., and Rodriguez, S., Phys. Rev. **B25**, 2681 (1982).

22. Picquart, M., Anzallag, E., Balkanski, M., Julien, C., Gebicki, W., and Nazarewicz, W., phys. stat. sol. (b) **99**, 683(1980).

23. Feng, Z.C. , Sudharsanan, R., Perkowitz, S., Erbil, A., Pollard, K.T., and Rohatgi, A., J. Appl. Phys. **64**, 6861 (1988).

24. Venugopalan, S., Kolodziejski, L.A., Gunshor, R.L., and Ramdas, A.K., Appl. Phys. Lett. **45**, 974 (1984).

25 Suh, E.-K., Bartholomew, D.U., Ramdas, A.K., Rodriguez, S., Venugopalan, S., Kolodziejski, L.A., and Gunshor, R.L., Phys. Rev. **B36**, 4316 (1987).

26. Arora, A.K., Bartholomew, D.U., Peterson, D.L., and Ramdas, A.K., Phys. Rev. **B35**, 7966 (1987).

27. Petrou, A., Peterson, D.L., Venugopalan, S., Galazka, R.R., Ramdas, A.K., and Rodriguez, S., Phys. Rev. **B27**, 3471 (1983); Phys. Rev. Lett. **48**, 1036 (1982).

28. Peterson, D.L., Bartholomew, D.U., Ramdas, A.K., and Rodriguez, S., Phys. Rev. **B31**, 7932 (1985).

29. Limmer, W., Bauer, S., and Gebhardt, W., Proc. Fourth Int. Conf. on II-VI Compounds, Berlin 1989.

30. Peterson, D.L., Petrou, A., Dutta, M., Ramdas, A.K., and Rodriguez, S.,

Solid State Commun. **43**, 667 (1982).

31. Peterson, D.L., Bartholomew, D.U., Debska, U., Ramdas, A.K., and Rodriguez, S., Phys. Rev. **B32**, 323 (1985).

32. Dietl, T. and Spalek, J., Phys. Rev. **B28**, 1548 (1983).

33. Suh, E.-K., Bartholomew, D.U., Ramdas, A.K., Bicknell, R.N., Harper, R.L., Giles, L.C., and Schetzina, J.F., Phys. Rev. **B36**, 9358 (1987).

34. Fleury, P.A., Phys. Rev. **180**, 591 (1969).

35. Perry, C.H., Anastassakis, E., and Sokolof, J., Indian Journal of Pure and Applied Physics, **9**, 930 (1971).

36. Grynberg, M. and Picquart, M., J. Phys. C: Solid State Phys. **14**, 4677 (1981).

37. Samarth, N., Luo, H., Furdyna, J.K., Quadri, S.B., Lee, Y.R., Ramdas, A.K., and Otsuka, N., Appl. Phys. Lett. **54**, 2680 (1989).

38. Dorhaus, R. and Giriat, W., in **Proc. 9th Int. Conf. on Raman Spectroscopy** (Tokyo, 1984), p. 388.

39. Arora, A.K., and Ramdas, A.K., Phys. Rev. **B35**, 4345 (1987).

40. Suh, E.-K., Arora, A.K., Ramdas, A.K., and Rodriguez, S. (unpublished).

41. Alonso, R.G., Suh, E.-K., Ramdas, A.K., Samarth, N., Luo, H., and Furdyna, J.K., Phys. Rev. **B40**, 3720 (1989).

42. Valakh, M.Ya. and Litvinchuk, A.P., Sov. Phys. Solid State **27**, 1176 (1985).

43. Nawrocki, M., Planel, R., Fishman, G., and Galazka, R., Phys. Rev. Lett. **46**, 735 (1981).

44. Heiman, D., Appl. Phys. Lett. **41**, 585 (1982).

45. Heiman, D., Wolff, P.A, and Warnock, J., Phys. Rev. **B28**, 4848 (1983).

46. Heiman, D., Shapira, Y., Foner, S., Khazai, B., Kershaw, R., Dwight, K., and Wold, A., Phys. Rev. **B29**, 5634 (1984).

47. Planel, R., Nhung, T.H., Fishman, G., and Nawrocki, M., J. Physique **45**, 1071 (1984). Also, Nawrocki, M., Planel, R., Mollot, F., and Kozielski, M.J., phys. stat. sol. (b) **123**, 99 (1984).

48. Bartholomew, D.U., Suh, E.-K., Rodriguez, S., Randas, A.K., and Aggarwal, R.L., Solid State Commun. **62**, 235 (1987).

49. Zigone, N., Beserman, R., and Balkanski, M., in **Light Scattering in Solids,** Eds. M. Balkanski et. al. (Flammarion, Paris, 1972), p. 61; Beserman, R., Zigone, N., and Balkanski, M., in **Phonons,** Ed. M.A. Nusimovici (Flammarion, Paris, 1972), p. 405.

50. Alov, D.L., Sov. Phys. Solid State **29**, 900 (1987).

51. Alov, D.L., Gubarev, S.I., and Timofeev, V.B., Sov. Phys. JETP **57**, 1052 (1983).

52. Heiman, D., Shapira, Y., and Foner, S., Solid State Commun. **45**, 899 (1983).

53. Douglas, K., Nakashima, S., and Scott, J.F., Phys. Rev. **B29**, 5602 (1984).

54. Alov, D.L., Sov. Phys. Solid State **30**, 555 (1988).

55. Alov, D.L., Gubarev, S.I., and Timofevv, V.B., Sov. Phys. JETP **59**, 658 (1984).

56. Valakh, M.Ya. and Litvinchuk, A.P., Sov. Phys. Solid State **25**, 1597 (1983). Olszewski, A., Wojdowski, W., and Nazarewicz, W., phys. stat. sol. (b) **104**, K155 (1981).

57. Oles, B. and von Schnering, H.G., J. Phys. C: Solid State Phys. **18**, 6289 (1985).

58. Arora, A.K., Suh, E.-K., Debska, U., and Ramdas, A.K., Phys. Rev. **B37**, 2927 (1988).

59. Lu, W., Liu, P.L., Shi, G.L., Shen, S.C., and Giriat, W., Phys. Rev. **B39**, 1207 (1989).

60. Anastassakis, E., J. Raman Spectrosc. **10**, 64 (1981). Also, Anastassakis, E., Pinczuk, A., Burstein, E., Cardona, M., and Pollak, F.H., Solid state Commun. **8**, 533 (1970).

61. Cerdeira, F., Buchenauer, C.J., Pollak, F.H., and Cardona, M., Phys. Rev. **B5**, 580 (1972).

62. Leiderer, H., Limmer, W., and Gebhardt, W., see conference of Ref. 29.

63. Heiman, D., Shapira, Y., and Foner, S., Solid State Commun. **51**, 603 (1984).

64. Beserman, R., Nusimovici, M.A., and Balkanski, M., phys. stat. sol. **34**, 309 (1969).

65. Zigone, M., Beserman, R., and Balkaniski, M., J. Phys. (Paris), Colloque C3 Suppl. **35**, 153 (1974).

66. Zigone, M., Kunc, K., Plumelle, P., and Vandevyver, M., in **Proc. 3rd. Int. Conf. on Lattice Dynamics**, Ed. M. Balkanski (Flammarions, Paris, 1978), p. 405.

67. Natadze, A.L., Pevnitskii, I.V., Ryskin, A.I., Khil'ko, G.I., Sov. Phys. Solid State **18**, 1125 (1976). Krol', A.V.,Levichev, N.V., Natadze, A.L., and Ryskin, A.I., Sov. Phys. Solid State, **20**, 85 (1978). Egorov, V.Yu., Krol, A.V., Levichev,N.V., and Ruskin, A.I., Opt. Spectrosc **44**, 354 (1978).

68. Anastassiadou, A., Liarocapis, E., and Anastassakis, E., Solid State Commun. **69**, 137 (1988).

69. Jahne, E., Goede, O., and Weinhold, V., phys. stat. sol. (b) **146**, K157 (1988).

70. Zigone, M., Beserman, R., and Fair, H.D., Jr., in **Proc. 3rd Int. Conf. on Light Scattering in Solids**, Eds. M. Balkanski et. al. (Flammarion, Paris, 1976), p. 597.

71. Zigone, M., Vandevyver, M., and Talwar, D.N., Phys. Rev. **24**, 5763 (1981), and references.

72. Anastassakis, E., Liarokapis, E., Stoyanov, S. and Anastassakis, E.,

273

Solid State Commun., **67**, 633 (1988).

73. Amirtharaj, P.M., Tiong, K.K., Parayanthal, P., Pollak, F.H., and Furdyna, J.K., J. Vac. Sci. Technol. **A3**, 226 (1985).

74. Gebicki, W, and Nazarewicz, W., phys. stat. sol. (b) **80**, 307 (1977).

75. Geyer, F.F. and Fan, H.Y., IEEE J. Qu. El. **E-16**, 1365 (1980).

76. Heiman, D., Petrou, A., Bloom, S.H., Shapira, Y., Isaaks, E.D., and Giriat, W., Phys. Rev. Lett. **60**, 1876 (1988).

MAGNETIC AND OPTICAL PROPERTIES OF

Fe-BASED SEMIMAGNETIC SEMICONDUCTORS

(The role of nonvanishing orbital momentum)

Andrzej Twardowski*

*   *Institute of Experimental Physics, Warsaw University*
    Hoza 69, 00681 Warsaw, Poland
    and
    Physics Department, Eindhoven University of Technology
    Eindhoven, The Netherlands

Table of contents

III. Fe-based SMSC (L=2, S=2)

I. INTRODUCTION

So far research on Semimagnetic Semiconductors (SMSC) has been focused mainly on SMSC containing Mn ions[1]. The magnetic behaviour of these systems shows common characteristics, which can be understood on the basis of a random array of localized magnetic moments coupled by (long ranged) isotropic antiferromagnetic interaction. The underlying microscopic mechanisms of this interaction are still subjects of investigations.

In that respect, substitutional Mn^{++}, with its 6-fold degenerate $^6A_1$ spin-only ground state (L=0, S=5/2), represents a rather simple, although theoretically attractive case, since all the phenomena which involve orbital momentum are absent.

In contrast, substitutional iron Fe^{++} (d^6) can serve as a much more general case, since it possesses both spin and orbital momenta (S=2 and L=2). The ground state of the Fe ion in SMSC is - due to spin-orbit interaction - a singlet [2,3]. Consequently, the situation for Fe is essentially different than it was for Mn: Fe ions reveal only a field induced magnetic moment, leading to typical Van Vleck-type paramagnetism. The properties of Fe-based SMSC are however, still relatively unexplored[4], although recently some experimental data were reported on these compounds (ZnFeSe, CdFeSe and HgFeSe)[4-12].

In view of the above, it seems worthwhile to examine the magnetic and magnetooptical properties of Fe-based SMSC in some detail. These properties will be related to those of Mn-based SMSC, in order to trace directly the influence of nonvanishing orbital momentum.

This chapter is arranged as follows: in sec.II we briefly revise magnetic and magnetooptical properties of Mn-based SMSC founding the base for comparison between Mn- and Fe-type materials. In this section we also recall the idea of Extended Nearest Neighbour Pair Approximation (ENNPA). In sec.III we discuss properties of Fe-based SMSC. Interpretation of the magnetic properties of these crystals takes into account the d-d exchange interaction between Fe ions. The High Temperature Expansion (HTE) series developed for susceptibility of Fe-type materials will be presented. We conclude in sec.IV.

II. Mn—BASED SMSC (L=0, S=5/2)

2.MAGNETIC PROPERTIES

For the Mn—based SMSC such as HgMnTe, CdMnTe, CdMnSe, ZnMnTe, $(CdMn)_3As_2$ or $(ZnMn)_3As_2$ rather typical magnetic behaviour is observed[1,13,14,15]. This behaviour can be characterized as follows:

— Curie—Weiss behaviour of the magnetic susceptibility at high temperatures indicating antiferromagnetic (AF) Mn—Mn interaction.

— A cusp or kink at low—temperature susceptibility, indicating spin—glass like transition at a temperature depending on the Mn concentration x.

— A magnetic contribution to the specific heat with a broad maximum shifting to higher T with increasing x.

— A field dependence of the magnetization indicating AF interaction, usually accompanied by steps at high fields (for x 0.05).

In this chapter we present examples of such typical behaviour, which will be used as a reference for Fe—based SMSC.

2.1 Experimental Results

A. Specific heat

The representative results for magnetic specific heat c_m are shown in Fig.1[16]. The overall behaviour of c_m is common for all Mn—based SMSC: at low temperatures (T=0.4—0.5K) a maximum is found, which shifts to higher temperatures with increasing x. In the case of small x this maximum is not observed since it is located at T<0.4K. In the presence of a magnetic field, the specific heat data show a shift of the maximum to higher

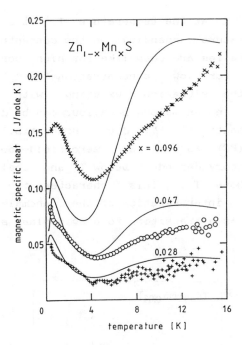

Fig.1 Magnetic specific heat of $Zn_{1-x}Mn_xS$; the solid lines represent ENNPA calculations with $J_{NN}/k_B = -16K$ and $J_1/k_B = -10/R_1^{7.6}K$. After Ref.16.

temperatures with increasing field, as shown in Fig.2[15].

B. High-temperature susceptibility

The high-temperature susceptibility results[14,17] show a Curie-Weiss behaviour, with a negative Curie-Weiss temperature indicating an AF interaction between Mn ions. Moreover Θ was found to be a linear function of x[17], suggesting a random distribution of Mn ions.

C. Low-temperature susceptibility

Representative results of a.c. susceptibility are shown in Fig.3[13]. One can clearly see an anomalous behaviour of χ at the temperature T_f (the so called freezing temperature) depending on the concentration of Mn ions. The anomalies are cusp-like for high concentrations and kink-like for low concentrations[14,18,13,15]. In contrast to the situation existing with regards to susceptibility, no anomalous behaviour can be detected in the specific heat[14,13,15]. d.c. susceptibility data, field-cooled (FC) as well as zero-field-cooled (ZFC) are cooling-history dependent below T_f and cooling-history independent above T_f. This characteristic behaviour supports the interpretation of the anomaly in the susceptibility as a transition to a spin-glass state.

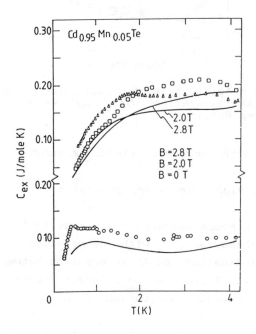

Fig.2 Magnetic specific heat of $Cd_{0.95}Mn_{0.05}Te$. The solid lines represent calculations with the ENNPA. After Ref.15.

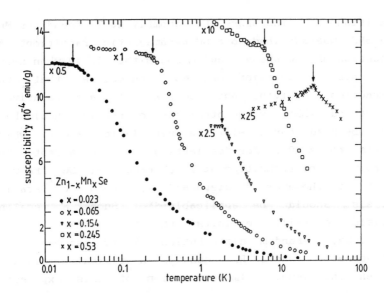

Fig.3. ac susceptibility of $Zn_{1-x}Mn_x Se$ as a function of temperature for different Mn compositions. The arrows indicate the freezing temperature T_f. After Ref.13.

D. Magnetization

Magnetization of Mn-type SMSC has been measured up to 42T and shows typical Brillouin-like behaviour[1], as examplified in Fig.4[19] for ZnMnSe. Magnetization per one Mn ion decreases with increasing x (Fig.4), indicating an AF interaction between Mn ions. At sufficiently high magnetic fields characteristic step-like structures have been observed[20].

2.2 Spin-glass Phase Transition

The existence of a cusp or a kink at T_f for a.c. susceptibility, history-dependent d.c. susceptibility below T_f, and the monotonic behaviour of c_m in the vicinity of T_f are three experimental pieces of evidence for the presence of a spin-glass (SG) phase

282

transition[21]. Therefore the anomaly at T_f found for Mn-type SMSC was ascribed to a SG-paramagnetic transition. An inspection of the phase diagram shows that $T_f ->0$ when $x->0$. From this observation one may conjecture that the interactions involving this SG transition are relatively long-ranged (LR), since otherwise no freezing should have been observed for x below the percolation limit, which amounts to $x_c=0.18$ for fcc lattice. If one accepts the nature of the transition at T_f as canonical SG freezing (which is however, disputed[21]), then a scaling analysis should be applicable[22]. Such a scaling analysis generally exploits the fact that for a continuous, random distribution it is assumed that $R_{i,j}^3 x=$const., where $R_{i,j}$ denotes a typical distance between the ions. Implementation of this expression

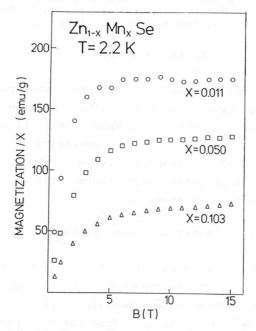

Fig.4 Magnetization per mole Mn of $Zn_{1-x}Mn_x$Se for x=0.011, 0.05 and 0.103. After Ref.19.

in a model of SG freezing, given a known functional form
for the radial dependence of the exchange interaction,
then yields a theoretical prediction of $T_f(x)$, which can
be compared with the experimental data. It was found[13,23]
that the experimental data obey the relation $\ln(T_f)\sim\ln(x)$
in the whole concentration range (Fig.5), what suggests
simple radial dependence of $J(R)$: $J(R)=J_o/R^n$. In Fig.5[13]
the available experimental data on $T_f(x)$ for SMSC are
plotted in the coordinates suitable for power dependence.
The exponent n deduced from Fig.5 is between n=5 (HgMnTe)
and n=7.6 (ZnMnS) — see Tab.2. We would like to stress
here that although it is clear that a power law yields

Table 1

E_d is Fe^{++} d-level position above the bottom of the
conduction band. E_g is the energy gap between conduction
and valence bands for x=0.

material	x range	structure	E_d (eV)[4]	E_g (eV)
ZnFeS	0.1		−2.3	3.8
ZnFeSe	0.2	cubic (fcc)	−1.6	2.8
ZnFeTe	0.01			2.4
CdFeSe	0.15	hexagonal	−1.2	1.8
CdFeTe	0.03	cubic	−1.45	1.6
HgCdFeSe	x<0.1	cub/hex	see Fig.7	
	y<0.5			
HgFeSe	0.1	cubic	+0.23	0

a better fit to the experimental data than an exponential decay in the concentration range from far below to far above the percolation limit, one should not exclude the possibility that the x dependence of T_f can be different above and below the percolation limit. Therefore one should be careful in drawing definitive conclusions from Fig.5.

2.3 Single Mn Ion and Mn-Mn Pair (theory)

We recall here some basic facts concerning the magnetic properties of an isolated Mn ion as well as Mn-Mn

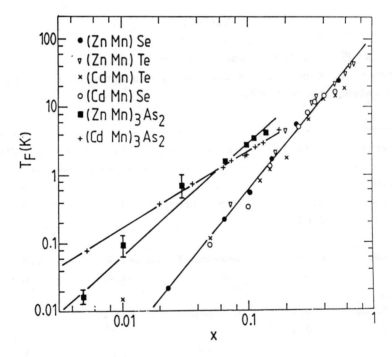

Fig.5 Freezing temperature T_f as a function of the Mn concentration x for various SMSC on logarithmic scale. The straight lines are fitted to the data yielding the power dependence $J(R) \sim R^{-n}$ as tabulated in Tab.2. After Ref.13.

Table 2

	d_{NN} (Å)	$-J_{NN}/k_B$ (K)	n	$N_0\alpha$ (eV)	$N_0\beta$ (eV)
ZnFeSe	4.0	22[11]	12[12]	0.22[8]	−1.74[8,103]
CdFeSe	4.28	19[11,37]		0.23[9]	−1.9[72]
HgCdFeSe		18[37]			
HgFeSe	4.3	18[37]			
ZnMnS	3.83	16[80]	7.6[80]		
ZnMnSe	4.00	12.6[80]	6.8[13]	0.26[19]	−1.31[19]
ZnMnTe	4.31	8.8[80]	6.8[81]	0.20[29]	−1.1[29]
CdMnS	4.12	10.5[82]		0.22[83]	−1.8[83]
CdMnSe	4.28	8.7[20]	6.8[18,84,81]		
		7.9[85]		0.23[75]	−1.26[75]
		10.6[17]			
CdMnTe	4.58	7[17]	6.8[14,18,81]		
		6.3[85]		0.22[28]	−0.88[28]
HgMnSe	4.30	11[17]	5[86]	0.9[87]	−1.5[87]
		6[88]		0.4[89]	−0.7[89]
HgMnTe	4.55	7.2[17]	5[14,1,90]	0.7[91]	−1.4[91]
		5.1[88]		0.4[89]	−0.6[89]
$(ZnMn)_3As_2$	2.94	100[92]	4.5[92]		
$(CdMn)_3As_2$	3.17	30[15]	3.5[15]		
PbMnSe	4.34	1.7[93]			
PbMnTe	4.56	1.4[94]	3[95]		

pairs. A substitutional Mn^{++} ion has d^5 electronic configuration. In a tetrahedral crystal field, its ground state is $^6A_1$, followed by $^4T_1$, $^4T_2$, $^4A_1$ and so on[1]. Since the first excited state ($^4T_1$) lies at about 2.4eV ($\sim 30000K$) above the ground state[24], all the excited states can be neglected in magnetic considerations. Consequently, magnetic properties of an isolated Mn ion are determined by the $^6A_1$ ground term, which is an orbital singlet (L=0) and a spin sextet (S=5/2). This state is a magnetically active one. The resulting magnetization is described by a Brillouin function and the low-field susceptibility exhibits typical Curie behaviour. On the other hand, in the absence of a magnetic field, there are no closely lying (few tens K) excited states which could be thermally populated, and such a system does not then contribute to the specific heat. Only if a magnetic field is applied, does a nonzero specific heat appear for a single Mn ion. The situation is essentially different for an Mn–Mn pair coupled by the exchange interaction The Hamiltonian for the pair is usually assumed in isotropic Heisenberg form

(1) $H = -2J S_1 S_2 + g\mu_B (S_1 + S_2) B$

where $J = J(R_{12})$.
In this case instead of a single sixfold degenerate level, one is dealing with a series of six levels. The ground state is now a magnetically inactive singlet, separated from the first excited level by $|2J|$. These excited states are sufficiently close to the ground state to be thermally populated, and then produce excess magnetic specific heat[14]. Application of a magnetic field lifts the degeneracy of the excited states. Since there is no field induced mixing between the ground and excited states, the ground state is crossed by the excited states for sufficiently high fields, which gives a step-like form of pair magnetization[20].

2.4 Extended Nearest Neighbour Pair Approximation (ENNPA)

The simplest magnetic systems discussed in the preceding section can be regarded as a basis for description of the relevant thermodynamic properties of an entire SMSC crystal with low Mn mole fraction. It follows from the data presented, that we are dealing with an array of magnetic ions immersed in a diamagnetic host lattice and coupled by an long-ranged (LR) interaction. Such a system can be described in the so called Extended Nearest Neighbour Pair Approximation (ENNPA), introduced by Matho[25] and recently successfully used for description of magnetic properties of Mn-based SMSC[15,13,16]. ENNPA is an approximative calculation method particularly useful for random, diluted systems with a LR interaction. It is based on the assumption that the partition function of the whole crystal can be factorized into contributions from pairs of ions. In this method, each ion is considered to be coupled by exchange interaction J_1 to its nearest magnetic neighbour only, and the latter may be located anywhere at a distance R_1. The statistical weights of pair configurations with different R_1 are given by a random distribution of magnetic ions. Therefore, any thermodynamic quantity can be expressed in the following form:

(2) $A = \sum_{l=1}^{\infty} A_1 (J_1) P_1 (x) / 2$

where A_1 stands for the pair specific heat, susceptibility or magnetization and depends on J_1. $P_1 (x)$ is the probability of finding at least one nearest Fe ion in the l-th shell and for a random distribution takes the form:

(3) $P_1 (x) = (1-x)^{n_{l-1}} - (1-x)^{n_l}$

where n_l is the number of lattice sites inside a shell of radius R_l. The calculations are performed numerically and summation over l shells in (3) is carried out up till $l=v$, for which:

$$\sum_{l=1}^{v} P_l(x) > 0.99$$

In practice $v=12$ was quite sufficient for all x. The only input parameters for this model are the exchange constants J_l for the different possible pairs. In Figs.1,2 we show example results of ENNPA calculations, where NN exchange constants were obtained from independent experiments (see also Ref.13,16).

3. OPTICAL PROPERTIES

Optical properties of SMSC were studied in the visible spectral range (1.5-3eV). Below the fundamental absorption edge typical intra-manganese absorption was observed, originating from $^6A_1 - \rangle^4T_1$, $^4T_2$, $^4A_1$ intra manganese transitions[1,24]. At higher photon energies, band to band transitions resulting in free exciton structures were also observed. From the exciton behaviour, especially in the presence of a magnetic field, one can infer data on conduction and valence bands, and consequently, about the s,p-d exchange interaction.

3.1 Free Exciton Spectroscopy

Well pronounced exciton reflectivity and absorption structures were observed in several SMSC[1]. In the presence of a magnetic field, large exciton splitting is

observed.   The spectra taken in the Faraday configuration
(i.e. magnetic field parallel to the incident light beam)
exhibit splitting into 4 components: two of them observable
in 6^+ polarization (strong A line and weak B line) and two
in 6^- one (weak C line and strong D line)[1,19,29]. The
6^+-6^- splitting of the strong components (E_D-E_A) tends to
saturation, in particular, for low concentrations.

3.2 s-d Exchange In SMSC

The   s-d exchange   interaction has   been   studied by
Liu[26]   and   for magnetic ions with vanishing   orbital
momentum can be written in the spirit of the mean field
theory   and   virtual   crystal   approximation[27]   in   the
following form:

(4) $H_{ex} = -\sum_n J(r-R_n) \langle S \rangle s = -JN_0 x \langle S \rangle s$

Assuming a magnetic field along the z axis,   we   obtain
energies of the exciton states[27]:

$$E_A = E_o + 3B - 3A$$
for 6^+ polarization

$$E_B = E_o + B + 3A$$

(5)

$$E_C = E_o - B - 3A$$
for 6^- polarization

$$E_D = E_o - 3B + 3A$$

where: E_o is the zero-field exciton energy ($E_o = E_g - E_{1s}$,
         exciton binding energy E_{1s} is assumed to be the
         same for A, B, C and D excitons),
         $A = 1/6(N_0\alpha)x\langle S_z \rangle$, $\alpha = \langle S|J|S \rangle$,
         $B = 1/6(N_0\beta)x\langle S_z \rangle$, $\beta = \langle X|J|X \rangle$ [27].

It follows from (5) that the ratio β/α can be evaluated
from the 6^+-6^- splitting as:

$$(6) \qquad \beta/\alpha = \frac{l+1}{l-1/3} \quad , \qquad l = \frac{E_C - E_B}{E_D - E_A}$$

The mean z component of an Mn ion spin $\langle S_z \rangle$, appearing in (5) can be easily related to macroscopic magnetization (per unit mass):

$$(7) \qquad M_m = (x/m) \ P$$

where $P = -\mu_B \langle L_z + g S_z \rangle$ is the magnetic moment of an ion and $m = (1-x)m_{Zn} + xm_{Mn} + m_{Se}$ is the mass of a SMSC molecule. In the Mn^{++} case $L=0$ and then $P = -\mu_B g \langle S_z \rangle$, and

$$(8) \qquad M_m = (x/m) g \mu_B |\langle S_z \rangle|$$

In this way, the exciton splittings (5) are seen to be proportional to macroscopic magnetization:

$$(9) \qquad \Delta E = E_D - E_A = (N_o \alpha - N_o \beta) \ (m/g\mu_B) \ M_m$$

It was demonstrated for different Mn-based SMSC[28,29,30,19] that, in fact, the exciton splitting is a linear function of macroscopic magnetization. A detailed discussion of s-d exchange parameters, as well as a comparison with other materials will be given later.

3.3 Bound Magnetic Polaron (BMP)

As a particular case of s-d exchange interaction we should mention so called BMP i.e. system of an electron localized on an impurity (donor or acceptor) interacting with some Mn ions[31]. Such a system exhibits many interesting effects, in particular zero field splitting (nonzero spin-flip energy) of donor or acceptor state due to nonvanishing local magnetization of Mn ions located inside impurity electron orbit. For more detailed discussion we refer to Ref.31.

III.Fe-BASED SMSC (L=2, S=2)

4. THE CRYSTALS

The Fe-based SMSC are grown by modified Bridgman technique (low or high pressure). The available crystals are tabulated in Tab.1. The energy gap variation with Fe concentration for ZnSe[8] and CdSe[32,33] is shown in Fig.6 , as well as for HgCdSe as a function of Cd content for low x (x<0.01) in Fig.7[35]. In Fig.7 we also show Fe (d^6) level position relatively to the energy bands: in zero-gap HgFeSe Fe d-level is located 0.2eV above the bottom of the conduction band, whereas in wide gap materials like CdFeSe or ZnFeSe it is located in the middle of the energy gap (see also Tab.1).

5.MAGNETIC PROPERTIES

5.1 Experimental Data

We start discussion of the magnetic properties by reviewing basic experimental data.

A. Specific heat

Specific heat data for Fe-based SMSC were obtained in the temperature range 0.4-20K. The representative results of heat capacity (c_p) are shown in Fig.8[36] for HgFeSe together with the data for pure HgSe. The problem of extracting magnetic contribution (c_m) to total c_p is not trivial[36,37]. Substitution of nonmagnetic cations (Hg, Cd

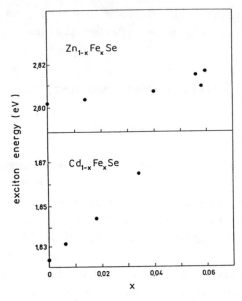

Fig.6 Free exciton energy (energy gap) for $Zn_{1-x}Fe_xSe$ and $Cd_{1-x}Fe_xSe$ as a function of Fe concentration. After Ref.8 and 32.

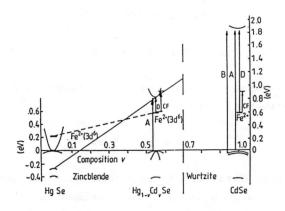

Fig.7 The schematic band structure of $Hg_{1-v}Cd_vSe$ at T=4.2K. The possible optical transitions are indicated by the arrows. The position of the Fe^{++} level in the crystals is also shown. After Ref.35.

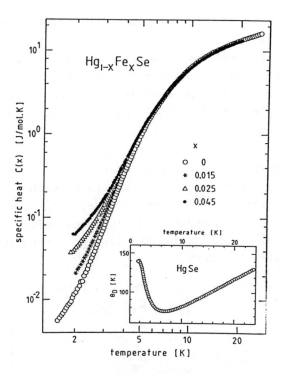

Fig.8 Specific heat (c_p) of $Hg_{1-x}Fe_xSe$. the inset shows the Debye temperature of pure HgSe. After Ref.36.

or Zn) by magnetic ions (Mn, Fe), which have smaller masses than Hg, Cd and slightly Zn, shifts low lying vibrational modes of the lattice to the higher frequencies. Thus reliable, positive excess specific heat can be observed only in low temperatures[36,37]. This situation is extreme for Hg compounds because of large Hg, Fe mass difference, whereas for Zn materials no great changes in vibrational spectrum are expected. In general one should obtain magnetic contribution to total c_p by:

$$c_m (A_{1-x} Fe_x B) = c_p (A_{1-x} Fe_x B) - c_{latt} (A_{1-x} M_x B)$$

where c_{latt} is specific heat of fictitious "nonmagnetic" $A_{1-x}Fe_xB$. In Ref.37 different methods of simulation of c_{latt} were proposed yielding practically the same c_m at low temperatures, but producing different results at high temperatures (T>10K). Nevertheless it seems that for all the investigated materials (ZnSe, CdSe, HgCdSe and HgSe) some general features of c_m can be observed:

- the overall behaviour of c_m resembles a Shottky-type anomaly but only for crystals with lowest Fe concentrations, for which a well pronounced, broad maximum is observed (Fig.9[6']) indicating energy gap between the ground and the excited states (this gap is estimated to 20K).

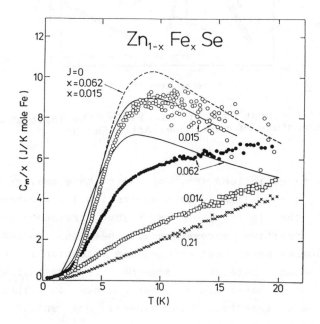

Fig.9 Magnetic specific heat per mole of Fe ions versus temperature for $Zn_{1-x}Fe_xSe$. Solid lines—ENNPA calculations with $J_{NN}=-22K$, dashed line — $J_{NN}=0$. $Dq=293cm^{-1}$, $\lambda=-95cm^{-1}$. After Ref.6.

- c_m practically does not depend on magnetic field ($B < 3T$) — Fig.10[37], which proves that the ground state is a singlet one. These two features are in contrast to Mn-type materials properties (see Sec.2).

- The decrease of c_m/x with x indicates the relevance of the Fe-Fe interaction. If the only contribution to c_m would be that coming from isolated ions, c_m would scale with x.

- The absolute values of c_m (at maximum) for CdSe (and HgSe) exceeds that of ZnSe considerably[37]. We will return to this point later.

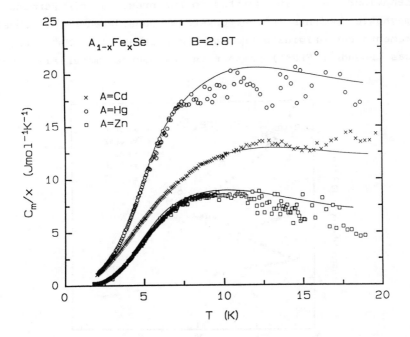

Fig.10 Magnetic specific heat per mole Fe (c_m/x) of $Cd_{1-x}Fe_xSe$ (x=0.034), $Hg_{1-x}Fe_xSe$ (x=0.024) and $Zn_{1-x}Fe_xSe$ (x=0.015) in the presence of an external magnetic field B=2.8T, each for one Fe concentration. The zero-field results are given by lines. After Ref.37.

B. Low temperature susceptibility

Some representative a.c. susceptibility data for ZnFeSe
measured in magnetic fields less of than 0.0001T, are
shown in Fig.11[12]. The data for low concentrations
(x<0.07) differ from susceptibility results reported
for Fe^{++}-doped crystals[38]: instead of temperature-
independent susceptibility at low temperatures (i.e.
typical Van Vleck-type paramagnetism), we observe a
monotonic increase of susceptibility and only a kink is
present at about 4K. Similar situation was also
encountered for CdFeSe[10] and HgFeSe[10]. This particular
behaviour of χ is ascribed to the presence of permanent
magnetic moments in the crystals considered[8,39]. These
moments can originate from the impurities (like Mn^{++}, which
was found in EPR[8]) present in the source materials used

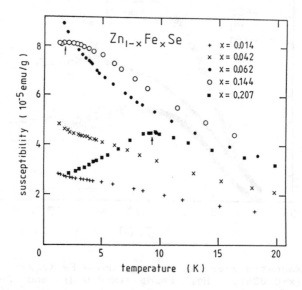

Fig.11 ac susceptibility of $Zn_{1-x}Fe_xSe$ as a function of
temperature. The arrows indicate anomaly discussed in the
text. After Ref.12.

for crystal growing (ZnFeSe, CdFeSe) or in the case of
HgFeSe, also from the selfionization process of Fe^{++} to
Fe^{+++} (which is, from the magnetic point of view,
equivalent to Mn^{++})[35,4]. In the latter case, the
concentration of Fe^{+++} increases with concentration of Fe^{++}
if $x<0.01$. For $x>0.01$ the concentration of Fe^{+++} stabilizes
at the level of $N=5*10^{18} cm^{-3}$ [35,4]. Another possible source
of permanent magnetic moments suggested recently is the
Bound Magnetic Polaron effect leading to s-d exchange-
interaction-induced magnetic moments localized on the
donors[32]. Although the amount of paramagnetic impurities
($10^{18}-10^{19} cm^{-3}$) is only very small in comparison with the
number of Fe^{++} ions, their Curie-type contribution to the
susceptibility can completely mask the temperature
independent susceptibility expected of Fe^{++} ions at low
temperatures. We should stress that although the above
interpretation is very probable, it still requires further
studies. In particular simultaneous susceptibility
measurements and independent determination of paramagnetic
impurity concentration would be very helpful in deriving
the final conclusions. At the moment one cannot exclude the
possibility that non-Van Vleck-type susceptibility reflects
a more complex situation of Fe ions than the assumed so
far.

For crystals with higher x one can observe a maximum in χ
(Fig.11). Since no specific heat anomalies occur for
these samples at the corresponding temperatures,
these maxima were tentatively ascribed to
paramagnetic-spinglass phase transition[10,12], in
analogy to the situation found in Mn systems. We should
stress, however, that this conjecture requires further
studies. In particular, zero-field-cooled and field-
cooled dc susceptibility should be measured, and a
wider composition range should be investigated. However,
strong improvement in crystal growing technology is
necessary to satisfy the latter condition. Nevertheless,

if one accepts the nature of the susceptibility anomaly
as spinglass freezing, the scaling analysis discussed in
Sec.2.2 should then be applicable. The result of a
similar procedure for Fe-based SMSC, is shown in
Fig.12[12]. The exponent n deduced from Fig.12 is $n \approx 12$,
much larger than that for Mn-type materials (see Tab.2).
However, one should be careful in ascribing this
difference solely to the difference in the
interaction ranges. As will be shown in Chap.5.2, the
energy level pattern of Fe^{++} is much more complicated than
that of Mn^{++}, and thus $T_f(x)$ does not necessarily reveal
the radial dependence of $J(R)$ in some simple way, as is
in the case for Mn (see discussion in Ref.12).

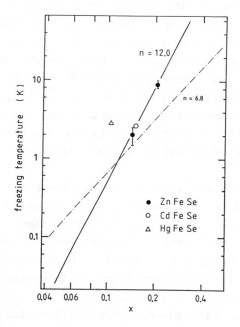

Fig.12 Freezing temperature as a function of Fe
concentration for ZnFeSe, CdFeSe and HgFeSe. The straight
lines indicate the radial decay of the exchange integral
$J \sim R^{-n}$. After Ref.12.

C. High temperature susceptibility

High temperature susceptibility of Fe-based SMSC reveals typical Curie-Weiss behaviour[10,11], as examplified in Figs.13 for HgFeSe and ZnFeSe. The Curie-Weiss temperatures are negative (suggesting antiferromagnetic interactions between Fe ions) and scale with x[10,11], similarly as for Mn-type SMSC.

D. Magnetization

Typical results of high field magnetization of ZnFeSe are shown in Fig.14[40,41] and 15[40]. Relatively strong magnetic field dependence is observed at low temperatures. In particular, saturation effects, present in Mn-based crystals (Chap.2.1), are observed only at very high magnetic fields[40,41] (Fig.14). The temperature dependence of magnetization is also different from that for Mn-type SMSC. Instead of being a monotonically decreasing function, the magnetization of ZnFeSe is practically temperature independent (or has a local maximum) between 2 and 6K, whereas for higher temperatures, it decreases monotonically with temperature[40] (Fig.15). It is worth noticing that magnetization per mole of iron decreases with increasing Fe content (Fig.14), as does c_m/x, also indicating the relevance of the Fe-Fe interaction. Preliminary high magnetic field data (Fig.14 - Ref.41) show gradual and smooth increase of magnetization with magnetic field[41]. No step-like structures are observed even at T=1.5K, in contrast to Mn-type DMS (see Sec.2.1 and 2.3). Another interesting fact is the anisotropy of magnetization found for cubic CdFeTe[42] and hexagonal CdFeSe[9].

5.2 Theoretical Description of the Magnetic Properties

We start this section by stressing the common overall
magnetic behaviour of all the Fe-based SMSC (including
HgFeSe) investigated so far. This observation can be
regarded as a strange one, in view of the different

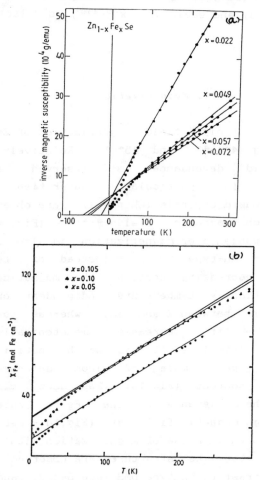

Fig.13 Inverse magnetic susceptibility of a). $Zn_{1-x}Fe_xSe$
and b). $Hg_{1-x}Fe_xSe$ as a function of temperature. The
straight lines show the Curie-Weiss law. After Ref.11, 10.

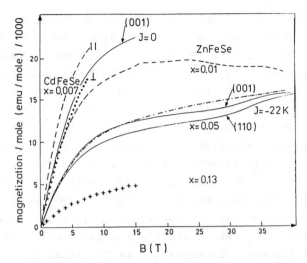

Fig.14 Magnetization per mole Fe of $Zn_{1-x}Fe_xSe$[41] and $Cd_{1-x}Fe_xSe$[7] (x=0.007, for B parallel and perpendicular to the crystal hexagonal axis). The solid lines show results of ENNPA calculations for B||(100) and (110); Dq=294cm^{-1}, λ =-95cm^{-1}.

locations of the Fe-energy level in relation to the band structure[35,4] (Fig.7). The weak Fe-level-dependence can be understood on the basis of recent calculations of CdMnTe band structure[43]. In Fig.16 we reproduce a schematic diagram for the d levels in II-VI compounds. In the case of the d^5 configuration (Mn-based SMSC), the five electrons occupy the e_+ and t_+ orbitals ($e_+^2t_+^3$ configuration) located deeply in the valence band. For Fe-type SMSC (d^6), the sixth electron must occupy an e_- orbital, yielding $e_+^2t_+^3e_-^1$ configuration. The reported variation of Fe level position with the host material[35] concerns precisely the electrons from e_- orbital[43,44]. These orbitals, however hybridize very weakly with p-bands[45] and therefore no first-order effects should be expected with variation of e_- level position relatively to the band structure.

The common magnetic behaviour of all the Fe-type SMSC suggests also the same theoretical description of all the crystals. Since in the previous section we evidenced that in Fe-based SMSC one is dealing with the system of interacting ions, which possess no permanent magnetic moments, one should start the discussion by reexamining the Hamiltonian describing such a situation. The general Hamiltonian for interacting ions possessing both spin and orbital momenta is rather complicated and contains a large number of unknown parameters[46]. However, in II-VI Fe-based SMSC the ^5T term is separated from the ground ^5E term by about 4000K[47], and the magnetic properties should therefore be determined mainly by the ^5E term, which has its orbital momentum largely quenched[3]. In that respect, it is reasonable to approximate the interaction of the ion system by an isotropic Heisenberg-type Hamiltonian:

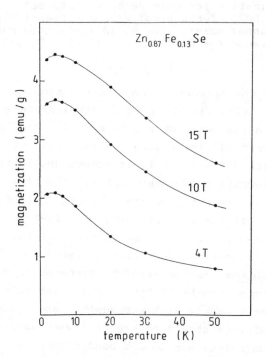

Fig.15 High-field magnetization of $Zn_{0.87}Fe_{0.13}Se$ as a function of temperature. the lines are to guide the eye only. After Ref.40.

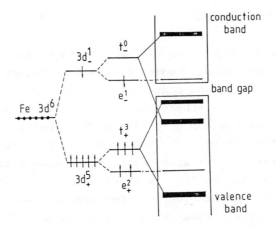

Fig.16 Schematic representation of the electronic Fe-levels
within the band structure of II-VI semiconductor, based on
calculations on Mn-containing CdTe[43].

(10) $H_{int} = -2 \sum_{ij} J_{ij} \mathbf{S}_i \mathbf{S}_j$

where the summation runs over all the Fe ions, $\mathbf{S} = (S_x, S_y, S_z)$
is the spin operator of an Fe ion and J is the exchange
integral. We should stress here that the main reason for
choosing the Hamiltonian in the form (10) is its simplicity
and the small number of parameters which must be inserted.
The full Hamiltonian of an Fe ion system would then read:

(11) $H = \sum_i \{ H_{cf} + \lambda \mathbf{S}_i \mathbf{L}_i + \mu_B (g\mathbf{S}_i + \mathbf{L}_i)\mathbf{B} \} - 2 \sum_{ij} J_{ij} \mathbf{S}_i \mathbf{S}_j$

where λ is the spin orbit parameter and H_{cf} is the crystal
field term[3], $\mathbf{L} = (L_x, L_y, L_z)$ is orbital momentum operator
and $\mathbf{B}$ is magnetic field. The Jahn-Teller effect of the ^5E
term is neglected here.
In order to understand magnetic situation in Fe-based SMSC
we discuss the simplest magnetic systems: single Fe ions
and Fe-Fe pairs (i.e. the systems which have been solved so
far).

5.2.1 Isolated Fe^{++} ion

The isolated Fe^{++} ion in a tetrahedral symmetry environment has been studied by Low and Weger[2]. They found that the ground state of the Fe^{++} (d^6) free ion (^{5}D) is split by a tetrahedral crystal field into a ^{5}E orbital doublet and a higher lying ^{5}T orbital triplet (separated from ^{5}E by $\triangle$ =10Dq, where Dq is crystal field parameter). The spin-orbit interaction splits the ^{5}E term into a singlet A$_1$, a triplet T$_1$, a doublet E, a triplet T$_2$ and a singlet A$_2$. Performing second order perturbation calculations within the ^{5}E term one finds that the energy separation between these states is

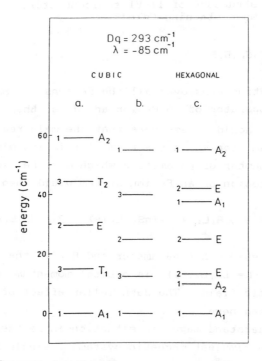

Fig.17 Energy levels of an isolated Fe^{++} ion in a cubic symmetry environment: a). Low and Weger model[2] b). Slack's model[3] as well as c). for hexagonal symmetry.

exactly equal to $6\lambda^2/10Dq$ (Fig.17a), where is the
spin-orbit parameter)[2]. The ground state is a
magnetically inactive singlet A_1, yielding Van Vleck-type
paramagnetism. It is worth noticing that in this model
admixtures of the ^5T term states to the ^5E term are
neglected, which results in vanishing orbital momentum of
the ^5E term. A better description of Fe^{++} was proposed by
Slack at al[3]. They have accounted exactly for ^5E-
^5T mixing, choosing appropriate zero-order basis
i.e. a basis simultaneously diagonalizing the crystal
field and spin-orbit Hamiltonians (Table IIIV and IX in
Ref.3). The energy levels obtained are shown in Fig.17b.
A slight difference can be noticed relative to the Low and
Weger model. Some other models[48,49] were based on either
the Slack one or the Low and Weger model. In hexagonal
case the main modification is splitting of triplets T_1 and
T_2 into singlets and doublets[38] (Fig.17c). In the presence
of magnetic field crystal hexagonality leads to anisotropy
of magnetization and susceptibility[9,38]. It is worth
mentioning that anisotropy effects can be also expected for
cubic crystals for high magnetic fields (therefore
magnetization should be anisotropic while susceptibility
should remain isotropic[48,38,42]).

5.2.2. Jahn-Teller coupling

In the above discussion of the Fe^{++} ion energetical
structure we have ignored Jahn-Teller (JT) effect. However
it is well known that for all the Transition Metal
impurities in II-VI or II-V compounds the JT coupling plays
very important role. In our case the JT coupling between
electronic E and T states and ϵ, τ phonon modes is
expected[50,52]. Roughly speaking this coupling leads to the
changement of internal structure of E, T terms and in
particular to lowering of the ground state energy (by JT

energy E_{JT})[55,49,50,52] . Many spectroscopic experiments
performed for Fe^{++} dopand have revealed the importance of
JT effect for ^5T term[52,55] . The structure of this term has
however minor importance for the magnetic properties, as we
noted before. On the other hand the available data have not
given any compelling evidence for the view that JT
coupling is significant in the ^5E term. Usually the ^5E
term energetical structure can be reasonably described by
the crystal field together with the spin—orbit
interaction[48,52,55] . However, one can not completely rule
out such a possibility. It was suggested that JT coupling
can be quite effective for phonons with energies
comparable to the spin—orbit splitting of the ^5E term[49] .
Such a situation was encountered for $CdTe:Fe$[49] . On the
other hand, if the phonon energies are appreciably larger
than the ^5E splittings, the most important effect[64] of
JT coupling is to reduce the intervals between the ^5E
states[52] . It is, however, difficult to distinguish
this reduction in the experimental data because it can
be accommodated in the "effective" spin—orbit parameter
(λ) value. For cubic ZnFeSe, phonon energies seem to be
large enough[53] to neglect the JT effect, as is the case
for $ZnS:Fe$[51,48] . In the case of hexagonal CdFeSe, the
situation is not so clear because of lower phonon
energies[54] . Too high absolute value of c_m mentioned in
Sec.5.1 may suggest importance of JT effect in this
material. The full consideration of JT effect (especially
for a Fe—Fe pair problem) is still to be worked out.

5.2.3 High Temperature Expansion series for Fe—based SMSC

In Secs.5.2.1 and 5.2.2 we considered only noninteracting
Fe ions. Using results for isolated ions some
information about ion—ion coupling can be gathered. It
was shown[11] that susceptibility for high temperatures,

can be expanded as a   Curie–Weiss law:

$$C(x)/(T-\Theta(x))$$

with the Curie constant

$$(12) \qquad C(x) = \frac{\mu_B 2N\langle M_z{}^2\rangle}{k_B V} x = C_o x$$

and a Curie–Weiss temperature

$$(13) \quad \Theta(x) = \frac{\langle M_z{}^2\rangle\langle E\rangle - \langle M_z{}^2 E\rangle}{\langle M_z{}^2\rangle k_B} + \frac{2\langle M_z S_z\rangle^2}{\langle M_z{}^2\rangle} x \sum_{p} z_p (J_p/k_B)$$

$$= A + \Theta_o x$$

where $\langle .. \rangle$ are quantum–mechanical averages of appropriate operators calculated for isolated Fe^{++} states[11]. Eqs.12 and 13 reduce to standard HTE for $L=0$[17]. If the distribution of magnetic ions is truly random both the Curie constant and the Curie–Weiss temperature should be linear functions of x. The values of the averages in (12) and (13) can be evaluated for a particular ion in a particular crystal field, if the ion eigenfunctions are known[11]. The NN exchange constants obtained in this way for several Fe–type SMSC are tabulated in Tab.2.

5.2.4 Theory of an Fe–Fe pair

A. Cubic case

The full Hamiltonian (11) reduces for an $Fe^{++}-Fe^{++}$ pair to the following expression[56,12]:

(14) $H = H_{cf} + H_{so} + H_B + H_{exch}$

where: H_{cf} is the crystal field term,

$H_{so} = \lambda S_1 L_1 + \lambda S_2 L_2$,

$H_B = (L_1 + L_2 + 2S_1 + 2S_2) \mu_B B$,

$H_{exch} = -2J_x S_{1x} S_{2x} - 2J_y S_{1y} S_{2y} - 2J_z S_{1z} S_{2z} =$

$= -2J S_1 S_2$

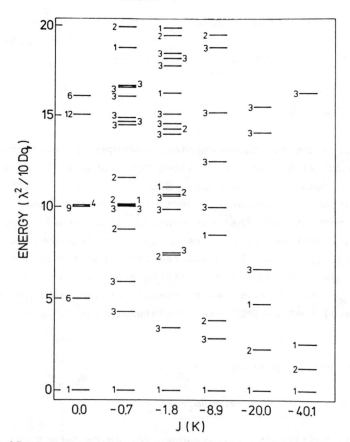

Fig.18 Low lying energy levels for Fe-Fe pair for example J values. $Dq = 293 cm^{-1}$, $\lambda = -85 cm^{-1}$. After Ref.6.

J_{ij} is the exchange constant and the indices i and j refer to ions in the pair considered. Since no information is provided about a possible anisotropy of J and then, in cubic case, H_{exch} is simplified to isotropic exchange. Numerical solution of Hamiltonian (14) leads to 100 energy levels. Lowest energy levels of a Fe—Fe pair are shown in Fig.18. It can be noticed that the exchange interaction does not significantly change the iron energy structure, in the sense that the ground state is still a singlet one. However, the first excited state approaches the ground state when J increases. In the presence of a magnetic field the degeneracy of the Fe—Fe pair is lifted (Fig.19). The effect of the field—induced coupling between the ground state and the excited states can also be noticed (Fig.19).

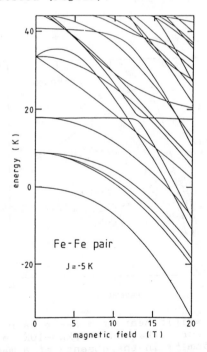

Fig.19 Splitting of the energy levels of a pair in a magnetic field (J/k_B=-5K, Dq=293cm⁻¹ and λ=-85cm⁻¹). After Ref.12.

The energy level scheme obtained can be used to calculate all the thermodynamic properties of an Fe-Fe pair. In Fig.20, we show the specific heat of a pair, for different exchange interactions. For all the J values presented, a well-pronounced maximum in the specific heat is observed, a feature similar to that seen for a single Fe ion (J=0). This maximum shifts to lower temperatures as the exchange interaction increases (this is a direct consequence of the reduction of the energy interval between the ground and the first excited states), and slightly changes its shape. Application of moderate magnetic fields influence the specific heat only slightly (less than 5% for B<6T), which is a consequence of the small changes of the energy level separations with magnetic field (cf Fig.19).

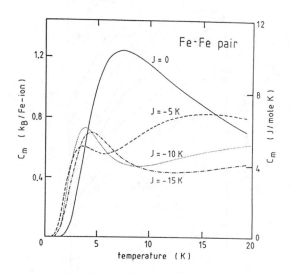

Fig.20 Magnetic specific heat of a Fe-Fe pair as a function of temperature, for $J/k_B=-5K$, $J/k_B=-10K$ and $J/k_B=-15K$ (Dq=293cm^{-1}, $\lambda =-85$cm^{-1}) in the absence of a magnetic field. Also shown, for comparison, is the noninteracting ions case (J=0). After Ref.12.

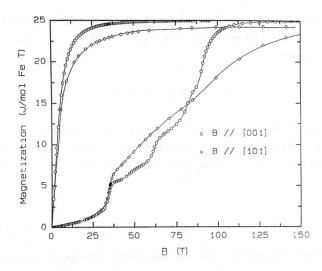

Fig.21 Calculated magnetization of an isolated Fe^{++} ion and Fe-Fe pair (J=-22K) for B||(100) and (110). After Ref.41.

In Fig.21 we show calculated magnetization of a Fe-Fe pair as a function of magnetic field[41]. The high field magnetization of both single Fe ion and Fe-Fe pair reveal pronounced anisotropy for different orientation of magnetic field in relation to the crystal axis. We notify characteristic step-like structure of the pair magnetization, which reflects crossings of the excited pair levels with the ground state, similarly as for Mn-Mn pair. However, in the present case the level crossing (and consequently steps in magnetization) depend strongly on interaction strength: it occurs only for rather strong interaction (like J=-22K, Fig.21) whereas it is absent if the interaction is rather weak (for instance J=-5K, Fig.19). In the latter case the exchange interaction is too weak to overcome the field-induced downward shift of the ground state, which therefore is not crossed by the excited states.

Finally for the calculated susceptibility the Curie-Weiss behaviour $(\chi \sim (T-\theta)^{-1})$ is predicted for the high temperature limit when the interaction is present in contrast to the Curie-type behaviour ($\chi \sim T^{-1}$) seen for a single Fe ion[3].

B. Hexagonal case

In hexagonal case no great changes are found[57,37]. The principal difference is additional strong anisotropy of magnetization and low field susceptibility for B||c and B⊥c[57].

5.3. Description of the Paramagnetic Phase of Fe-based SMSC

Using the results of the preceding sections (5.2.1, 5.2.4) one can now describe the relevant thermodynamic properties, such as specific heat, susceptibility and magnetization of, a "real" Fe-based SMSC. This was done for cubic ZnFeSe in ENNPA (described in Chap.3.4).

The specific heat calculated, together with the experimental data, are shown in Fig.9 for x=0.015 and for x=0.062. It is obvious that if the exchange interaction between the Fe ions is taken into account, a much better description of the experimental data can be obtained than is the case for the noninteracting ion model (J=0 in Fig.9). The latter description is reasonable for x=0.015, whereas substantial discrepancy is found for x=0.062. We should note, however, that the dilute limit, which is essential in ENNPA, may not be reached for x=0.06. Actually, for Mn-based SMSC only crystals containing less than $x \approx 0.05$ magnetic ions could be properly described by ENNPA[16,13]. Similar calculations for CdFeSe show that experimental c_m is always substantially larger than that calculated even for very low x

$(x=0.008)$[37]. These exceptionally large values of c_m were ascribed to Jahn-Teller effect, which supplies energetical spectrum with additional, vibrational levels and therefore increases c_m absolute value. Since JT effect was not taken into account in the pair calculations, the detailed comparison between ENNPA and experimental data for CdFeSe seems to be meaningless. Similar problem concerns HgFeSe for which experimental uncertainty at high temperatures is too large to perform reasonable comparison with the theory. Nevertheless we should point out that the low temperature behaviour of c_m for CdFeSe and HgFeSe is well recovered by ENNPA[37].

High field magnetization results are shown in Fig.14 for $Zn_{0.95}Fe_{0.05}Se$. Also in this case, it is evident that the ion-ion interaction has to be taken into account, although the model predicts steps which are not observed in experiment[41].

Magnetic susceptibility calculations are presented in Fig.22 for $Zn_{0.978}Fe_{0.022}Se$ and $Zn_{0.928}Fe_{0.072}Se$[6,11]. The characteristic curvature of the calculated susceptibility , which is indeed observed in experimental data, is due to exchange interaction and is neither observed for semi-isolated Fe ions[38], nor reproduced by calculations for isolated ions as shown in Fig.22 (J=0). Again, we observe increasing discrepancy between the calculations and experimental data as the concentration increases.

We summarize this section notifying that the theoretical model based on ENNPA and incorporating a solution of the pair problem provides a reasonable, simultaneous description of all the magnetic properties. This description is obviously not perfect, and limited to crystals with low Fe contents. At the present stage it is difficult to ascertain whether the discrepancy between the experimental data and the theory is solely due to limitations of the pair approximation, or rather,

314

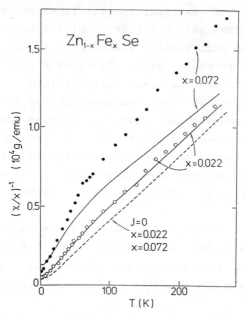

Fig.22 Inverse susceptibility per mole of Fe ions versus temperature for $Zn_{1-x}Fe_xSe$. Solid lines—ENNPA calculations with $J_{NN}=-22K$, dashed line — $J_{NN}=0$. $Dq=293cm^{-1}$, $\lambda=-95cm^{-1}$. After Ref.6.

to the simplicity of the Hamiltonian used. Generally speaking anisotropic terms (in both exchange and single ion terms) are expected in the Hamiltonian. However, data to support this conjecture are not available, and therefore these effects were not included in discussion. Nevertheless, the presented model seems to be the most complete one up to date.

5.4 The Nature of d–d Interaction

Finally we discuss the possible nature of the exchange interaction. We notice that although the exchange interaction is stronger for Fe than for Mn, it is of the same sign (AF) and of the same order of magnitude in both systems (Tab.2).

Another useful information would be knowledge about the interaction range in Fe-type DMS but, as we noted above, one can hardly derive pertinent conclusions about it on the basis of the data collected up to date.

Considering the different possible mechanisms we notice that the *dipole-dipole* interaction can be effective only at milikelvin temperatures. *RKKY* interaction[58] is also ineffective because of too low concentration of free carriers in Fe-type DMS (similarly as for Mn-type materials). The *Blombergen-Rowland* mechanism[59,60] would yield a strong dependence of exchange on the energy gap of the crystal. In contrast, we observe practically the same strength of exchange in the series HgFeSe (zero-gap material), HgCdFeSe (narrow gap material), CdFeSe as well as ZnFeSe (both wide gap crystals) - cf. Tab.2. Therefore, it seems very likely that also for an Fe ion system *superexchange*[61] is the dominant mechanism of interaction (at least for NN), as it was well established in Mn-type DMS[62,17]. This hypothesis is further supported by the fact that for both Fe and Mn ion systems, the major contribution to the exchange comes from the same electrons i.e. electrons on t_+ orbitals (Fig.16).

The chemical trends in d-d exchange could be discussed in the so called "3-level" model of superexchange interaction. This model is based on kinetic superexchange and was originally developed for Mn-based SMSC[62] but can be easily modified for the Fe case. Although "3-level" model provides simple analytical expressions for d-d and p-d exchange constants, it requires some material parameters to be inserted[62]. Current knowledge about these parameters for Fe-type DMS is rather poor and therefore we are not able to perform an analysis similar to that performed for Mn-based DMS.

The above observations are generally in agreement with the situation observed for the well-known Mn- and Fe-based antiferromagnets (Tab.3). These materials crystallize mainly in the fcc rock-salt (RS) structure or the zinc-blende (ZB) structure. Below the Neel temperature (T_N) they exhibit AF ordering. Since superexchange is the driving mechanism responsible for this ordering[61], the RS structure M–anion–M interaction paths for NN and NNN correspond to 90° ($pd\pi$–$pd\sigma$) and 180° ($pd\pi$–$pd\pi$) coupling, respectively. Therefore, $J_{NN} \approx J_{NNN}$ is expected, in agreement with the J values (Tab.3) and the observation of type II AF ordering. On the other hand, for ZB structures, only for the NN ions is the efficient M–anion–M (109°) superexchange possible, whereas for the NNN, inclusion of several anions is necessary, yielding $J_{NNN} \ll J_{NN}$. A similar situation is encountered for SMSC, where magnetic ions are tetrahedrally coordinated (cubic or hexagonal structure): for all the materials presented in Tab.2, $J_{NNN} \ll J_{NN}$ (J_{NNN}/J_{NN} varies from 3.4 for n=3.5 to 14 for n=7.6). Moreover the values of the exchange integrals are comparable for antiferromagnets (Tab.3) and SMSC (Tab.2). In particular, ZnFeSe and CdFeSe can be compared to FeS, since in all these materials Fe^{++} is tetrahedrally coordinated and the Fe^{++} ions have similar energy level patterns (in the RS structure, Fe^{++} is octahedrally coordinated and the energy diagram is then inverted, i.e. ^5T term is the ground term). We find it remarkable that both Mn or Fe chalcogenides and the SMSC, which in most cases have to be regarded as dilute or very dilute Mn or Fe chalcogenides, reveal similar magnetic behaviour. We believe this must result from the same (superexchange) interaction mechanism. More pertinent conclusions about interactions in Fe-based SMSC could be derived only if more experimental data were provided, especially for other Fe-type materials.

Table 3

	structure	T_N (K)	$-J_{NN}/k_B$ (K)	$-J_{NNN}/k_B$ (K)	type	
MnO	RS	116	7.2	3.5	II	[96]
FeO	RS	198	7.8	8.2	II	[96]
αMnS	RS	154	4.4	4.5	II	[96]
βMnS cub	ZB	155	10.5	7.2	III	[96]
		100	12.4		III	[97]
FeS	ZB	234[98]	15[99]		III?	
MnSe	RS	150[100]	7.9	12.8[101]	II	
MnSe	ZB	110[102]			III	

6. OPTICAL PROPERTIES

6.1 Far Infrared Spectroscopy

The first FIR experiments on crystals containing Fe^{++} were performed for ZnS:Fe[51,48] and CdTe:Fe[3], where the structure of ^5E term was investigated. Recently similar study was done for real mixed crystals of ZnFeSe, CdFeTe and CdFeSe[6,7,63]. In Fig.23 we show transmission spectra of ZnFeSe in the absence of magnetic field. Two well pronounced absorption lines are attributed to intra single Fe ion A_1-T_1 and A_1-T_2 transitions[6]. The results of laser FIR spectroscopy in the presence of magnetic field corroborate this assignment[63]. We should notice that this assignment is correct for low Fe concentrations. In the crystals with higher Fe content, one can also expect transitions between pair levels. In particular the differences observed between the spectra of $Zn_{0.985}Fe_{0.015}Se$ and $Zn_{0.938}Fe_{0.062}Se$ may be due to increasing contribution of intra pair transitions.

318

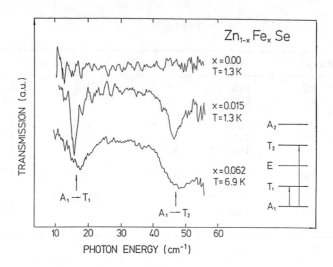

Fig.23 Transmission versus energy for $Zn_{1-x}Fe_xSe$, x=0.00, 0.015and 0.062. After Ref.6.

6.2 IR Intra Ion Transitions

Except of FIR transitions within ^5E term, IR transitions $^5E-^5T$ were observed for both doped[47,52] and mixed crystals[55,65] (Fig.24). These transitions have been recently re-analysed[55] and the role of JT effect has been pointed out.

We mention that direct observation of intra ion transitions within ^5E term together with the $^5E-^5T$ transitions (in particular Zero Phonon Line) allows one to determine crystal field (Dq) and spin-orbit (λ) parameters (see Tab.4).

6.3 Free Exciton Spectroscopy

Band to band optical transitions were first observed for zero-gap HgFeSe in FIR spectroscopy[44]. For open-gap materials, in the visible spectral range (1.5-3eV), band to band transitions, resulting in free exciton

Table 4

Parameters of the crystal field model[35]. Parameters v, v' characterize hexagonal symmetry[32,37].

material	Dq	λ	v	v'
ZnFeSe	293	−95	0	0
CdFeSe	257	−93	30	35
HgCdFeSe	280	−80	0	0
HgFeSe	293	−85	0	0

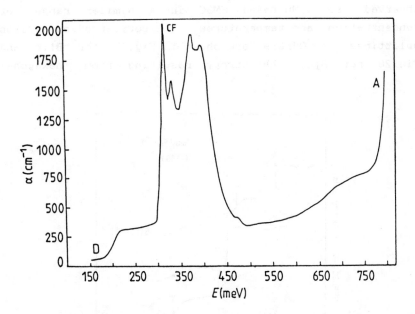

Fig.24 Absorption of $Hg_{0.40}Cd_{0.55}Fe_{0.05}Se$ at T=4.2K. The tresholds ascribed to D transitions, $^5E \rightarrow ^5T$ transition (CF) and absorption edge (A) are indicated. After Ref.35.

320

structures, were observed for ZnFeSe[67,8,68,69,70], CdFeSe[71,72], CdFeTe[34] and ZnFeTe[73]. A representative exciton reflection spectrum is shown in Fig.25. In the presence of a magnetic field, ZnFeSe and CdFeSe show a large splitting of the exciton structure (Figs.26, 27 and 28). For cubic materials the reflection spectra split into 4 components (lines A,B in 6^+ and C,D in 6^- polarization — Fig.26), in analogy with the case of Mn-type SMSC (Sec.3.1). The 6^+–6^- splitting ($\Delta E=E_D-E_A$) is shown in Fig.26b for several different ZnFeSe samples. It should be noticed that these splittings do not exhibit the tendency to saturation with the magnetic field observed for Mn-based SMSC in a similar range of concentration and temperatures. The corresponding exciton splittings of CdFeSe are shown in Fig.27 for $B \parallel c$ and Fig.28 for $B \perp c$. Anisotropy resulting from hexagonal

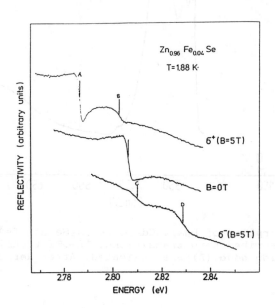

Fig.25 Exciton reflectance spectra of $Zn_{0.96}Fe_{0.04}Se$ at T=1.88K. After Ref.8.

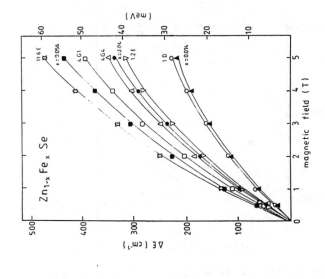

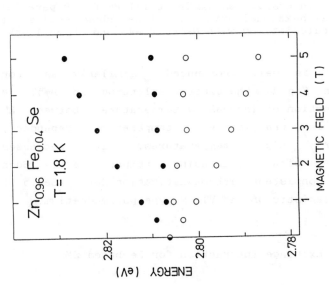

Fig.26 a). Energies of the 6^+ (lines A and B) and 6^- (lines C and D) components of the exciton in $Zn_{0.96}Fe_{0.04}Se$ versus the magnetic field at, T=1.8K. b). Exciton splitting ($\Delta E = E_D - E_A$) for different x at T=1.8K. The solid lines are guide to the eye only. After Ref.8.

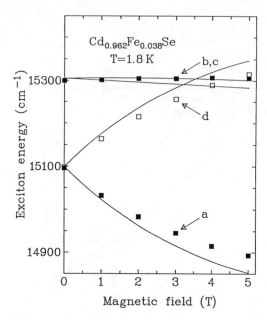

Fig.27 Energies of the exciton lines (a,b — 6$^+$; c,d — 6$^-$) (reflectivity) in Cd$_{0.962}$Fe$_{0.038}$Se at T=1.8K for B parallel to the crystal hexagonal axis. The lines show results of theoretical calculations with N$_0\alpha$=0.23eV and N$_0\beta$=-1.9eV. After Ref.72.

symmetry[74,30] is well pronounced, similarly as for CdMnSe[75]. In Fig.29 the exciton splitting in B=5T is shown as a function of increasing temperature. Between 2K and 6K $\triangle$E is practically temperature-independent, whereas for higher temperatures, it decreases monotonically with increasing temperature[8]. Such behaviour is consistent with magnetization data (Sec.5.1) and is characteristic of Van Vleck-type paramagnetism.

6.4 s-d Exchange Interaction for Fe-based SMSC

The band splittings in Fe-based SMSC resulting from the s-d exchange interaction can, in general, be treated in a way similar to that adopted for Mn-based materials

(Sec.3.2). Only the exchange Hamiltonian (4) has to be revised. This hamiltonian is the correct one when the interacting states are simple multiplets[26], as for materials containing Mn^{++}. Although for the iron case the situation is more complicated (all the five multiplets resulting from the ^5E term should be considered, as well as the influence of the higher lying ^5T term), we assume for further analysis an exchange Hamiltonian in the form (4) having in mind similar exciton behaviour in Mn- and Fe-based SMSC. The mean ion spin $\langle S_z \rangle$ in Mn^{++} case was easily expressed by macroscopic magnetization (Eq.8). In Fe^{++} this problem requires more careful study. The mean iron spin $\langle S_z \rangle$ can be calculated using isolated Fe^{++} ion wavefunctions for the ^5E term[3].

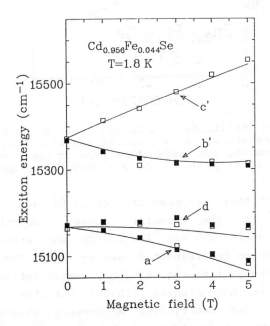

Fig.28 Energies of the exciton lines (a,b – 6^+; c,d – 6^-) (reflectivity) in $Cd_{0.956}Fe_{0.044}Se$ at T=1.8K for B perpendicular to the crystal hexagonal axis. The lines show results of theoretical calculations with $N_0\alpha=0.23eV$ and $N_0\beta=-1.9eV$. After Ref.72.

324

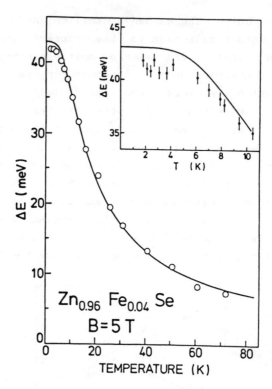

Fig.29 Exciton splitting ($\Delta E = E_D - E_A$) for $Zn_{0.96}Fe_{0.04}Se$ versus the temperature, in a constant magnetic field B=5T. The inset shows the low temperature part in more details. The solid line is a guide to the eye only. After Ref.8.

It was found[8] that the mean spin of an Fe ion $\langle S_z \rangle$ is proportional to the magnetic moment $\langle M_z \rangle = \langle L_z + 2S_z \rangle$: $\langle S_z \rangle = k \langle M_z \rangle$, where k=0.447 for ZnFeSe and k=0.444 for CdFeSe[32], whereas for spin-only case k=1/2[76]. The smaller k value (in respect to the spin-only case) reflects the contribution of nonvanishing orbital momentum to the magnetization. The coefficient k depends slightly on the temperature as well as on the magnetic field but this variation is far below experimental accuracy of exciton splitting, and we therefore assume constant k value in further considerations.

Recently similar analysis was performed for Fe-Fe pair[77]. It was found that also in this case pair spin is proportional to the pair magnetic moment with the coefficient k=0.462 (ZnFeSe) and k=0.469 (CdFeSe). Therefore, in the spirit of ENNPA, one can assume that macroscopic magnetization measures $\langle S_z \rangle$ (at least for not very concentrated crystal, for which ENNPA apply) and that the proportionality factor k is the same as for a single ion. Under this assumption we get for the magnetization:

(15) $M_m = |\langle M_z \rangle| (\mu_B x/m) = |\langle S_z \rangle| (\mu_B /k)(x/m)$

where $m=(1-x)m_{Zn}+ xm_{Fe}+ m_{Se}$ is, as in (4.2.7) and (4.2.8), mass of a MeFeSe molecule. Exciton splitting E can thus be expressed in the same way as for Mn-type SMSC (Eq.9):

(16) $\Delta E = E_D - E_A = (N_o\alpha - N_o\beta)$ mk M_m/μ_B

Thus plotting ΔE versus mkM_m/μ_B one should obtain a straight line with a slope $(N_o\alpha - N_o\beta)$. This was done only for ZnFeSe[8] (Fig.30) and CdFeSe[72] so far. For both materials similar values of s,p-d exchange are found (Tab.2).

The s-d exchange for the conduction band ($N_o\alpha$) is practically the same for Fe and Mn SMSC, a result which could have been expected[78]. The exchange interaction for the valence band ($N_o\beta$) seems to be stronger for Fe-based materials (suggesting stronger hybridization of the Fe level with the valence band and/or different location of the Fe level in the valence band), although a detailed comparison is difficult because the data for materials other than ZnFeSe or CdFeSe are lacking.

The p-d exchange integrals could be used for discussion of chemical trends in d-d exchange interaction of Fe-type SMSC (see Sec.5.4). However as we noticed the available data are still too poor for comprehensive analysis.

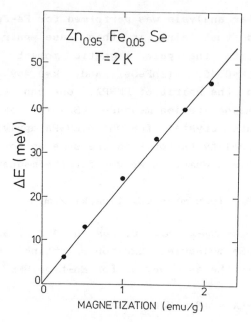

Fig.30 Splitting of the exciton ($\Delta E = E_D - E_A$) versus M_n. The straight line is the theoretical dependence for $N_o(\alpha-\beta) = 1.96eV$[103]. After Ref.8.

6.5   Polaron Effects

The first data on BMP in Fe-type SMSC are provided by Raman spectroscopy performed on CdFeSe[32]. Although no characteristic zero field splitting was observed in this case, this does not exclude the existence of BMP. the calculations assuming isotropic Heisenberg-type s-d exchange[32] shows that in fact first order correction to the complex's energy vanishes at B=0 leading to the absence of zero field splitting. However second order effects lower energy of the ground state (by 0.2cm for x=0.02[32]) as well as mixes wavefunctions of some states, yielding noncrossing effects. Such noncrossings of Raman lines were observed in experiment[32]. It was also shown that s-d exchange results in nonvanishing local spin polarization ($<S>=0.06$ at the center of polaron). Such a local spin polarization can produce effects similar to the paramagnetic impurities and therefore can also contribute significantly to the low temperature susceptibility masking its Van Vleck-type character (see Sec.5.1).

IV. CONCLUDING REMARKS

In this chapter we presented results of the study of basic magnetic and optical properties of bulk Fe-based SMSC. It follows from the data presented above that the nonvanishing orbital momentum of Fe^{++} significantly influences the magnetic properties of Fe-based SMSC: the specific heat, susceptibility and magnetization of these materials differ appreciably from those of Mn-based SMSC. On the other hand, the optical properties, and in particular the band splittings, seem to follow macroscopic magnetization like analogous quantities for the Mn-type materials and, in this sense, are influenced indirectly by the orbital momentum of magnetic ions.

We also discussed some attempts to interpret these data. Although some problems seem to be solved, there are still some fundamental questions (problems) to be worked out:

- The leading terms in the Hamiltonian describing the Fe ion system seem already to be taken into account. However, some additional terms should probably be added in order to provide better ENNPA description of experimental data.

- The existence of a spinglass-paramagnetic phase transition should be established by both experimental studies (in a wide Fe concentration range, new experimental methods like Mossbauer[79] spectroscopy e.t.c.) and understanding microscopic situation of freezing field-induced magnetic moments. Proper elaboration of this problem can be helpful in a study of the interaction range.

- The problem of the range of interaction between Fe ions is important in establishing the nature of the exchange interaction. It is less relevant from the point of view of description of experimental data, since a weakly coupled Fe-Fe pair behaves in nearly the same way as an isolated Fe ion. Therefore, inclusion of long ranged interaction is not as crucial as for Mn-based materials.

- There are some experimental suggestions about the important role of the Jahn-Teller effect for the ^5E term. This aspect should be investigated, in particular for the pair problem.

- The possible interaction between bivalent Fe^{++} and Fe^{+++} subsystems (in HgFeSe) should also be considered.

- Very recently quantum well (QW) structures based on ZnFeSe/ZnSe were fabricated and interesting properties of this system were found[70]. In particular large magnetic field induced p-d splittings of the ZnFeSe valence band yields simultaneous existence of type I and type II QW depending on valence band spin state. One can expect many new results on this subject in the future.

Acknowledgement

I'm very greatful to Professors W.J.M. de Jonge,
J.K.Furdyna, R.R.Galazka, A.M.Mycielski and J.A.Gaj for
numerous discussions.

330

References

1. see: Semiconductors and Semimetals, **25**, "Diluted Magnetic Semiconductors" ed. J.K.Furdyna and J.Kossut, Academic Press (1988) and rewiew paper: Furdyna,J.K., J.Appl.Phys. **64** (4), R29 (1988)

2. Low, W. and Weger, M. Phys.Rev. **118**, 1119 (1960)

3. Slack, G.A., Roberts, S. and Wallin, J.T., Phys.Rev. **187**, 511 (1969)

4. see review paper: Mycielski,A., J. Appl. Phys. **63**, 3279 (1988) and references therein.

5. Dobrowolski,W., Dybko,K., Skierbiszewski,C., Suski,T., Litwin-Staszewska,E., Miotkowska,S., Kossut,J. and Mycielski,A., in Proc. of 19th Int. Conf. on the Phys. of Semiconductors (Warsaw 1988) ed. W.Zawadzki, Institute of Physics, Polish Academy of Sciences, p.1247 (Wroclaw 1989)

6. Twardowski,A., Swagten,H.J.M. and de Jonge,W.J.M., in Proc. of 19th Int. Conf. on the Phys. of Semiconductors (Warsaw 1988) ed. W.Zawadzki, Institute of Physics, Polish Academy of Sciences, p.1543 (Wrocław 1989)

7. Witowski,A.M., Twardowski,A., Pohlmann,M., de Jonge,W.J.M., Wieck,A., Mycielski,A. and Demianiuk,M., Solid State Commun. **70**, 27 (1989)

8. Twardowski,A., Glod,P., de Jonge,W.J.M., Demianiuk,M., Solid State Commun. **64**, 63 (1987)

9. Heiman,D., Petrou,A., Bloom,S.H., Shapira,Y., Isaacs,E.D. and Giriat,W., Phys. Rev. Letters, **60**, 1876 (1988) D.Heiman, E.D.Isaacs, P.Becla, A.Petrou, K.Smith, J.Marsella, K.Dwight and A.Wold, in Proc. of 19th Int. Conf. on the Phys. of Semiconductors (Warsaw 1988) ed. W.Zawadzki, Institute of Physics, Polish Academy of Sciences, p.1539 (Wrocław 1989)

10. A.Lewicki, J.Spalek and A.Mycielski, J.Phys. C, **20**, 2005 (1987)

11. A.Twardowski, A.Lewicki, M.Arciszewska, W.J.M.de Jonge, H.J.M.Swagten, M.Demianiuk, Phys. Rev. **B38**, 10749 (1988)

12. H.J.M.Swagten, A.Twardowski, W.J.M. de Jonge and

M.Demianiuk, Phys.Rev. **B39**, 2568 (1989)

13. A.Twardowski, H.J.M.Swagten, W.J.M.de Jonge and
    M.Demianiuk, Phys.Rev. **B36**, 7013 (1987) and references
    therein.

14. R.R.Galazka, S.Nagata , P.H.Keesom, Phys. Rev. **B22**,
    3344 (1980); S.Nagata , R.R.Galazka, D.P.Mullin,
    H.Arbarzadeh, G.D.Khattak, J.K.Furdyna, P.H.Keesom,
    Phys. Rev. **B22**, 3331 (1980)

15. C.J.M.Denissen, H.Nishihara, J.C.van Gool, W.J.M. de
    Jonge, Phys. Rev. **B 33**, 7637 (1986)
    C.J.M.Denissen and W.J.M. de Jonge, Solid State
    Commun. **59**, 503 (1986)

16. H.J.M.Swagten, A.Twardowski, W.J.M.de Jonge,
    M.Demianiuk, J.K.Furdyna, Solid State Commun. **66**, 791
    (1988)

17. J.Spalek, A.Lewicki, Z.Tarnawski, J.K.Furdyna,
    R.R.Galazka, Z.Obuszko, Phys. Rev **B33**, 3407 (1986)

18. M.A.Novak, O.G.Symko, D.J.Zheng, S.Oseroff,
    J.Appl.Phys. **57**(1),3418 (1985)
    M.A.Novak, O.G.Symko, D.J.Zheng, S.Oseroff, Phys.
    Rev.**B33**, 6391 (1986)

19. A.Twardowski, M. von Ortenberg, M.Demianiuk,
    R.Pauthenet, Solid State Commun. **51**, 849 (1984)

20. R.L.Aggarwal, S.N.Jasperson, Y.Shapira,
    S.Foner,T.Sakibara, T.Goto, N.Miura, K.Dwight and
    A.Wold, Proc. of the 17th Int. Conf. on the Phys. of
    Semiconductors, San Francisco 1984, p.1419 (ed.
    J.D.Chadi and W.A.Harrison) Springer, New York (1985)
    Y.Shapira, S.Foner, D.H.Ridgley, K.Dwight and A.Wold,
    Phys. Rev. **B30**,4021 (1984)
    J.P.Lascaray, M.Nawrocki, J.M.Broto, M.Rakoto and
    M.Demianiuk Solid State Commun. **61**, 401 (1987)

21. J.A.Maydosh in Hyperfine Interactions, Vol.31 of
    Lecture Notes in Physics (Springer, New York 1986),
    p.347

22. J.Souletie, J. de Phys. **C2**, 3 (1978)

23. W.J.M.de Jonge, A.Twardowski and C.J.M.Denissen
    in Material Research Society Fall Meeting (Boston
    '86), Symposium Proceedings **89**, 156 (1987), Diluted
    Magnetic (Semimagnetic) Semiconductors, ed. S.von
    Molnar, R.L.Aggarwal and J.K.Furdyna

24. J.E.Morales Toro, W.M.Becker, B.I.Wang, U.Debska and J.W.Richardson, Solid State Commun. **52**, 41 (1984) B.Oczkiewicz, A.Twardowski and M.Demianiuk, Solid State Commun., **64**, 107 (1987)

25. K.Matho, J. Low Temp. Physics **35**, 165 (1979)

26. S.H.Liu, Phys. Rev. **121**, 451 (1960)

27. J.A.Gaj, J.Ginter and R.R.Galazka, Phys.Stat.Sol. (b) **89**, 655 (1978)

28. J.A.Gaj, R.Planel, G.Fishman, Solid State Commun., **29**,435 (1979)

29. A.Twardowski, P.Swiderski, M. von Ortenberg, R.Pauthenet, Solid State Commun. **50**, 509 (1984)

30. R.L.Aggarwal, S.N.Jasperson, J.Stankiewicz, Y.Shapira, S.Foner, B.Khazai and A.Wold, Phys. Rev. **B28**, 6907 (1983)

31. see review paper: P.A.Wolf, in Semiconductors and Semimetals, **25**, "Diluted Magnetic Semiconductors" ed. J.K.Furdyna and J.Kossut, Academic Press (1988)

32. D.Scalbert, J.A.Gaj, A.Mauger, J.Cernogora C.Benoit a la Guillaume, Phys. Rev. Lett. **62**, 2865 (1989) D.Scalbert, J.Cernogora, A.Mauger and C.Benoit a la Guillaume and A.Mycielski, Solid State Commun.**69**, 453 (1989)

33. J.A.Gaj, private communication

34. A.Mycielski, private communication

35. A.Mycielski, P.Dzwonkowski, B.Kowalski, B.A.Orlowski, M.Dobrowolska, M.Arciszewska, W.Dobrowolski and J.M.Baranowski, J.Phys. C, **19**, 3605 (1986)

36. H.J.M.Swagten, F.A.Arnouts, A.Twardowski, W.J.M.de Jonge and A.Mycielski, Solid State Commun. **69**, 1047 (1989)

37. A.Twardowski, H.J.M.Swagten and W.J.M. de Jonge, SPECIFIC HEAT OF THE DILUTED MAGNETIC SEMICONDUCTOR $Hg_{1-x-y}Cd_yFe_xSe$, to be published

38. J.Mahoney, C.Lin, W.Brumage and F.Dorman, J.Chem.Phys. **53**,4286 (1970)

39. M.Arciszewska, A.Lenard, T.Diel, W.Plesiewicz, T.Skoskiewicz and W.Dobrowolski, MAGNETIC

SUSCEPTIBILITY OF HgFeSe, Acta Physica Polonica, in press (1990)

40. A.Twardowski, M.von Ortenberg and M.Demianiuk, J.Cryst.Growth **72**, 401 (1985)

41. H.Draaisma, H.J.M.Swagten, W.J.M.de Jonge, A.Twardowski, unpublished

42. C.Testelin, A.Mauger, C.Rigaux, M.Guillot and A.Mycielski, Solid State Commun. in press (1989)

43. Su-Huai Wei and A.Zunger, Phys. Rev. **B35**, 2340 (1987)

44. B.A.Orlowski, B.J.Kowalski, A.Sarem, A.Mycielski, B.Velicki and V.Chab, in Proc. of 19th Int. Conf. on the Phys. of Semiconductors (Warsaw 1988) ed. W.Zawadzki, Institute of Physics, Polish Academy of Sciences, p.1267 (Wrocław 1989)
B.J.Kowalski, V.Chab, B.A.Orlowski, J.Majewski, A.Sarem and A.Mycielski, Acta Physica Polonica **A73**, 455 (1988)
A.Kisiel, M.Piacentini, F.Antonangeli, N.Zema and A.Mycielski, Solid State Commun. **70**, 693 (1989)

45. Weak hybridization with s-type conduction band (cation-type) can be also expected. This hybridization should be however weak because of small overlap of Fe and nonmagnetic cation s-orbitals.

46. D.Bloor, G.M.Copland, Rep. Prog. Phys. **35**, 1173 (1972)
P.M.Levy, Phys. Rev. **177**, 509 (1969)

47. J.M.Baranowski, J.W.Allen and G.L.Pearson, Phys. Rev. **160**,627 (1967)

48. J.T.Vallin, G.A.Slack and C.C.Bradley, Phys. Rev. **B2**, 4406 (1970)

49. J.T.Vallin, Phys. Rev **B2**, 2390 (1970)

50. F.S.Ham, Phys. Rev. **166**, 307 (1968)

51. G.A.Slack, S.Roberts and F.S.Ham, Phys. Rev. **155**, 170 (1967)

52. F.S.Ham and G.A.Slack, Phys. Rev. **B4**, 777 (1971)

53. E.Carnell, Bull. Am. Phys. Soc.**40**, 1086 (1965)
B.Hanion, F.Moussa, G.Pepy and K.Kunc, Phys. Lett. **36A**, 376 (1971)

54. R.Geich, C.H.Perry and S.S.Mitra, J. Appl. Phys. **37**,

1994 (1966)
A.K.Arara and A.K.Ramdas, Phys. Rev. **B35**, 4345 (1987) and references therein.

55. G.A.Slack, F.S.Ham and R.M.Chrenko, Phys. Rev. **152**,376 (1966)
     M.A.de Oure, J.Rivere—Iratchet and E.Vogel, J.Crystl Growth **86**, 28 (1988)
     J.Rivere—Iratchet, M.A.de Oure and E.Vogel, Phys. Rev. **B34**, 3992 (1986)
     E.Vogel, J.Rivere—Iratchet and M.A.de Oure, Phys. Rev. **B38**, 3556 (1986)
     E.Vogel, J.Rivere—Iratchet and M.A.de Oure, Phys. Rev. **B22**, 4511 (1980)

56. A.Twardowski, H.J.M.Swagten, T.F.H.v.d.Wetering and W.J.M.de Jonge, Solid State Commun. **65**, 235 (1988)

57. A.Twardowski, Solid State Commun. **68**, 1069 (1988)

58. M.A.Ruderman and C.Kittel, Phys. Rev. **96**, 99 (1954)
     K.Yoshida, Phys. Rev. **106**,893 (1957)
     T.Kasuya, Prog. Theor. Phys. **16**, 45 (1956)

59. N.Blombergen and T.J.Rowland, Phys. Rev. **97**, 1679 (1955)

60. C.Lewiner and G.Bastard, J.Phys. C13, 2347 (1980)
     C.Lewiner, J.A.Gaj and G.Bastard, J.Phys.(Paris) **41**, C5-289 (1980)

61. P.W.Anderson, in Solid State Physics **14**, 99 (1963), ed. F.Seitz and D.Turnbull, Academic Press
     G.A.Sawatzky, W.Geertsma and C.Hass, J. of Magn. and Magn. Materials **3**, 37 (1976)
     C.E.T.Goncalves da Silva and L.M.Falicov, J.Phys. C5, 63

62. B.E.Larson, K.C.Hass, H.Ehrenreich and A.E.Carlsson, Phys. Rev. **B37**, 4137 (1988)
     H.Ehrenreich, K.C.Hass, B.E.Larson and N.F.Johnson in Material Research Society Fall Meeting (Boston '86), Symposium Proceedings **89**, 159 (1987), Diluted Magnetic (Semimagnetic) Semiconductorsuctors, ed. S.von Molnar, R.L.Aggarwal and J.K.Furdyna
     B.E.Larson, K.C.Hass, H.Ehrenreich and A.E.Carlsson, Solid State Commun. **56**, 347 (1985)

63. M.Hauseman, A.Twardowski, C.L.Claessen, A.Wittlin, M.von Ortenberg, W.J.M.de Jonge and P.Wyder, Solid State Commun, in press

64. We notice that the magnetic properties of the ion are

determined by the relative energy differences between the ground and the excited states. Therefore the magnitude of JT energy E_{JT} is of less importance.

65. M.Arciszewska, unpublished

66. Y.Guldner, C.Rigaux, M.Menant, D.P.Mullin and J.K.Furdyna, Solid State Commun. **33**, 133 (1980) H.Serre, G.Bastard, C.Rigaux, J.Mycielski and J.K.Furdyna in: Proc.Int.Conf. on Narrow Gap Semiconductors, Linz '81, Lecture Notes in Physics, Vol.**152** (Springer, Berlin 1982), p.321

67. A.V.Komarov, S.M.Ryabchenko and O.V.Terletskii, Phys.Stat.Sol. (b) **102**, 603 (1980)

68. B.T.Jonker, J.J.Krebs, S.B.Quadri and G.A.Prinz, Appl.Phys.Lett. **50**, 848 (1987)

69. X.Liu, A.Petrou, B.T.Jonker, G.A.Prinz, J.J.Krebs and J.Warnock, Appl. Phys. Lett. **53**, 476 (1988)

70. X.Liu, A.Petrou, J.Warnock, B.T.Jonker, G.A.Prinz and J.J.Krebs, submitted to Phys. Rev. Lett.

71. A.Petrou, X.Liu, G.Waytena, J.Warnock and W.Giriat, Solid. State Commun. **61**, 767 (1987)

72. A.Twardowski, K.Pakula, M.Arciszewska and A.Mycielski, MAGNETOSPECTROSCOPY OF FREE EXCITON IN CdFeSe SEMIMAGNETIC SEMICONDUCTOR, to be published

73. A.Twardowski and M.Demianiuk, unpublished

74. G.L.Bir and G.E.Pikkus, Symmetry and Strain-Induced Effects in Semiconductorsuctors, Wiley, New York (1974)

75. M.Arciszewska and M.Nawrocki, J.Phys.Chem.Solids **47**, 309 (1986)

76. In the Low and Weger model[2], one finds that the mean spin of an Fe ion $\langle S_z \rangle$ is proportional to the magnetic moment with $k=1/2$. This k value is exactly the same as that for a Mn^{++} ion and results from the fact that in the Low and Weger model the admixture of the ^5T term to the ^5E term is neglected in the wavefunctions, and, consequently, the orbital momentum of the ^5E term is totally quenched. Therefore, magnetization of this term is produced solely by the spin momentum.

77. A.Twardowski, unpublished

78. A.K.Bhattacharjee, G.Fishman and B.Coqblin, Physica **117B** and **118B**, 449 (1983)

79. I.Nowik, E.R.Bauminger, A.Mycielski and H.Szymczak, Physica **B153**, 215 (1988)
    A.Gerard, P.Imbert, H.Prange, F.Varret and M.Winterberger, J.Phys. Chem. Solids **32**, 2091 (1971)

80. T.M.Giebultowicz, J.J.Rhyne, J.K.Furdyna J.Appl.Phys. **61**, 3537 (1987); **61**, 3540 (1987)

81. A.Twardowski, C.J.M.Denissen, W.J.M. de Jonge, A.T.A.M. de Waale, M.Demianiuk, R.Triboulet, Solid State Commun. **59**, 199 (1986)

82. Y.Shapira, N.F.Oliveira, Phys.Rev. **B35**, 6888 (1987)

83. M.Nawrocki, J.P.Lascaray, D.Coquillat and M.Demianiuk, in Material Research Society Fall Meeting (Boston '86), Symposium Proceedings **89**, 65 (1987), Diluted Magnetic (Semimagnetic) Semiconductors, ed. S.von Molnar, R.L.Aggarwal and J.K.Furdyna

84. C.D.Amarasekara, R.R.Galazka, Y.Q.Yang, P.H.Keesom, Phys. Rev. **B27**, 2868 (1983)

85. B.E.Larson, K.C.Hass, R.L.Aggarwal, Phys.Rev. **B33**, 1789 (1986)

86. R.R.Galazka, W.J.M. de Jonge, A.T.A.M. de Waale, J.Zeegers, Solid State Commun. **68**, 1047 (1988)

87. S.Takeyama, R.R.Galazka, Phys. Stat. Sol. (b), **96**, 413 (1979)

88. R.R.Galazka, W.Dobrowolski, J.P.Lascaray, M.Nawrocki, A.Bruno, J.M.Broto and J.C.Ousset, J.of Mag. and Mag. Mat. **72**, 174 (1988)

89. M.Dobrowolska, W.Dobrowolski, M.Otto, T.Dietl, R.R.Galazka, J.Phys. Japan **49**, Suppl.A, 815 (1980)

90. A.Mycielski, C.Rigaux, M.Menant, T.Dietl, M.Otto, Solid State Commun. **50**, 257 (1984)

91. G.Bastard, C.Rigaux, Y.Guldner, J.Mycielski, A.Mycielski, J.Physique **39**, 87 (1978)

92. C.J.M.Denissen, Sun Dakun, K.Kopinga, W.J.M.de Jonge, Phys. Rev. **B36**, 5316 (1987)

93. M.Gorska, J.R.Anderson, Solid State Commun. **63**, 1055 (1987)

94. J.R.Anderson, M.Gorska, Solid State Commun., **52**, 601 (1984)
    M.Bartkowski, A.H.Reddoch, D.F.Williams, G.Lemarche, Z.Korczak, Solid State Commun. **57**, 185 (1986)

95. M.Escorne, A.Mauger, R.Triboulet, J.L.Tholence, Phys. Rev. **B29**, 6306 (1984)

96. J.S.Smart in "Magnetism" ed. G.Rado and H.Shul, **3**, New York-London, (1963) — Tab.7

97. L.Jansen, R.Ridler and E.Lombardi, Physica **71**, 425 (1974)

98. M.Winterberger, B.Srour, C.Meyer, F.Hartmann-Boutron and Y.Gros, J. de Physique **39**, 965 (1978)

99. This value was estimated from Neel temperature T_N assuming type III fcc AF ordering and then $J_{NNN} \ll J_{NN}$. Mean field formula $T_N = -(8J_{NN} - 4J_{NNN})S(S+1)/3k_B$ was used.

100. E.D.Jones, Phys. Letters **18**, 98 (1965); Phys.Rev. **151**, 315 (966)

101. R.Shanker and R.A.Singh, Indian J. Pure Appl. Phys. **12**, 589 (1974)

102. R.MacLaren Murray, B.C.Forbes and R.D.Heyding, Can.J.Chem. **50**, 4059 (1972)

103. The value $N_0\beta = -1.57eV$ reported in Ref.8 resulted from $k = 1/2$, obtained in Low & Weger model[2], in better, Slack's model[3] one obtains $k = 0.447$ and consequently $N_0\beta = -1.74eV$.

DILUTED MAGNETIC IV-VI COMPOUNDS

G. Bauer* and H. Pascher**

* Semiconductor Physics Group, Johannes-Kepler-Universität Linz,
A-4040 Linz, Austria

** Experimentalphysik I, Universität Bayreuth, D-8580 Bayreuth,
Federal Republic of Germany

1.0   INTRODUCTION

Diluted magnetic cubic IV-VI compounds which crystallize in the rock salt structure have attracted considerable interest in recent years [1]. In PbTe, PbSe, PbS the group IV element has been replaced by Mn^{2+}, Eu^{2+} (or gadolinium). These IV-VI compounds have, for comparatively small concentrations of the magnetic ions, a direct narrow gap at the L-point of the Brillouin zone. The electrons and hole masses are more or less mirror images of each other, the mass anisotropies $K = m_l/m_t$ vary from 10 (PbTe) to about 2 (PbSe) and about 1(PbS). The incorporation of magnetic ions causes a more (Eu) or less (Mn) rapid increase of the energy gap with composition x. Since narrow gap IV-VI materials are attractive for mid infrared laser applications, especially the systems $Pb_{1-x}Eu_xTe$ [2] and $Pb_{1-x}Eu_xSe$ [3] have found a considerable interest, even though in these applications just the tuning of the energy gap with composition x is of interest.

As far as the magnetic interaction between Mn^{2+}- or Eu^{2+}- ions in the rock salt IV-VI host lattices is concerned, already the early investigations have established that their antiferromagnetic interaction is considerably weaker than that found in a comparable narrow gap II-VI compound: $Hg_{1-x}Mn_xTe$ [4,5]. The Eu-Eu exchange effects are even weaker in comparison to those of Mn-Mn, as expected [6].

An interesting feature in dilute magnetic IV-VI compounds is the observation of ferromagnetic ordering like in $Sn_{1-x}Mn_xTe$ (Escorne et al. [7]) in $Ge_{1-x}Mn_xTe$ (Cochrane et al. [8] an in PbGeMnTe (Hamasaki [9]). Recently a carrier concentration induced paramagnetic-ferromagnetic transition was found in $Pb_{1-x-y}Sn_xMn_yTe$ (x≈0.7, y≈0.03) [10]. Thus in the magnetic phase diagram of this compound the carrier concentration appears as an additional and tunable parameter.

Apart from the study of the evolution of magnetic interactions the main interest in dilute magnetic semiconductros results from the consequences of the exchange interaction between the mobile carriers in the conduction and valence bands and the localized magnetic moments [11]. Until about 2 years ago the situation was rather unclear in the IV-VI compounds with indications for negligibly small exchange induced ef-

fects for the electrons in $Pb_{1-x}Mn_xTe$ [12-14] and $Pb_{1-x}Mn_xS$ [15,16].
From interband magnetooptical data [14], however, hints were obtained
that the exchange induced correction for the g-factors might be much
larger for holes than for electrons . Using direct measurements of the
electron and hole g-factors by a coherent resonant Raman scattering
techniques in external fields, Pascher et al [17] established the impor-
tantance of exchange induced effects in $Pb_{1-x}Mn_xTe$ and for the first
time the exchange parameters were deduced with reasonable accuracy.

In this review we first describe the magnetic properties (Sec. 2).
We then proceed with the principal concepts of the calculation of Landau
states within the mean field approach for the many valley situation encountered in cubic IV-VI compounds (Sec. 3). In Sec. 4 results on magnetotransport, in Sec. 5 on optical interband transitions are described.
In Sec. 6 the coherent Raman scattering technique is treated in detail
because of its great importance for the investigation of diluted magnetic narrow gap semiconducturs. Results especially on $Pb_{1-x}Mn_xTe$ and
$Pb_{1-x}Eu_xSe$ are presented. In Sec. 7 the results on far infrared spectroscopy are given. In Sec. 8 the calculated results of the mean field theory are compared with experimental data and are discussed. Finally we
conclude by a summary of our present knowledge on dilute magnetic IV-
VI's and emphasize the need for further band strucutre investigations,
especially for determination of the Mn^{2+} and Eu^{2+} levels and their hybridization, necessary for a deeper theoretical understanding of diluted
magnetic IV-VI materials.

2.0 MAGNETIC PROPERTIES

Magnetic properties of dilute magnetic lead compounds were studied
by low field susceptibility, high field magnetization and electron paramagnetic resonance experiments.

2.1 Susceptibility

Susceptibility measurements were performed on $Pb_{1-x}Mn_xS$ [5]
$Pb_{1-X}Mn_xSe$ [18], $Pb_{1-x}Mn_xTe$ [18], $Pb_{1-x}Eu_xTe$ [6] as well as $Pb_{1-x}Gd_xTe$

342

[18]. The low field χ-data have been fitted to a Curie-Weiss-expression
of the form [18]:

$$\chi = \frac{P_1}{T+\theta} + \chi_{dia} \tag{1}$$

where χ_{dia} represents the diamagnetic contribution from the host lattice, θ is the Curie-Weiss parameter and P_1 a further fitting parameter
which is used to obtain the effective number of ion spins $\bar{x}$ [18]:

$$\bar{x} = \frac{m_A + m_B}{S(S+1)\ (g\mu_B)^2 N_A/(3kP_1) + m_A - m_M} \tag{2}$$

where m_A and m_B are the cation and anion atomic masses respectively, m_M
the magnetic ion atomic mass, S the spin of the magnetic ion, g its g
factor (usually g = 2) and N_A the Avogadro constant.

For all materials investigated, the high temperature extrapolation
of χ^{-1} vs T yields small negative values, implying weak antiferromagnetic interaction. The Curie-Weiss-parameter θ (K) is always less than 5
K. From $\theta_0 = \theta/\bar{x}$ the nearest neighbour exchange interaction J can be extracted from the relation

$$\frac{2J}{k} = \frac{3\theta_0}{S(S+1)z} \tag{3}$$

where z is the number of nearest neigbours on cation sites. For the NaCl
lattice z = 12 (J_1); the number of next nearest neigbours on cation
sites is 6 (J_2). In order to compare the different compounds (S = 5/2
for Mn, S = 7/2 for Eu and Gd) in Table I the susceptibility parameters
are given, including $\theta_s = \theta/S(S+1)z$. It should be pointed out, that the
exchange parameter 2J/k is about a factor of 10 lower for these narrow
gap materials in comparison to the II-VI DMS (like $Hg_{1-x}Mn_xTe$).

2.2 High Field Magnetization

High field magnetization studies were carried out so far on
$Pb_{1-x}Mn_xS$ (Karczewski et al. [5]), $Pb_{1-x}Mn_xTe$ (Pascher et al. [14,19],
Anderson et al. [20]), $Pb_{1-x}Gd_xTe$ (Bruno et al. [21] Anderson et al.

Table I Susceptibility parameters

Compound	x	$\bar{x}$	θ(K)	$\theta_s = \theta/(S+1)S \cdot z$ (K)	$2J/k_B$ (K)
$Pb_{1-x}Mn_xS$ [a]	0.012	0.0169	2.045	1.95×10^{-2}	2.92
	0.02	0.0218	1.848	1.76×10^{-2}	2.26
	0.04	0.0378	2.822	2.69×10^{-2}	3.46
$Pb_{1-x}Mn_xSe$ [b]	0.03	0.022	3.0	2.68×10^{-2}	3.9
	0.077	0.045	4.4	4.19×10^{-2}	2.8
$Pb_{1-x}Mn_xTe$ [c]	0.01	0.009	0.8	7.62×10^{-3}	2.41
	0.022	0.024	1.2	1.14×10^{-2}	1.43
	0.04	0.03	1.3	1.24×10^{-2}	1.24
$Pb_{1-x}Eu_xTe$ [d]	0.042	0.037	0.6	3.18×10^{-3}	0.26
	0.096	0.076	1.0	5.29×10^{-3}	0.21
	0.316	0.320	4.6	2.43×10^{-2}	0.23
$Pb_{1-x}Gd_xTe$ [e]	0.033	0.033	1.7	8.99×10^{-3}	0.82
	0.056	0.053	1.8	9.52×10^{-3}	0.54
	0.064	0.066	2.6	1.38×10^{-2}	0.63

Refs: a): [5]; b,c,e): [18], d): [6]

[20]) and $Pb_{1-x}Eu_xTe$ (Braunstein et al. [6]). Recently $Pb_{1-x}Mn_xSe$ and $Pb_{1-x-y}Sn_yMn_xSe$ were investigated by Anderson et al [20]. Traditionally a modified Brillouin function has been used to fit the data [11]:

$$M_s = g\mu_B N_0 S\bar{x} \ B_s(\xi) \tag{4}$$

$$\xi = \frac{Sg\mu_B H}{k(T+T_0)} \tag{5}$$

$$B_s(\xi) = \frac{2S+1}{2S} \coth\left(\frac{2S+1}{2S}\xi\right) - \frac{1}{2S} \coth\left(\frac{\xi}{2S}\right) \tag{6}$$

In this case the fitting parameters are $\bar{x}$, the effective occupation probality of a cation site by an Mn^{2+}- (Eu^{2+}-, Gd^{2+}-) ion and T_0 represents a parameter for the description of the exchange interaction.

Anderson et al. [20] have recently performed 3 parameter fits to the magnetization by fixing $T_0 = 0$ K and adding a further term to M, namely M_p which represents the pair exchange according to Bastard and Lewiner [22]:

$$M_p = \frac{1}{2} g\mu_B \ N_0 \bar{x}_2 \frac{\displaystyle\sum_{S=0}^{S_{max}} \exp\left(\frac{J_p}{kT}(S(S+1))\right) S\left(\sinh\left(\frac{2S+1}{2S}\xi_p\right)\right) B_s(\xi_p)}{\displaystyle\sum_{S=0}^{S_{max}} \exp\left(\frac{J_p}{kT}(S(S+1))\right) \sinh\left(\frac{2S+1}{2S}\xi_p\right)} \tag{7}$$

In this way $\bar{x}_1$, $\bar{x}_2$ and J_p are fitting parameters. In Table II some results on the modified Brillouin function fit are summarized as well as the three paramter fit [20].

In the three parater fit J_p/k should directly represent the contribution of pairs to the exchange and $\bar{x}_1$ the number of individual Mn^{2+}-ions in contrast to $\bar{x}_2$ the number of Mn^{2+}- ion pairs.

Apart from those fits Karczewski et al. [5] have applied a selfconsistant cluster model of interacting magnetic ions to the analyses of their $Pb_{1-x}Mn_xS$ (x = 0.0208) magnetisation data which takes into ac-

Table II: MAGNETIZATION PARAMETERS

Compounds	x	y	Two paramter Fit		Three parameter Fit		
			$\bar{x}$	T_0 (K) (T=4.2K)	$\bar{x}_1$	$\bar{x}_2$	J_p/k_B (K)
$Pb_{1-x}Mn_xSe$	0.03	–	0.038	2.6	0.022	0.015	-0.93
	0.07	–	0.057	4.0	0.027	0.028	-1.03
$Pb_{1-x}Mn_xTe$	0.01	–	0.013	1.9	0.009	0.004	-1.97
	0.022	–	0.026	1.2	0.019	0.006	-0.88
	0.04	–	0.033	1.7	0.022	0.010	-0.92
$Pb_{1-x-y}Sn_yMn_xSe$	0.007	0.02	0.0071	0.8	0.006	0.001	-1.21
	0.015	0.03	0.014	1.2	0.011	0.003	-0.89
	0.017	0.05	0.015	1.06	0.012	0.003	-0.81
$Pb_{1-x}Gd_xTe$	0.035	–	0.040	1.6	0.026	0.014	-0.39
	0.055	–	0.054	2.0	0.023	0.031	-0.27
	0.07	–	0.074	4.4	0.037	0.036	-0.69
$Pb_{1-x}Eu_xTe$	0.042	–	0.041	0.7			
	0.096	–	0.077	1.1			
	0.316	–	0.32	4.2			

After Refs. 5,6,20,21

count the contribution of isolated Mn^{2+}- ions, the contribution of two types of interacting Mn pairs as well as further the contribution of open triples. It turned out that with $J_1/k \approx J_2/k \approx 0.5K$ and $J_3/k \approx 0.06K$ reasonable fits were obtained.

In Fig.1 we present a fit on magnetization data on $Pb_{1-x}Mn_xTe$ (x = 0.01) for two different temperatures $T_1 = 2.5$ K and $T_2 = 11$ K. The simple modified Brillouin function was used with the parameters as given in the figure caption.

So far all magnetization data were taken at temperatures of $T \geq 2K$ and above. Due to the comparatively small values of J, it is not astonishing, that under these conditions no steps in the magnetization as a function of external field H [23] were found. These steps in M(H) are a direct consequence of the existence of antiferromagnetically coupled nearest neighbour pairs and the magnetic fields at which these steps are expected to occur would provide a direct measure of J. Further investigations at much lower temperatures are needed.

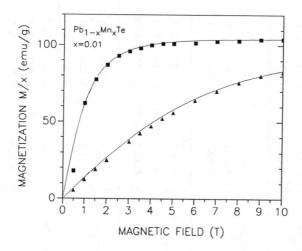

Fig.1: Magnetization vs. temperature for $Pb_{1-x}Mn_xTe$, x=0.01. Experimental data for T=2.5K (■) and T=11K (▲).
Full lines: results of calculation using modified Brillouin function (T_o(2.5K) = 0.7K, S_o = 2.25; T_o(11K) = 0.15K, S_o = 2.25)

2.3 Spin Glass Phase

At sufficiently low temperatures, the magnetic properties of $Pb_{1-x}Mn_xTe$ [24] but also other compounds like $Sn_{1-x}Mn_xTe$ [25] turn out to be determined by a spin glass phase. In $Pb_{1-x}Mn_xTe$ this behaviour was observed which is characterized by a thermoremanent magnetization which has been studied as a function of magnetic field. A cusp in the reversible part of the magnetic susceptibility is used to define the freezing temperature T_g. Escorne et al [24] argue that the spin glass freezing at low manganese concentrations is due to the onset of a comparatively long range indirect interaction which becomes relevant due to the fact that PbMnTe is not only a narrow gap semiconductor but has four extrema at the L-points of the Brillouin zone.

Although, experimental evidence for what commonly is used as a characteristic for "spin-glass" behaviour has been found, further theoretical and experimental studies are needed to clarify the origin of the observed phenomena and to rigorously classify the magnetic interactions involved.

2.4 Free Carrier Induced Ferromagnetism

One of the most fascinating magnetic properties of the diluted magnetic lead salt alloys found so far, has been described by Story et al. [10] and has since been quantitavely explained by several authors [26-28]. In $Pb_{1-x-y}Sn_xMn_yTe$ (for $x \approx 0.7$, $y \approx 0.03-0.08$) a ferromagnetic ordering has been observed above a critical hole concentration of about $3 \times 10^{20} cm^{-3}$. The transition temperature (0-5K) increases first with increasing hole concentration as shown in Fig.2. The ferromagnetic phase has been unambiguously identified by magnetization, magnetic susceptibility and specific heat measurements and experiments under hydrostatic pressure. The interesting feature is the follwing: for such large hole concentrations a second set of valence levels (Σ-band) becomes populated apart form the L-levels [26]. The position of the Σ levels relatively to the L-levels strongly depends on chemical composition and hydrostatic

348

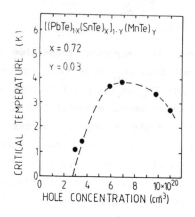 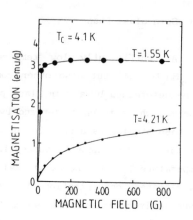

Fig. 2: Paramagnetic- ferromagnetic phase transition in PbSnMnTe.

Left hand side: dependence of T_c on hole concentration.

Right hand side: magnetization exhibits rapid saturation for $T<T_c$ measured saturation value is close to the theoretical one (M_{theor} = 3.1emu/g). After [10]

pressure as shown in Fig.3. The interaction which mediates this ferromagnetic transition is the Ruderman-Kittel-Kasuya-Yosida (RKKY) interaction which was originally developed for parabolic bands. Swagten et al [26] have extended this RKKY interaction to the two valence band situation (L-Σ) using the following assumptions:

i) intraband RKKY interaction between the magnetic moments

ii) mean field approximation for the spin-system

iii) finite mean free path for the holes

iv) L and Σ bands are reprensented by two parabolic valence bands with two different effective masses (m_L = 0.05 m_0, m_Σ = m_0).

Using for the Curie-Weiss parameter the expression:

$$\theta = \frac{2S(S+1)x}{3k} \sum_{i=1}^{2} \sum_{j\geq 1} J_i^{eff}(R_j) \tag{8}$$

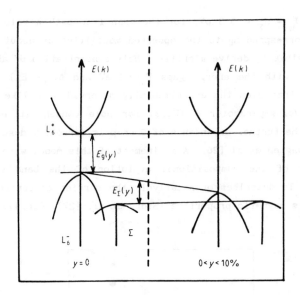

Fig.3: Evolution of the band structure of $Pb_{1-x-y}Sn_xMn_yTe$ with Mn- content for x=0.72. After [27]

where

$$J_i^{eff} (R_j) = \left(\frac{m_i}{m_0} J_i^2 \right) \frac{m_0}{128\pi^3\hbar^2} \left(\frac{a_0^2}{j^2} \right) \exp(-R_j/\lambda) \; F(z_j^i) \tag{9}$$

with

$$F(z_j^i) = \left(\sin(z_j^i) - z_j^i \cos(z_j^i) \right) \tag{10}$$

where $\quad z_j^i = 2k_F^i R_j$ $\qquad$ (11)

k_F^i: Fermi wave number in $i = L$ or $i = \Sigma$ band

R_j: distance between two lattice points in the NaCl lattice

$S = 5/2$ for Mn^{2+}

J_i: exchange constant for L or Σ band

a_0: lattice parameter

Using for $J_L = J_\Sigma = 0.3$ eV, for the mean free path λ values between 7 and 13 Å corresponding to the observed mobilities of about 20 cm^2/ Vs) it was possible to derive a critical hole concentration of about p = 2.4 x 10^{20} cm^{-3} with the energy gaps E_g=0.15eV and $\Delta(E_L - E_\Sigma)$ = 0.6eV. It turned out that even the experimentally observed step like increase of the Curie-Weiss-parameter θ (Eq.8) for hole concentrations p > p$_{crit}$ (θ>0) and the further dependence of θ on p is correctly described by the model of Swagten et al [26]. A refinement of this model with the correct description of the compositional evolution of the band structure in Pb$_{1-x-y}$Sn$_x$Mn$_y$Te describes quanitatively the carrier concentration dependence of its magnetic properties (Story et al. [27], Fig.4).

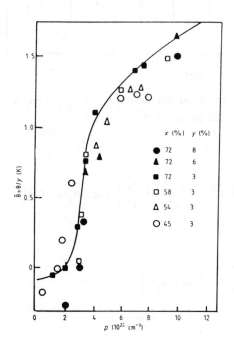

Fig.4: Normalized paramagnetic Curie- Weiss parameter $\bar{\theta}$=θ/y for Pb$_{1-x-y}$Sn$_x$Mn$_y$Te with different compositions x and y versus free hole concentration. After [27]

The explanation of the ferromagnetic ordering in $Pb_{1-x-y}Sn_xMn_yTe$ due to the subtleties of the exchange due to RKKY interaction including 2 valence bands is a beautiful demonstration of the effects of large free carrier concentrations on magnetic properties of dilute magnetic semiconductors.

2.5 Magnetic Properties of Layered IV-VI Diluted Magnetic Semiconductors

The study of magnetic properties in quasi twodimensional systems is of particular interest both for the aspects of magnetism as well as of fundamental importance for an understanding of the electronic proper-ties. The first experiment of this kind was a susceptibility study on PbTe/ EuTe short period superlattices [29]. PbTe is diamagnetic, EuTe paramegnetic and below 9.58 K antiferromagnetic. A systematic study on MBE grown material [2] was undertaken using three samples consisting of $(EuTe)_m/(PbTe)_n$ layer structures with (m,n) = (1,3); (2,6) and (4,4) and periods between 100 and 400. Susceptibility measurements were performed using a SQUID magnetometer and applying external fields both parallel and perpendicular to the layers. Results on the (4/4) and (2/6) samples are shown in Fig.5. χ^{-1}(T) follows a Curie-Weiss-law with anisotropic extrapolated Curie-Weiss-parameters. For the (4/4) sample the antiferro-magnetic phase tranistion is indeed observed at a Néel temperature of T = 8.5 K, only slightly below the bulk value for EuTe. No phase transi-tion was observed for the (2/6) and (1/3) samples where the Eu atoms do not have all of their 6 nearest neighbours in contrast to the (4/4) sam-ple. The anisotropy of θ was explained by involving effects of strain $(\varepsilon \approx 10^{-2})$ on the nearest and next nearest neighbour interactions. The disappearance of the antiferromagnetic properties in the ultrathin (two and one monolayer EuTe) samples is at least consistent with the tendency and the predictions of the Mermin-Wagner-theorem [11].

3.0 CALCULATION OF LANDAU STATES: MEAN FIELD THEORY FOR IV-VI COMPOUNDS

The $\vec{k} \cdot \vec{p}$ band model for the lead compounds has been derived by seve-ral authors. In the following we use the notation of Adler et al. [30]

352

who treat exactly the interaction of the upper most valence and the low-est conduction band level whereas the interaction with two more distant conduction and valence levels is treated in k^2- approximation. The 4x4 matrix Hamiltonian for the calculation of the Landau states resulting from this procedure is given by [30]:

$$H = \begin{bmatrix} h_v & h_{vc} \\ h_{cv} & h_c \end{bmatrix}$$ (12)

with

$$h_c = \left(E_c + \frac{k_3^2}{2m_l^-} + \frac{k_1^2 + k_2^2}{2m_t^-} \right) I_2 + \left(g_l^- \sigma_z \, B_3 + g_t^- (\sigma_x B_1 + \sigma_y B_2) \right) \mu_B / 2$$ (13)

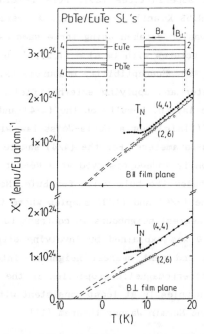

Fig.5: Magnetic properties of short- period PbTe/EuTe superlattices: invers susceptibility vs temperature for B perpendicular (bottom) and parallel (top) to the layers. The (4/4) sample exhibits an antiferromagnetic phase transition whereas the (2/6) sample re-mains paramagnetic. After [29]

$$h_v = \left(E_v - \frac{k_3^2}{2m_l^+} - \frac{k_1^2 + k_2^2}{2m_t^+} \right) I_2 - \left(g_l^+ \sigma_z B_3 + g_t^+ (\sigma_x B_1 + \sigma_y B_2) \right) \mu_B / 2 \tag{13a}$$

$$h_{cv} = P_{\parallel} k_3 \sigma_z + P_{\perp} (k_1 \sigma_x + k_2 \sigma_y) \tag{14}$$

$$h_{vc} = h_{cv}^+ \tag{14a}$$

I_2 is the 2x2 unit matrix, σ are the Pauli spin matrices (σ_z diagonal), μ_B is the Bohr magneton and the indices 1,2,3 refer to the valley axes system with axis 3 parallel to the valley main axis. $P_{\parallel}$ and $P_{\perp}$ are the momentum matrix elements for the two band interaction perpendicular and parallel to <111>, $m_{l,t}^{+-}$ are the far band contributions to the conduction band (-) and valence band (+), parallel (l) and perpendicular (t) <111>, $g_{l,t}^{+-}$ the same for the g-factors. The symmetry of the band edge functions for the conduction band is given by [31]:

$$L_6^-(L_2^-)_\alpha = -\sin\theta^- \, Z \uparrow - \cos\theta^- \, X_+ \downarrow$$
$$L_6^-(L_2^-)_\beta = \sin\theta^- \, Z \downarrow - \cos\theta^- \, X_- \uparrow \tag{15}$$

and for the valence band:

$$L_6^+(L_1^+)_\alpha = i\cos\theta^+ \, R \uparrow + \sin\theta^+ \, S_+ \downarrow$$
$$L_6^+(L_1^+)_\beta = i\cos\theta^+ \, R \downarrow + \sin\theta^+ \, S_- \uparrow \tag{15a}$$

The relative mixing of the states by spin orbit coupling is determined by the parameters θ^+, θ^- [32]. R, Z, X_{+-} and S_{+-} denote functions which are derived from functions with s,p,d symmetry, respectively and are explicitly given in Ref.31. The system of coordinates is chosen such that $x \parallel [\bar{1}\bar{1}2]$, $y \parallel [1\bar{1}0]$ and $z \parallel [111]$.

The secular equation for the calculation of Landau states is given by [33]:

$$\begin{vmatrix} -E_g/2-\hbar\tilde{\omega}_c^+(n+1/2) \\ -\hbar^2 k_B^2/(2m_B^+)+ & 0 & [E_g/(2m_B)]^{1/2}\hbar k_B & (E_g\hbar\tilde{\omega}_c)^{1/2}\sqrt{n+1} \\ \tilde{g}_1^+\mu_B B/2 \\ \\ & -E_g/2-\hbar\tilde{\omega}_c^+(n+3/2) \\ 0 & -\hbar^2 k_B^2/(2m_B^+)- & (E_g\hbar\tilde{\omega}_c)^{1/2}\sqrt{n+1} & -[E_g/(2m_B)]^{1/2}\hbar k_B \\ & \tilde{g}_1^+\mu_B B/2 \\ \\ & & E_g/2+\hbar\tilde{\omega}_c^-(n+1/2) \\ [E_g/(2m_B)]^{1/2}\hbar k_B & (E_g\hbar\tilde{\omega}_c)^{1/2}\sqrt{n+1} & +\hbar^2 k_B^2/(2m_B^-)+ & 0 \\ & & \tilde{g}_1^-\mu_B B/2 \\ \\ & & & E_g/2+\hbar\tilde{\omega}_c^-(n+3/2) \\ (E_g\hbar\tilde{\omega}_c)^{1/2}\sqrt{n+1} & -[E_g/(2m_B)]^{1/2}\hbar k_B & 0 & +\hbar^2 k_B^2/(2m_B^-)- \\ & & & \tilde{g}_1^-\mu_B B/2 \end{vmatrix}$$

$$(16)$$

with

$$\lambda_3^2 = \cos^2\Phi \;\; ; \;\; \lambda_1^2 = \sin^2\Phi \;\; ; \;\; v_l = P_\parallel/m_0 \;\; ; \;\; v_t = P_\perp/m_0$$

$$v_B^2 = (\lambda_3^2 v_t^4 + \lambda_1^2 v_t^2 v_l^2)^{1/2} \;\; ; \;\; \sin\gamma = \lambda_1 v_l v_t/v_B^2 \;\; ; \;\; \cos\gamma = \lambda_3 v_t^2/v_B^2$$

$$\tilde{\omega}_c^+ = eB\left[\lambda_3^2/(m_t^+ m_t^+) + \lambda_1^2/(m_t^+ m_l^+)\right]^{1/2}$$

$$\tilde{\omega}_c^- = eB\left[\lambda_3^2/(m_t^- m_t^-) + \lambda_1^2/(m_t^- m_l^-)\right]^{1/2}$$

$$m_B^+ = \lambda_3^2 m_l^+ + \lambda_1^2 m_t^+ \;\; ; \;\; m_B^- = \lambda_3^2 m_l^- + \lambda_1^2 m_t^-$$

$$\tilde{g}_1^+ = \frac{1}{v_B^2}\left[\lambda_3^2 v_t^2 g_l^+ + \lambda_1^2 v_l v_t g_t^+\right]$$

$$\tilde{g}_1^- = \frac{1}{v_B^2} \left[\lambda_3^2 v_t^2 g_1^- + \lambda_1^2 v_1 v_t g_t^- \right]$$

$$m_B = \frac{E_g v_B^4}{2v_1^2 v_t^4} \quad ; \qquad \tilde{\omega}_c = 2 \frac{eBv_B^2}{E_g}$$

where E_g denotes the energy gap, Φ is the angle between $\vec{B}$ and [111].

For diluted magnetic semiconductors with rather small contents x of the paramegnetic ions we assume that the sequence of energy levels is not altered with respect to the host crystal. In the spirit of the mean field approximation [34] (MFA) an exchange Hamiltonian is added to the 4x4 matrix Hamiltonian [12,13,14,17]. It describes the interaction between the electron spin $\vec{\sigma}$ and spin $\vec{S}$ the paramegnetic ions is given by [34]:

$$H_{exch} = J(|\vec{r}-\vec{R}|) \; \vec{\sigma}\cdot\vec{S} \tag{17}$$

J is the exchange integral.

The $\vec{k}\cdot\vec{p}$ Hamiltonian is finally obtained for $S=5/2$:

$$H_{exch} = 0.5 \times S_0 B_{5/2} \left[\frac{5\mu_B |\vec{B}|}{k \; (T+T_0)} \right]$$

$$\times \begin{bmatrix} A\sigma_z \alpha_3 + a_1 (\sigma_x \alpha_1 + \sigma_y \alpha_2) & 0 \\ 0 & B\sigma_z \alpha_3 - b_1 (\sigma_x \alpha_1 + \sigma_y \alpha_2) \end{bmatrix} \tag{18}$$

(For the Eu- compounds $S=5/2$ has to be replaced by $S=7/2$).

$\alpha_{1,2,3}$ are the direction cosines of $\vec{B}$ in the valley axes system, $B_{5/2}$ is the modified Brillouin function describing the paramagnetic orientation of the magnetic moments [11], S_0 and T_0 are adjustable parameters, determined from magnetization data, A, a_1, B, b_1 are the four exchange parameters for the valence and conduction bands, respectively [14]. This 4x4 exchange Hamiltonian contains off diagonal terms which cause a strong mixing of Landau states with different spin and Landau quantum numbers. In order to calculate the energies of the 0^+ and 0^- Landau states of the conduction and valence bands a 12x12 secular determinant $\tilde{\mathcal{H}} = \tilde{\mathcal{X}} + \tilde{\mathcal{Y}}$ is used where $\tilde{\mathcal{X}}$ has three 4x4 blocks according to Eq.(16) along the main diagonal.

$$\tilde{\mathcal{X}} = \begin{vmatrix} \begin{vmatrix} Eq.(1) \\ n = -1 \end{vmatrix} & & \\ & \begin{vmatrix} Eq.(1) \\ n = 0 \end{vmatrix} & \\ & & \begin{vmatrix} Eq.(1) \\ n = 1 \end{vmatrix} \end{vmatrix} \qquad (19)$$

The secular determinant $\tilde{\mathcal{Y}}$ describing the exchange is given by:

$$\tilde{\mathcal{Y}} = \frac{1}{2} x S_0 B_{5/2} \left[\frac{5\mu_B B}{k\ (T+T_0)} \right] \cdot \begin{vmatrix} +a & o & o & o & o & o & o & o & o & o & o & o \\ o & -a & o & o & +c & o & o & o & o & o & o & o \\ o & o & +b & o & o & o & o & o & o & o & o & o \\ o & o & o & -b & o & o & +d & o & o & o & o & o \\ o & +c & o & o & +a & o & o & o & o & o & o & o \\ o & o & o & o & o & -a & o & o & +c & o & o & o \\ o & o & o & +d & o & o & +b & o & o & o & o & o \\ o & o & o & o & o & o & o & -b & o & o & +d & o \\ o & o & o & o & o & +c & o & o & +a & o & o & o \\ o & o & o & o & o & o & o & o & o & -a & o & o \\ o & o & o & o & o & o & o & +d & o & o & +b & o \\ o & o & o & o & o & o & o & o & o & o & o & -b \end{vmatrix} \qquad (20)$$

The matrix elements a, c and b, d are functions of four exchange parameters as defined in Refs. 12, 13 for the valence (A, a_1) and conduction band (B, b_1), respectively [14,17]:

$$a = A \cdot \cos\Phi\ \cos\gamma + a_1 \cdot \sin\Phi\ \sin\gamma \qquad (21)$$

$$c = -A \cdot \cos\Phi \, \sin\gamma + a_1 \cdot \sin\Phi \, \cos\gamma \qquad (21a)$$

$$b = B \cdot \cos\Phi \, \cos\gamma - b_1 \cdot \sin\Phi \, \sin\gamma \qquad (22)$$

$$d = -B \cdot \cos\Phi \, \sin\gamma - b_1 \cdot \sin\Phi \, \cos\gamma \qquad (22a)$$

where

$$A = a_1 - a_2 = \alpha \cdot \cos^2\theta^+ - \delta \cdot \sin^2\theta^+$$

$$B = b_1 - b_2 = \beta_\| \cdot \sin^2\theta^- - \beta_\perp \cdot \cos^2\theta^- \qquad (23)$$

For the rotation of the reference frame the angle γ is given by [30]:

$$\cos\gamma = \left[\cos^2\Phi \, \frac{P_\perp^2}{P_\|^2} \middle/ \left(\cos^2\Phi \, \frac{P_\perp^2}{P_\|^2} + \sin^2\Phi \right) \right]^{1/2} \qquad (24)$$

Due to the symmetry properties of the wave functions at the L-point of the Brillouin zone instead of two exchange integrals as in zincblende semiconductors now four are needed where the first two are for the holes and the second two for the electrons [12,13,14]:

$$\alpha = (R|J|R)/\Omega, \qquad \delta = (S_\pm|J|S_\pm)/\Omega,$$

$$\beta_\| = (X_\pm|J|X_\pm)/\Omega, \qquad \beta_\perp = (Z|J|Z)/\Omega \qquad (25)$$

For the calculation of higher Landau states the schemes for $\tilde{x}$ and $\tilde{y}$ have to be extended accordingly. It should be noted that the approach outlined above is only suitable for the momentum $k_B = 0$.

Using the band parameters of Table III and the exchange parameters of Table IV in Figs.6a,b Landau levels for the conduction and valence band are shown calculated for two temperatures, $\Phi = 90°$ for $Pb_{1-x}Mn_xTe$. At T=1.8K the exchange induced corrections alter drastically the magnetic field dependence of the Landau states causing even a positive slope of the 0^- valence band level at small magnetic fields. For higher temperatures the influence of the exchange interaction decreases and the magnetic field dependence of the Landau states resembles nearly the diamagnetic host material. In $Pb_{1-x}Eu_xSe$ (Fig.7), the exchange induced effects are weaker but of different sign and even a crossing of the n^- and n^+ levels occurs in the valence band.

Table IV: Exchange parameters

Mn-content	T(K)	A(meV)	a_1(meV)	B(meV)	b_1(meV)
x=0.010	1.8	−182±15	−288±15	−33±10	27±5
x=0.008	1.8	−192±15	−315±15	−66±10	55±5
x=0.006	1.8	−225±15	−314±15	--	50±5
x=0.006	3.5	−142±15	−279±15	−41±10	50±5
x=0.006	4.4	−124±15	−279±15	−51±10	51±5
x=0.006	12.0	−51±15	−288±15	--	59±5
Eu-content					
x=0.045	1.7	102±10	96±10	21±5	24±5
x=0.008	1.7	78±10	65±10	19±5	8±5
x=0.008	6.0	77±10	59±10	25±5	3±5
x=0.008	12.0	70±10	56±10	31±5	7±5

Table V: Exchange integrals

Mn-content	T(K)	α(meV)	δ(meV)	$\beta_{\parallel}$(meV)	$\beta_{\perp}$(meV)
x=0.010	1.8	−305	−2460	109	80
x=0.006	1.8	−305	−1700	225	--
x=0.006	3.5	−305	−3290	225	120
x=0.006	4.4	−305	−4040	225	140
x=0.006	12.0	−305	−5900	225	140
Eu-content					
x=0.045	1.7	98	−144	73	4
x=0.008	1.7	61	−433	18	−19
x=0.008	6.0	61	−408	18	−28
x=0.008	12.0	61	−240	18	−37

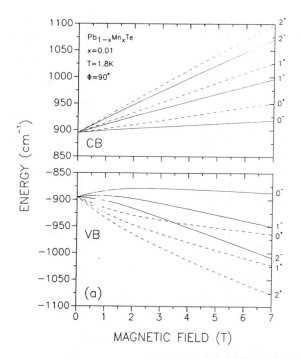

Fig.6a: Dependence of Landau states in the conduction band of $Pb_{1-x}Mn_xTe$ (for $\Phi=90°$, Φ denotes angle between $\vec{B}$ and valley axis) and in the valence band on the magnetic field for T=1.8K.

4.0 MAGNETOTRANSPORT

In order to study quantum oscillations of the magnetoresistance, high quality samples are necessary. In Fig.8 the Hall mobility as a function of temperature is shown for three $Pb_{1-x}Mn_xTe$ epitaxial films containing from x=0.007 to x=0.028 manganese.. In general, with increasing Mn-content, the low temperature mobility levels off at lower levels. This fact seems to indicate that the number of lattice defects increases with increasing Mn-concentration. In this context it should be noted, that in order to increase the Mn-Content in the epitaxial films during HWE [35] or MBE [36] growth, more and more tellurium has to be

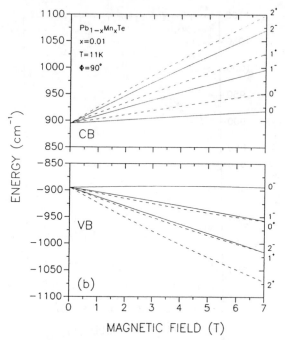

Fig.6b: as Fig.6a but for T=11K.

Table III: Band parameters.

Sample	PbTe	$Pb_{1-x}Mn_xTe$	$Pb_{1-x}Mn_xTe$	$Pb_{1-x}Mn_xTe$
x	0	0.006	0.01	0.012
E_g (meV)	189.7	209.8	221.9	225.7
$2P_\perp^2/m_o$ (eV)	6.02	5.77	5.51	--
$P_\perp/P_\parallel$	3.42	3.52	3.74	--

Sample	PbSe	$Pb_{1-x}Eu_xSe$	$Pb_{1-x}Eu_xSe$	$Pb_{1-x}Eu_xSe$
x	0	0.0035	0.0045	0.008
E_g (meV)	146.3	181.1	189.1	219.0
$2P_\perp^2/m_o$ (eV)	3.6	3.6	3.6	3.6
$P_\perp/P_\parallel$	1.35	1.40	1.40	1.40

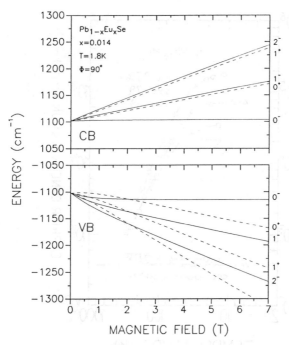

Fig.7: as Fig.6a but for Pb$_{1-x}$Eu$_x$Se.
Note the different curvature of valence band spin levels compared
to Pb$_{1-x}$Mn$_x$Te

offered. Thus, most probably the total number of lattice defects increases. The low temperature mobilities in the n-type PbMnTe samples with x
less than 3%, range from 30.000 cm^2/Vs to values of up to 300.000 cm^2/Vs
[37] at liquid helium temperatures.

In the narrow gap II-VI diluted magnetic semiconductors an anomalous temperature dependence of the Shubnikov- de Haas oscillations has
been found, deviating considerably from the x/sinh(x) behaviour
(x=$2\pi^2$kT/$\hbar\omega_c$).

The amplitude of the Shubnikov- de Haas oscillations depends also
on a factor cos($\pi g^* m^*/(2m_0)$). If g* depends on temperature due to the
temperature dependence of the magnetization quite startling phenomena,
like a disappearance of the oscillotions at a certain temperature may
occur [11].

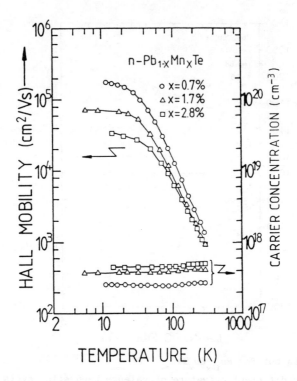

Fig.8: Hall mobility and carrier concentration of electrons vs temperature in epitaxial $Pb_{1-x}Mn_xTe$ films.

In Fig.9 the magnetic field dependence of the transverse magnetoresistance of n- $Pb_{1-x}Mn_xTe$ (x=0.012) are shown with temperature as a parameter [13]. Due to biaxial tensile strain in the $Pb_{1-x}Mn_xTe$ epitaxial film, the degeneracy of the four L-Valleys is lifted. The $\Phi=0°$ valley is shifted downwards by about 4meV for this sample. Due to the low carrier concentration of about $1\times10^{17}/cm^3$ at higher magnetic fields only this valley contributes to the observed oscillations.

The spin-splitting is well resolved, but the amplitude of the oscillations follows roughly the behaviour which is expected for a non-magnetic material. This is due to the fact, that the exchange parameters

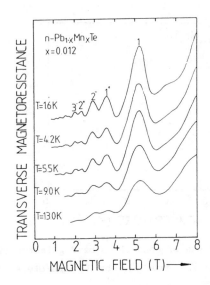

Fig.9: Temperature dependence of Shubnikov de- Haas oscillations in n-Pb$_{1-x}$Mn$_x$Te

are small in the conduction band. In Fig.10 the g-factor for B=3.5T and Φ=0° as a function of temperature is calculated with the parameters listed in Tables II andIV. There is almost no dependence, in accordance with the experimental findings of Fig.9 where the spin splitting (maxima at about 3.5T and 5.2T) also does not depend on temperature.

5.0 OPTICAL INTERBAND TRANSITIONS

Interband transition energies can be determined either by photoluminescence or by absorption spectroscopy. With thin epitaxial films it is very difficult to interpret measurements of the transmission as a function of wavelength due to Fabry Perot like interferences of the radiation reflected from both sides of the samples. This difficulty is avoided with magnetooptical interband absorption experiments where the transmission is registered as a function of an external magnetic field. Transmission minima occur for magnetic fields where the incident photon energy is equal to the transition energy of an allowed interband transi-

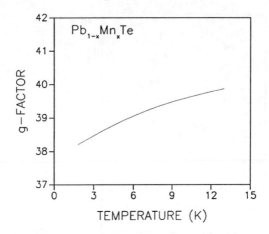

Fig.10: Calculated dependence of conduction band g factor ($\Phi=0°$) on temperature for B=3.5T, x=0.012, band parameters according to Tables III and IV.

tion between Landau levels. However, with this method no experiments with very low magnetic fields are possible. At B=0 the energy gap must be determined from photoluminescence experiments.

5.1 Photoluminescence Without Magnetic Field

Photoluminescence spectra were taken in back scattering geometry with a Q- switched CO- or Nd:YAG laser, respectively depending on the band gap of the investigated material. The samples were immersed in superfluid He in a bath cryostat.

Figures 11a-c reproduce the photoluminescence spectra of $Pb_{1-x}Mn_xTe$ (x=0.01), $Pb_{1-x}Eu_xTe$ (x=0.012) and $Pb_{1-x}Eu_xSe$ (x=0.013). The spectra show narrow single lines with no indication of a zero field splitting, i.e. the occurence of two different gap energies at zero external megnetic field. From the energy gaps which can be found out from the low energy side of the lines the x- values can be calculated in a very easy and accurate way by the following equations, valid for He- temperatures and small concentrations x [14,38,2]:

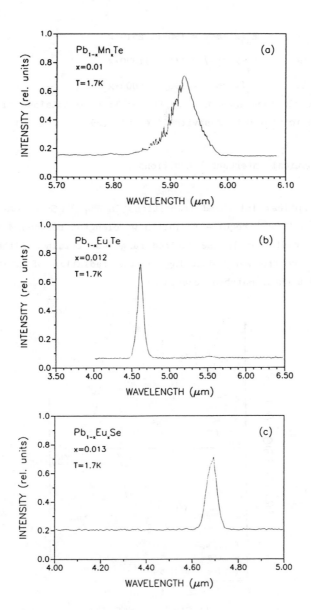

Fig. 11: Photoluminescence spectra of: a) $Pb_{1-x}Mn_xTe$, b) $Pb_{1-x}Eu_xTe$, and c) $Pb_{1-x}Eu_xSe$ epitaxial films on (111) BaF_2.

$$Pb_{1-x}Mn_xTe: \quad E_g(x)/meV = 190 + 2510 \cdot x \quad\quad (26)$$

$$Pb_{1-x}Eu_xSe: \quad E_g(x)/meV = 114 + 11370 \cdot x \quad\quad (27)$$

$$Pb_{1-x}Eu_xTe: \quad E_g(x)/meV = 190 + 6000 \cdot x \quad\quad (28)$$

Recently photoluminescence data on $Pb_{1-x}Mn_xTe$ were obtained also in external magnetic fields by Zasavitskii et al. [39].

5.2 Magnetooptical Interband Transitions

Magnetooptical interband transitions in $Pb_{1-x}Eu_xSe$ ($x \leq 0.01$) and $Pb_{1-x}Mn_xTe$ ($x \leq 0.02$) require radiation wavelengths between 4.5 and 7.5 μm. Coherent radiation in the desired range was obtained either from a CO-laser or by frequency doubling of the radiation of a Q-switched CO_2-laser in a phase matched Te-crystal.

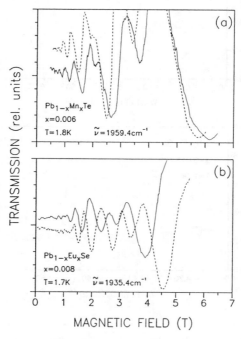

Fig.12: Interband magnetotransmission: a) $Pb_{1-x}Mn_xTe$, b) $Pb_{1-x}Eu_xSe$ in Faraday configuration. Full curves: σ^+ and broken curves σ^- circular polarization.

The transmission spectra were observed in Faraday configuration $\vec{B} \| \vec{k} \| [111]$ with circularly polarized σ^+ and σ^- radiation and in Voigt configuration $\vec{B} \perp \vec{k}$, $\vec{B} \| [1\bar{1}0]$, $\vec{E} \| \vec{B}$.

Figure 12a shows the interband transmission of a $Pb_{1-x}Mn_xTe$ sample with x=0.006 as a function of the magnetic field for the two circular polarizations in Faraday configuration, Fig 12b the same for $Pb_{1-x}Eu_xSe$, x=0.008. It is evident that the resonant magnetic fields are different for the two polarizations in contrast to the PbTe and PbSe case, respectively. Whereas in $Pb_{1-x}Mn_xTe$ the σ^+ resonances occur at higher magnetic fields than σ^-, in $Pb_{1-x}Eu_xSe$ it is vice versa. From this fact one can conclude that the spin splitting of the valence band is diminished in $Pb_{1-x}Eu_xSe$ by the exchange interaction of free carriers with localized spins and increased in $Pb_{1-x}Mn_xTe$. From a comparison of Faraday- and Voigt data it follows that this interaction mainly affects holes in the valence band.

From a large number of such recordings taken with various infrared laser frequencies fan charts for $Pb_{1-x}Mn_xTe$ were obtained as shown in Fig.13 for the Faraday configuration and in Fig.14 for the Voigt configuration. The full lines correspond to calculated transitions based on the equations given in Sec.3. with the selection rules $\Delta n=0$, $\Delta s=\pm 1$ (Faraday geometry) and $\Delta n=0$, and $\Delta s=0$ ($\vec{E} \| \vec{B}$). We would like to point out that the mean field theory of Sec.3. is capable of explaining nearly all of the experimental data, even such delicate features like the crossing of several transitions as a function of B in Voigt geometry (see Fig.14).

In Voigt geometry (Fig.14) one relatively strong series of interband transitions is observed which does not obey the selection rules $\Delta n=0$, $\Delta s=0$. It turns out that a transition $0^+(vb) \to 0^-(cb)$, i.e. $\Delta s=-1$ is observed, within the $\Phi=35,26°$ valleys. The allowed transitions $0^+(vb) \to 0^+(cb)$ for $\Phi=35.26°$ also appears, whereas $0^-(vb) \to 0^-(cb)$ is blocked due to the position of the Fermi energy. Experiments were performed on several $Pb_{1-x}Mn_xTe$ samples with x values up to x=1.7%. The band parameters from the fits are given in Table III, the exchange parameters in Table IV.

In Faraday geometry at higher fields in excess of about 2T the transitions $n^+(vb) \to n^-(cb)$ as well as $n^-(vb) \to n^+(cb)$ have nearly iden-

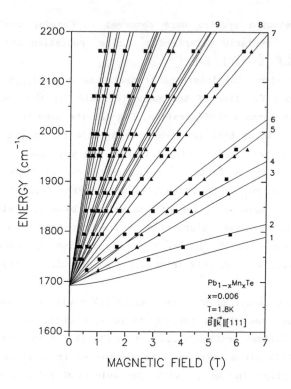

Fig.13: Fan chart for interband magnetooptical transitions in Faraday
geometry: ■, ▲ experimental data (σ^-, σ^+);
full lines: calculated data; identification:
1: $0^- \to 0^+ (\Phi=70.53°)$; 2: $0^+ \to 0^- (\Phi=70.53°)$; 3: $0^- \to 0^+ (\Phi=0°)$;
4: $0^+ \to 0^- (\Phi=0°)$; 5: $1^- \to 1^+ (\Phi=70.53°)$; 6: $1^+ \to 1^- (\Phi=70.53°)$;
7: $2^- \to 2^+ (\Phi=70.53°)$; 8: $2^+ \to 2^- (\Phi=70.53°)$; 9: $1^- \to 1^+ (\Phi=0°)$; etc.

tical slope. In PbTe these transitions almost coincide with each other.
In Fig.15 the energy difference between these two kinds of transitions,
ΔE, is plotted versus the content x of Mn^{2+}- ions. With increasing x, ΔE
increases nearly linearly.

Figures 16 and 17 show the fan charts for one of the $Pb_{1-x}Eu_xSe$
samples. Again the reasonable agreement between experimental points and

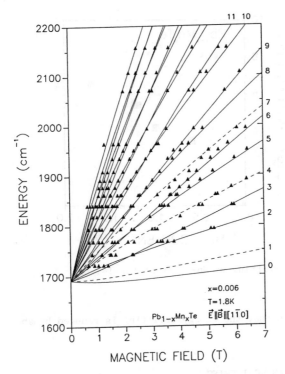

Fig.14: As Fig.4 but for Voigt geometry $\vec{B} \| [1\bar{1}0]$.

Experimental data: in $\vec{E} \| \vec{B}$ polarization;

full lines: calculated data for $\Phi=90°$, dashed lines: $\Phi=35.26°$.

0: $0^- \to 0^- (\Phi=90°)$; 1: $0^- \to 0^- (\Phi=35.26°)$; 2: $0^+ \to 0^+ (\Phi=90°)$;

3: $1^- \to 1^- (\Phi=90°)$; 4: $0^+ \to 0^- (\Phi=35.26°)$; 5: $1^+ \to 1^+ (\Phi=90°)$;

6: $2^- \to 2^- (\Phi=90°)$; 7: $0^+ \to 0^+ (\Phi=35.26°)$; 8: $2^+ \to 2^+ (\Phi=90°)$;

9: $3^- \to 3^- (\Phi=90°)$; 10: $3^+ \to 3^+ (\Phi=90°)$; 11: $4^- \to 4^- (\Phi=90°)$;

theoretical curves, calculated by the molecular field approach is noticed. The modulation of the transmitted intensity with B in general is weaker in $Pb_{1-x}Eu_xSe$ than in $Pb_{1-x}Mn_xTe$, particularly in Voigt configuration. This may be caused by the poorer mobilities of the $Pb_{1-x}Eu_xSe$ samples. Therefore only the strongest transitions can be observed and transitions which are forbidden in PbSe are not observed in $Pb_{1-x}Eu_xSe$. In Fig.18 ΔE is plotted versus the Eu^{2+}- content x. In comparison to

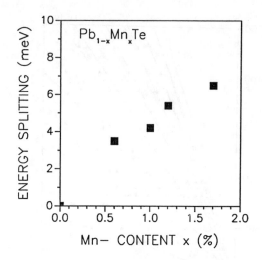

Fig.15: σ^{+}- σ^{-} splitting vs Mn- content for T=1.8K

Fig.15 it is seen that the energy splitting is reduced by about a factor of two.

6.0 COHERENT RAMAN SCATTERING

In nonmagnetic semiconductors the cyclotron resonance is infrared active, whereas the spin resonance of the free carriers is Raman active. However, Raman spectra of narrow gap semiconductors are very difficult to be obtained by conventional Raman methods since visible light is strongly absorbed by interband transitions. Spontaneous Raman scattering in the infrared is hardly observable due to the poor sensitivity of infrared detectors and due to the frequency dependence of scattering cross sections.

With the coherent Raman methods which are based on optical four-wave mixing the scattered light leaves the sample as a collimated beam, by far easier to be detected than the spontaneously scattered light which is emitted into the whole solid angle. This makes coherent Raman scattering applicable to narrow gap semiconductors [40] and allows an

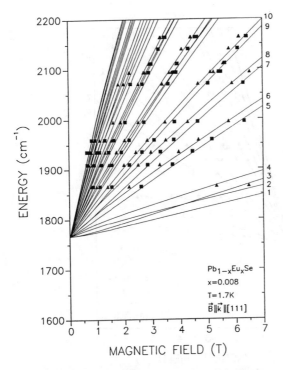

Fig.16: Fan chart for interband magnetooptical transitions in Faraday geometry: ■, ▲ experimental data (σ^-, σ^+); full lines: calculated data; identification:

1: $0^+ \to 0^- (\Phi=70.53°)$; 2: $0^- \to 0^+ (\Phi=70.53°)$; 3: $0^+ \to 0^- (\Phi=0°)$;

4: $0^- \to 0^+ (\Phi=0°)$; 5: $1^+ \to 1^- (\Phi=70.53°)$; 6: $1^- \to 1^+ (\Phi=70.53°)$;

7: $1^+ \to 1^- (\Phi=0°)$; 8: $1^- \to 1^+ (\Phi=0°)$; 9: $2^+ \to 2^- (\Phi=70.53°)$; etc.

observation of spin flip transitions of electrons and holes in a magnetic field. The resulting highly precise data on the effective g factors are of special importance with diluted magnetic semiconductors.

6.1 Theory.

6.1.1 Classical approach.

The nonlinear Raman effects are produced by the last term:

372

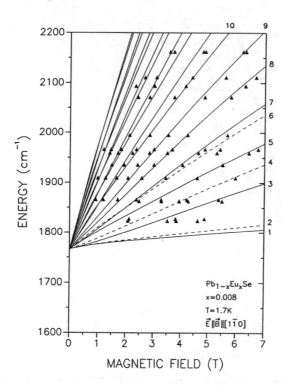

Fig.17: As Fig.4 but for Voigt geometry $\vec{B} \parallel [1\bar{1}0]$.
Experimental data: ▲ in $\vec{E} \parallel \vec{B}$ polarization;
full lines: calculated data for $\Phi=90°$, dashed lines: $\Phi=35.26°$.
1: $0^- \rightarrow 0^-(\Phi=90°)$; 2: $0^- \rightarrow 0^-(\Phi=35.26°)$; 3: $0^+ \rightarrow 0^+(\Phi=90°)$;
4: $0^+ \rightarrow 0^-(\Phi=35.26°)$; 5: $1^- \rightarrow 1^-(\Phi=90°)$; 6: $1^- \rightarrow 1^-(\Phi=35.26°)$;
7: $1^+ \rightarrow 1^+(\Phi=90°)$; 8: $2^- \rightarrow 2^-(\Phi=90°)$; 9: $2^+ \rightarrow 2^+(\Phi=90°)$;
10: $3^- \rightarrow 3^-(\Phi=90°)$; etc.

$$P_i^{NL} = \varepsilon_o \chi_{ijkl}^{(3)} E_j E_k E_l \qquad (29)$$

in a power series of the polarization $\vec{P}$, which reads for the ith component [41]:

$$P_i = \varepsilon_o (\chi_{ij}^{(1)} E_j + \chi_{ijk}^{(2)} E_j E_k + \chi_{ijkl}^{(3)} E_j E_k E_l + \ldots) \qquad (30)$$

(The meaning of the symbols is as usual)

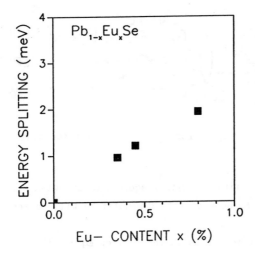

Fig.18: $\sigma^+- \sigma^-$ splitting vs Eu- content for T=1.8K

We write the electric field of a wave as

$$\vec{E}(\vec{r},\omega,t) = \frac{1}{2}\left[\vec{E}(\vec{r},\omega)\cdot\exp(i(\vec{k}\vec{r}-\omega t)) + cc.\right] \tag{31}$$

From Maxwell's relations the wave equation is deduced:

$$ik_x\frac{\partial E_i(x,\omega)}{\partial x}\cdot\exp(i(k_x x-\omega t)) = -\frac{1}{2}i\omega\mu_o\sigma E_i(x,\omega)\cdot\exp(i(k_x x-\omega t))+\mu_o\frac{\partial^2 P_i^{NL}}{\partial t^2}$$

$$\tag{32}$$

This equation holds under the following assumptions:

    a) The waves propagate in x-direction.

    b) Amplitudes do not change significantly within one wavelength.

    c) Power transfer between the waves is neglected.

We consider the experimental situation as schematically drawn in Fig.19. Two laser beams with frequencies ω_L and ω_S are collinearly focused onto a sample with length L. The radiation leaving the sample is fed into a grating monochromator and either the intensity I at the anti-Stokes frequency $\omega_{AS}= 2\omega_L - \omega_S$ or at the second Stokes frequency $\omega_{2S}= 2\omega_S - \omega_L$ is registered. These attestations are according to incoherent Raman scatte-

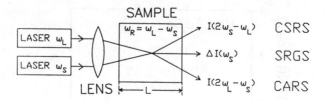

Fig.19: Schematic diagram for four wave mixing techniques.

ring of pump laser radiation with frequency ω_L and with a Raman shift of $\omega_R = \omega_L - \omega_S$. The usefulness of this nomenclature will be seen in the subsequent section.

With the electric fields in the form of Eq.31, the polarization according to Eq.29 has components, oscillating at the different frequencies $\omega_S, \omega_{2S}, \omega_{AS}$. Which one is observed is decided by setting the spectrometer. Let us assume, the monochromator is set to ω_{AS}. The term of the polarisation, oscillating with the correct frequency is:

$$P_i^{NL}(\omega_{AS}, x, t) = \frac{3\varepsilon_0}{8} \cdot \chi_{ijjl}(-\omega_{AS}, \omega_L, \omega_L, -\omega_S) \cdot E_j(x, \omega_L) \cdot E_j(x, \omega_L) \cdot E_l^*(x, \omega_S)$$

$$\times \exp\left[i(2k_L - k_S)x\right] \cdot \exp\left[-i(2\omega_L - \omega_S)t\right] + cc. \qquad (33)$$

This polarization is replaced in Eq.(32) and it is solved for the intensity at the end of the sample. The result is [42,45]:

$$I_{AS}(L) = L^2 \frac{9}{16} \cdot \frac{\omega_{AS}^2}{c^4 n_{AS} n_L^2 \varepsilon_0^2} |\chi^{(3)}|^2 I_L^2 I_S \cdot \left(\frac{\sin(\Delta\vec{k}\vec{x}_0 L/2)}{\Delta\vec{k}\vec{x}_0 L/2} \right)^2 \qquad (34)$$

($\Delta\vec{k} = 2\vec{k}_L - \vec{k}_S - \vec{k}_{AS}$, $\vec{x}_0$: vector of unit length in growth direction)

In thin epitaxial films where the length L is small compared to the coherence length $\pi/\Delta k$ the phase factor can be approximated by 1. Due to the proportionality of the intensity I_{AS} to the length, squared the signals are difficult to be observed with samples thinner than about $3\mu m$.

The informations on the investigated material are contained in the nonlinear susceptibility, particularly in its dependence on the frequencies.

6.1.2 Nonlinear susceptibility in semiconductors.

The susceptibility $\chi^{(3)}$ has non resonant and resonant contributions. Wolff [43] has demonstrated that the non resonant part is mainly
caused by the non parabolicity of the bands and is proportional to the
number N_o of free carriers and inversely proportional to the energy gap
E_g and the square of the effective mass m^*. Thus it is very large in
doped narrow gap semiconductors.

Using the energy level scheme of Fig.20 the susceptibility can be
calculated by time dependent perturbation theory. The result is reproduced in Table A2 of Ref.44. The susceptibility is a sum over the states
of the scattering medium as intermediate states $|b\rangle$, $|c\rangle$ and $|d\rangle$ of 24
terms. One of them which is responsible for CARS is reproduced in
Eq.(35):

$$\chi^{(3)} = \frac{\mu_{jkli}}{(\omega_{ba} - \omega_L - i\Gamma_{ba})(\omega_{ca} - (\omega_L - \omega_S) - i\Gamma_{ca})(\omega_{da} - \omega_{AS} - i\Gamma_{da})} \qquad (35)$$

The abbreviation μ_{ijkl} stands for a product of four dipole matrix ele-
ments $\mu_{ab}\mu_{bc}\mu_{cd}\mu_{da}$ with the field polarizations i,j,k,l. ω_{12} is equal to
$(E_1-E_2)/\hbar$. Γ_{12} is the homogeneous linewidth of the transition $|1\rangle \rightarrow |2\rangle$.
Identifying ω_{ba} with the energy gap it is seen from Eq.(35) that the
susceptibility is resonantly increased if $\hbar\omega_L \approx E_g$.

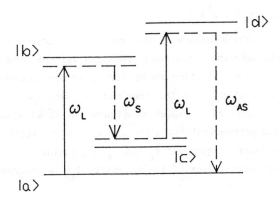

Fig.20: Energy level diagram.

376

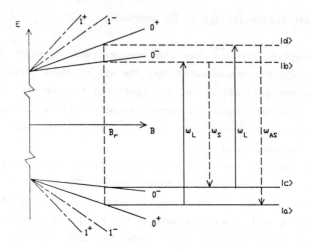

Fig.21: Selection rules for CARS transitions. Polarization of radiation
with frequency: $\omega_L:\vec{E}{\perp}\vec{B}$, $\omega_S:\vec{E}\|\vec{B}$, $\omega_{AS}: \vec{E}\|\vec{B}$.

For a semiconductor exposed to a magnetic field B the most important Raman resonances are due to spin flip transitions if

$$\hbar\,\omega_{ca} = \hbar\,(\omega_L - \omega_S) = g^*\mu_B B. \qquad (36)$$

In Fig.21 a correspondence of the energy level scheme (Fig.20) with
spin split Landau levels in the conduction and valence band is assumed.
From the polarization selection rules for interband transitions it is
seen that the product μ_{ijkl} of the four matrix elements is not zero if
the sample is irradiated in Voigt configuration ($\vec{B}{\perp}\vec{k}$) with the laser
with ω_L polarized perpendicular and the one with ω_S parallel to $\vec{B}$.

With fixed laser frequencies ω_L and ω_S and tuning ω_{ca} by a magnetic
field, a resonant enhancement of the intensity I_{AS} is observed. From the

resonant magnetic field B_r the effective g-factor can be calculated by Eq.36.

For magnetic fields near B_r the susceptibility is a complex number (see Eq.35). This leads to complicated lineshapes caused by interferences between real and imaginary parts of the susceptibility. Either intensity maxima or points of inflection correspond to the resonant magnetic fields. Whether extrema or points of inflection have to be evaluated depends on the relative strength of resonant in comparison to nonresonant contributions. The lineshapes are discussed in detail elsewhere [42,44].

6.3 Experimental Results and Discussion.

The experiments were performed with two simultaneously Q-switched CO_2 lasers. The peak powers were attenuated as much as possible to avoid power broadening of the lines and saturation of the spin resonances. For $\hbar\omega_L \approx E_g$ powers of the order of 50mW are sufficient, for $\hbar\omega_L \approx E_g/2$ 500W to 2kW are necessary. The pulse lengths are about 100ns. Therefore the energy per pulse is low enough to avoid heating of the crystal lattice. For collinear adjustment of the laser beams the optimum sample thickness is equal to the coherence length $L_c = \pi/|\Delta\vec{k}|$. With thin epitaxial films $L \ll L_c$ and the phase factor can be approximated by 1. Due to the proportionality $I_{AS} \propto L^2$ in very thin films the signals are difficult to observe without taking advantage of the band gap resonance $\hbar\omega_L \approx E_g$. In this case using a double monochromator and a sensitive MCT photovoltaic detector a sample thickness of at least 3μm is preferable.

In $Pb_{1-x}Mn_xTe$ as well as $Pb_{1-x}Eu_xSe$ with $x \leq 0.012$ there are CO_2 laser lines with $\hbar\omega_L > E_g/2$. Using these lines the laser radiation produces a considerable number of minority carriers by two photon absorption. Therefore electron and hole resonances can be observed in one sample independent of its doping.

6.3.1 Results: $Pb_{1-x}Mn_xTe$

Figure 22a shows experimental data on the CARS intensity for $\Delta\omega$ =

378

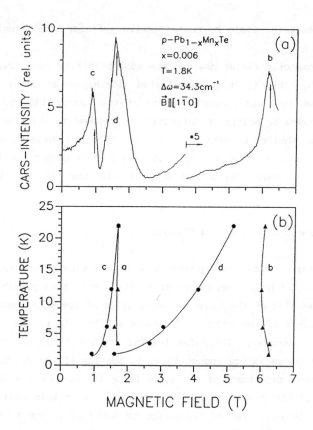

Fig.22: a) CARS intensity vs. magnetic field for $\Delta\omega=34.3cm^{-1}$ at T=1.8K.
   The transitions marked b), d), c) correspond to spin flip of
   electrons ($\Phi=90°$) and of holes ($\Phi=90°$, $\Phi=35.26°$), respectively. Arrows indicate resonance positions.
   b) temperature dependence of CARS intensity resonances for both,
   electrons and holes in the $\Phi=35.26°$ and $\Phi=90°$ valleys (transition a: electrons $\Phi=35.26°$).

$\omega_L-\omega_S=$ 34.3 cm^{-1} as a function of magnetic field. The arrows indicate the resonance positions. (b) corresponds to the spin flip of photoexcited electrons in the $\Phi=90°$ valley. Both of the spin flip transition of holes ((d), $\Phi=90°$), (c), $\Phi=35.26°$) are observable. The spin flip resonance of photoexcited electrons for the $\Phi=35.26°$ valleys, (a) is not observed at T=1.8 K since it coincides almost with that of the holes (d), which is much stronger in the p-type sample. However, as the temperature is changed (a) is observable. In Fig.22b the resonance positions are shown for the identical $\Delta\omega=34.3$ cm^{-1} as a function of T. Apparently, the spin flip resonances of the electrons ((a), (b)) shift only slightly with increasing temperature to somewhat smaller magnetic fields whereas the situation is opposite for the valence band. The spin flip resonance for the holes ((c) and (d)) shift drastically to higher magnetic fields as the temperature is increased. This shift is particularly pronounced for the $\Phi=90°$ valley (resonance (d)).

Recordings as plotted in Fig.22a were taken for a large number of laser frequencies with different $\omega_L-\omega_S=\Delta\omega$. In Figs.23a-c results on three samples with x=0.006, x=0.008 and x=0.01 at T=1.8 K are summarized. From an inspection of the data it follows that: (i) the contributions to the spin flip transition energies induced by the exchange interaction are much larger for the valence band than for the conduction band (ii) for fields in excess of about 2 T, the difference between the spin flip energies for electrons and holes for the oblique $\Phi=35.26°$ valleys as well as for the $\Phi=90°$ valleys changes only slightly with field (arrows in Figs.23a-c) for fields above 2T. This energy difference is the same within experimental error which is observed for interband magnetooptical transitions in $\vec{B}\|\vec{k}$ -geometry if energy differences for the σ^+ and σ^- polarized radiation $(0^-(vb) \rightarrow 0^+(cb), 0^+(vb) \rightarrow 0^-(cb))$ are derived from the experimental data (see Fig.15).

The calculated transition energies are presented by the full lines in Figs.23a-c using the model outlined in Sec.3. Actually, the four exchange parameters listed in Table IV as well as T_o were determined from the fits shown in Fig.23a, whereas the two band parameters $P_\perp$, $P_\|$ and the energy gap were obtained from the magnetooptical interband transition energies. The far band parameters were assumed to be identical to

380

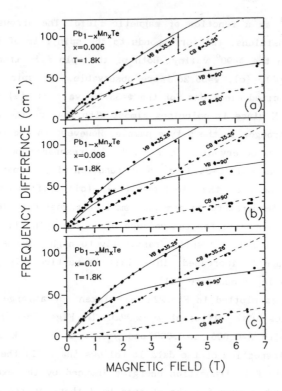

Fig.23: Results of CARS measurements (frequency differences) as a function of magnetic field for x=0.006 (a), x=0.008 (b) and x=0.01 (c) for $\vec{B}\|[1\bar{1}0]$ and T=1.8K. Experimental data: •, calculated data for holes: full lines for $\Phi=35.26°$ and $\Phi=90°$ valleys; electrons: broken lines for $\Phi=35.26°$ and $\Phi=90°$ valleys.

those of PbTe.

The temperature dependence of these transition energies is quite dramatic, as shown in Figs.24a-f for experiments performed up to 22K for a sample with a Mn-content of x=0.006. The temperature dependence is particularly pronounced for the hole spin flip transitions. With increasing lattice temperature the sequence of spin flip transition energies becomes more and more PbTe-like apart from the fact that the absolute

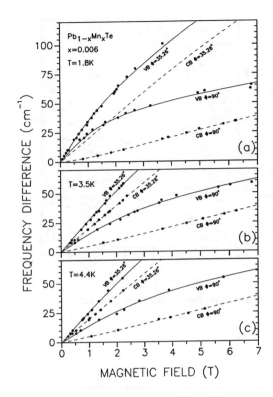

Fig.24a-c: Results of CARS measurements on Pb$_{1-x}$Mn$_x$Te (x=0.006) for three temperatures: 1.8K, 3.5K and 4.4K. Experimental data ●, calculated data: full lines for holes (Φ=90°, Φ=35.26°) and dashed lines for electrons (Φ=90°, Φ=35.26°).

values of the g-factors are smaller due to the larger gap. In Fig.25 the electron and hole g-factors (calculated from the experimental spin flip energies at B = 0.05 T using Eq.(36) are shown as a function of temperature in the range 1.8 K - 22 K. For $\vec{B}\|[1\bar{1}0]$ two conduction and two valence band g-factors are shown. The most dramatic temperature shifts exhibits the g-factor for the Φ=90° VB-valleys.

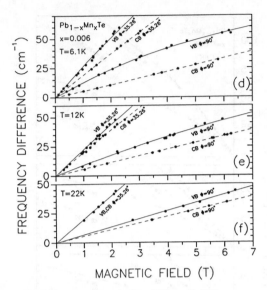

Fig.24d-f: As Fig.24a-c but for 6.1K, 12K and 22K.

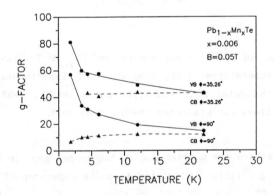

Fig.25: Temperature dependence of electron (▲) and hole g-factors (●)
for a magnetic field of B=0.05T as derived from CARS experiments. The full and dashed lines are just a guide to the eye.

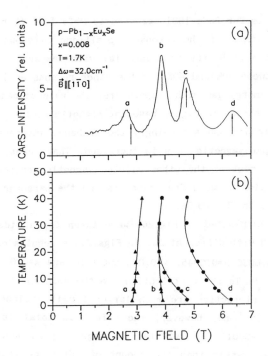

Fig.26: a) CARS intensity vs. magnetic field for $\Delta\omega=32.0\text{cm}^{-1}$ at T=1.7K.
The transitions marked b), a), d), c) correspond to spin flip
of electrons ($\Phi=90°$, $\Phi=35.26°$) and of holes ($\Phi=90°$,
$\Phi=35.26°$), respectively. Arrows indicate resonance positions.
b) temperature dependence of CARS intensity resonances for both,
electrons and holes in the $\Phi=35.26°$ and $\Phi=90°$ valleys.

6.3.2 Results: $Pb_{1-x}Eu_xSe$

In Fig.26a an experimental registration of the CARS intensity vs.
magnetic field is reproduced for a $Pb_{1-x}Eu_xSe$ sample with x = 0.008 at a
temperature of 1.7 K. (a) and (b) are the conduction band spin resonances in the valleys with Φ = 35.26° and 90°, respectively, (c) and (d)
belong to the valence band [46,55]. The anisotropy both of the effective
masses and the effective g-factors is much smaller in $Pb_{1-x}Eu_xSe$ than in

$Pb_{1-x}Mn_xTe$ as it is for PbSe [53] compared to PbTe [42] (Table III). The temperature dependence of the resonance positions is shown in Fig.26b. The small increase of B_r for resonance (a) with temperature as well as the increase between 30K and 40K for the other resonances is due to the increase of the energy gap with temperature. The valence band resonances (c) and (d) are shifted to higher resonant magnetic fields when the temperature is diminshed and accordingly the exchange interaction between free holes and paramagnetic ions is increased. That means, in contrast to $Pb_{1-x}Mn_xTe$ and most of the other diluted magnetic semiconductors the exchange interaction between free carriers and the paramagnteic ions is antiferromagnetic in $Pb_{1-x}Eu_xSe$.

Recordings as plotted in Fig.26a were taken for a wide variety of laser frequencies with different $\Delta\omega$. In Figs.27a-e results obtained with samples with x=0.0035 x=0.0045, x=0.008 x=0.013, and x=0.014 are summarized. As in $Pb_{1-x}Mn_xTe$ the contributions to the spin splittings induced by the exchange interaction are much larger for the valence band than for the concuction band, however, the effect in total is smaller in $Pb_{1-x}Eu_xSe$. With x=0.014 for B<4T the exchange induced decrease of the spin splitting is larger than the amount of splitting which would be present without Eu. Spin up and spin down levels are exchanged below 4T. However, by the CARS method only the absolute value of the effective g factor can be determined experimentally. The conclusions concerning the sign are drawn from the comparison between experimental values and theory.

The lines in Figs.27a-e and 28a-c present the calculated transition energies. Figure 28 shows experimentally observed as well as calculated spin splittings for three temperatures. As in $Pb_{1-x}Mn_xTe$ the model outlined in Sec. 3 is used with the exchange parameters and T_o fitted to the CARS data, the two band parameters to the magnetooptical interband data and the far band parameters assumed to be the same as in PbSe.

In analogy to Fig.25 in Fig.29 the temperature dependence of electron and hole g factors is plotted for $Pb_{1-x}Eu_xSe$ for fixed magnetic field. The hole g factors change sign at about 5K, a fact caused by the spin level crossing. Particularly impressive is the strong temperature dependence of the hole g factors ($\Phi=90^o$) for the samples with highest

Eu- contents (Fig.30). It reflects the temperature dependence of the exchange interaction. The results are obtained with fixed frequency differences $\omega_L - \omega_S$ of about 8 cm^{-1}.

6.4 Effective Electron and Hole g factors

The spin resonant four wave mixing experiments provide an ideal
means for the study of the magnetic field dependence of the spin splitting between the $0^- - 0^+$ states both in the conduction and in the valence

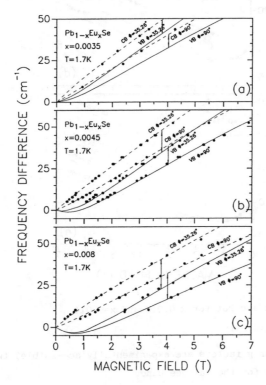

Fig.27a-c: Results of CARS measurements (frequency differences) as a
function of magnetic field for x=0.0035 (a), x=0.0045 (b) and
x=0.008 (c) for $\vec{B} \parallel [1\bar{1}0]$ and T=1.8K. Experimental data: •,
calculated data for holes: full lines for $\Phi=35.26°$ and $\Phi=90°$
valleys; electrons: broken lines: $\Phi=35.26°$ and $\Phi=90°$ valleys.

bands. In Fig.31 experimentally derived g-factors of $Pb_{1-x}Mn_xTe$ (x=0.01) are plotted as a function of magnetic field together with calculated data based on Sec. III and Tables III and IV. Figure 32 shows the analogous plot for $Pb_{1-x}Eu_xSe$ (x=0.008). For an orientation of the magnetic

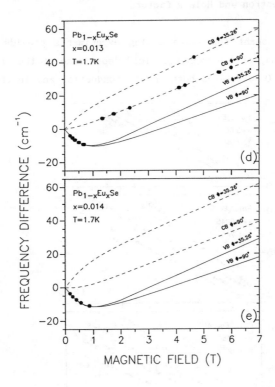

Fig.27d,e: As Fig.27a-c but for x=0.013 and x=0.014

field $\vec{B}//[1\bar{1}0]$ four g factors are experimentally accessible, two for the conduction and two for the valence band.

With increasing magnetic field the absolute values of the g-factors (i) decrease in the valence band of $Pb_{1-x}Mn_xTe$ for both kinds of valleys. The relative change is larger for the $\Phi=90°$ valleys than for the $\Phi=35.26°$ valleys In $Pb_{1-x}Eu_xSe$ the valence band g factors increase. If x is large enough, at small magnetic fields the ordering of the spin levels is inverted; g^* goes through zero.

(ii) In the conduction bands the deviation from the case without exchange interaction is small for both materials. There is a small increase in the conduction band of $Pb_{1-x}Mn_xTe$ for both kinds of valleys,
whereas in $Pb_{1-x}Eu_xSe$ the g-factor for the $\Phi=35°$ valleys decreases and
for $\Phi=90°$ increases.

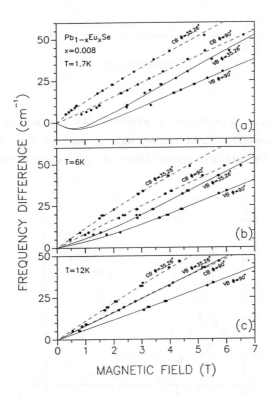

Fig.28a-c: Results of CARS measurements on $Pb_{1-x}Eu_xSe$ (x=0.008) for
three temperatures: 1.7K, 6.0K and 12K. Experimental data ●,
calculated data: full lines for holes ($\Phi=90°$, $\Phi=35.26°$) and
dashed lines for electrons ($\Phi=90°$, $\Phi=35.26°$).

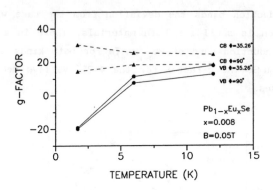

Fig.29: Temperature dependence of electron (▲) and hole g-factors (●)
for a magnetic field of B=0.05T as derived from CARS experiments. The full and dashed lines are just a guide to the eye.

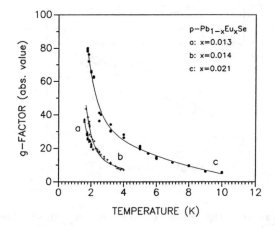

Fig.30: Temperature dependence of hole g factors (Φ=90°) for three
$Pb_{1-x}Eu_xSe$ samples.

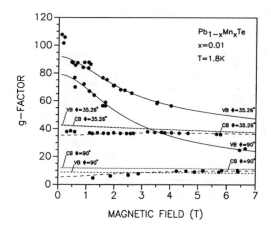

Fig.31: Magnetic field dependence of g-factors derived from CARS experiments in $\vec{B}\|[1\bar{1}0]$ geometry with theoretical data: full lines for holes, dashed lines for electrons; dotted lines: calculated without exchange interaction. $Pb_{1-x}Mn_xTe$, x=0.01 at T=1.8K.

The mean field approach yields an excellent overall agreement for all gfactors involved. In order to demonstrate the effect of the exchange interaction on the magnetic field dependence of the g factors we note that nonparabolicity decreases the absolute values of the g factors with increasing field for electrons and holes for all types of valleys. In Figs.31,32 for comparison also theoretical values of the magnetic field dependence of the g factors are plotted without exchange interaction (i. e. based on Eq.16 or on Eq.19), for $0^- \rightarrow 0^+$ transitions. With increasing field the electron g factors in $Pb_{1-x}Mn_xTe$ approach the values of the host material from <u>below</u> whereas the actual hole g-factors approach the corresponding g factors without taking into account the exchange interaction from <u>above</u>. In $Pb_{1-x}Eu_xSe$ it is vice versa (apart from $\Phi=90^\circ$, CB).

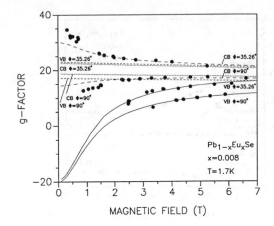

Fig.32: As Fig.31 but for Pb$_{1-x}$Eu$_x$Se.

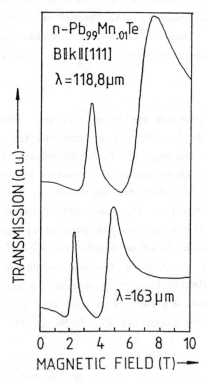

Fig.33: FIR magnetotransmission of n- Pb$_{1-x}$Mn$_x$Te in Faraday geometry for two laser wavelengths.

7.0 FAR INFRARED SPECTROSCOPY

Magnetooptical intraband transitions in Faraday geometry were investigated in a number of n- and p-type samples to get independent information on the effective masses and their anisotropy by cyclotron resonance absorption. Earlier investigations on n- and p-Pb$_{1-x}$Mn$_x$Te were
presented in Refs.13,37,47,48. For $\vec{B} \| [111]$ two resonant frequencies as-
sociated with cyclotron transitions in the [111] (ω_{c1}, $\Phi=0°$) and in the

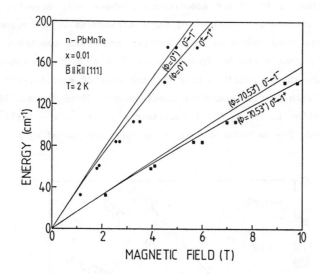

Fig.34: Intraband magnetooptical transition energies as derived from
cyclotron resonances for infrared experiments in Faraday geometry. Two kinds of resonances associated with the $\Phi=0°$ (●) and
$\Phi=70.53°$ (■) valleys are observed. Full lines: calculated data

three obliquely oriented <111> ($\Phi=70.53°$) valleys are observed (Fig.33).
In samples with high mobility and consequently long relaxation times
with FIR laser spectroscopy both $n^- \to (n+1)^-$, $n^+ \to (n+1)^+$ transitions can
be observed as shown in Fig.34 for a n-type sample with x=0.01. The
identification of the resonance frequencies is generally based on oscillator fits using a model dielectric function [33,49] since for laser

energies $\hbar\omega_L$ below about 14 meV, $\hbar\omega_{TO} < \hbar\omega_L < \hbar\omega_{LO}$ where ω_{TO} and ω_{LO} de-note the TO and LO optic mode phonon frequencies. Thus, in this range, the magnetotransmission experiments are performed in the reststrahlen region where the resonances are accompanied by dielectric anomalies, as shown in Fig.33 for two infrared laser wavelengths λ=163μm and λ=118.8μm. In addition, for transition energies close to $\hbar\omega_{LO} + E_F(B)$, where $E_F(B)$ denotes the magnetic field dependent Fermi energy, energy shifts due to polaron effects occur [49.50].

In addition to FIR laser spectroscopy, where only magnetic field dependent transitions are observable, magnetotransmission and magnetore-flectivity data were taken using a Fourier transform spectrometer at various fixed magnetic fields. In Fig.35 for Faraday geometry results on p-Pb$_{1-x}$Mn$_x$Te are shown, obtained from the analysis of magnetoreflectivi-ty data taken with Fourier transform spectroscopy. Since the mobility of the p-type samples is lower, the transitions $n^+\rightarrow (n+1)^+$, $n^-\rightarrow (n+1)^-$ can-not be resolved. The calculated transition energies, based on the para-

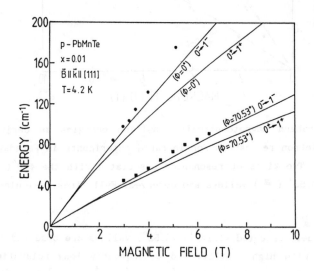

Fig.35: As Fig.34 for a p-type Pb$_{1-x}$Mn$_x$Te (x=0.01) sample. Due to lower mobilities experimentally the two different transitions $1^-\rightarrow 0^-$, $1^+\rightarrow 0^+$ are not resolved.

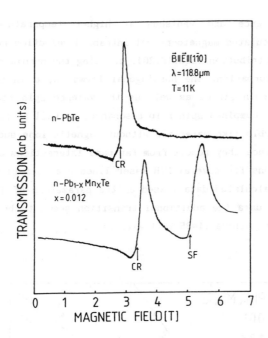

Fig.36: Far infrared transmission in Voigt geometry, $\vec{E}\|\vec{B}\|[1\bar{1}0]$:
In PbTe just the oblique valley resonance ($\Phi=35.26°$) is observed
whereas in $Pb_{1-x}Mn_xTe$ an additional resonance identified as spin
flip resonance occurs.

meters from Table III and IV which were obtained from optimum fits to
the interband and the g-factor data are in reasonable agreement with the
observed ones.

In Voigt geometry, $\vec{E}\|\vec{B}$, experiments were performed again on n- and
p-type samples with $\vec{B}\|[1\bar{1}0]$, i.e. $\Phi=35.26°$ and $\Phi=90°$. For the PbTe band-
structure just the oblique valley cyclotron resonance ($\Phi=35.26°$) is
allowed whereas no cyclotron transition is possible for the two $\Phi=90°$
valleys. In Fig.36 FIR transmission experiments on n-PbTe and
n-$Pb_{1-x}Mn_xTe$ are compared with each other. A strong second resonance is
observed at fields higher than the cyclotron resonance transition. As
already shown in Refs.13,37, the second resonance exhibits a strong tem-

perature dependence and vanishes at higher temperatures. Zawadzki
[13,51] has calculated magnetooptical intraband selection rules based on
the exchange contribution (Eqs.18,20), treating the nondiagonal exchange
terms as a perturbation. The nondiagonal terms (c, d in Eqs.18,20 mix
the conduction spin states as well as the valence spin states and enhance additional combined spin flip resonances as well as pure spin flip
resonances. In PbTe like materials without magnetic ions such resonances
are very weak since they result from far band interactions only. The observed transitions for several FIR laser lines are summarized in Fig.37
together with calculated data (based on the parameters of Table III and
IV). From these data the additional transition seems to be a spin flip
resonance in the oblique (Φ=35.26°) valleys.

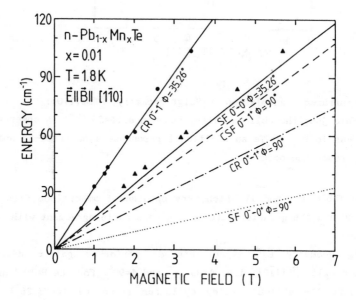

Fig.37: Experimental (●, ▲) and calculated resonance positions in
Voigt geometry ($\vec{E} \| \vec{B} \| [1\bar{1}0]$)] for n-Pb$_{1-x}Mn_x$Te together with iden-
tification: CR: cyclotron resonance, SF: spin resonance, CSF:
combined spin flip resonance.

From the transmission data alone, due to the dielectric anomalies associated with both the CR and SF transitions, no direct conclusion on the oscillator strengths of both transitions can be drawn. In addition the data were obtained on thin biaxially strained epitaxial samples, where the tensile strain at low twmperatures enhances the SF transition as well.

Similar FIR results in Voigt geometry $\vec{E}\|\vec{B}\|[1\bar{1}0]$ for p-type $Pb_{1-x}Mn_xTe$ (x = 0.01) also exhibit two resonances but are qualitatively different as shown in Fig.38. Again a cyclotron resonance of the holes of the oblique valley ($0^- \to 1^-$, $\Phi=35.26°$) is observed together with an additional resonance which is only observable for fields in excess of 2.5 T by Fourier transform spectroscopy and may be hidden for lower fields due to the dielectric anomaly associated with the cyclotron resonance $0^- \to 1^-$. In Fig.38 several transitions are calculated as indicated.

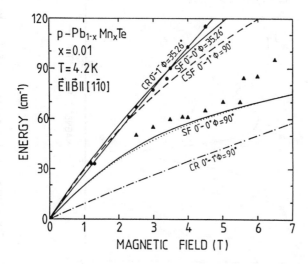

Fig.38: As Fig.37 but for p-type $Pb_{1-x}Mn_xTe$ (x=0.01) at T=4.2K.
Experimentally observed transitions correspond to cyclotron resonance (●) in the oblique valleys and to spin flip resonance (▲). The bowing is due to strong effect of exchange interaction.

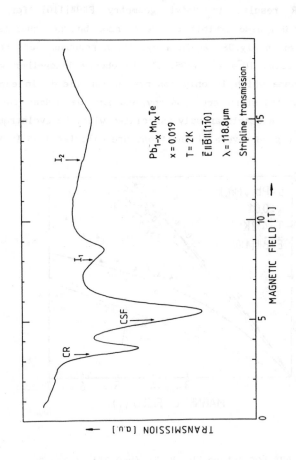

Fig.39: FIR stripline transmission of $Pb_{1-x}Mn_xTe$ vs B. Resonance positions marked by arrows.

A spin flip transition $(0^- \to 0^+)$ of the oblique $(\Phi=35.26^\circ)$ valleys would be indistinguishable from their cyclotron resonance, a combined spin flip (CSF $0^- \to 1^+$) for the $\Phi=90^\circ$ valleys deviates at higher fields. Therefore the additional resonance in p-type $Pb_{1-x}Mn_xTe$ is likely to be a spin flip resonance of the $\Phi=90^\circ$ valleys [54].

As already shown by von Ortenberg [48,52] the sensitivity of FIR spectroscopy towards the detection of rather weak transitions can be enhanced using strip- line transmission transmission spectroscopy. In Fig.39 such data are shown for $\vec{E} \| \vec{B} \| [1\bar{1}0]$ on n- $Pb_{1-x}Mn_xTe$ at T=2K. In addition to the CR and SF transitions, further transitions are observed which are absent in n-PbTe. In Fig.40 the corresponding fan chart is

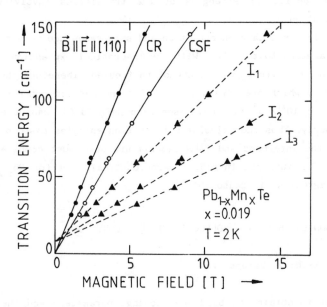

Fig.40: Fan chart of intraband transitions in Voigt configuration. Full lines: calculated CR and SF transitions, dashed lines: guide for the eye for transitions I_1 to I_3.

shown, where the full lines represent calculated data and the dashed lines for the I_1, I_2 and I_3 transitions are just a guide for the eye. Such additional transitions were observed for Mn- contents ranging from 1% to 1.9% in n-type material. Whether the observed transitions I_1, I_2 and I_3 are related to impurities or defects is not yet established.

Cyclotron resonance experiments on bulk $Pb_{1-x}Mn_xTe$ using strip-line techniques were performed by Gorska et al. [47]. Due to the rather moderate mobilities the resonances are broadened considerably and only limited informations can be extracted.

Further experimental investigations, especially on the temperature dependence of the observed resonances in Voigt geometry in p-type $Pb_{1-x}Mn_xTe$ as well as in $Pb_{1-x}Eu_xSe$ are necessary. In addition calculations on the oscillator strengths of the transitions involving spin flips are required.

The far infrared data on cyclotron and spin resonance were not used as input parameters to the fits. With the theoretical expressions as given in Eqs.16-20 finite k_B-values cannot be treated. Therefore the occupation effects, which are always present for the carrier concentrations of the order of $10^{17}/cm^3$ and above (see also Refs.33,49) cannot be considered properly. Thus the calculated transition energies based on parameters from the interband and CARS experiments were just compared with experimental far infrared intraband magnetooptical data for $Pb_{1-x}Mn_xTe$. The overall agreement is reasonable.

8.0 COMPARISON EXPERIMENT - MOLECULAR FIELD THEORY

8.1 Band and Exchange Parameters

In order to obtain the band and exchange parameters and their dependences on Mn- and Eu- content, respectively, which are listed in Tables III and IV, the following procedure was performed: to determine the two band parameters like the minimum energy gap , the two band momentum matrix elements $2P_\perp^2/m_o$ and $2P_\parallel^2/m_o$, fits to the magnetooptical interband transitions were made with the far band parameters $m_t^+, m_l^+, g_l^+, g_t^+$ assumed to be identical to those of PbTe (PbSe), which were taken from

Refs.42,([53]). This assumption seems to be valid for the rather small x- values (x<0.02) of the samples under investigation. It means that neither the band ordering nor the energy separation for the two higher conduction and two lower valence levels at the L-point of the Brillouin zone changes considerable when the paramagnetic ion is introduced into the host lattice, for such small x-values. From the fits to the inter-band data it follows that with increasing Mn-content $2P_\perp^2/m_o$ decreases whereas the anisotropy value $P_\perp^2/P_\parallel^2$ increases (see Table III). The two band parameters for $Pb_{1-x}Eu_xSe$ remain even constant, except for the energy gap [55].

The exchange induced coefficients A, a_1, B, b_1 and the parameter T_0 in the modified Brillouin function (see Eqs.4-6) are obtained from least square fits to the experimentally observed $0^- \to 0^+$ transitions, i. e. from the CARS experiments which determine these spin splittings most ac-curately. Since in these experiments for $\vec{B}\|[1\bar{1}0]$ two kinds of valleys, (i.e. $\Phi=35.26°$ and $\Phi=90°$) yield two g-factors both for the conduction as well as for the valence band all band and exchange parameters can be de-termined unambiguously. S_0 (Eq.18) is taken from fits to the experimen-tal magnetization data for $Pb_{1-x}Mn_xTe$ which yield also values for T_0. The latter are in agreement with those obtained from the CARS data. For $Pb_{1-x}Eu_xSe$ no magnetization data are available so far. We assumed S_0 to be 7/2, T_0 derived from the CARS data turned out to be OK.

Using Eqs.21-23 the actual exchange constants α, δ (cb) and $\beta_\parallel$, $\beta_\perp$ (vb) are derived and given in Table V as a function of composition and temperature. In order to calculate these values the spin orbit mixing parameters $\Theta^\pm$ as given by Bernick and Kleinman are used [32]. The large absolute value of δ for $Pb_{1-x}Mn_xTe$ is most probably an artifact caused by the uncertainty of $\sin\Theta^+$(Ref.32) and a small change of this parameter drastically affects the value of δ. Unfortunately, even the general un-certainty in the spin orbit mixing parameters $\Theta^\pm$ is too large to derive reliable values for α, δ (vb) and $\beta_\parallel$, $\beta_\perp$ (cb). Nevertheless Zasavitskii et al. [56] have tried to obtain exchange integrals from luminescence data.

For PbSe the spin orbit mixing parameters $\Theta^\pm$ as given by Bernick and Kleinman are more accurate than for PbTe since the g- factors and masses deduced in Ref.32 agree much better with experimental data for PbSe [53]. We conclude that thus in $Pb_{1-x}Eu_xSe$ also α, δ (vb) and $\beta_\parallel$, $\beta_\perp$ (cb) are more reliable than in $Pb_{1-x}Mn_xTe$.

The exchange induced effects in the valence band are larger than in the conduction band and therefore we restrict the following discussion to these, more precisely determined parameters. For $Pb_{1-x}Mn_xTe$ the parameter $|A|$ decreases with increasing temperature whereas for $Pb_{1-x}Eu_xSe$ it stays approximately constant in the same temperature interval from 1.7 to 12 K. In addition with increasing x in both materials, $|A|$ and $|a_1|$ decrease. Such a drastic variation of exchange parameters is a hint for the necessity for further improvements in the theoretical description of diluted magnetic IV-VI compounds in external fields. It demonstrates that the temperature dependence of the hole g-factors does not reflect the temperature dependence of the macroscopic magnetization. In contrast to the diluted magnetic II-VI compounds [11], the physical origin of the s,p,d- d or s,p,d- f exchange constants in the IV-VI compounds has not been analyzed by ab initio calculations. In wide gap materials like ZnMnTe the exchange constants $N_o\alpha$ and $N_o\beta$ are insensitive to the Mn- content. Since in the IV-VI's the minimum gap is at the L-point instead of Γ and furthermore already the band edge functions contain admixtures of d- like functions, the problem of the exchange interaction with the localized moments is more delicate than in the II-VI's.

8.2 Selection Rules

In the magnetooptical experiments in Faraday geometry the selection rules are the same in the diluted magnetic as in the non magnetic compounds [33]. In Voigt geometry (with $\vec{E}\|\vec{B}\perp\vec{k}$, $\vec{B}\|[1\bar{1}0]$, $\vec{k}\|[111]$) new transitions were found: in the interband data shown in Fig.14 a new spin flip transition $0^+(vb) \to 0^-(cb)$ was found induced by the Mn^{2+}- ions [19]. In the intraband data again spin flip transitions were found both in the conduction band as well as in the valence band (Figs.37 and 38) besides the tilted orbit cyclotron resonance in the $\vec{E}\|\vec{B}$ Voigt geometry [19]. These observations demonstrate a modification of the selection rules brought about by the interaction between the localized Mn^{++}-spins and the free electrons as well as holes.

9.0 CONCLUSION:

Despite the fact that the diluted magnetic IV-VI compounds have been studied for about a decade, the amount of knowledge which has been acumulated is much less than compared to the situation in $A_{1-x}^{II} Mn_x B^{VI}$ or $A_{1-x}^{II} Fe_x B^{VI}$ alloys. The complications due to the many valley band structure, the fact that the band edge functions are composed of a combination of s, p, and d - like functions as well as general material properties make this class more difficult to deal with.

As far as the magnetic properties are concerned, the susceptibility data follow a Curie-Weiss-behaviour with rather small paramagnetic Curie-Weiss-temperatures which indicate weak antiferromagnetic exchange coupling between the magnetic ions. The exchange constants obtained from $\chi(T)$ are generally an order of magnitude lower than those found in narrow gap II-VI diluted magnetic semiconductors.

Following the interpretation as given by Gorska and Anderson [18] superexchange dominates in the IV-VI's like in the II-VI's despite the rock salt crystal structure. The NaCl structure, however, has the important consequence, that the cation-anion separation is a/2 instead of $a3^{1/2}/4$ (for zinc blende structure) which turns out to lead to an order of magnitude difference in the superexchange parameter J if estimations based on the Anderson theory of superexchange and the interatomic matrix elements according to Harrison and Straub [57] are used. Also the trends in exchange parameters within the series of lead salts containing manganese and varying the anions (S, Se, Te) as well as those if rare earth elements (Eu, Gd) replace Mn are reproduced. The exchange parameter $2J_B/k$ decreases from $Pb_{1-x} Mn_x S$ (2.57 K) to $Pb_{1-x} Mn_x Te$ (1.69 K) and decreases further for $Pb_{1-x} Eu_x Te$ (0.23 K).

Highfield magnetization data for various Mn compounds are available as well. These data were usually interpreted with a modified Brillouin function but also more elaborate approaches were used. So far a step-like behaviour of the magnetization as a function of the magnetic field has not been observed. Such steps would provide a direct measure of J.

The most striking phenomenon observed in the magnetic properties of IV-VI compounds with large hole concentrations is the existence of a ferromagnetic phase. In PbSnMnTe for hole concentrations in excess of 3×10^{20} cm^{-3} an abrupt transition from the paramagnetic to the ferromagnetic phase has been found with transition temperatures from 0 K to about 5 K. On the basis of the RKKY interaction and details of the band structure (population of L and Σ-valleys) a quantitative explanation for the carrier concentration dependence of the magnetic properties was found.

Another important feature is the occurence of a spin glass phase found in Pb$_{1-x}$Mn$_x$Te as inferred from a cusp of the low field susceptibility at a temperature T_g. Tentatively the existence of long range interband exchange interaction was made responsible for the presence of this phase.

As far as the exchange effects between the mobile electrons or holes and the localized 3d and 4f carriers are concerned, for a relatively long period progress in understanding was comparatively slow. Bulk materials are rather inadequate for such studies due to their high carrier concentrations (10^{18} cm^{-3}) and low mobilities. Progress in epitaxial growth has yielded Pb$_{1-x}$Mn$_x$Te samples (with $x \leq 0.04$) and Pb$_{1-x}$Eu$_x$Te samples (entire range of composition) as well as Pb$_{1-x}$Eu$_x$Se with rather low carrier concentrations ($\approx 10^{17}$ cm^{-3}) and sufficiently high mobilities.

Experimental data on intraband magnetooptical properties and magnetotransport data on n-type samples have given negligible exchange contributions. Early magnetooptical interband experiments both on Pb$_{1-x}$Mn$_x$Te as well as on Pb$_{1-x}$Mn$_x$S were interpreted as giving evidence for a so-called zero field spin-splitting, i. e. the existence of two spin split energy gaps at $B_{ext} = 0$. However, this picture was based on the extrapolation of magnetooptical interband transitions, obtained both in Faraday and Voigt-geometries towards B$\rightarrow$0, and it turned out to be not correct. The direct experimental determination of the effective g-factors of electrons and holes in Pb$_{1-x}$Mn$_x$Te in magnetic fields down to 0.05 T as a function B, as well as of temperature yielded a clear evidence for exchange induced corrections to g*:

a) these exchange induced corrections to the effective g factors are larger for holes than for electrons

b) in the compounds $Pb_{1-x}Mn_xTe$ and for $Pb_{1-x}Eu_xSe$ they are of different sign

c) the main features can be explanied within a mean field model for the exchange interaction. This model also explains the magnetooptical interband as well as intraband data quantitatively

d) within the experimental uncertainty of less than 0.5 meV there is no evidence for a zero field spin splitting of the energy gap in $Pb_{1-x}Mn_xTe$.

e) due to the symmetry properties of the band edge wave functions at the L-points of the Brillouin zone, which are made up of linear combinations of two terms both in the conduction and in the valence band four exchange integrals, two for the electrons and two for the holes are needed. The appreciable amount of spin-orbit interaction causes these pecularities.

f) The two exchange parameters for the holes in $Pb_{1-x}Mn_xTe$ as well as in $Pb_{1-x}Eu_xSe$ turn out to be dependent on composition x, a feature which is at present not understood. In $Pb_{1-x}Mn_xTe$ the hole exchange parameters even decrease with increasing temperature.

These facts demonstrate, that despite of the progress which has been made, primarily due to the direct determination of g factors using CARS spectroscopy, further investigations are necessary. In order to study theoretically the physical origin of the spd-d and spd-f interactions, information on the hybridization of the $3d^5$ and $4f^7$ levels with the s, p, d like band electrons would be extremely helpful. Such data could be provided by photoemission and inverse photoemission experiments on the diluted magnetic IV-VI's. Such investigations would yield information on the location of occupied and unoccupied states of the 3d (4f) orbitals with respect to the host band structure levels.

Studies on the diluted magnetic semiconductor quantum wells and superlattices of IV-VI compounds are in their infancy. The magnetism of quasi 2 D layers in $(EuTe)_m/(PbTe)_n$ has been studied by Heremans and Partin [29]. In order to observe antiferromagnetic coupling strong

404

enough that it leads to a finite Néel temperature. It turned out that four monolayers of EuTe sandwiched between PbTe layers are necessary in order to observe the antiferromagnetic phase. The carrier-ion exchange interaction in PbTe/Pb$_{1-x}$Mn$_x$Te quantum well structures is now being studied and new effects like the influence of the exchange interaction on confined carriers in diamagnetic wells begin to emerge. Due to the extension of the wave functions of the confined carriers into the barriers where the Mn ions are located, exchange induced corrections appear for electrons and holes in the diamagnetic wells [58-60]. These phenomena are examples for exciting new opportunities.

ACKNOWLEDGEMENTS:

A large number of data presented in this review is taken from the Ph.D. thesis of P.Röthlein. We gratefully acknowledge this contribution. Furthermore experimental results were obtained by E.J.Fantner, S.Gerken, G.Meyer and M.von Ortenberg. Samples were grown by L.Palmetshofer, H.Clemens and M.Tacke. For helpful discussions on the theoretical aspects we thank W.Zawadzki, M.Kriechbaum and E.Bangert. The work was supported by the Deutsche Forschungsgemeinschaft (Bonn) and Fonds zur Förderung der wissenschaftlichen Forschung (Vienna).

REFERENCES:

[1] G. Bauer, Mat. Res. Soc. Proc. 89, 107 (1987).
     (eds. R. L. Aggrarwal, J. K. Furdyna and S. von Molnar).
[2] D. L. Partin, IEEE J.Quantum Electronics QE 24, 1716 (1988).
[3] P. Norton and M. Tacke, J. Crystal Growth 81, 405 (1987).
[4] J. R. Anderson and M. Gorska, Solid State Commun. 52, 601 (1984).
[5] G. Karczewski, M. von Ortenberg, Z. Wilamowski, W. Dobrowolski, and J. Niedwodniczanska-Zawadzka, Solid State Commun. 55, 249 (1985).
[6] G. Braunstein, G. Dresselhaus, J. Heremans, and D. L. Partin, Phys. Rev. B35, 1969 (1987).
[7] M. Escorne, A. Ghazalli, P. Leroux Hugon, Proc 12th Int. Conf. Phys. Semicond., ed. M. H. Pilkuhn (Teubner, Stuttgart 1974)

[8] R. W. Cochrane, M. Plischke, J.O. Strom-Olsen, Phys. Rev. B9, 3013 (1974).

[9] T. Hamasaki, Solid State Commun. 32, 1069 (1979)

[10] T. Story, R. R. Galazka, R. B. Frankel, P. A. Wolff, Phys. Rev. Lett. 56, 777 (1986)

[11] For a recent general review on diluted magnetic II-VI compounds see: J. K. Furdyna, J. Appl. Phys. 64 R29 (1988) and references cited therein.

[12] J. Niedwodniczanska-Zawadzka, J. Kossut, A. Sandauer, W. Dobrowolski, Lecture Notes in Physics 133, 245 (1980). (Springer, Berlin, Heidelberg and New York).

[13] J. Niedwodniczanska-Zawadzka, J. G. Elsinger, L. Palmetshofer, A. Lopez-Otero, E. J. Fantner, G. Bauer, W. Zawadzki, Physica B+C 117 and 118B, 458 (1983).

[14] H. Pascher, E. J. Fantner, G. Bauer, W. Zawadzki, M. v.Ortenberg, Solid State Commun 48, 461 (1983).

[15] G. Karczewski and L. Kowalczyk, Solid State Comm. 48, 653 (1983).

[16] G. Karczewski and M. von Ortenberg, in Proc. 17th Int. Conf. on the Physics of Semiconductors, San Francisco 1984, ed. by J.D. Chadi and W. A. Harrison (Springer New York 1985), p. 1435.

[17] H. Pascher, P. Röthlein, G. Bauer, L. Palmetshofer, Phys. Rev. B36 9395, (1987).

[18] M. Gorska and J. R. Anderson, Phys. Rev. B38, 9120, (1988).

[19] H. Pascher, P. Röthlein, G. Bauer, M. von Ortenberg, Phys. Rev. B40, 10469 (1989).

[20] J. R. Anderson, G. Kido, Y. Nishina, M. Gorska, L. Kowalczyk, Z. Golacki, Phys. Rev. B41, 1014 (1990).

[21] S. Bruno, J. P. Lascaray, M. Averous, J. M. Broto, J. C. Ousset, J. F. Dumas, Phys. Rev. B35, 2068 (1987).

[22] G. Bastard and C. Lewiner, J. Phys. C13, 1469 (1980).

[23] Y. Shapira, S. Foner, D. H. Ridgley, K. Dwight, A. Wold, Phys. Rev. B30, 4021 (1984).

[24] M. Escorne, A. Mauger, J. L. Tholence, R. Triboulet, Phys. Rev. B29, 6306 (1984).

[25] A. Mauger and M. Escorne, Phys. Rev. B35, 1902 (1987).

406

[26] H. J. M. Swagten, W. J. M. de Jonge, R. R. Galazka, P. Warmenbol, J. T. Devreese, Phys. Rev. B37, 9907 (1988).

[27] T. Story, G. Karczewski, L Swierkowski, M. Gorska, R. R. Galazka, Semicond. Sci. Technol. 5 , S138 (1990).

[28] W. J. M. de Jonge, H. J. M. Swagten. S. J. E. Eltink, N. M. J. Stoffels, Semicond. Sci. Technol. 5, S131 (1990).

[29] J. Heremans and D. L. Partin, Phys. Rev. B37, 6311 (1988).

[30] M. S. Adler, C. R. Hewes and S. D. Senturia, Phys. Rev. B7, 186 (1973).

[31] D. L. Mitchell and R. F. Wallis, Phys. Rev. 151, 581 (1966).

[32] R. L. Bernick and L. Kleinman, Solid State Comm. 8, 569 (1970).

[33] G. Bauer in Narrow Gap Semiconductors, Physics and Applications, Vol.133 of Lecture Notes in Physics, ed. by W. Zawadzki (Springer Berlin 1980).

[34] R. R. Galazka and J. Kossut, in Narrow Gap Semiconductors, ed. W. Zawadzki, in Lecture Notes in Physics 133, 245 (1980).

[35] G. Elsinger, L. Palmetshofer and A. Lopez-Otero, Nuovo Cimento 2D, 1869 (1983).

[36] H. Clemens, P. C. Weilguni, U. Stromberger, G. Bauer, J. Vac. Sci. Technol. A7, 3197 (1989)

[37] G. Bauer, in Physics of Semiconducting Compounds ed. R. R. Galazka (Ossolineum, Warszawa 1983) p. 62.

[38] R. Roseman, A. Katzir, P. Norton, K.-H. Bachem, H. M. Preier, IEEE J. Quant. El. QE-23, 94 (1987).

[39] I. I. Zasavitskii and A. V. Sazonov, Fiz. Tverd. Tela 30, 1669 (1988) [Sov. Phys. - Solid State 30, 962 (1988)].

[40] C. K. N. Patel, R. E. Slusher, P. A. Fleury, Phys. Rev. Lett. 17, 1011 (1966).

[41] N. Bloembergen, Nonlinear Optics, Benjamin, New York, 1965

[42] H. Pascher, Appl. Phys. B34, 107 (1984);
H. Pascher, Semicond. Sci. Technol. 5, S141 (1990).

[43] P. A. Wolff, G. A. Pearson, Phys. Rev. Lett. 17, 1015 (1966).

[44] G.L.Eesley, Coherent Raman Spectroscopy (Pergamon Press, Oxford (1981)

[45] R. L. Aggarwal, in Physics of High Magnetic Fields, ed. by
S. Chikazumi, N. Miura, Springer Ser. Solid-State Sci. 24 (Springer, Berlin, Heidelberg 1981) p. 105.

[46] P. Röthlein, H. Pascher, G. Bauer, M. Tacke, Semicond. Sci. Technol. 5, S147 (1990).

[47] M. Gorska, T. Wojtowicz, W. Knap, Solid State Commun. 51, 115 (1984)

[48] M. von Ortenberg, G. Bauer and G. Elsinger Proc. Int. Conference Millimeter Waves, Marseille 1985,

[49] H. Burkhard, G. Bauer and W. Zawadzki, Phys. Rev. B19, 5149 (1979).

[50] P. Vogl and P. Kocevar, Proc. Int. Conf. Phys. Semicond., Edinburgh, 1978, Inst. of Physics Conf. Series ed. R. H. Wilson (I.O.P. London 1978) p.1317.

[51] W. Zawadzki, to be published.

[52] M. von Ortenberg, in Infrared an Millimeter Waves, edited by K. Button (Academic, New York, 1980), Vol. III, p. 275.

[53] H. Pascher, G. Bauer, R. Grisar, Phys. Rev. B38, 3383 (1988)

[54] H. Pascher, P. Röthlein, I. Roschger, G. Bauer, Proc. 19th Int. Conf. on the Physics of Semiconductros, ed. W. Zawadzki, (Institute of Physics, Polish Academy of Sciences 1988), S1535

[55] H. Pascher, P. Röthlein, M. Tacke, idid p. 1591

[56] I. I. Zasavitskii, L. Kowalczyk, B. N. Matsonashvili, A. V. Sazonov, Fiz. Tekh. Poluprovodn. 22, 2188 (1988) [Sov. Phys.-Semicond. 22, 1338 (1988)].

[57] W. A. Harrison and G. K. Straub, Phys. Rev. B36, 2695 (1987).

[58] H. Krenn, K. Kaltenegger, N. Frank, G. Bauer, H. Pascher, M. Kriechbaum, to be published.

[59] H. Pascher, P. Röthlein, M. Kriechbaum, N. Frank, G. Bauer, to be published.

[60] E. T. Heyen, M. Magerott, A. V. Nurmikko, D. L. Partin, Appl. Phys. Lett. 54, 653 (1989).

MAGNETIC PROPERTIES OF CO-BASED DILUTED MAGNETIC SEMICONDUCTORS

A.Lewicki[a,b], A.I.Schindler[a], J.K.Furdyna[c], and T.M.Giebultowicz[c,d]

[a] Department of Physics, Purdue University, West Lafayette, Indiana 47907, USA.

[b] Permanent address: Department of Solid State Physics, Akademia Górniczo-Hutnicza (AGH), PL-30-059 Kraków, Poland.

[c] Department of Physics, University of Notre Dame, Notre Dame, Indiana 46556, USA.

[d] National Institute of Standards and Technology, Gaithersburg, MD 20899, USA.

1. Introduction.

The most extensively studied diluted magnetic semiconductors (DMS's), known also as semimagnetic semiconductors, are alloys based on II-VI semiconductor compounds, in which a fraction x of cations has been randomly replaced by Mn^{2+} (Furdyna, 1988b; Furdyna and Kossut, 1988c; Galazka and Kossut,1982) or Fe^{2+} (Mycielski, 1988; Twardowski, 1990) ions. Recently, DMS alloys containing Co^{2+} ions have also been successfully prepared.

Writing a review article on such a new topic as Co-based DMS's is very attractive, but also very difficult. Many problems remain unsolved or are currently being investigated. For instance, we still do not know the maximum possible concentrations of Co^{2+} ions in some host compounds. Furthermore, there are no experimental data concerning the narrow-gap DMS's such as $Hg_{1-x}Co_xSe$ or $Hg_{1-x}Co_xTe$. In spite of these difficulties, basic magnetic properties of Co-based DMS's seem to be quite well understood. The theoretical model describing the properties of the Co^{2+} ion in a crystal field was introduced many years ago, and is applicable (with minor modifications) to the DMS materials. This model is based on a simple spin Hamiltonian, and may serve as a starting point for calculations of thermodynamic properties such as magnetic susceptibility, magnetization, and specific heat.

The Co-based DMS's are isostructural with their nonmagnetic counterparts, and provide us with an opportunity to study their magnetic properties from the very dilute limit to $x \approx 0.14$ (in $Zn_{1-x}Co_xS$).

Preliminary results of the x-ray powder diffraction analysis suggest no change (or at most a very small change) of the lattice parameter **a** as a function of the concentration x of Co^{2+} ions in $Zn_{1-x}Co_xS$ and $Zn_{1-x}Co_xSe$ (Lewicki *et al.*,1989). This observation is in contrast with the results for Zn-based DMS's containing Mn, where the lattice parameters vary sharply with x (and, by the use of Vegard's law can be successfully exploited for determining the concentration of Mn). Such differences between the Zn-based alloys containing Co and Mn can be related to the difference in the covalent (tetrahedral) radii of Zn, Co, and Mn. Namely, the changes of the lattice constant of $Zn_{1-x}Mn_xS$ and $Zn_{1-x}Mn_xSe$ with increasing x result from the fact that the covalent radius of Mn (r_{Mn}=1.326 Å) is larger than that of Zn (r_{Zn}=1.225 Å , Yoder-Short *et al.*, 1985). Therefore, from the powder diffraction results for the Co-based alloys, one can infer that the covalent radius of Co is smaller than that of Mn and similar to that of Zn.

The position of the ground state of the Co^{2+} ion relative to the valence and conduction bands may also influence the magnetic properties. Unfortunately, there are large discrepancies between the results available from the literature. From the infrared luminescence measurements Radlinski (1979) concludes that the $^4A_2$ ground state of the Co^{2+} ion is about 0.3 eV to 0.69 eV below the conduction band edge for various wide-gap II-VI semiconductors. Noras *et al.*(1981) report much larger values - 2.5 eV for $Zn_{1-x}Co_xS$ and 2.2 eV for $Zn_{1-x}Co_xSe$. Robbins *et al.*(1980), on the other hand, claim that for $Zn_{1-x}Co_xSe$ this level lies very close to the valence band. In our opinion, the optical properties of the Co-based DMS's should be reexamined.

Single crystal samples of DMS's containing Co^{2+} ions were obtained by the Bridgman method ($Cd_{1-x}Co_xS$, $Cd_{1-x}Co_xSe$) or by chemical (iodine) vapor transport ($Zn_{1-x}Co_xS$ and $Zn_{1-x}Co_xSe$). The former technique usually leads to larger and more homogeneous crystals. However, in some cases application of the Bridgman method is very difficult and does not succeed, whereas successful results have been obtained using the chemical vapor transport.

In this article we intend to review the magnetic properties of Co-based DMS's as well as some other features closely related to the presence of Co^{2+} ions, i.e., their contribution to the specific heat and inelastic neutron scattering. We shall also point out problems which require further investigations.

The material is presented in the following sequence. In Sec.2 we present the theoretical model that describes the properties of a single Co^{2+} ion and a pair of Co^{2+} ions in a crystal field, for both the zinc blende and the wurtzite structures. We then review experimental results of magnetic susceptibility, neutron scattering and specific heat, along with their analysis in Sec.3. Finally, in concluding remarks, we comment on current and future investigations of Co-based DMS's.

2. Co^{2+} ions in a tetrahedral crystal field.

2.1. Single ion case.

It is well known that the crystal field plays an important role in determining the properties of transition metal ions. In Co-based DMS's, which crystallize in the zinc blende or the wurtzite structures, each Co^{2+} ion is surrounded by four anions, forming a regular tetrahedron. Wurtzite materials, however, crystallize with a small distortion, which leads to an additional term in the crystal-field potential and has an important influence on the observed properties of these crystals. We shall first discuss the simpler zinc blende structure.

The point charge model of the crystal-field theory expresses the energy-level diagram of the Co^{2+} ion in a tetrahedral crystal-field in terms of two parameters: the crystal-field amplitude Dq, and the spin-orbit coupling λ. Formally, the problem consists of finding the eigenvalues of a Hamiltonian operator, which is a sum of several terms. This can be performed analytically using a perturbation expansion. The details of the solution of the crystal-field Hamiltonian have been published by several authors (e.g., Weakliem,1962; Koidl *et al.*,1973; Ryskin *et al.*,1973; Villeret,1989; Villeret *et al.*, 1990a; Villeret *et al.*, 1990b). Therefore, we are going to give only a brief outline of the final results.

The energy-level diagram for a Co^{2+} ion in a crystal field is shown in Fig.1 (one should remember that for a zinc blende structure $V_{trig} = 0$, and the splitting at the extreme right in Fig.1 is not present). Under the action of a tetrahedral (V_{tet}) crystal field, the free-ion ^4F ground term splits into a $^4A_2$ $(^4\Gamma_2)$ orbital singlet and two orbital triplets, $^4T_2$ $(^4\Gamma_5)$ and $^4T_1$ $(^4\Gamma_4)$. The first excited $^4T_2$ state is separated from the ground state by an amount $\Delta=10Dq\cong3700$ cm^{-1}(Weakliem, 1962). The next term $^4T_1$ is about 18Dq above the $^4A_2$ state. In the presence of spin-orbit interaction ($\lambda L \cdot S$ in the Fig.1), both orbital triplets split into four energy levels each, whereas the ground orbital singlet remains unsplit. Only the application of a magnetic field (H) removes the degeneracy of the ground state and splits $^4A_2$ into four levels ($S_z = -\frac{3}{2}$, $-\frac{1}{2}$, $\frac{1}{2}$, $\frac{3}{2}$). In order to calculate the g-factor for these levels, one has to include the spin-orbit coupling terms connecting the ground states and the excited states. This coupling "mixes in" a small amount of the excited state wave function into the ground state, thus modifying the magnetic properties. The second order perturbation correction for the energy adds a small shift of the energy levels (the same for all S_z values, with no zero-field splitting), and gives the following g-factor correction (Carlin, 1965)

$$g = 2 - \frac{8\lambda}{\Delta} \quad , \tag{1}$$

where λ is the spin-orbit coupling constant. Equation (1) is not exact, since here only mixing between $^4A_2$ and $^4T_2$ was taken into account. However, from Eq.(1), and with negative λ, one can predict an increase of the g-factor from the spin-only value, g=2. This increase was confirmed by electron paramagnetic resonance experiments (Sec.3.1).

Energy levels are given by

$$E_m = g\mu_B mH + GH^2 \quad ,m = -\frac{3}{2} , -\frac{1}{2} , \frac{1}{2} , \frac{3}{2} , \tag{2}$$

and

$$G = -\frac{4\mu_B^2}{\Delta} \quad ,$$

where μ_B is the Bohr magneton, m is the S_z eigenvalue, H is the magnetic field, and g and Δ were defined earlier. The second term in Eq.(2) is very small and usually may be neglected. The value of Δ is much larger than the thermal energy at room temperature, and thus the occupation of all excited states is much smaller than that of the ground-state level. The magnetic susceptibility is therefore determined almost entirely by the properties of the orbital singlet $^4A_2$ state, and takes the form of a Curie law plus a small temperature-independent term

$$\chi_S = \frac{g^2\mu_B^2 S(S+1)}{3k_B T} - 2G \quad . \tag{3}$$

One should remember that χ_S in Eq.(3) was calculated for an isolated Co^{2+} ion, in the absence of interactions with its magnetic neighbors. Moreover, we have neglected the hyperfine interactions. The energy level splittings produced by these interactions are of the order of 1 mK, and therefore are important only at milikelvin temperatures.

The case of DMS's with the wurtzite (hcp) lattice structure is somewhat more complicated. The hexagonal structure possesses a polar direction (c-axis), corresponding to the trigonal symmetry axis. As we mentioned above, wurtzite materials crystallize with a small c-axis distortion, i.e., the c/a ratio is not exactly equal to the value $\sqrt{8/3}$ predicted by an ideal hcp arrangement. In this case the crystal field potential consists of two terms. The first — and dominant — is the tetrahedral symmetry V_{tet} term. The second, and much smaller V_{trig} term, results mainly from the above lattice distortion and has a trigonal symmetry. The basic difference between this and the previously discussed zinc blende structure is that in the presence of the trigonal term and the spin-orbit interaction the energy levels undergo further splitting. The fourfold degenerate $^4A_2$ ground state divides into two Kramers doublets. At zero magnetic field the doublet states are separated by a value 2D, as shown in Fig.1.

In the absence of the c-axis distortion, the four anions surrounding a Co^{2+} cation would form a regular tetrahedron, and hence they would not contribute to V_{trig}. The remaining contribution to V_{trig} comes from the electrostatic potential of more distant ions, which is much weaker and partially screened. Hence, the V_{trig} term and the energy level splitting would be significantly smaller in that case.

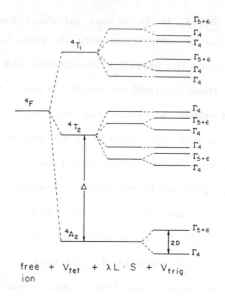

free + V_{tet} + $\lambda L \cdot S$ + V_{trig}
ion

Fig.1. Schematic diagram of the energy levels of a Co^{2+} ion in a tetrahedral (V_{tet}) and trigonal (V_{trig}) crystal field. For the sake of clarity, distances between the levels are not in scale [after Villeret, 1989].

Energy levels can be described by the following Hamiltonian (Bleaney and Stevens, 1953; Bowers and Owen, 1955; Villeret, 1989; Villeret *et al.*, 1990a; Villeret *et al.*, 1990b)

$$H_S = D\left(S_z^2 - \frac{5}{4}\right) + g_\perp \mu_B S_x H\sin\theta + g_\parallel \mu_B S_z H\cos\theta + G_\perp(H\sin\theta)^2 +$$

$$+ G_\parallel(H\cos\theta)^2 \quad , \tag{4}$$

where $D = \frac{\lambda}{2}(g_\parallel - g_\perp)$, $G_\perp = \frac{\mu_B^2}{\lambda}(0.5g_\perp - 1)$, $G_\parallel = \frac{\mu_B^2}{\lambda}(0.5g_\parallel - 1)$,

S_x, S_z are the components of spin, θ is the angle between the magnetic field $\vec{H}$ and the c-axis, and $g_\perp$ and $g_\parallel$ are the g-factors for $\vec{H} \perp \hat{c}$ and $\vec{H} \parallel \hat{c}$, respectively. The two quadratic terms arise from the admixture of $^4T_2$ with the $^4A_2$ state and, as in the zinc blende case, may be neglected in many cases.

A straightforward calculation now gives the energy eigenvalues for $\vec{H} \parallel \hat{c}$:

$$E_{1,2} = -D \pm \frac{1}{2} g_\parallel \mu_B H + G_\parallel H^2 \quad , \tag{5}$$

$$E_{3,4} = D \pm \frac{3}{2} g_\parallel \mu_B H = G_\parallel H^2 \quad , \tag{6}$$

and for $\vec{H} \perp \hat{c}$:

$$E_{1,2} = \frac{1}{2} g_\perp \mu_B H \pm \left(g_\perp^2 \mu_B^2 H^2 - g_\perp \mu_B HD + D^2 \right)^{1/2} + G_\perp H^2 \quad , \tag{7}$$

$$E_{3,4} = -\frac{1}{2} g_\perp \mu_B H \pm \left(g_\perp^2 \mu_B^2 H^2 + g_\perp \mu_B HD + D^2 \right)^{1/2} + G_\perp H^2 \quad , \tag{8}$$

An example of the energy levels splitting as a function of magnetic field for these two orientations is shown in Fig.2. When the magnetic field is parallel to the c-axis and $D>0$, the lower Kramers doublet corresponds to $S_z = \pm\frac{1}{2}$, and the upper to $S_z = \pm\frac{3}{2}$. For $\vec{H} \perp \hat{c}$, all states are mixed and therefore cannot be indexed by the S_z number.

The magnetic susceptibility can be readily calculated by the standard Van Vleck formula (Van Vleck, 1932), which yields for $\vec{H} \parallel \hat{c}$:

$$\chi_{S_\parallel} = \frac{g_\parallel^2 \mu_B^2}{4 k_B T} \frac{9 + \exp(2D/k_B T)}{1 + \exp(2D/k_B T)} - 2G_\parallel \quad , \tag{9}$$

and for $\vec{H} \perp \hat{c}$:

$$\chi_{S_\perp} = \frac{g_\perp^2 \mu_B^2}{k_B T} \frac{1}{1 + \exp(-2D/k_B T)} + \frac{3 g_\perp^2 \mu_B^2}{4D} \tanh(D/k_B T) - 2G_\perp \quad . \tag{10}$$

418

From Eqs.(9) and (10) it is clear that in wurtzite crystals the magnetic susceptibility is anisotropic at low temperatures. The source of this single–ion anisotropy is the zero–field splitting 2D. In the absence of the zero–field splitting (D=0) the g–factor becomes isotropic ($g_{\parallel}=g_{\perp}=g$), and both formulas for $\chi_{S_{\parallel}}$ and for $\chi_{S_{\perp}}$ converge to the expression already obtained for the susceptibility χ_S of the Co^{2+} ion in a zinc blende crystal, given by Eq.(3).

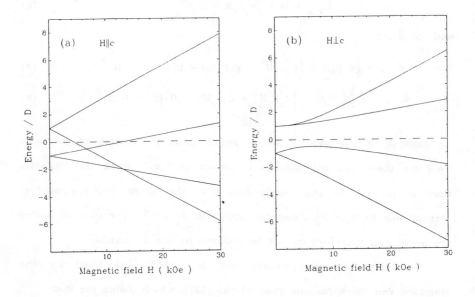

Fig.2. Splittings of the ground state energy levels (Γ_4 and Γ_{5+6}) of a Co^{2+} ion in a wurtzite lattice as a function of magnetic field H; (a) is for $\vec{H}\|\hat{c}$, and (b) is for $\vec{H}\perp\hat{c}$; D is the zero–field splitting parameter.

In the high-temperature limit, the above expressions for χ reduce to Curie-law expressions for $S=\frac{3}{2}$ minus a temperature independent term $2G$ (which is negative), as follows

$$\chi_{S\parallel} = \frac{5g_\parallel^2\mu_B^2}{4k_BT} - 2G_\parallel \quad , \tag{11}$$

$$\chi_{S\perp} = \frac{5g_\perp^2\mu_B^2}{4k_BT} - 2G_\perp \quad , \tag{12}$$

Because the difference between $g_\perp$ and $g_\parallel$ is small the magnetic susceptibility is nearly isotropic in this limit. Typically, $\chi_\perp/\chi_\parallel \simeq g_\perp^2/g_\parallel^2 \simeq 1.01$.

At very low temperatures, i.e., $k_BT \ll 2D$, and $\vec{H}\parallel\hat{c}$, the upper doublet $(S_z=\pm\frac{3}{2})$ will not be occupied, and one should expect to obtain a Curie-law behavior with $S = \frac{1}{2}$. The direct calculation from Eq.(9) gives exactly that result:

$$\chi_{S\parallel}(T\to0) = \frac{g_\parallel^2\mu_B^2}{4k_BT} \quad . \tag{13}$$

When $\vec{H}\perp\hat{c}$, the interpretation is not so obvious, because of the mixing of states corresponding to various S_z. However, from Eq.(10) one can obtain

$$\chi_{S\perp}(T\to0) = \frac{g_\perp^2\mu_B^2}{k_BT} \quad , \tag{14}$$

which also has a Curie-law form, but with the magnetic moment twice as large as for $\vec{H}\parallel\hat{c}$.

If the angle θ between the magnetic field $\vec{H}$ and the c-axis is not equal to $0°$ $(\vec{H}\parallel\hat{c})$ or to $90°$ $(\vec{H}\perp\hat{c})$, then the energy levels depend both on the value and the direction (determined by θ) of the magnetic field $\vec{H}$. In order to calculate these energy levels, one has to find all matrix elements and eigenvalues of the spin Hamiltonian given by Eq.(4).

Expansion of the secular determinant gives the following equation (Orton, 1969)

$$E^4 - 2\left(D^2 + \frac{5}{4}g^2\mu_B^2H^2\right)E^2 + 2g^2\mu_B^2H^2D\left(1 - 3\cos^2\theta\right)E +$$
$$+ \left(D^2 + \frac{3}{4}g^2\mu_B^2H^2\right)^2 - g^2\mu_B^2H^2D^2\left(1 + 3\cos^2\theta\right) = 0 \quad , \qquad (15)$$

which is valid for $g_\perp \cong g_\parallel \cong g$. An analytical solution of the biquadratic Eq.(15) is rather complicated, but the eigenvalues can be easily obtained by a numerical procedure.

In spite of the fact that the magnetization data for Co-based DMS's are not yet available, we would like to briefly outline the spin Hamiltonian predictions. In this connection, Eq.(2) may be rewritten as

$$E_m = \mu_B H\left(mg - \frac{4\mu_B H}{\Delta}\right) \simeq g\mu_B m H \quad . \qquad (16)$$

Note that even for H=100 kOe, the second term (because of large Δ) is less than 1% of the first. Hence, at magnetic fields available under laboratory conditions and at low temperatures ($k_B T \ll \Delta$), we can neglect the quadratic term. With this approximation, a straightforward calculation leads to the well known Brillouin function

$$M = g\mu_B S B_S\left(\frac{g\mu_B S H}{k_B T}\right) \quad , \qquad (17)$$

where M is the magnetization per ion for the zinc blende structure.

Analogous approximation for the wurtzite materials and for $\vec{H}\|\hat{c}$ leads to

$$M_\parallel = \frac{1}{2}g\mu_B \frac{\sinh(y/2) + 3\exp(-2D/k_B T)\times\sinh(3y/2)}{\cosh(y/2) + \exp(-2D/k_B T)\times\cosh(3y/2)} \quad , \qquad (18)$$

where $y = \dfrac{g_\parallel \mu_B H}{k_B T}$ .In the absence of the zero-field splitting (D=0) the above equation reduces to Eq.(17). For the second case, i.e., for $\vec{H}\perp\hat{c}$ it should be noted that some results from earlier articles (Figgis, 1960; Brown *et al.*, 1977) are incorrect. These authors follow a standard method based on an expansion of the energy as a power series of H. Usually, it is sufficient to terminate this series after the quadratic term. However, for $\vec{H} \perp \hat{c}$ the convergence of the series is quite slow, and therefore the standard procedure is insufficient. Instead, one has to calculate the derivatives

$$\mu_i = -\frac{\partial E_i}{\partial H} \quad , \tag{19}$$

where E_i is given by Eqs.(7) and (8) with $G_\perp = 0$, and then substitute them into the general expression

$$M = N \frac{\sum_i \mu_i \exp\left(-E_i / k_B T \right)}{\sum_i \exp\left(-E_i / k_B T \right)} \quad . \tag{20}$$

The resulting equation is rather complicated and we will not present it here in explicit form. Needles to say, inclusion of the exchange interaction between Co^{2+} ions would lead to even more complex expressions for the magnetization (see next section).

For completeness, we should mention that some authors (Ryskin *et al.*, 1973; Koidl *et al.*, 1973) point out the importance of the dynamic Jahn-Teller effect in explaining the optical absorption spectra of Co^{2+} ions in tetrahedrally coordinated crystals (specifically in $Zn_{1-x}Co_xS$). This effect indeed leads to additional splittings of the $^4T_2$ and $^4T_1$ levels, but leaves the $^4A_2$ state unchanged. It therefore does not influence significantly the magnetic properties of Co-based DMS alloys (which involve primarily the $^4A_2$ state of the Co^{2+} ion), and we shall not concern ourselves with this effect.

2.2. Co^{2+} ion pairs.

In this section we will consider a pair of Co^{2+} ions, which is the simplest interacting magnetic system. It is also the starting point for including exchange interaction in the calculation of the magnetic properties.

Co^{2+} ions in DMS's (similarly to Mn^{2+} ions) are localized. Hence, an interaction between these ions may be expressed by the Heisenberg Hamiltonian

$$\mathbb{H}_P = \mathbb{H}_{S1} + \mathbb{H}_{S2} - 2J \, \vec{S}_1 \vec{S}_2 \quad , \tag{21}$$

where $\mathbb{H}_{S1}$, $\vec{S}_1$, $\mathbb{H}_{S2}$, $\vec{S}_2$ represent the single-ion Hamiltonian and spin of the first and second ions, respectively, and J is the value of the spin-spin exchange integral.

For zinc blende DMS's Eq.(21) may be rewritten in the form

$$\mathbb{H}_P = g\mu_B \left(S_{z1} + S_{z2} \right) H_z - 2J \, \vec{S}_1 \vec{S}_2 \quad . \tag{22}$$

The eigenvalues of the above Hamiltonian can be calculated as follows (Kambe, 1949):

$$E_m = -J\left(S'(S'+1) - 2S(S+1) \right) - g\mu_B m H_z = -J\left(S'(S'+1) - \frac{15}{2} \right) - g\mu_B m H_z \quad , \tag{23}$$

where S' is the value of the total spin $\vec{S}' = \vec{S}_1 + \vec{S}_2$, $S' = 0, 1, ..., S_1 + S_2$ ($S' = 0, 1, 2, 3$ for a pair of Co^{2+} ions), and $|m| \leq S'$. According to Eq.(23) the diagram of the energy levels of a Co^{2+}- Co^{2+} pair (shown in Fig.3) is similar to that for a Mn^{2+}- Mn^{2+} pair (Nagata et al., 1980).

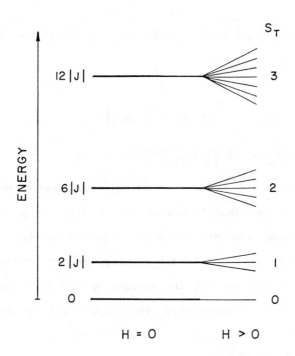

Fig.3. Energy levels for a pair of Co^{2+} ions in a zinc blende lattice. S_T stands for the total spin of a pair.

Because of the zero-field splitting, the Hamiltonian for a Co^{2+}-Co^{2+} pair in a wurtzite lattice depends on two parameters: J and D. It may also be written in the form of Eq.(21), but with single ion terms H_{S1} and H_{S2} given by Eq.(4). If the absolute value of the exchange integral $|J|$ is much larger than the zero-field splitting 2D of the ground orbital state, then again, the pair is most conveniently described in terms of the total spin $\vec{S}' = \vec{S}_1 + \vec{S}_2$. This Hamiltonian can then be explicitly written as (Owen, 1961)

$$H_P = W_{S'} + g_{\perp}\mu_B H_x S_x' + g_{\parallel}\mu_B H_z S_z' + \beta_{S'} D\left(S_z'^2 - \frac{1}{3}S'(S'+1)\right) + 2G_{\perp}H_x^2 +$$

$$2G_{\parallel}H_z^2 \quad , \tag{24}$$

where

$$W_{S'} = -J\left(S'(S'+1) - 2S(S+1)\right) \quad , \tag{25}$$

$$\beta_{S'} = \frac{3S'(S'+1) - 4S(S+1) - 3}{(2S'-1)(2S'+3)} \quad , \tag{26}$$

and $S'=0,1,2,3$; $S=\frac{3}{2}$; $2D \ll J$. One can find the eigenvalues of the above Hamiltonian by an analytical calculation, but we will not specify those complicated expressions in explicit form in the present article.

In the general case, i.e., for the exchange integral $|J|$ comparable to the zero-field splitting $2D$, the solution of the pair Hamiltonian (Eq.(21)) is much more troublesome. The simplest way to obtain the energy levels is to calculate all the matrix elements, and then perform a numerical diagonalization (see Sec.3.4).

3. Results and analysis.

3.1. Electron paramagnetic resonance.

Electron paramagnetic resonance (EPR) is the most direct method for the determination of the g-factor, which in turn is necessary for the interpretation of magnetic properties. The experimental data for Co-doped $A^{II}B^{VI}$ semiconductors with zinc blende structure, i.e., ZnS, ZnSe, ZnTe, and CdTe, have been analyzed (Ham et al., 1960; Hall and Hayes, 1960; Henning et al., 1966) using the following spin Hamiltonian

$$\mathcal{H} = g\mu_B \vec{H}\cdot\vec{S} + A\vec{S}\cdot\vec{I} - g_N\mu_N\vec{I}\cdot\vec{H} + uS^3H + US^3I \quad , \tag{27}$$

where A is the hyperfine splitting constant, $\vec{I}$ and g_N are the nuclear spin and the nuclear magneton, respectively, and S^3H and S^3I denote higher order terms (for explicit form of these terms, see Ham *et al.*, 1960). Because the hyperfine splitting constant is very small ($A\simeq 0.001$ cm^{-1}) and $\mu_N/\mu_B = 5.45\times 10^{-4}$, the first term in Eq.(27) is the dominant one. The values of the g-factor obtained for various DMS's are given in Table I. The higher order terms in Eq.(27) lead to a weak angular variation of the g-factor, given by

$$g_a(\theta) = g - \frac{9}{5}up \quad , \tag{28}$$

with

$$p = 1 - \frac{5}{4}\sin^2\theta(1 + 3\cos^2\theta) \quad , \tag{29}$$

where θ is the angle between the magnetic field H and the [001] crystallographic direction, and u is a multiplicative coefficient of the higher order term in Eq.(27). The experimental results confirmed the above prediction (Ham *et al.*, 1960; Henning *et al.*, 1966), but the coefficient u is so small ($u\leq 0.004$) that the anisotropy of the magnetic susceptibility caused by the angular variation of the g-factor would be extremely hard to detect. Actually, as we will see in the next section, such an anisotropy has not been observed in the zinc blende DMS's (within the present experimental accuracy of 2%).

For $Cd_{1-x}Co_xS$ and $Cd_{1-x}Co_xSe$, which have the wurtzite structure, the following spin Hamiltonian was used for the interpretation of the EPR spectra (Morigaki, 1963)

$$\mathcal{H} = g_\parallel \mu_B H_z S_z + g_\perp \mu_B (H_x S_x + H_y S_y) + D(S_z^2 - 5/4) + AI_z S_z + B(I_x S_x + I_y S_y) , \tag{30}$$

TABLE I.

Values of the g-factor and other spin Hamiltonian parameters for various $A_{1-x}^{II} Co_x B^{VI}$ semiconductors.

$A_{1-x}^{II} Co_x B^{VI}$	g	D/k_B (K)	u	$\|A\|$ 10^{-4} cm^{-1}	$\|U\|$ 10^{-4} cm^{-1}
$Zn_{1-x} Co_x Te$ (a)	2.2972	—	+0.00078	17.5	0.65
$Zn_{1-x} Co_x S$ (a)	2.248	—	-0.0048	1.8	—
$Zn_{1-x} Co_x Se$ (a)	2.270	—	-0.005	—	—
$Zn_{1-x} Co_x Se$ (b)	2.2742	—	-0.0038	7.7	1.2
$Zn_{1-x} Co_x Se$ (f)	2.27	—	—	—	—
$Cd_{1-x} Co_x Te$ (a)	2.3093	—	+0.0016	23.4	0.85
$Cd_{1-x} Co_x Te$ (c)	2.30	—	—	23.0	—
$Cd_{1-x} Co_x S$ (d)	$g_\| =2.269$ $g_\perp =2.286$	0.96	—	—	—
$Cd_{1-x} Co_x Se$ (e)	$g_\| =2.295$ $g_\perp =2.303$	1.11	—	—	—

(a) Ham et al., 1960.

(b) Henning et al., 1966.

(c) Hall and Hayes, 1960.

(d) Morigaki, 1964.

(e) Hoshina, 1966.

(f) Jonker et al., 1988.

where A and B are the hyperfine splitting constants, and other symbols were defined previously. One should note that this Hamiltonian is equivalent to the spin Hamiltonian discussed in Sec.2.1 (Eq.(4)), with additional very small hyperfine terms (A,B $\leq$ 0.001 cm^{-1}). The values of the g-factors for the wurtzite systems are listed in Table I for the two principal orientations of the magnetic field, $\vec{H}\|\hat{c}$ and $\vec{H}\perp\hat{c}$. An additional shift of the resonance lines associated with the higher order terms $S^3 H$ and $S^3 I$ was not found for either $Cd_{1-x}Co_x S$ or $Cd_{1-x}Co_x Se$. It is possible, however, that the much stronger angular variation of the EPR lines due to the zero-field splitting hides the weak effect of these higher-order terms.

The zero-field splitting parameter D was also calculated from the EPR data: for $Cd_{1-x}Co_x S$ it was found to be D/k_B=0.96±0.02 K (Morigaki, 1963; Morigaki, 1964), and for $Cd_{1-x}Co_x Se$, the result is D/k_B=1.11±0.03 K (Hoshina, 1966). An example of the angular dependence of the positions of the EPR lines for $Cd_{1-x}Co_x S$ is shown in Fig.4. According to the relation given by the crystal field theory (see Eq.(4)),

$$\lambda = \frac{2D}{g_\| - g_\perp} \quad , \tag{31}$$

one can estimate the value of the spin-orbit coupling parameter λ. For $Cd_{1-x}Co_x S$ this parameter is significantly smaller than the free ion value (λ=-180 cm^{-1}), whereas for $Cd_{1-x}Co_x Se$ it is comparable the free ion case. Hoshina (1966) ascribes this difference to the higher degree of covalency for CdSe than for CdS.

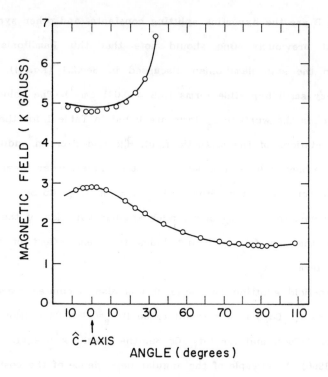

Fig.4. Position of EPR lines of Co^{2+} ions in $Cd_{1-x}Co_xS$ for the magnetic field in the $(10\bar{1}0)$ plane at 4.2 K and 9.202 GHz. The solid lines represent theoretical values of resonance fields [after Morigaki, 1964].

3.2. Magnetic susceptibility.

3.2.1. DMS's with the zinc blende lattice: $Zn_{1-x}Co_xS$ and $Zn_{1-x}Co_xSe$.

In this Section we will review the magnetic susceptibility results for $Zn_{1-x}Co_xS$ and $Zn_{1-x}Co_xSe$ in the temperature range from 4.2 to 300 K. Both materials crystallize in the zinc blende structure. Magnetic susceptibility measurements at high (T≃300 K) temperatures usually require samples with a moderate (x≥0.02) concentration of magnetic ions. For other Co-based zinc blende systems ($Cd_{1-x}Co_xTe$, $Hg_{1-x}Co_xTe$, $Hg_{1-x}Co_xSe$, etc.), the above requirement has so far not been successfully fulfilled. On the other hand, as we mentioned in the Introduction, the maximum concentration of Co^{2+} ions in these compounds has not been determined. The magnetic susceptibility of these systems therefore still remains as an open field for further study.

The magnetic susceptibility measurements (Lewicki *et al.*, 1989) were performed for x≤0.145 for $Zn_{1-x}Co_xS$, and x≤0.05 for $Zn_{1-x}Co_xSe$. The concentration of Co^{2+} ions in these cases is too high to neglect exchange interactions, and is even (for $Zn_{1-x}Co_xS$) beyond the limit of applicability of the NNPA procedure (see Sec.3.4). However, in the high-temperature regime (i.e., $k_BT > J$), exchange interactions can be readily incorporated into the susceptibility by using the mean-field approximation or the high-temperature expansion for a dilute Heisenberg antiferromagnet (Spalek *et al.*, 1986). Using this approximation and assuming a random distribution of Co^{2+} ions over the cation sublattice, one can express the susceptibility in the form of the Curie-Weiss law

$$\chi(T) = \frac{C(x)}{T - \Theta(x)} \quad , \tag{32}$$

with the Curie constant per mole given by

$$C(x) = xN_A(g\mu_B)^2 S(S+1)/3k_B \quad , \tag{33}$$

and the Curie-Weiss temperature

$$\Theta(x) = 2xS(S+1)zJ_{eff}/3k_B \quad , \tag{34}$$

where z is the number of cations in the first coordination sphere (z=12 for the zinc blende structure), and N_A, k_B, μ_B, and S have their usual meaning, and S=3/2.

The inverse molar susceptibility obtained for $Zn_{1-x}Co_xS$ is shown in Fig.5. The straight lines represent the least-squares fit of the Curie-Weiss law to the experimental data in the high-temperature range. Below approximately 130 K, the magnetic susceptibility does not follow the Curie-Weiss law. This deviation is qualitatively similar to that observed in all Mn-based DMS's. However, the occurrence of the deviation from the straight-line behavior at much higher temperatures than is observed for the Mn-based DMS's, suggests a larger value of the exchange integral for $Zn_{1-x}Co_xS$ than for $Zn_{1-x}Mn_xS$. A similar behavior is observed for $Zn_{1-x}Co_xSe$ (see Fig.6).

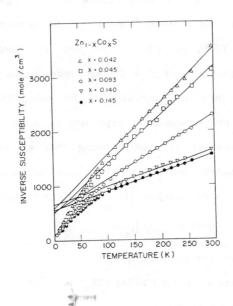

Fig.5. Inverse molar magnetic susceptibility of $Zn_{1-x}Co_xS$ alloys as a function of temperature for various concentrations of Co^{2+} ions x [after Lewicki et al., 1989].

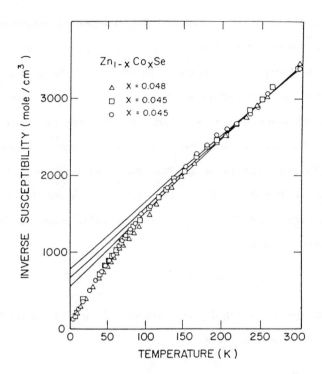

Fig.6. Inverse molar susceptibility of $Zn_{1-x}Co_xSe$ as a function of temperature for x=0.045 (two samples) and x=0.048 [after Lewicki et al., 1989].

A least squares fit of the Curie-Weiss law makes it possible to analyze the data quantitatively. Using Eqs.(33) and (34) one can determine two microscopic parameters of the systems studied : the magnetic moment of the Co^{2+} ion, and the effective exchange integral J_{eff} between nearest-neighbor Co^{2+} ions. The results of J_{eff} are given in Table II. It should be remarked that the value of J_{eff} obtained from the Curie-Weiss law is not exactly equal to the exchange integral between nearest neighbors (that is why we call it an *effective* exchange integral), but rather is an upper limit for the nearest-neighbor value, since it also contains contributions from interactions between second-, third-, and higher-order neighbors. These contributions are usually small, but not entirely negligible (Lewicki et al., 1988). Furthermore, the small temperature-independent term in the magnetic susceptibility predicted by the crystal field theory (Eq.(3)) was not included in the analysis of the experimental data. This may lead to a small ($\leq$ 10%) over-estimate of the J_{eff}. Since this correction is smaller than the experimental error (see Table II) of the available data, it does not affect the final conclusions.

TABLE II.

Effective exchange integrals J_{eff}/k_B between Co^{2+} ions determined by the high-temperature susceptibility.

$A^{II}_{1-x} Co_x B^{VI}$	Ref.	J_{eff}/k_B (K)
$Zn_{1-x}Co_xSe$	(a)	$- 54 \pm 8$
$Zn_{1-x}Co_xS$	(a)	$- 47 \pm 6$
$Cd_{1-x}Co_xSe$	(b)	$- 38 \pm 5$
$Cd_{1-x}Co_xS$	(c)	$- 32 \pm 6$

(a) Lewicki et al., 1989.

(b) Lewicki et al., 1990a.

(c) Lewicki et al., 1990b.

The negative value of the exchange integral J_{eff} shows that the interactions between Co^{2+} ions are antiferromagnetic, which is consistent with the general trends predicted by the theory of superexchange (Spalek *et al.*, 1986). On the other hand, in sharp contrast to the Mn-based DMS's, the value of J_{eff} increases for larger anion size. A more definite conclusion about the possible contributions of other mechanisms of exchange (e.g., the Bloembergen-Rowland interaction or the two-electron processes; see Larson *et al.*, 1985) will be possible after measurements of J_{eff} for narrow band-gap DMS's, such as $Hg_{1-x}Co_xSe$ or $Hg_{1-x}Co_xTe$, become available.

One should also note that the values of the exchange integrals for Co-based DMS's are about three to four times larger than those for their manganese counterparts. Theoretical interpretation of this distinct and striking effect has not yet been undertaken.

3.2.2. DMS's with the wurtzite lattice: $Cd_{1-x}Co_xS$ and $Cd_{1-x}Co_xSe$.

The review of the magnetic susceptibility of $Cd_{1-x}Co_xS$ and $Cd_{1-x}Co_xSe$ will be divided into two parts - one for low, and one for high temperatures. Here the term "low" and "high" refers to the ratio of the temperature to the value of the dominant (NN) exchange integral.

The most distinctive feature differentiating the Co-based wurtzite DMS's from their zinc blende counterparts is the occurrence of anisotropy of the magnetic susceptibility at low temperatures. This anisotropy is a result of the small distortion of the hexagonal lattice structure along the c-axis. The theoretical formulas for the anisotropic susceptibility have been presented in Sec.2.1. We wish to emphasize that underlying this anisotropy is the behavior of the *isolated* Co^{2+} ion in a wurtzite lattice, and that the effect does not imply any ordering of Co^{2+} ions (i.e., the magnetically anisotropic sample remains in the paramagnetic phase).

As we mentioned earlier, the lattice distortion leads to an axial symmetry of the physical properties of the crystals. It is well known that in the presence of such symmetry (and with $\chi_\perp > \chi_\parallel$) the susceptibility $\chi_\theta \equiv \chi(\theta)$ obeys the following angular dependence:

$$\chi_\theta - \chi_\parallel = (\chi_\perp - \chi_\parallel)\sin^2\theta \quad , \tag{35}$$

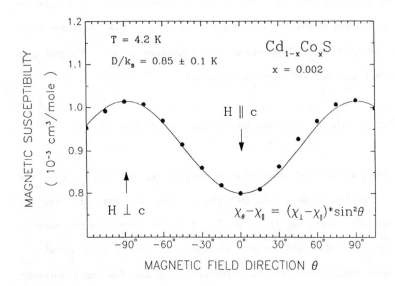

Fig.7. Angular dependence of the magnetic susceptibility of $Cd_{1-x}Co_xS$, x=0.002, at a fixed temperature T=4.2 K [after Lewicki *et al.*, 1990a].

where θ is the angle between the direction of the magnetic field $\vec{H}$ and the c-axis. This dependence was observed for $Cd_{1-x}Co_xS$ and $Cd_{1-x}Co_xSe$. As an example, the angular dependence of the magnetic susceptibility of $Cd_{1-x}Co_xS$, x=0.002, at a fixed temperature T=4.2 K is shown in Fig.7 (Lewicki *et al.*, 1990a).

The temperature dependence of the magnetic susceptibility of $Cd_{1-x}Co_xSe$, x=0.0065 for two orientations of the magnetic field, $\vec{H}\|\hat{c}$ and $\vec{H}\perp\hat{c}$, is presented in Fig.8 (Lewicki *et al.*, 1990a). Continuous curves represent the best fits of the theoretical expression for the molar susceptibility which include isolated ion and nearest-neighbor (NN) pair contributions, given by

$$\chi = xN_A\left(P_S(x)\chi_S + \frac{1}{2}P_P(x)\chi_P\right) \quad , \tag{36}$$

where χ_S is the single ion contribution (Eqs.(9) and (10)), χ_P is the NN pair contribution obtained from the standard formula for magnetic susceptibility (Van Vleck, 1932) calculated for energy levels given by Eqs.(24)-(26), and $P_S(x)$ and $P_P(x)$ are probabilities for the occurrence of single ions and NN ion pairs, respectively. The expression for the total susceptibility contains four parameters: the concentration x of Co^{2+} ions, the zero-field splitting D, the exchange integral J, and the g-factors $g_{\|}$ or $g_{\perp}$. In our calculation, the latter two parameters were taken from the high-temperature susceptibility (J) and from the EPR results ($g_{\|}$ and $g_{\perp}$). The remaining two parameters were adjusted during the fitting process, and the best agreement with the experimental data was obtained for D/k_B=1.0±0.1 K and x=0.0065.

Similar analysis of the experimental data for $Cd_{1-x}Co_xS$, x=0.002 yields the value of the zero-field splitting equal to D/k_B=0.85±0.10 K.

436

One should note, that the difference between $g_\perp$ and $g_\parallel$ is very small (Sec.3.1): for example $(g_\perp - g_\parallel)/g_\perp \approx 3.5 \times 10^{-3}$ for $Cd_{1-x}Co_x Se$. Thus, in the high-temperature regime $\chi_\perp \cong \chi_\parallel$ within the experimental accuracy (see Eqs.(11) and (12)), and there is no need to specify the orientation of the magnetic field for that temperature range.

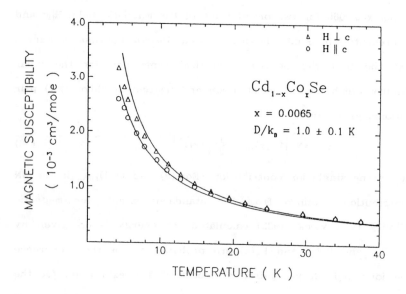

Fig.8.   Molar susceptibility of $Cd_{1-x}Co_x Se$, x=0.0065, versus temperature for two orientations of the magnetic field, $\vec{H} \| \hat{c}$ and $\vec{H} \perp \hat{c}$ [after Lewicki et al., 1990a].

The magnetic susceptibility of $Cd_{1-x}Co_x Se$ at high temperatures were analyzed using the same equations as for the materials with zinc blende structure, i.e. Eqs.(32)-(34), with an additional temperature-independent term predicted by the crystal field theory (Eqs.(11) and (12)). This small term was estimated from

$$\chi_\infty \simeq -2GxN_A = xN_A \frac{\mu_B^2}{2D}\left(g_\perp - g_\parallel\right)\left(g - 2\right) \quad , \tag{37}$$

where $g = \frac{1}{2}(g_\perp + g_\parallel)$. The inverse molar susceptibility of $Cd_{1-x}Co_xSe$ (after subtraction of the above constant term) versus temperature is shown in Fig.9. Here again, the straight lines represent the least-squares fit of the Curie-Weiss law to the experimental data in the high-temperature regime. The values obtained for the effective exchange integral are: $J_{eff}/k_B = -37\pm5$ K (for $Cd_{1-x}Co_xSe$, x=0.062) and $J_{eff}/k_B = -38\pm5$ K (for x=0.085). Recently, the value of J_{eff} was also determined for $Cd_{1-x}Co_xS$ and found to be equal to $J_{eff}/k_B = -22\pm6$ K (Lewicki, 1990b).

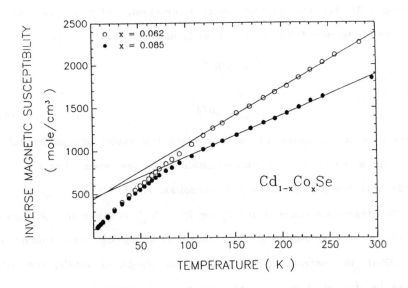

Fig.9. Inverse molar susceptibility of $Cd_{1-x}Co_xSe$ as a function of temperature for x=0.062 and x=0.085 [after Lewicki et al., 1990a].

3.2.3. The spin-glass state.

The transition from the paramagnetic phase to the spin-glass state in DMS's was first observed for Mn-based semiconductors (for a review, see Furdyna, 1988b). Although an empirical model has been proposed (Furdyna and Samarth, 1987) which presumes an interlocking process of different clusters (with short range intracluster order, but no long range order), the nature of the spin-glass state in DMS's is not fully understood. Twardowski *et al.*(1986) argued that the transition temperature T_f gives information about the radial dependence of the exchange interaction, as represented by a radial-dependent exchange integral $J(R)$. For the superexchange interaction, which is dominant in the case of Mn-based DMS's, it is usually assumed that

$$J(R) \propto R^{-n} \quad , \tag{38}$$

and

$$T_f(x) \propto x^{n/3} \quad , \tag{39}$$

where n is a parameter which determines the radial dependence of J (n>0). The majority of the Mn-based DMS's follow the dependence given by Eq.(39) with the parameter n=6.8 (Twardowski *et al.*, 1987).

The transition temperature T_f for $Zn_{1-x}Co_xS$ is given in Fig.10 as a function of concentration of Co^{2+} ions, on a log-log scale (Shand *et al.*, 1990a). In contrast to the results for Mn-based DMS's, the data cannot be described by a simple power law behavior (i.e., by a linear dependence on a log-log scale). A crossover from the superexchange to the dipolar coupling has been suggested to explain the observed behavior. Because of the long-range character of the dipolar coupling ($J(R) \propto R^{-3}$) compared to superexchange, this form of exchange may be important at sufficiently low concentrations (such as those involved in Fig.10).

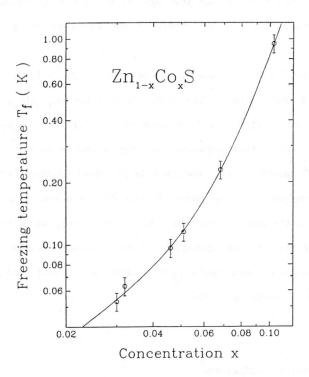

Fig.10. Freezing temperature T_f as a function of cobalt concentration x in $Zn_{1-x}Co_xS$. The solid curve is a theoretical fit as discussed in the text [after Shand *et al.*, 1990a].

Considering both the dipolar coupling (the first term in Eq.(40)) and the superexchange (the second term in Eq.(40)), the freezing temperature T_f can be described by

$$T_f(x) = Ax + Bx^{n/3} \quad . \tag{40}$$

The line shown in Fig.10 represent the mean-square fit of Eq.(40) to the experimental data obtained for A=1.7, B=3.1x10^4, and n=14. The difference between the values of the freezing temperature for $Zn_{1-x}Co_xS$ and for the Mn-based DMS's (for the same x) are generally smaller than might be expected from the difference of the exchange integrals (see Sec.3.2.1).

One additional comment should be made about the value of n. For $Zn_{1-x}Co_xS$, n is much larger than that obtained for $Zn_{1-x}Mn_xS$ (n≅6.8, Twardowski *et al.*, 1987), and rather close to that for $Zn_{1-x}Fe_xSe$ (n≅12, Swagten *et al.*, 1989). This parameter is very sensitive to the freezing temperature of samples with relatively large concentration (x≈0.1) of Co^{2+} ions. Therefore, any inaccuracy in the determination of x or any inhomogeneity of the sample leads to a large uncertainty in n. Recently, experimental studies of the spin-glass state in other Co-based DMS's have been undertaken (Shand and Crooker, 1990b), and should shed further light on this fundamental problem.

3.3. Inelastic neutron scattering.

One of the best ways of determining the exchange parameters in magnetically diluted systems is by studying the excitation levels of small isolated clusters of magnetic ions (such as pairs, triads, etc.). This can be readily seen from the fact that for a pair of spins with the Heisenberg-type antiferromagnetic interaction, the energy levels E_m are related to the exchange integral J as given by Eq.(23). While various macroscopic measurements may be used to obtain J (e.g., the high-temperature susceptibility (Sec.3.2) or the step-like features on the high-field magnetization of DMS's (Shapira *et al.*,1984; Shapira, 1990), the most accurate and straightforward method is by neutron scattering.

In inelastic scattering processes, the neutron energy loss or gain is equal to the difference between two adjacent levels (the selection rules permit $\Delta S_T = \pm 1$, see Fig. 3), thus providing a direct measure of J. The theoretical background of this method is given in detail in a review paper by Furrer and Gudel, 1979.

The strength of the antiferromagnetic interactions for the new Co-based DMS's appears to be several times larger than that for their Mn-based counterparts. Hence, the technique of magnetization steps cannot be applied to these materials, because the magnitude of J_{NN} would in this case require magnetic field intensities that are not available. In contrast, the neutron scattering method is particularly well suited for measuring energy transfers up to several tens of meV, thus offering a perfect tool for studying the excitation levels of NN Co-Co pairs.

As the pairs involve only a small fraction of the total number of atoms in the sample, the observed scattering intensity is usually relatively weak - a shortcoming which can be serious when phonon scattering peaks occur in the same energy range. However, pair scattering can be distinguished from phonon scattering, as follows:

(i) The energy transfer in pair scattering does not depend on the momentum transfer $\mathbf{Q}$. In contrast, phonon lines usually exhibit pronounced dispersion, as illustrated schematically in Fig. 11a. The fact that pair spectra do not depend on $\mathbf{Q}$ makes it possible to use both single crystal and powder samples for pair scattering studies.

(ii) Pair scattering spectra exhibit a characteristic temperature behavior, which is illustrated in Fig. 11b. At $T \ll J/k_B$, essentially all

442

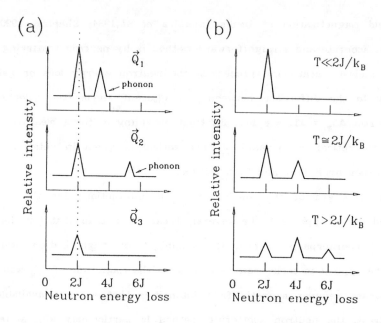

Fig.11. Basic features of neutron scattering on magnetic ion pairs.

(a) A diagram illustrating the behavior of pair scattering and phonon lines with the change of neutron momentum transfer Q.

(b) A diagram illustrating the temperature behavior of pair scattering spectra (see text).

pairs are in the E=0 state, so that one observes only a single $|0\rangle \rightarrow |1\rangle$ transition line for neutron energy loss $\Delta E=2J$ (see Fig.3). As T is raised, the occupation of the excited states increases, giving rise to a $|1\rangle \rightarrow |2\rangle$ peak at 4J, a $|2\rangle \rightarrow |3\rangle$ peak at 6J, etc. At the same time, the intensity of the first peak decreases. This latter effect is again in sharp contrast to the normal behavior of phonon lines.

(iii) Although the energy transfer is **Q**-independent, the peak *intensity* does vary with **Q** in a characteristic way. In the case of a single crystal specimen the $I(\mathbf{Q})$ dependence can be written as (Furrer and Gudel, 1979)

$$I(\mathbf{Q}) \propto f^2(Q) \exp[-2W(Q)] \sum_{i,j} \left(1 + (-1)^{\Delta S_T} \cos \mathbf{Q} \cdot \mathbf{R}_{ij}\right) \quad , \qquad (41)$$

where $\mathbf{R}_{ij} = \mathbf{R}_j - \mathbf{R}_i$, $Q = |\mathbf{Q}|$, $f(Q)$ is the magnetic formfactor, W is the Debye–Waller factor, $\mathbf{R}_i$ and $\mathbf{R}_j$ are the positions of the two ions comprising a given pair, and the sum in Eq. (41) runs over all pairs in the crystal. Due to the oscillating term $\cos \mathbf{Q} \cdot \mathbf{R}_{ij}$, the intensity vs. $\mathbf{Q}$ data yield direct information about the ion–ion distance, thus enabling a precise identification of the cluster type (e.g., distinguishing NN pairs from NNN pairs). For powders or polycrystalline materials, Eq.(41) has to be averaged in $\mathbf{Q}$ space, which leads to

$$I(Q) \propto f^2(Q) \exp\left(-2W(Q)\right) \left(1 + (-1)^{\Delta S_T} \frac{\sin(QR)}{QR}\right) \quad , \qquad (42)$$

where $R = |\mathbf{R}_j - \mathbf{R}_i|$. Although the oscillating character of the Q dependence for polycrystalline specimens is less pronounced than that for single crystals, it is sufficient to enable an identification of pair spectra in a fcc lattice.

Experimental studies of inelastic neutron scattering from Co–Co pairs in polycrystalline samples of $Zn_{1-x}Co_xS$, $Zn_{1-x}Co_xSe$, and $Zn_{1-x}Co_xTe$ with x ranging from 0.01 to 0.06 have been carried out at the 20 MW NBSR reactor at the National Institute of Standards and Technology. (Giebultowicz et al., 1990a; Giebultowicz et al., 1990b). An example of the inelastic neutron scattering spectrum for $Zn_{0.94}Co_{0.06}S$ is displayed in Fig.12. In addition to the two distinct pair transition lines which exhibit the expected temperature behavior, the data show a broad feature at about 11 meV which has been identified as phonon scattering (the inset in Fig.12 shows the phonon density of states in ZnS which exhibits a pronounced maximum at about the same energy). The intensity of the $|0>\rightarrow|1>$ transition vs. Q obtained from measurements for two different $Zn_{1-x}Co_xS$ specimens (with x=0.05 and x=0.06) is displayed in Fig.13, showing good agreement with the theoretical curve calculated for NN pairs from Eq.(42).

The inelastic neutron scattering data for $Zn_{1-x}Co_xS$, $Zn_{1-x}Co_xSe$ and $Zn_{1-x}Co_xTe$ gave the values of J_{NN}/k_B for these systems as −47.5±0.6, −50±1, and −36±2 K, respectively. These values are, indeed, 3 to 4 times larger than those for Mn-based alloys (see Table III). It should be

TABLE III.

Antiferromagnetic exchange integrals J_{NN}/k_B for $Zn_{1-x}Mn_xB^{VI}$ and $Zn_{1-x}Co_xB^{VI}$ (B^{VI} = S, Se, Te) determined by neutron scattering.

Host lattice	Mn-Mn pairs	Co-Co pairs
ZnS	− 16.1 ± 0.2 K (a)	− 47.5 ± 0.6 K (b)
ZnSe	− 12.3 ± 0.2 K (a)	− 49.5 ± 1.0 K (c)
ZnTe	− 9.35 ± 0.05 K (c)	− 38.0 ± 2.0 K (c)

(a): Giebultowicz *et al.*, 1987.

(b): Giebultowicz *et al.*, 1990a.

(c): Giebultowicz *et al.*, 1990b.

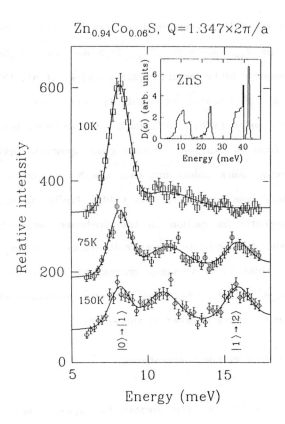

Fig.12. Inelastic scattering spectra obtained from triple axis spectrometer measurements on a polycrystalline $Zn_{1-x}Co_xS$ sample at various temperatures, showing the $|0\rangle \rightarrow |1\rangle$ and $|1\rangle \rightarrow |2\rangle$ pair transition lines. For clarity, the spectra are shifted upward on the intensity axis [after Giebultowicz *et al.*, 1990a]. The broad maximum emerging at E=11 meV is identified as phonon scattering. The inset shows the phonon density of states in ZnS [after Kunc, *et al.*, 1975].

noted that there is a very good agreement between neutron scattering data and the J_{NN} values for $Zn_{1-x}Co_xS$ and $Zn_{1-x}Co_xSe$ obtained from magnetic susceptibility measurements (Lewicki *et al.*, 1989). Taking into account that the magnetic susceptibility data contain contributions from other Co-Co pairs (NNN, NNNN, etc.), and neutrons selectively measure the effect from NN pairs only, such a good agreement between the results of both experiments points out that the NN coupling, which plays an overwhelming role in the Mn-based DMS family, is also the dominant antiferromagnetic interaction in the Co-based materials. Surprisingly, however, the data indicate stronger interactions in $Zn_{1-x}Co_xSe$ than in $Zn_{1-x}Co_xS$, which is in striking contrast to the regularity observed in the $A_{1-x}^{II}Mn_xB^{VI}$ alloys, where the interaction strength always decreases with increasing atomic number of the anion.

3.4. Specific heat at low temperatures.

In this Section we will discuss the specific heat results of the two Co-based wurtzite DMS's — $Cd_{1-x}Co_xS$ and $Cd_{1-x}Co_xSe$ (Lewicki *et al.*, 1990c). There is one important difference between the specific heat results for Co- and Mn-based wurtzite alloys. The zero-field splitting of the ground $^6A_1$ state of the Mn^{2+} ion is much smaller ($D/k_B \simeq 2$ mK, Title, 1963) than that for the Co^{2+} ion. Therefore, in contrast with Co^{2+} ions, isolated Mn^{2+} ions do not contribute to the specific heat in the temperature range investigated (0.4 - 4.0 K), unless a magnetic field is present. The contribution to the specific heat of the ground state of isolated Co^{2+} ions, on the other hand, must be taken into account.

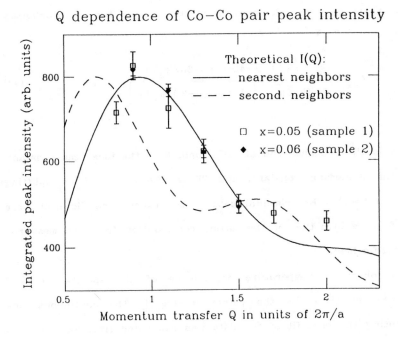

Fig.13. Integrated intensities of the $|0\rangle \to |1\rangle$ pair transition lines in $Zn_{1-x}Co_xS$ plotted versus Q. The data were obtained from experiments on polycrystalline samples with x=0.05 and x=0.06 prepared by two different techniques. The solid and dashed curves show the theoretical integrated intensities calculated, respectively, for NN and NNN Co-Co pairs in $Zn_{1-x}Co_xS$ using Eq.(42) [after Giebultowicz et al., 1990a].

In the dilute limit, i.e., neglecting the exchange interactions, and in the absence of an external magnetic field, we have to consider only two doubly degenerate levels (see the spin Hamiltonian in Sec.2.1). Thus, the Co^{2+} ion contribution C_{ex} to the molar specific heat (excess specific heat) may be calculated using the simple two-level Schottky formula,

$$C_{ex}(T) = \frac{xR}{(k_B T)^2} \frac{4D^2 \exp(-2D/k_B T)}{\left[1 + \exp(-2D/k_B T) \right]^2} \quad , \tag{43}$$

where x is the concentration of Co^{2+} ions, R is the ideal gas constant, k_B is the Boltzmann constant, and 2D is the energy level splitting defined in Sec.2.1. Experimental data of the excess specific heat were obtained by subtraction of the lattice contribution from the measured total specific heat.

The observed temperature dependence of the specific heat of $Cd_{1-x}Co_x Se$ is given by the points in Fig.14. The continuous line representing the best fit of the data was calculated from Eq.(43), with D/k_B=0.72±0.03 K, and x=0.00115 . Experimental results for $Cd_{1-x}Co_x S$ are shown in Fig.15. Here a good fit was achieved for D/k_B=0.97±0.03 K and x=0.00155 . The above results demonstrate that in the dilute limit the excess specific heat follows the Schottky formula quite well, and can be used for an exact determination of the zero-field splitting 2D.

The specific heat in nonzero magnetic field can also be described using the spin Hamiltonian discussed in Sec.2.1 (Eq.(4)). From Eq.(15) one can see that the energy eigenvalues of the spin Hamiltonian depend both on the value of the magnetic field H and the angle θ between $\vec{H}$ and the c-axis. Hence, the excess specific heat in nonzero magnetic field is a function of H, θ, and zero-field parameters D and x. The results for $Cd_{1-x}Co_x S$ in various magnetic fields are presented in Fig.16. Since, the determination of θ during the experiment was not possible, θ was used as a fitting parameter, and found that the best fit for all the data (for all values of H) was obtained with $\theta = 18°$. Other parameters, i.e., x and D were taken from the zero-field measurements and fixed. From Fig.16 it is clear that the specific heat strongly depends on the value of magnetic field H. This occurs because the magnetic field splitting of energy levels is comparable (or even larger) than the zero-field splitting 2D.

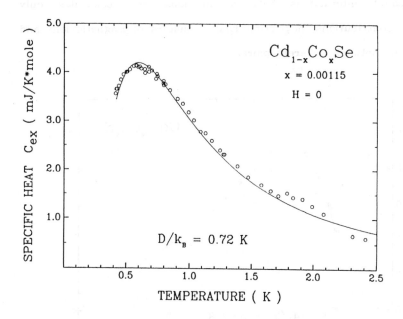

Fig.14. Excess specific heat of $Cd_{1-x}Co_xSe$ as a function of temperature in zero magnetic field [after Lewicki et al., 1990c].

The temperature dependence of the excess specific heat of $Cd_{1-x}Co_xS$, x=0.0077 for H=0 is shown in Fig.17. For this sample, with its larger concentration of Co^{2+} ions, an attempt to fit Eq.(43), shown by the dashed curve, was unsuccessful. Because of the higher cobalt concentration, it was necessary to include in the calculation the exchange interactions between Co^{2+} ions. The nearest-neighbor pair correlation approximation (NNPA) was applied for this purpose. This approximation is based on the assumption that the partition function of a long-range interaction may be factored into contributions of pairs containing nearest neighbors (NN), next nearest neighbors, etc. (Matho, 1979; Denissen and de Jonge, 1986). In other words, instead of taking into account interactions between all ions, one calculates only pair-wise contributions, neglecting larger clusters of magnetic ions and interactions between different pairs.

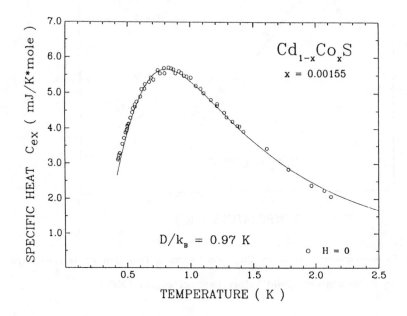

Fig.15. Excess specific heat of $Cd_{1-x}Co_xS$ versus temperature in zero magnetic field [after Lewicki *et al.*, 1990c].

The NNPA procedure consists of the following steps. First, a value of the exchange integral J_i is calculated for specific pairs of ions. Here again a simple power law is assumed,

$$J_i(R_i) = J_1 \left(\frac{R_i}{R_1} \right)^{-n} , \quad i=1,2,..., \tag{44}$$

where J_i and R_i are the exchange integral and the distance between NN, NNN, etc., and n is a parameter which determines how rapidly J_i decreases as a function of R_i. The NN exchange integral (J_1/k_B = -19 K) was estimated from the effective exchange integral J_{eff}/k_B = -22 K, which, in turn, was obtained from high-temperature susceptibility measurements (Lewicki, 1990b),

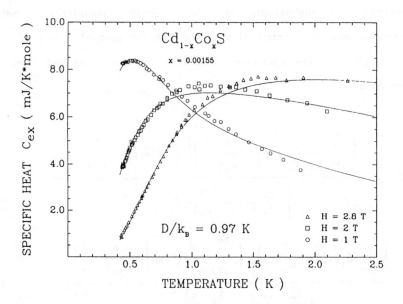

Fig.16. Temperature dependence of the excess specific heat of $Cd_{1-x}Co_xS$ for three values of applied magnetic field [after Lewicki *et al.*, 1990c].

$$J_{eff}N_1 = \sum_{i=1}^{k} N_i J_i = J_1 \sum_{i=1}^{k} N_i \left(\frac{R_i}{R_1} \right)^{-n} , \tag{45}$$

where N_i is the number of cation sites in the i-th coordination sphere.

Next, the energy levels for a single $Co^{2+}-Co^{2+}$ pair are calculated. The general pair Hamiltonian (Eq.(21)) may be written in zero magnetic field as

$$\mathbb{H}_P = \mathbb{H}_{S1} + \mathbb{H}_{S2} - 2J_i \vec{S}_1 \cdot \vec{S}_2 = D \left(S_{z1}^2 + S_{z2}^2 \right) - \frac{5}{2} D - $$
$$J_i \left(S_{+1}S_{-2} + S_{-1}S_{+2} + 2S_{z1}S_{z2} \right) , \tag{46}$$

where $S_{+1}, S_{-1}, S_{+2}, S_{-2}$ are spin-shift operators for the first and second ions of the pair, respectively. Sixteen eigenvalues were found by a numerical diagonalization of the above Hamiltonian matrix representation. The contribution to the specific heat of a pair of Co^{2+} ions coupled by the exchange integral J_i is then computed from the expression

$$C_i = \frac{\left(\sum_{i=1}^{16} \frac{E_i^2}{k_B T^2} \exp\left(-\frac{E_i}{k_B T} \right) \right) \left(\sum_{j=1}^{16} \exp\left(-\frac{E_j}{k_B T} \right) \right)}{\left(\sum_{i=1}^{16} \exp\left(-\frac{E_i}{k_B T} \right) \right)^2} - $$
$$\frac{\left(\sum_{i=1}^{16} E_i \exp\left(-\frac{E_i}{k_B T} \right) \right) \left(\sum_{j=1}^{16} \frac{E_j}{k_B T^2} \exp\left(-\frac{E_j}{k_B T} \right) \right)}{\left(\sum_{i=1}^{16} \exp\left(-\frac{E_i}{k_B T} \right) \right)^2} , \tag{47}$$

where E_i and E_j are the energy levels.

The excess specific heat C_{ex} is finally obtained by the summation of all pair contributions C_i multiplied by the probability $P_i(x)$ of finding a magnetic neighbor in the i-th coordination sphere,

$$C_{ex} = \sum_{i=1}^{\infty} \frac{1}{2} x N_A P_i(x) C_i \quad , \quad (48)$$

with

$$P_i(x) = \left(1 - x\right)^{m_{i-1}} - \left(1 - x\right)^{m_i} \quad , \quad (49)$$

where $m_i = \sum_{j=1}^{i} N_j$, $m_0 = 0$, x is the concentration of Co^{2+} ions, and N_j is the number of sites in the j-th coordination sphere of the cation sublattice in the hcp structure. The above calculation was performed for the first 25 coordination spheres, and because coupling with the remaining (more distant) ions is very weak ($J_{26}/k_B < 0.005$ K), they were treated like isolated (noninteracting) ions.

The NNPA theoretical curve is shown in Fig.17 (continuous curve). The value of the zero-field splitting parameter D was taken from results for $Cd_{1-x}Co_xS$ in the dilute limit, and then the best fit was obtained using n=6.0±1.0 . For illustrative purpose, a dashed line calculated for the same concentration x, but neglecting interaction between Co^{2+} ions (Eq.(43)) is also drawn. One can also see from the figure that inclusion of the exchange between the magnetic ions leads to a broadening of the calculated specific heat curve, a behavior which is confirmed by the experimental data.

The zero-field splitting obtained from the specific heat may be compared with the results obtained by other techniques. In the case of $Cd_{1-x}Co_xS$, this value is in rather good agreement with the value obtained from the anisotropic magnetic susceptibility (D/k_B=0.85±0.10 K, Sec.3.2.2), and matches exactly the EPR value (D/k_B=0.96±0.02 K,

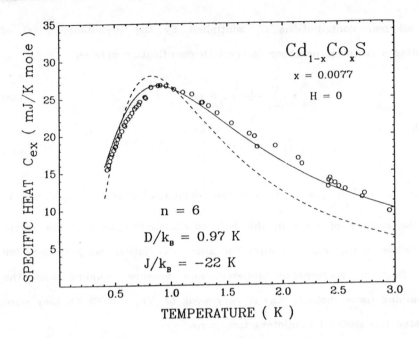

Fig.17. Temperature dependence of the excess specific heat of $Cd_{1-x}Co_xS$, x=0.0077, in zero magnetic field. The continuous curve represents the best fit of the NNPA model, whereas the dashed curve depicts the result of calculation which does not include exchange interactions [after Lewicki et al., 1990c].

Sec.3.1). For $Cd_{1-x}Co_xSe$, on the other hand, the differences between the corresponding results are somewhat larger, the value of D/k_B from the specific heat being 0.72±0.03 K, while that from the susceptibility is 1.0±0.1 K (Sec.3.2.2), and the value of D/k_B from EPR is 1.11±0.03 K (Sec.3.1). One possible explanation of the above differences is that the magnetic susceptibility measurements were carried out on a sample with a larger concentration of Co^{2+} ions (x=0.0065). Exchange interaction between more distant ions, i.e. NNN, NNNN, etc., which were not included in the interpretation of the susceptibility data may lead to an apparent increase of D. The discrepancy with the EPR data remains unexplained.

Finally, the value of the exponent n in the expression for $J_i(R_i)$ (Eqs.(38) and (45)) merits a comment. The specific heat measurements gave a value of n=6.0±1.0, which is close to that obtained for Mn-based DMS's (Twardowski *et al.*, 1987). On the other hand, the transition temperatures to the spin-glass state for zinc blende $Zn_{1-x}Co_xS$ suggest a larger value of n (Sec.3.2.3). It is not yet clear whether the difference between the wurtzite and the zinc blende structures could cause a significant change in the radial dependence of the exchange integrals. The proposed dipolar coupling invoked in the discussion of the spin-glass transition (Sec.3.2.3) was not included in the present specific heat analysis, but it is too weak to affect the specific heat results in the temperature range investigated (T>0.4 K). More studies focused on this problem are clearly needed.

3.5. Other experimental results.

While the primary purpose of this article has been to provide a review of the magnetic properties of Co-based DMS's, for the sake of completeness a brief description of other investigations of these materials, which have not been included in the preceding sections, will now be given:

1. An investigation of electronic Raman scattering in $Cd_{1-x}Co_xSe$, x=0.035 and x=0.082, has been carried out by Bartholomew *et al.*, (1989). Specifically two effects are observed: Raman-EPR of Co^{2+} ions, and

spin-flip Raman scattering of donor-bound electrons. The Raman-EPR scattering yields an isotropic g-factor equal to 2.32±0.04 . This result is not contradictory to the anisotropic g-factor from EPR (see Sec.3.1), since the difference between $g_\perp$ and $g_\parallel$ is smaller than the sensitivity of the Raman scattering method. The spin-flip Raman scattering associated with donors, observed at low temperatures, provides a measure of the sp-d exchange coupling between the spins of the donor electrons and those of the Co^{2+} ions. The value of the sp-d exchange constant was found to be αN_o=320 meV , i.e., larger than that for $Cd_{1-x}Mn_xSe$. Analysis of such spin-flip Raman data determines magnetization of the Co^{2+} system, yielding an effective temperature $(T+T_{AF})$. The value of T_{AF} found for $Cd_{1-x}Co_xSe$ indicates that the exchange interaction between Co^{2+} ions is stronger than that for $Cd_{1-x}Mn_xSe$. This result remains in qualitative agreement with the magnetic susceptibility measurements (Lewicki et al., 1990a).

2. $Zn_{1-x}Co_xSe$ films were successfully grown by the molecular beam epitaxy on GaAs substrates (Jonker et al., 1988). Samples with Co concentrations up to x=0.075 were grown to a typical thickness of 1-1.5 μm. The EPR measurements of the g-factor in such films gave the same value (g=2.27) as for bulk material (see Sec.3.1). The observed temperature dependence of the magnetic moment is that characteristic for a paramagnetic material. The magnitude of the measured moment, however, is reduced by a factor of $\approx$ 0.4 from that expected from the total number of magnetic ions in the sample. This is due to the antiferromagnetic coupling between Co^{2+} ions. On the basis of the value of this reduction factor, the authors have suggested that a significant clustering of Co^{2+} ions occurs in epitaxial films of $Zn_{1-x}Co_xSe$. This conclusion has not

been supported by any quantitative analysis, and has not yet been confirmed.

3. A magnetoreflectivity study performed on the same type of films (Liu *et al.*, 1989) yields a value of the sp-d exchange parameter $N_o(\alpha-\beta)=2420\pm40$ meV, significantly larger than the corresponding value observed for either Fe- or Mn-based DMS's.

4. Finally, from the nuclear magnetic resonance (NMR) experiment accomplished for the single crystal of CdS doped with Co, nuclear spin-lattice relaxation time T_1 was obtained (Look and Locker, 1972). Minima in T_1 vs. temperature dependence were observed for various frequencies (ν=2-15 MHz), and the absolute value of the effective electron relaxation time τ_e was calculated from the data ($2\pi\nu\tau_e\simeq1$). However, it should be noted that the τ_e deduced from the NMR T_1 data is an "effective" value, and cannot be assigned to a specific electronic transition, such as ($-\frac{1}{2} \leftrightarrow \frac{1}{2}$). Furthermore, below T = 5 K, τ_e is dominated by a resonant Orbach process involving the two ground-state Kramers doublets (see Sec.2.1). The value of the zero-field splitting was estimated from these data to be equal to 4±1 K, which is twice as large as the more recent results described in Secs.3.2.2 and 3.4.

4. Concluding remarks.

Investigations of the magnetic properties of Co-based DMS's, which have been reviewed in this article, represent a part of a systematic study of magnetic effects due to various transition metal ions in a nonmagnetic (i.e., diamagnetic) matrix.

The most important conclusions from the results presented above

458

are:

(i)   the presence of an antiferromagnetic coupling between Co^{2+} ions;

(ii) the large values of the exchange integral J (about three to four times larger than those for Mn-based DMS's);

(iii) the difference between the dependence of the freezing temperature T_f on concentration x for the Co- and for the Mn-based DMS's;

(iv) the zero-field splitting of the Co^{2+} ion ground orbital state in the wurtzite lattice structure, which leads to an anisotropy of the magnetic susceptibility and to the Schottky anomaly in the specific heat at low temperatures.

Let us finally mention two issues that have not been resolved either in the Co-based systems, or even in the simpler case of Mn-based DMS's, and still await a satisfactory explanation. First is the nature of the spin-glass state in DMS's. Transition to this state has been frequently attributed to frustration of the antiferromagnetic interactions between nearest neighbors in an fcc lattice. Observation of the spin-glass state for samples with the concentration of magnetic ions well below the percolation threshold of $x \simeq 0.17$ (e.g. Novak et al., 1984; Twardowski et al., 1986; Shand et al., 1990a) proved the importance of interactions between more distant ions in this context. Although an empirical model of the spin-glass state formation has been proposed (Furdyna and Samarth, 1987), we are still far from a full understanding of this process. In particular, because of the lack of any quantitative model, we still do not know which of the microscopic parameters determine the transition temperature T_f. Moreover, from ac susceptibility measurements, Geschwind et al., 1988, suggest that a transition observed for $Cd_{1-x}Mn_xTe$, $x \geq 0.40$, may be a dynamically inhibited transition to a type–III antiferromagnetic state rather than a spin–glass formation.

The second "old" problem that should be mentioned is the lack of an exact expression for the magnetic susceptibility for low temperatures and large concentrations of magnetic ions. The high–temperature expansion (Spalek *et al.*, 1986), which may be improved by taking into account higher order terms, must fail for temperature $T < J/k_B$. A more promising approximation is the NNPA model and its extensions. However, in this model, at sufficiently large concentration of magnetic ions (i.e. $x>0.1$) the probability of finding triplet and even larger clusters becomes significant, and the results of the NNPA method become inaccurate. Unfortunately, the inclusion of larger clusters into the calculation leads to an enormous increase in mathematical complexity. The above problem is common for many magnetic materials with localized moments, and a satisfactory solution has not yet been found.

Further studies are also needed to explain several problems which are specific to Co–based DMS's themselves. For instance, the position of the Co^{2+} ion ground state energy relative to the valence and conduction bands has not been definitely established. The results concerning the values of the exchange integrals for NNN and more distant ions (parameter n in Eq.(38)) are not yet consistent between themselves. Thin film semiconductors containing Co still remain almost unexplored, only one example of such materials ($Zn_{1-x}Co_xSe$), having been investigated so far. The physical origin of the large value of the effective exchange integral J_{eff} observed in Co–based DMS's has not been explained theoretically. Although the superexchange appears to be the likely dominant mechanism of $Co^{2+}-Co^{2+}$ interactions, other contributions should also be considered. Finally, the total absence of measurements for the narrow gap semiconductors, e.g. $Hg_{1-x}Co_xSe$ and $Hg_{1-x}Co_xTe$, is especially conspicuous at this time.

Acknowledgments.

We are grateful to Professors A.K.Ramdas and S.Rodriguez for very useful discussions. We also wish to acknowledge the support of NSF Grant DMR-89-13706.

REFERENCES.

Bartholomew,D.U., Suh,E-K., Ramdas,A.K., Rodriguez,S., Debska,U., and Furdyna,J.K. (1989). *Phys.Rev.* **B39**,5865.

Bleaney,B., and Stevens,K.W.H. (1953). *Repts.Progr.Phys.* **16**,108.

Bowers,K.D., and Owen,J. (1955). *Repts.Progr.Phys.* **18**,304.

Brown,D.B., Crawford,V.H., Hall,J.W., and Hatfield,W.E. (1977). *J.Phys.Chem.* **81**,1303.

Carlin,R.L. (1965). *Transition Metal Chemistry* **1**,1.

Denissen,C.J.M., and de Jonge,W.J.M. (1986). *Solid State Commun.* **59**,503.

Figgis,B.N. (1960). *Trans.Faraday Soc.* **56**,1553.

Furdyna,J.K., and Samarth,N. (1987). *J.Appl.Phys.* **61**,3526.

Furdyna,J.K., Samarth,N., Frankel,R.B., and Spalek,J. (1988a). *Phys.Rev.* **B37**,3707.

Furdyna,J.K. (1988b). *J.Appl.Phys.* **64**,R29.

Furdyna,J.K. and Kossut,J. (1988c). *Semiconductors and Semimetals*, Vol.25 (Academic, Boston).

Furrer,A., and Gudel,H.U. (1979). *J.Magn.Magn.Mater.* **14**,256.

Galazka,R.R. and Kossut,J. (1982) in *Landolt-Börnstein New Series, Group III*, Vol.17b, Springer, Berlin, p.302.

Geschwind,S., Ogielski,A.T., Devlin,G., and Hegarty,J. (1988). *J.Appl.Phys.* **63**,3291.

Giebultowicz,T.M, Klosowski,P., Rhyne,J.J., Udovic,T.J., Furdyna,J.K., and Giriat,W. (1990). *Phys.Rev.* **B41**,504.

Giebultowicz,T.M, Rhyne,J.J., and Furdyna,J.K. (1987). *J.Appl.Phys.* **61**, 3537.

Giebultowicz,T.M, Rhyne,J.J., Furdyna,J.K., and Klosowski,P. (1990). *J. Appl.Phys.* - in press.

Hall,T.P.P, and Hayes,W. (1960). *J.Chem.Phys.* **32**,1871.

Ham,F.S., Ludwig,G.W., Watkins,G.D., and Woodbury,H.H. (1960). *Phys.Rev. Lett.* **5**,468.

Henning,J.C.M., van den Boom,H., and Dieleman,J. (1966). *Philips Res.Repts.* **21**,16.

Hoshina,T. (1966). *J.Phys.Soc.Japan* **21**,1608.

Jonker,B.T., Krebs,J.J., and Prinz,G.A. (1988). *Appl.Phys.Lett.* **53**,450.

Koidl,P., Schirmer,O.F., and Kaufmann,U, (1973). *Phys.Rev.* **B8**,4926.

Kunc,K., Balkanski,M., and Nusimovici,M.A. (1975). *Physica Status Solidi (b)* **72**,229.

Larson,B.E., Hass,K.C., and Ehrenreich,H. (1985). *Solid State Commun.* **56**,347.

Lewicki,A. Spalek,J., Furdyna,J.K., and Galazka,R.R. (1988). *Phys.Rev.* **B37**,1860.

Lewicki,A., Schindler,A.I., Furdyna,J.K., and Giriat,W. (1989). *Phys.Rev.* **B40**,2379.

Lewicki,A., Schindler,A.I., Miotkowski,I., and Furdyna,J.K. (1990a).

Phys.Rev. B41,4653.

Lewicki,A. (1990b) – unpublished.

Lewicki,A., Schindler,A.I., Miotkowski,I., Crooker,B.C., and Furdyna,J.K. (1990c). *Phys.Rev.* B. – in press.

Liu,X., Petrou,A., Jonker,B.T., Prinz,G.A., Krebs,J.J., and Warnock,J. (1989). *Appl.Phys.Lett.* 55,1023.

Look,D.C., and Locker,D.R. (1972). *Phys.Rev.* B6,713.

Matho,K. (1979). *J.Low Temp.Phys.* 35,165.

Morigaki,K. (1963). *J.Phys.Soc.Japan* 18,1558.

Morigaki,K. (1964). *J.Phys.Soc.Japan* 19,2064.

Mycielski,A. (1988). *J.Appl.Phys.* 63,3279.

Nagata,S., Galazka,R.R., Mullin,D.P., Akbarzadeh,H., Khattak,G.D., Furdyna,J.K., and Keesom,P.H. (1980). *Phys.Rev.*B22,3331.

Noras,J.M., Szawelska,H.R., and Allen,J.W. (1981). *J.Phys.* C14,3255.

Novak,M.A., Symko,O.G., Zheng,D.J., and Oseroff,S. (1984). *Physica* B126,469.

Orton,J.W. (1969). *Electron Paramagnetic Resonance*, Gordon and Breach, New York, p.55.

Owen,J. (1961). *J.Appl.Phys.* 32,213S (Suppl.).

Radlinski,A.P. (1979). *J.Phys.* C12,4479.

Robbins,D.J., Dean,P.J., Glasper,J.L., and Bishop,S.G. (1980). *Solid State Commun.* 36,61.

Ryskin,A.I., Natadze,A.L., and Kazanskii,S.A. (1973). *Zh.Eksp.Teor.Fiz.* 64,910.; [*Sov.Phys.-JETP* 37,462.]

Shand,P.M., Lewicki,A., Crooker,B.C., Giriat,W., and Furdyna,J.K. (1990a). *J.Appl.Phys.* 67,5246. (Proc. 34-th MMM Conference).

Shand,P.M., and Crooker,B.C. (1990b) – private communication.

Shapira,Y., Foner,S., Ridgley,D.H., Dwight,K., and Wold,A. (1984). *Phys.Rev.* **B30**,4021.

Shapira,Y., (1990). *J.Appl.Phys.* **67**,5090. (Proc. 34-th MMM Conference).

Spalek,J. Lewicki,A., Tarnawski,Z, Furdyna,J.K., Galazka,R.R., and Obuszko,Z. (1986). *Phys.Rev.* **B33**,3407.

Swagten,H.J.M., Twardowski,A., de Jonge,W.J.M., and Demianiuk,M. (1989). *Phys.Rev.* **B39**,2568.

Title,R.S. (1963). *Phys.Rev.* **131**,2503.

Twardowski,A., Denissen,C.J.M., de Jonge,W.J.M., de Waele,A.T.A.M., Demianiuk,M., and Triboulet,R. (1986). *Solid State Commun.* **59**,199.

Twardowski,A., Swagten,H.J.M., de Jonge,W.J.M., and Demianiuk,M. (1987). *Phys.Rev.* **B36**,7013.

Twardowski,A. (1990). *J.Appl.Phys.* **67**,5108. (Proc. 34-th MMM Conference); see also: Twardowski,A. - "*Magnetic and Optical Properties of Fe-based Semimagnetic Semiconductors*" in this book.

Van Vleck,J.H. (1932). *The Theory of Electric and Magnetic Susceptibilities,* Oxford University Press, p.182.

Villeret,M. (1989). Ph.D. Thesis, Purdue University.

Villeret,M., Rodriguez,S., and Kartheuser,E. (1990a). *Phys.Rev.* **B41**, 10028.

Villeret,M., Rodriguez,S., and Kartheuser,E. (1990b). *Physica* **B162**,89.

Weakliem,H.A. (1962). *J.Chem.Phys.* **36**,2117.

Yoder-Short,D.R., Debska,U., and Furdyna,J.K. (1985). *J.Appl.Phys.* **58**,4056.

DILUTED MAGNETIC SEMICONDUCTOR HETEROSTRUCTURES

Gilmar Eugenio Marques

Departamento de Física, Universidade Federal de São Carlos

13560, São Carlos, SP, Brazil.

I - INTRODUCTION.

The pseudobinary II-VI diluted magnetic semiconductors (DMS) of type $A_{(1-x)} M_{(x)} B$ grow, basically, in the zincblende (cubic) or in the wurtzite (hexagonal) crystalline[1] lattices. In every material, the element A from the group II, located in a given site of the Bravais sublattice, is randomly replaced by chemical elements, M, with high atomic magnetic moment such as Mn, Fe or Eu. For example, $Cd_{(1-x)} Mn_{(x)} Te$ in its zincblende crystal phase, displays the ions Mn^{++} randomly distributed in the fcc sublattice of the ions Cd^{++} and their half-filled $3d^5$ atomic shell are highly localized at each replaced Bravais site. The five paired spin components form a local magnetic moment which presents an antiferromagnetic coupling with their closest neighbors. In virtue of this tendency of antiparallel alignment of spins this family is also known as semimagnetic semiconductors.

Due to thermal agitation, a DMS is paramagnetic above a certain temperature $T_{sg}(x)$, the spin-glass freezing temperature which depends on the Mn concentration and on the host II-VI material[2] and is antiferromagnetic below $T_{sg}(x)$. This phase transition is shown clearly as a kink in the magnetic susceptibility[3] however, for DMS heterostructures such as $CdTe$-$Cd(Mn)Te$ superlattice, the spin-glass phase may be totally suppressed[4] below approximately 20 $\overset{o}{A}$ of the semimagnetic layer thickness. The substitution of the element of group II for a magnetic ion makes the alloy to exhibit a vast range of properties[6] new to the physics of semiconductors and, most of them, *are originated from this antiferromagnetic coupling between neighbor local atomic magnetic moments.*

One of the most interesting aspect of these materials is the fact that the magnetization causes strong changes on their electronic, magneto-optical, transport and thermal properties. Therefore, the family of DMS displays extremely large electronic gyromagnetic factor, giant magnetoresistance[5] and Faraday rotation, anomalous Shubnikov-de Haas oscillations[6], and a sharp kink in the magnetic susceptibility[5] even though the specific heat[5] shows just a smooth dependence on the temperature.

In the recent years, the development of epitaxial growth techniques for II-VI materials, made it possible to fabricate DMS heterostructures such as $Cd_{(1-x)}Mn_{(x)}Te$-$CdTe$[6] or $Zn_{(1-x)}Mn_{(x)}Te$-$ZnTe$[7]. Some special aspects of such magnetization on the electronic subband structure, Landau levels and magnetic polaron will be reviewed briefly here. Before we will review the theoretical model and the necessary approximations that are frequently used to calculate the subband structure in these heterolayered materials, in special the quantum wells (QW).

II - THE SUBBAND STRUCTURE.

In figure 1 we show the change of a typical subband structure in the bulk of a zincblende semiconductor, near the Γ-point, with increasing concentration of Mn. Here, we will consider only the case of "normal" materials were the Γ_6 conduction band has s-symmetry and the Γ_8 valence band has p-symmetry of the T_d^2 crystal group. Also, these DMS materials usually exhibit a large spin-orbit energy, thus, the Γ_7 split-off branch in the valence band will be neglected in the following discussion of the empirical $\vec{k}.\vec{p}$ method which is commonly used for the calculation of the electronic subband structure of heterostructures.

The $\vec{k}.\vec{p}$ method, which gave origin to the so called envelope-function approximation, was first developed by Kane[8] and

Dresselhaus[9], *has been proved to be a rigorous basis for the determination of band structure in crystalline solids. The method can only be applied in the vicinity of an extremum within the Brillouin zone. As such, its validity is limited to values of linear momentum, $|\vec{k}|$, close to that extremum, however, its greatest convenience resides*

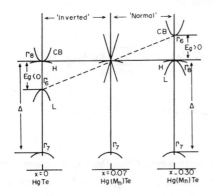

Fig. 1.- Qualitative change in the bulk band structure of $Hg_{(1-x)}Mn_{(x)}Te$ showning the "inverted" (Hg - rich) and the "normal" (Mn - rich) structures of DMS with zincblende symmetry at the Γ-point. Inversion occurs for approximately x = 7 % of Mn. Above x = 30% Mn the crystal exhibit islands of wurtzite and zincblende symmetries. The DMS families, such as Cd(Mn)Te and Zn(Mn)Te, display only "normal" zincblende symmetry and good crystal structures up to x = 70% Mn.

on simplicity, accuracy and unique capability to handle, with very little effort, the long range non-periodic potentials, electric and magnetic fields and internal strains. The method is based on a very small set of parameters (band-gap, gyromagnetic factor, spin-orbit energy and effective-masses) determined directly from magneto-optical absorption experiments.

The general bulk Hamiltonian can be derived directly from the Bloch theorem. Let $u_{nk}(\vec{r})$ be the periodic part of the Bloch wavefunction for an electron in a crystalline potential $V_C(\vec{R})$ and $H_C = T + V_C(\vec{R})$ the crystal Hamiltonian with the periodicity determined by $V_C(\vec{R})$. The Schrödinger equation for the periodic part of the wave function is given by

$$H_k \ u_{nk}(\vec{r}) \ = \ E_{nk} \ u_{nk}(\vec{r}) \tag{1}$$

where the set of eigen-values, $\{E_{nk}\}$ gives the bulk band structure and the Hamiltonian H_k being expressed as

$$H_k = e^{-i\vec{k}.\vec{r}} H_c e^{+i\vec{k}.\vec{r}} = H_C - i\vec{k}.[\vec{r}, H_C] - \frac{1}{2!} \sum_{\mu,\upsilon} k_\mu \, k_\upsilon \, [r_\mu, [r_\upsilon, H_C]]$$

$$+ \frac{i}{3!} \sum_{\mu,\upsilon,\lambda} k_\mu \, k_\upsilon \, k_\lambda \, [r_\mu, [r_\upsilon, [r_\lambda, H_C]]] + \ ..... \tag{2}$$

If H_C is the Hartree Hamiltonian, then the first commutator in eq.(2) gives $(i\frac{\hbar \vec{P}}{m})$, the second gives $(-\frac{\hbar^2}{m} \delta_{\mu\upsilon})$ and all terms of higher order are zero. Thus, the Hamiltonian in eq.(2) can be written as $H_k = H_C + \frac{\hbar \vec{k}.\vec{P}}{m} + \frac{\hbar^2 k^2}{2m}$. This is usually referred as the $\vec{k}.\vec{p}$ Hamiltonian for obvious reason.

The inclusion of spin-orbit interaction through the Hartree-Fock Hamiltonian $H_{SO} = (\vec{S} \times \vec{\sigma}).\vec{P}$, where $\vec{S} = \frac{\hbar}{4m^2c^2}(\vec{\nabla} V_C(\vec{R}))$, introduces two new terms in H_k, namely: $\vec{S}.\vec{\sigma} + (\vec{S} \times \vec{k}).\vec{\sigma}$, where the first term is the normal spin-orbit coupling and the last term is frequently referred as the k-linear spin-orbit contribution to the Bloch states in a crystalline potential $V_c(\vec{R})$.

In general there are an infinite number of solutions for the H_k Hamiltonians listed above. To simplify and make possible to find the most important solutions, the next step one should choose a finite number of Bloch states close to a given extremum, all of then compatible with the local symmetry, and calculate the matrix elements *of H_k in this finite number of states (first-order perturbation theory). There will appear a finite number of non-zero matrix elements in this context and these parameters are, then, determined from the local curvatures (effective-masses) of all branches close to the chosen extremum. The success of such method resides on the empirical determination of the values for these matrix elements.*

The effect of all the other neglected states can be introduced in second-order perturbation, as described by Löwdin[10]. These contributions give extra curvature to the valence bands and causes the splitting of the Kramer's doublet for semiconductors displaying inversion asymmetry, such as most of the binary III-V and II-VI compounds with cubic symmetry. We will restrict, from now on, our attention to materials having zincblende symmetry at the Γ-point.

The most important states there, compatible with the local symmetry of T_d^2 group (zincblende materials) can be written as

$$|U_1> = |\tfrac{1}{2}, +\tfrac{1}{2}> = |S> \uparrow,$$

$$|U_2> = |\tfrac{3}{2}, +\tfrac{3}{2}> = \frac{i}{\sqrt{2}} |X + iY> \uparrow, \tag{3}$$

$$|U_3> = |\tfrac{3}{2}, +\tfrac{1}{2}> = \frac{i}{\sqrt{6}} [|X + iY> \downarrow - 2|Z> \uparrow],$$

and the other three twin states can be obtained from (3) by an application of the time-reversal operator $\hat{K} = -i\,\hat{\sigma}_y\,\hat{C}\,\hat{J}$ for for zincblende symmetry, where $\hat{C}$ is the complex conjugation, $\hat{\sigma}_y$ is the Pauli matrix which flips the spin components and $\hat{J}$ is the inversion operators. They are ordered in the same way as in eq.(3) or, $|U_4> = \hat{K}|U_1>$, $|U_5> = \hat{K}|U_2>$ and $|U_6> = \hat{K}|U_3>$. The first and the fourth states have s-symmetry and correspond to the $|\tfrac{1}{2}, \pm\tfrac{1}{2}>$ (electrons) s-states with the total angular momentum $J=\tfrac{1}{2}$ ($L=0$, $S=\tfrac{1}{2}$). The second and fifth, the third and sixth are respectively the $|\tfrac{3}{2}, \pm\tfrac{3}{2}>$ (heavy-holes) and $|\tfrac{3}{2}, \pm\tfrac{1}{2}>$ (light-holes) p-states with angular momentum $J=\tfrac{3}{2}$ ($L=1$, $S=\tfrac{1}{2}$).

The full $\vec{k}.\vec{p}$ Hamiltonian, calculated in the Bloch states at the Γ-point, is written as a 6x6 matrix as a function of the momentum

$\vec{k} = (k_x, k_y, k_z)$, energy band-gap, E_g, and the spin-orbit energy, Δ, and the Kane's parameters (matrix elements). Its diagonalization gives three doubly degenerate eigen-values, which correspond to the conduction, heavy-hole and light-hole energy bands in the bulk, and the numbers referring to the order of the set of Bloch states in eq. (3) above.

Let the z-axis be perpendicular to the direction of the hetero-structure and consider atomic units. In the absence of an applied magnetic field, we can write an extended version of the $\vec{k} \cdot \vec{p}$ Hamiltonian in the set of Bloch states given in eqs.(3) as

$$
H = \begin{bmatrix}
D_{EL} & -\sqrt{3}\,A_+ & \sqrt{2}\,R & 0 & 0 & -A_- \\
 & D_{HH} & \sqrt{2}\,L & 0 & 0 & -W \\
 & & D_{LH} & A_+^* & W & 0 \\
 & & & D_{EL} & -\sqrt{3}\,A_- & \sqrt{2}\,R \\
 & & & & D_{HH} & \sqrt{2}\,L^* \\
 & & & & & D_{LH}
\end{bmatrix} \tag{4}
$$

where, the matrix elements are defined in terms of combinations of the linear momentum components, $k_x^2 + k_y^2 = k^2$, $k_\pm = (k_x \pm ik_y) = k\,exp(\pm i\Theta)$ and k_z, times the Kane's parameters[9] as

$$
D_{EL} = E_g + E_v^c + (F + \tfrac{1}{2})\,k^2 + [\,\hat{k}_z (F + \tfrac{1}{2})\,\hat{k}_z\,]\ ,
$$

$$
D_{HH} = D_+ \quad and \quad D_{LH} = D_-\ ,
$$

$$
D_\pm = E_v^v - \tfrac{1}{2}(\gamma_1 \pm \gamma_2)\,k^2 - \tfrac{1}{2}[\,\hat{k}_z (\gamma_1 \mp 2\gamma_2)\,\hat{k}_z\,]\ ,
$$

$$A_\pm = \sqrt{1/6}\ P\ \hat{k}_z \pm \sqrt{2/3}\ \{G, \hat{k}_z\}\ k_\mp\ ,$$

$$R = \sqrt{1/3}\ \{P, \hat{k}_z\} - 2i\ \sqrt{1/3}\ G\ k_x\ k_y\ ,$$

$$L = \sqrt{3/2}\ (\gamma_3, \hat{k}_z)\ k_-\ ,$$

$$W = \sqrt{3}\ [\bar{\gamma}\ k_-^2 - \mu\ k_+^2\ ],\quad \bar{\gamma} = \tfrac{1}{2}(\gamma_2 + \gamma_3),\quad \mu = \tfrac{1}{2}(\gamma_3 - \gamma_2),$$

For layered systems in the flat-band condition[11,12], the z-component of the linear momentum, perpendicular to the interface, becomes a differential operator, $\hat{k}_z = -i\frac{d}{dz}$ whereas in the bulk it is a number[8,12]. Also, in these expressions, $\{\hat{A}, \hat{B}\} = \frac{1}{2}(\hat{A}\hat{B} + \hat{B}\hat{A})$ is an average of the anticommutator; E_g is the smallest band-gap of the two materials in the heterostructure; P (F) is the first-order (second-order) Kane's parameter for the Γ_6 band; γ_1, γ_2 and γ_3 are equivalent to the Luttinger[13] parameters for the Γ_8 band; Δ is the spin-orbit energy which splits the Γ_7 and the Γ_8 multiplets and E_v^i is the band offset for the ith branch in a given interface.

The parameter μ, in the definition of the matrix element W, gives the warping of the valence subbands in normal materials and the inversion asymmetry terms T for the Γ_8 band and G for the Γ_6 band are responsible for the spin-splitting of the Kramer's doublets in zincblende materials and, since they give very small contribution[14] to the subband structure, we will neglect them in the future applications.

For bulk states the parameters P, F, γ_1, γ_2, γ_3, T and G are independent of the coordinate z and E_v^i is zero but for heterostructures they depend on z. Although we use the z-dependence of all parameters as two different constants for each side of the heterostructure they, in principle, can change smoothly from one side of the interface to the other (graded interface). Also, we choose the zero of energy at the top of the Γ_8 energy branch, and for this choice, the Γ_6 electrons have energies larger than E_g, the Γ_8 holes have negative energies.

The expansion of the eigen-values of eq. (4), up to quadratic order in k, determines the effective-masses at the Γ-point as a function of the Kane's parameters. For $\vec{k}$ in the [1,0,0] direction we obtain the effective-masses for each one of the three branches as (1 → electron, 2 → heavy-hole, 3 → light-hole)

$$\frac{1}{m_1} = 1 + 2F + \frac{4P^2}{3\,E_g}\,[\,1 + \frac{E_g}{2\,(E_g + \Delta\,)}\,],\tag{5}$$

$$\frac{1}{m_2} = (\,\gamma_1 - 2\,\gamma_2\,),\tag{6}$$

$$\frac{1}{m_3} = (\,\gamma_1 + 2\,\gamma_2\,) + \frac{4\,P^2}{3\,E_g},\tag{7}$$

For the $\vec{k}$ in the [111] direction, the heavy-hole effective-mass is found as

$$\frac{1}{m'_3} = (\gamma_1 - 2\,\gamma_3).\tag{8}$$

For heterostructures, the motion of carriers along the interface (xy-plane) is quasi-2D free motion whereas, in the perpendicular direction (z-axis) becomes quantized. This can be taken into account by setting the momentum component k_z into a differential operator, i.e. $\hat{k}_z = -\,i\,\frac{d}{dz}$. On the other hand, the effective-masses in each side of an interface are different, therefore, the anticommutators in the off-diagonal terms were a necessary symmetrization in order to make the full Hamiltonian for a heterojunction a hermitian operator which satisfies the time-reversal operator. The same justification holds for the diagonal quadratic terms of the form $\hat{k}_z\,\frac{1}{m(z)}\,\hat{k}_z$.

Let the envelope wave-functions for each 6-component spinor be written as

$$\Psi_{\vec{k}}(x, y, z) = \begin{bmatrix} A_1(x,y,z, \vec{k}) \\[6pt] A_2(x,y,z, \vec{k}) \\[6pt] A_3(x,y,z, \vec{k}) \\[6pt] A_4(x,y,z, \vec{k}) \\[6pt] A_5(x,y,z, \vec{k}) \\[6pt] A_6(x,y,z, \vec{k}) \end{bmatrix} \tag{9}$$

Since present heterojunctions, grown by epitaxial techniques, have interfaces of high quality, the translational invariance along the xy plane can be assumed without lost of reality, therefore, the explicit dependence of any component in eq.(5) on the parallel momentum can be separated as a plane wave in the xy plane and an explicit dependence on z and $\vec{k}$, in the form

$$A_j(x,y,z, \vec{k}) = e^{i\,\vec{k} \cdot \vec{r}}\, A_j(z, \vec{k}), \tag{10}$$

where $\vec{k}$ and $\vec{r}$ are two-dimensional vectors in the xy plane and j=1, 8.

If we consider the magnetic field applied in the z-direction, $\vec{B}=(0,0,B)$, the nearly free motion of electrons in the xy-plane will be further quantized into closed orbits and, therefore, the plane wave in eq.(10) will become closed Landau orbits of radius, R_c, and expressed by the Hermite polynomials, $|N> = H_N(\vec{r}/R_c)$. Let us define the cyclotron frequency, $\omega_c = (eB/mc)$, the cyclotron energy of free electrons, $E_c = \hbar\omega_c$, the Landau radius of a cyclotron orbit, $R_c^2 = (\hbar c/eB)$, and set the components of the momentum in the xy-plane in terms of the raising $(\hat{a}^+ |N> = \sqrt{N+1}\,|N+1>)$ and the lowering $(\hat{a}|N> = \sqrt{N}\,|N-1>)$ harmonic oscil-

lator operators with commutation relation for fermions, $[\hat{a}^+, \hat{a}] = (\hat{a}^+ a - \hat{a}\,\hat{a}^+) = 1$, as

$$\hat{k}_x = \frac{1}{R_c \sqrt{2}} (\hat{a} + \hat{a}^+), \qquad \hat{k}_y = \frac{i}{R_c \sqrt{2}} (\hat{a} - \hat{a}^+). \qquad (11)$$

Therefore, because the raising and lowering operators, the later good quantum numbers, that is, the components of the linear momentum, will change into operators such as: $\hat{k}^2 \rightarrow (2\hat{a}^+ \hat{a} + 1)/R_c^2$, $\hat{k}_- \rightarrow \sqrt{2}\,\hat{a}/R_c$, etc. Secondly, because the magnetic dependence of the total Hamiltonian, the envelope-function components will depend differently on the Landau quantum numbers, N, as imposed by the $\vec{k} \cdot \vec{p}$ method, as

$$\Psi_N(\vec{r}, z) = \begin{bmatrix} H_{N-1}(\vec{r}/R_c) & A_{1,N}(z) \\[2mm] H_{N-2}(\vec{r}/R_c) & A_{2,N}(z) \\[2mm] H_{N-1}(\vec{r}/R_c) & A_{3,N}(z) \\[2mm] H_{N}(\vec{r}/R_c) & A_{4,N}(z) \\[2mm] H_{N+1}(\vec{r}/R_c) & A_{5,N}(z) \\[2mm] H_{N}(\vec{r}/R_c) & A_{6,N}(z) \end{bmatrix} \qquad (12)$$

The matrix elements of eq.(4), will change their dependence on the quantum number $\vec{k}$ to the dependence on magnetic quantities such as the order of the Landau number, N, and cyclotron energy, E_c. Before showing this explicit dependence we must introduce the interaction of Bloch electrons, at the position $\vec{r}$ and having spin $\vec{s}$, with the local atomic spin $\vec{S}_i$ at the lattice site $\vec{R}_i$. The motion of Bloch electrons in this magnetized environment is well described by an one-particle Heisenberg exchange interaction in the form

$$H_{xc} = - \sum_{R_i} J(\vec{r} - \vec{R}_i) \; \vec{S}_i \cdot \vec{s} \qquad (13)$$

where $J(\vec{r} - \vec{R}_i)$ is the spin-spin exchange coupling constant and the sum running only over the sites occupied by the magnetic ions. This interaction Hamiltonian certainly cannot be treated one the same ground as the crystal Hamiltonian H_k given before since it lacks the perfect periodicity of the crystal. However, the highly localized nature of the $3d^5$ atomic states, the antiferromagnetic coupling between the neighbors atomic spins and the extension of the Bloch electrons wave-functions over the entire crystal, cause the band electrons, in reality, experience an average atomic spin from the magnetic ions in the sample. Therefore it is plausible to use two approximations in the treatment of the Hamiltonian in eq.(12). For a magnetic field applied in the z-direction, the mean-field approximation(MFA) replaces the z-component of the local spin $\vec{S}_i$ by a position independent thermal average throughout the crystal, $<S_z(B,T)>$. Also, the so called virtual crystal approximation(VCA) replaces the exchange coupling, $J(\vec{r} - \vec{R}_i)$, by an effective exchange spin-spin coupling, $x\,J(\vec{r} - \vec{R})$, with $\vec{R}$ running over the complete Bravais lattice and x being the concentration of the atomic magnetic ion in the sample. After these approximations we can rewrite the non-periodic Hamiltonian in eq.(13) as a periodic magnetic Hamiltonian

$$\bar{H}_{xc} = - x \sum_{R} J(\vec{r} - \vec{R}) \; <S_z(B,T)> \; s_z \qquad (14)$$

which now can be treated in the same set of Bloch states given in eq.(1). The matrix elements $<u_i|\tilde{H}_{xc}|u_j>$ gives only two non-zero values, namely: the s-component of the envelope-function gives, $\alpha = <S|J(\vec{R})|S>$, and the p-component, $\beta = \frac{1}{3}<Z|J(\vec{R})|Z>$.

The thermal average of the atomic spin can be very well fitted by a modified Brillouin function, $B_S[s,y] = \frac{2s+1}{2s} \coth(\frac{2s+1}{2s}y) - \frac{1}{2s}\coth(\frac{1}{2s}y)$, if we introduces an effective spin saturation, $S_o(x)$, and an effective temperature, $T_o(x)$. For example, for the semimagnetic semiconductor family $Cd_{(1-x)}Mn_{(x)}Te$, where $S=\frac{5}{2}$ is the Mn total atomic spin, we can write the thermal average of spins in the form[2,14]

$$<S_z(B,T)> = S_o(x)\, B_{\frac{5}{2}}\left[\frac{5}{2}, \frac{g_{Mn}\mu_B B}{K_B(T-T_o(x))}\right] \qquad (15)$$

where $g_{Mn} = 2$ is the Mn gyromagnetic factor, μ_B is the Bohr magneton and K_B is the Boltzmann constant. The magnetization of this sample can be determined directly from this thermal average of spins as

$$<M> = (x\,N_o)\,\mu_B\,g_{Mn}<S_z(B,T)> \qquad (16)$$

where (xN_o) is the fraction of Mn ions per unit of volume in a crystalline cell of the zincblende structure.

There are several modifications in the matrix elements of $(H_k+\tilde{H}_{xc})$, in the presence of the magnetic field $\vec{B}=(0,0,B_o)$. Firstly, we will neglect the asymmetry terms, T and G, and the warping term μ in eq.(4). Secondly, there are three new parameters for the magnetic case, namely: κ is the g-factor[13] for the Γ_8 band, q is the parameter arising from the spin-orbit splitting[13] after the inclusion (second-order perturbation) of the higher bands in the $\vec{k}.\vec{p}$ method which contributes further to the warping of the valence band, and finally, N_1 is an extra contribution to the g-factor[15] of the Γ_6 band. For the branches with spin $\vec{s}$ having eigen-values $s_z = \pm \frac{1}{2}$ in the conduction and $s_z = \pm \frac{3}{2}, \pm \frac{1}{2}$ in the valence bands, we find the following expressions for the matrix elements in eq.(4), in the presence of a magnetic field in the z-direction

$$D^{\pm}_{EL} = E_g + E^c_v + E_c \left\{ (F+\tfrac{1}{2})(2N \mp 1) \pm (N_1+1) \right\} - \frac{d}{dz}[F + \tfrac{1}{2}]\frac{d}{dz} \pm \tfrac{1}{2} Q_{cm} \,,$$

$$D^{\pm}_{HH} = E^v_v - E_c \left\{ (\gamma_1 + \gamma_2)(N \mp \tfrac{3}{2}) \pm (\kappa + \tfrac{9}{4} q) \right\} + \tfrac{1}{2}\frac{d}{dz}[\gamma_1 - 2\gamma_2]\frac{d}{dz} \pm \tfrac{3}{2} Q_{vm},$$

$$D^{\pm}_{LH} = E^v_v - E_c \left\{ (\gamma_1 - \gamma_2)(N \pm \tfrac{1}{2}) \mp \tfrac{1}{2}(\kappa + \tfrac{1}{4} q) \right\} + \tfrac{1}{2}\frac{d}{dz}[\gamma_1 + 2\gamma_2]\frac{d}{dz} \mp \tfrac{1}{6} Q_{vm},$$

$$A_{\pm} = P\sqrt{E_c(N \mp 1)}\, e^{i\Theta}, \qquad\qquad Q_{cm} = N_o\, \alpha\, <S_z(B_o,T)>,$$

$$R = i\sqrt{\tfrac{2}{3}} \{\tfrac{d}{dz}, P\}, \qquad\qquad Q_{vm} = N_o\, \beta\, <S_z(B_o,T)>,$$

$$L_{\pm} = E_c\, e^{i2\Theta}\, \bar{\gamma}\sqrt{3N(N \mp 1)}, \qquad W_{\pm} = -i\, e^{i\Theta}\{\tfrac{d}{dz}, \gamma_3\}\sqrt{6E_c(N \pm 1)},$$

$$(17)$$

where the quantities were already defined, but here, the $\pm$ refers to the signal of the components of the total spin on the z-axis. Notice that the semimagnetic terms will act as magnetic potential barriers, Q_{cm} and Q_{vm}, and their influence on the quantum well levels will be shown later.

The expansion of the bulk eigen-values for the Γ_6 band up to first order in the applied magnetic field, B_o, gives the gyromagnetic factor[15,16] for the conduction band as

$$g_c = 2 + 4 N_1 - \frac{4P^2}{3 E_g} [1 - \frac{E_g}{(E_g + \Delta)}] \qquad (18)$$

which shows clearly the net contribution of the parameter N_1 to the g-factor of Bloch electrons. For free electrons we have, $N_1 = 0$, $P=0$ and, as expected, $g_c = 2$.

The experimental values of the g-factor, the band-gap, the spin-orbit, the effective-masses in eqs.(5,6,7,8) and the split-off effective-mass, m_{so} (as for example shown in table I), will determine the set of parameters for the magnetic Hamiltonian in the presence of a magnetic field, F, P, γ_1, γ_2, γ_3, κ, q, N_1. Also, the experimental values for the magnetic parameters $(N_o \alpha)$ and $(N_o \beta)$ are easily determined from peaks of the σ^+ and σ^- magneto-optical absorption of excitons[6,7] in DMS samples.

The motion of electrons and holes in heterostructures is determined from the solutions of the Schrödinger equation for the components, A_j, in the form $(H + V) \Psi = E \Psi$, where V is 6x6 matrix which may include external electric field, internal potentials due to charge redistributions (self-consistent Hartree and exchange-correlation potentials) and strains (hydrostatic and axial). The hydrostatic strain only changes the band-gap but the axial strain splits the Γ_8 states at the center of the Brillouin zone[16]. In the following we will discuss the subband structure of quantum wells of Cd(Mn)Te-CdTe.

III - SUBBAND STRUCTURE OF $Cd_{(1-x)} Mn_{(x)} Te - Cd Te$

This material has "normal" zincblende band structure for Mn concentration[2] below $x_m = 30$ but x greater than 0.7. Above x_m the crystal structure starts to show poor quality and large $(MnTe_n)$ islands displaying wurtzite symmetry. For values of x, $0.7 \leq x \leq x_m$, the substitution of Cd by Mn mainly increases the fundamental band-gap and Cd(Mn)Te acts as a magnetic barrier in the heterostructure. In this sense the Mn has a similar effect as Al in GaAs-Ga(Al)As heterostructures, however the quantum well here has a larger spin-orbit energy. *Also, in the presence of a magnetic field, the influence of the magnetization, through Q_{cm} and Q_{vm}, will change the barrier height differently for each kind of particle and this is an unique feature of the DMS family.*

The value of the band-gap, E_g, as a function of the Mn concentration[17] is given by

$$E_g = 1595 + 1592\,x\,, \qquad (meV) \qquad\qquad (19a.)$$

Since MnTe has wurtzite symmetry, the band parameters for Cd(Mn)Te are not simple to find. We have chosen to get all masses at a given concentration where the alloy displays zincblende symmetry and, for $0.7 < x \leq x_m$, use linear interpolation or extrapolation to calculate the other parameters. They are given[17,18] in table-I below and, when necessary, we have used the values of CdTe. Therefore, this set of numbers is certainly subjected to changes once new experimental findings on its band structure are reported.

TABLE - I *The set of parameters for* CdTe(1) *and* $Cd_{(0.9)}Mn_{(0.1)}Te(2)$. *For the alloy use linear interpolation. The effective-masses, in units of free electron mass, were taken from refs.[6,7,18].*

	$m_c^*(001)$	$m_{hh}^*(001)$	$m_{lh}^*(001)$	$m_{hh}^*(111)$	$m_{so}^*(001)$	$\Delta(meV)$
(1)	0.0960	0.6600	0.1000	2.7000	0.2800	910.
(2)	0.0960	0.3200	0.0800	0.3200	0.1500	910.

The lattice mismatch in the heterostructure induces a strain in the CdTe layer, therefore, the potential V should play a role. We will discuss strain effects later, since another important parameter in any method of subband structure, the band-offset E_v^v, is not precisely known for this heterojunction. It is accepted, however, that a small fraction of the band-gap difference can be accomodated in the valence band[6,7]. Since this yet is an opened point for the system we have chosen to used the rule 20%(vb)-80%(cb) to determine each band-offset in the following results. For this rule we obtain the band offset for the valence band of $Cd_{(1-x)} Mn_{(x)} Te$,

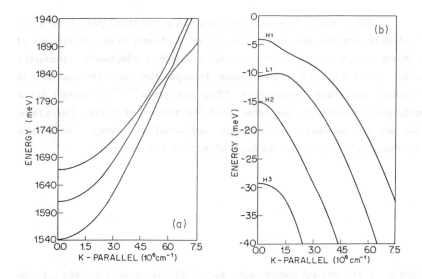

Fig. 2.- Subband structure of $Cd_{(0.90)} Mn_{(0.10)} Te$ - CdTe semimagnetic quantum well with L=100 Å, along the direction [100] in the Brillouin zone. Fig.-a:- Electrons. Notice the strong non-parabolicity, the subband mixing at large values of k-parallel and different curvatures (effective-masses) close to the bottom of each subband. Fig.- b:- Holes. The small subband mixing in this material is mostly due to large spin-orbit energy (910 meV) and heavy-hole mass (m_{hh} = 0.66 m_o).

$$E^v_v = - \; 318. \; x \qquad (meV). \qquad (19.b)$$

Figure 2a shows the subband structure for electrons and for holes in $Cd_{(0.94)} \; Mn_{(0.06)} \; Te$ - $CdTe$ quantum well with width L = 100 $\overset{o}{A}$. The electron subband structure is strongly non-parabolic and shows many bound states inside the well. In spite of the large band-gap, there is an important coupling between light-holes and electrons producing this strong subband mixing (mini-gaps) and a different effective-masses (curvatures) for the motion in the xy-plane in each bound state.

In figure 2b we show the valence subband structure for 100 $\overset{o}{A}$ quantum well with $Cd_{(0.90)} \; Mn_{(0.10)} \; Te$-$CdTe$. As should be expected, there is very little band admixture in the energy dispersions since we are dealing with a material having wide-gap and large spin-orbit. Also, due to the fact that the heavy-hole mass is almost three times heavier than the same mass for GaAs case, the smaller band-offset and the spin-orbit energy considerably larger, we observe very small number of light-hole states inside the CdTe quantum well. Moreover, since heavy-hole states are only weakly coupled to the other states the subband mixing only appears at very small values of k-parallel and, somewhat far away from the zone center, the dispersions approach rapidly to the parabolic case.

In figure 3 we show the valence subband structure for $Cd_{(0.94)} \; Mn_{(0.06)} \; Te$-$CdTe$ quantum well with L = 50 $\overset{o}{A}$. Here also we observe a strong non-parabolicity close to the bottom of the subband and away from the center they become parabolic. Also, due to the small band offset, only two states are bound to the well and show small mixing by the same reasons explaned above. The conduction band for this quantum well is similar to the fig.(2a) and will not be shown here since no new effect can be observed.

These features will certainly affect the optical absorption, joint density of states, excitons and magnetic polarons in DMS quantum wells. However, a much more drastic effect on these states are refered to the magnetic field. In the next section we will discuss the very special role played by the magnetic field on these states located inside the quantum well.

Fig. 3. - Hole subband struc-
ture of Cd$_{(0.94)}$ Mn$_{(0.06)}$ Te-CdTe
semimagnetic QW with L =
50 Å, along the direction
[100] in the Brillouin zone.
Again, the smaller subband
mixing has the same explanation as given in fig.(2) plus
the smaller band offset,
E_{v}^{V} = - 19.08 meV.

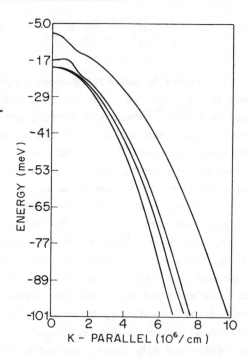

IV - THE LANDAU LEVELS OF Cd$_{(1-x)}$ Mn$_{(x)}$ Te - Cd Te

The Landau levels of particles in quantum wells is strongly
affected by the magnetization of the semimagnetic barriers. The
magnetic[6] exchange constants for Cd(Mn)Te are $(N_o \alpha)$ - 220 meV
and $(N_o \beta)$ = - 880 meV. The other new magnetic parameters
associated to the conduction and valence bands g-factors as well as to
the valence band warping for this system are calculated from bulk
quantities and are used independent of the concentration as N_1 - 1.37,
κ = 1.27 and q = 0.05. For small concentrations of Mn this is not too
drastic assumption.
        The complex coupling between the Landau levels and the
different dependence of each envelope-function component in eq.(12)
imposed by the $\vec{k}.\vec{p}$ method determine four types of Landau ladder
for magnetic quantum wells.

The first type of solution has the spinor as a totally decoupled singlet heavy-hole state coming from the component, $|\frac{3}{2}, -\frac{3}{2}>$, with $N = -1$. For this type of Landau level we should solve the simple parabolic equation

$$\left\{ \frac{d}{dz} [\gamma_1 - 2\gamma_2] \frac{d}{dz} \right\} A_{5,-1}(z) = \left\{ E - E^v + \frac{3}{2} E_C [(\gamma_1 + \gamma_2) + (\kappa + \frac{9}{4} q)] \right\} A_{5,-1}(z) \quad (20)$$

which gives a linear dependence of the energy with the z-component of the magnetic field, B_o. For this singlet solution the spinor, $\Psi_{N=-1}(\vec{r}, z)$, of equation (12) has every component zero except the fifth which is given by, $H_o(\vec{r}/R_C) A_{5,-1}(z)$.

In the second type, the solution is a triplet involving only the spin-down components, $|\frac{1}{2}, -\frac{1}{2}>$, $|\frac{3}{2}, -\frac{3}{2}>$ and $|\frac{3}{2}, -\frac{1}{2}>$. Therefore, the spinor, $\Psi_{N=0}(\vec{r}, z)$, for these solutions have all three spin-up components zero and only appears the three lowest

$$\Psi_{N=0}(\vec{r}, z) = \begin{bmatrix} 0 \\ 0 \\ 0 \\ H_0(\vec{r}/R_C) \, A_{4,0}(z) \\ H_1(\vec{r}/R_C) \, A_{5,0}(z) \\ H_0(\vec{r}/R_C) \, A_{6,0}(z) \end{bmatrix} \quad (21)$$

The three components A's are obtained by solving the following 3×3 matrix

$$
\begin{bmatrix} D^-_{EL} & -\sqrt{3}A_- & \sqrt{2}R \\ & D^-_{HH} & \sqrt{2}L^*_- \\ & & D^-_{LH} \end{bmatrix} \begin{bmatrix} A_{4,0}(z) \\ A_{5,0}(z) \\ A_{6,0}(z) \end{bmatrix} = E \begin{bmatrix} A_{4,0}(z) \\ A_{5,0}(z) \\ A_{6,0}(z) \end{bmatrix} \tag{22}
$$

where the matrix elements were already defined above but they must be calculated from the set of eqs.(17), with $N = 0$. For future reference let us label as the Landau levels $N = 0$, the two states in the valence and the state in the conduction (ground-state for electrons) bands since this number is attached to the each envelope-function components in eq.(22) or to the spinor in eq.(21).

The next set of solutions for the Landau levels appear as quintuplet involving all but the spin-up heavy-hole component. The spinor is determine from eq.(12) for $N = 1$ and has the form

$$
\Psi_{N=1}(\vec{r}, z) = \begin{bmatrix} H_0(\vec{r}/R_c)\, A_{1,1}(z) \\ 0 \\ H_0(\vec{r}/R_c)\, A_{3,1}(z) \\ H_1(\vec{r}/R_c)\, A_{4,1}(z) \\ H_2(\vec{r}/R_c)\, A_{5,1}(z) \\ H_1(\vec{r}/R_c)\, A_{6,1}(z) \end{bmatrix} \tag{23}
$$

Notice that there are two states in the conduction and three states in the valence bands and, for any future reference, we will label them as the Landau levels $N = 1$.

The five spinor components are determined by a 5x5 matrix equation where the row and the column defined by the D_{HH}^+ are not present and can be written as

$$\begin{bmatrix} D_{EL}^+ & \sqrt{2}R & 0 & 0 & -A_- \\ & D_{LH}^+ & A_+^* & W_+ & 0 \\ & & D_{EL}^- & -\sqrt{3}A_- & \sqrt{2}R \\ & & & D_{HH}^- & \sqrt{2}L_-^* \\ & & & & D_{LH}^- \end{bmatrix} \begin{bmatrix} A_{1,1}(z) \\ A_{3,1}(z) \\ A_{4,1}(z) \\ A_{5,1}(z) \\ A_{6,1}(z) \end{bmatrix} = E \begin{bmatrix} A_{1,1}(z) \\ A_{3,1}(z) \\ A_{4,1}(z) \\ A_{5,1}(z) \\ A_{6,1}(z) \end{bmatrix} \qquad (24)$$

Finally, the last type of solutions for the Landau levels will mix all six components of the spinor and we will label them as the Landau levels $N = 2$. The matrix equation is identical to eq.(12) after the proper substitution for the magnetic matrix elements in eq.(17).

It is worthwhile to point out that there exists strong interaction only between the levels in a given Landau ladder but, since the solutions of each of the four types of ladder are founded in separate matrix equations, two levels of different type cannot inter-act with each other.

Figure 4 shows the Landau levels associated with the spinors, $\Psi_{N=0}(\vec{r},z)$ and $\Psi_{N=1}(\vec{r},z)$ in the conduction band of the same quantum well described in figure 2. Notice that the tighter bound states to the quantum well show a linear dependence of the energy levels on the applied magnetic field but the top ones shows non-linear dependence for fields up to 5 Tesla. This can be understood since the envelope-functions for the lower states penetrate into the lateral semimagnetic barriers much less than the top ones. On the other hand,

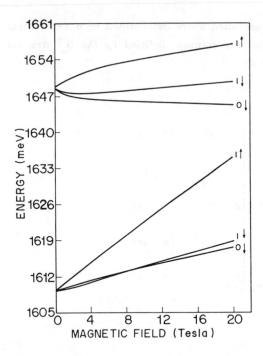

Fig. 4. Electron Landau levels for $N=0$ and $N=1$ in the semimagnetic QW $Cd_{(0.94)}Mn_{(0.06)}Te$ - $CdTe$ with L = 50 $\overset{o}{A}$. The spin splitting in the set $N=1$ is due to the parameter $N_1 = 1.37$ (atomic units) for CdTe. Notice the inversion of the normal ground state $N=0\downarrow$ to $N=1\downarrow$, at magnetic fields below 8 Tesla.

the low temperature experimental values of the thermal average of local spins, $<S_z(T,B)>$, shows clearly a spin saturation[2,13,18], $S_o(x)$, for magnetic fields above approximately 8 Tesla and, therefore, the magnetic potentials, Q_{cm} and Q_{vm}, will also shows the same saturation.

Figure 5 shows top five valence Landau levels for the spinor $\Psi_{N=0}(\vec{r},z)$ in the same quantum well of figure 4. The effect of the magnetic potential, Q_{vm}, is dominant in these states due to the large value of $(N_o\beta)$, the small value of the valence band-offset and the stronger subband mixing for valence states at small values of B_o. In order to give a clear example of the dominant effect of the magnetic potential on these levels, see in figure 6 the same top five valence states associated with the spinor $\Psi_{N=0}(\vec{r},z)$ in the same quantum well but now calculated with the band-offset only, that is, without the magnetic potential, Q_{vm}. It is apparent from fig.[4] and fig.[6] that the magnetic potentials makes the levels close to the continuum more tightly bound to the quantum well as B increases. Then, for values of

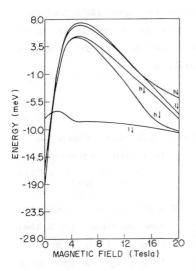

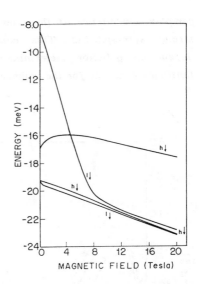

Fig. 5. - The top five hole Landau levels for N = 0 in the semimagnetic QW $Cd_{(0.94)} Mn_{(0.06)} Te - Cd Te$ with L = 50 Å.

Fig. 6. - The same top five hole Landau levels for N = 0 in the QW of fig.(5), only calculated with the crystal band offset or this sample, $E^V_v = -19.08$ meV.

magnetic field above the saturation of the thermal spin average, the barrier height remains constant and the penetration of the envelope-function components into these lateral barriers will not change further with increasing B. The complex and peculiar dependence of the Landau levels on B represent a clear competition between the localization due to the quantization in the z-direction and the localization due to the quantization into closed Landau orbits on the xy-plane.

The valence Landau ladder for the set of solutions associated with the spinor $\Psi_{N=2}(\vec{r}, z)$ is shown in figure 7. Notice the

488

complex dependence of these levels on the magnetic field and their strong spin-splitting. This complex dependence produces a B-dependent g-factor determined approximately by an effective Luttinger constant for the valence subbands as

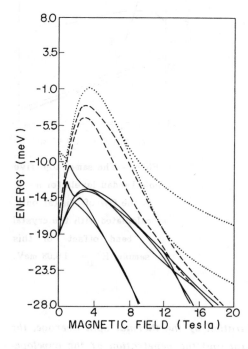

Fig. 7.- Hole Landau levels for $N = 2$, in the semimagnetic $Cd_{(0.94)} Mn_{(0.06)} Te - CdTe$ with L 50 $\overset{o}{A}$. Here, the dominant effect due to the potential, $V_m(B, T)$, is enhanced by the spin splitting of each level. The unusual g-factor for the hole Landau levels is caused by the subband mixing and different magnetic potential experienced by one heavy-hole and one light-hole state.

$$\kappa_{E_{ff}}(B,T) = \kappa - \frac{(N_o \beta) <S_z(B,T)>}{6 \mu_B B} \qquad (25)$$

where all terms were already defined.

In virtue of some uncertainties in the bulk parameters for Cd(Mn)Te, a direct comparison between the present calculation and the experimental values for transitions between these Landau levels should be taken carefully. However, figure 8 shows the calculated energy transitions from the top valence to the first electron Landau states satisfying the selection rules for the σ^+ and σ^- polarizations.

The experimental magnetooptical transitions for Cd(Mn)Te-CdTe multiple quantum wells reported by Warnock[6] et al. shows a quite similar dependence on the magnetic field and this agreement suggests that our present theoretical model for the calculation of the Landau levels in DMS heterostructures takes into account the most important features of the system which can be assigned to the antiferromagnetic spin-spin interaction between the ions and to the Heisenberg interaction of the spin of Bloch electrons with the average magnetization of the sample.

Fig.8.- Calculated optical transition from the Landau levels satisfying the selection rules for σ^+ polarization ($\Delta M_j = +1$) and for σ^- polarization ($\Delta M_j = -1$). The striking "features" at low fields, due to magnetic potential in our $\vec{k}.\vec{p}$ model, shows good agreement with magneto-exciton energies in Cd(Mn)Te-CdTe multiple quantum wells measured by Warnock[6] et al.

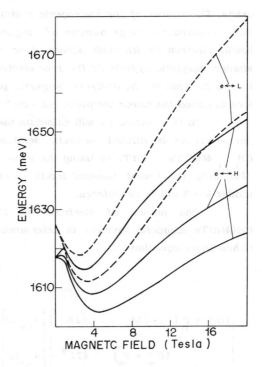

V - MAGNETIC POLARON EFFECT

The most striking feature ascribed to the magnetization of DMS samples is the magnetic polaron effect which is manifested by the motion of the Bloch electrons through the crystal even in the absence of an applied magnetic field. This so called free magnetic polaron (FMP) effect appears as the energy required by the Bloch electron to flip its spin from the alignment dictated by the atomic spin in a given site to the opposite alignment (antiferromagnetism) dictated by the spin of the nearest neighbor magnetic ion. This effect is not directional once there is a randomized substitution of the magnetic ion in that sublattice.

If there is some donor impurity at some other site of the crystal, the Bloch electron can bind to this impurity into hydrogen-like orbits. The radius of the hydrogenic orbits in DMS is very large and may encompass a large number of magnetic ions. The motion of the bound electron in its orbit across these magnetic ions presents the similar magnetic effects as the free electron and the energy required to flip its spin by the antiferromagnetic polarization of the magnetic ions is called the bound magnetic polaron (BMP) energy.

In this section we will calculate the MP and the BMP energies for electrons in diluted magnetic semiconductor quantum wells of $Cd_{(1-x)} Mn_{(x)} Te$ - $CdTe$ by using the same $\vec{k}.\vec{p}$ model discussed above. Also, only the lowest Landau levels coming from the solutions of ladder $N = 0$ will be considered.

The motion of electrons in CdTe quantum wells with Cd(Mn)Te magnetic barriers is determined from the solution of the Schrödinger equation

$$
\begin{bmatrix}
[D_{EL}^- + V_{EL}] & -\sqrt{3}A_- & \sqrt{2}R \\
& [D_{HH}^- + V_{HH}] & \sqrt{2}L_-^* \\
& & [D_{LH}^- + V_{LH}]
\end{bmatrix}
\begin{bmatrix}
A_{4,N}(z) \\
A_{5,N}(z) \\
A_{6,N}(z)
\end{bmatrix}
= E
\begin{bmatrix}
A_{4,N}(z) \\
A_{5,N}(z) \\
A_{6,N}(z)
\end{bmatrix}
\tag{26}
$$

where the zero of energy is the same as before, that is, we set it at the top of the bulk valence band of CdTe.

The potential terms in the diagonal include the magnetic potential, $V_m(B,T)$, the hydrostatic (δH) and the uniaxial (δU) strains produced by the lattice mismatch and, for the BMP case, the z-dependent part of the Coulomb potential of an electron bound to a donor impurity placed in the position z_i in the quantum well, $V_c(z,z_i)$. These terms are written as

$$V_{EL} = V_c(z,z_i) - \tfrac{1}{2} S_o(x) (N_o \alpha) V_m(B,T) + (1-\sigma) \delta H$$

$$V_{HH} = V_c(z,z_i) - \tfrac{1}{2} S_o(x) (N_o \beta) V_m(B,T) + \tfrac{1}{2} \sigma \delta H - \tfrac{1}{2} \delta U \qquad (27)$$

$$V_{LH} = V_c(z,z_i) - \tfrac{1}{6} S_o(x) (N_o \beta) V_m(B,T) + \tfrac{1}{2} \sigma \delta H + \tfrac{1}{2} \delta U$$

For quantum wells of width L, lateral magnetic barriers with concentration x, having spin saturation $S_o(x)$ and effective temperature $T_o(x)$ and, finally, if we include the local corrections, $B_x(q,z)$, to the external magnetic field, $\vec{B}=(0,0,B_o)$ then, the magnetic potential can be written as

$$V_m(B,T) = \Theta(|z| - \tfrac{L}{2}) \int_0^\infty dq \; B_S \left[S, \frac{g_{ion} \mu_B (B_o + B_x(q,z))}{K_B (T+T_o(x))} \right] \qquad (28)$$

where $\Theta(z)$ is the step-function, $B_S[S,Y]$ is the Brillouin function, as described before, and g_{ion} is the atomic g-factor of the magnetic ion. For an electron, moving in the xy-plane and bound to a quantum well with envelope-function component $A_{4,N}(z)$ and spin component $s_z = -\tfrac{1}{2}$, the local correction to the magnetic field is given by[20]

$$B_x(q, z) = - \frac{\alpha}{2} \frac{q}{g_{\text{Ion}} \mu_B} |A_{4,N}(z)|^2 \qquad (29)$$

It is important to notice that the local correction can be either symmetric or asymmetric in the z-coordinate measured from the center of the quantum well, which will depend on the geometry of the sample. Also, the magnetic potential acts differently for each spin component as, for example, it lowers the barrier for Bloch electrons with $s_z = -\frac{1}{2}$ and raises the barrier for those with $s_z = +\frac{1}{2}$. Therefore, there will be seen a clear spin splitting of the eigen-states.

There is experimental evidence of fluctuations in the concentration of magnetic ions along the interface of the heterostructure and also, due to the strong lattice mismatch, in the axial strains. These fluctuations will produce local changes in the total height of the lateral magnetic barriers and, certainly, may drive the carriers bound to the well closer to a certain interface region.

The z-dependent Coulomb potential is easily derived from the total potential seen by the carrier

$$V(\vec{r}, z-z_i) = - \frac{e^2}{\epsilon_0} \frac{1}{\sqrt{|\vec{r}|^2 + (z - z_i)^2}}. \qquad (30)$$

Since the envelope-function components have the associated harmonic oscillator states, $|N> = H_N(\vec{r}/R_c)$, the z-dependent Coulomb potential is obtained from the average of eq.(30) over the xy-plane, as $\int d\vec{r} <N|V(\vec{r}, z-z_i)|N>$. For the ground state of the harmonic oscillator, $|N=0>$, we find

$$V_c(z, z_i) = - \frac{e^2}{\epsilon_0 R_c} \left\{ 1 - erf(|\frac{z - z_i}{R_c}|) \right\} \qquad (31)$$

where $erf(x)$ is the error-function and ϵ_o is the low frequency limit for the dielectric function of the material forming the quantum well, that is, in the present calculation we will consider the CdTe.

  The solutions of the non-linear set of differential equations (26) with the local correction for the magnetic field is rather involved due to the dependence of $V_m(B,T)$ on the envelope-function components. They must be solved self-consistently, starting with a zero-order input potential solved for $B_x(q,z) = 0$. The output envelope functions are fed back into eq.(29) and its corresponding counterparts for holes. The new magnetic potentials are calculated as in eqs.(28) and, then, eq.(26) gives three new output eigen-values. The new output eigen-values are compared with the ones in the previous solution and the procedure continues until the eigen-values in the jth iteration differs from the eigen-values in the $(j-1)^{th}$ iteration by less than a priorly established accuracy. Usually the self-consistency takes less than 10 iterations for any good accuracy.

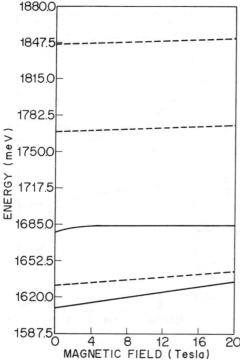

Fig.9.- Dependence on the applied field of the FMP energies of electrons in $Cd_{(1-x)}Mn_{(x)}Te$ - CdTe QWs of $L = 50$ Å. The solid lines are the FMP energies for $x = 6\% Mn$ and temperature $T = 2$ °K, the dashed lines are FMP energies for $x = 30\% Mn$ and temperature $T = 30$ °K.

Figure 9 *shows the lowest free magnetic polaron* (FMP) *energy as a function of the external magnetic field* B_o, *for electrons in* $Cd_{(0.94)}Mn_{(0.06)}Te$ - CdTe *quantum well with width* $L = 50 \overset{o}{A}$ *and at temperature* $T = 2.0 °K$ *(solid lines) and for electrons in* $Cd_{(0.70)}Mn_{(0.30)}Te$ - CdTe *quantum well with the same width L but at temperature* $T = 4.2 °K$ *(dashed lines). It apparent that the FMP effect in the conduction band of DMS heterostructures is weak. This is due to the large band offset (80%) where the envelope-function components can only penetrate less than* $10 \overset{o}{A}$ *into the lateral semimagnetic barriers. On the other hand, the spin-spin exchange for electrons,* α, *is a small parameter thus, the magnetic potential, even for large Mn concentrations and high magnetic fields, amounts to just few meVs. If compared to the barrier height due to the band offset* ($E_v^c \doteq 1200$ meV) *the contribution of the magnetic potential,* $V_m(B,T)$ *to the total potential seen by electrons,* V_{EL}, *in the DMS heterostructure is almost negligible.*

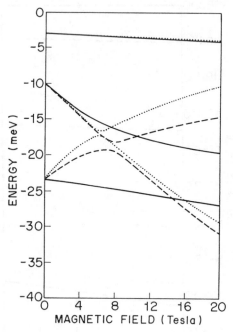

Fig.10. - Dependence on the applied field of the FMP and BMP energies of holes in $Cd_{0.94}Mn_{0.06}Te$ - Cd Te QW of $L = 50 \overset{o}{A}$ and $T = 2 °K$. The solid lines are three top FMP energies for heavy-hole and light-hole. The BMP energies for holes bound to a hydrogenic impurity placed in the center, $z_i = 0$, (dotted lines) and at $z_i = + 20 \overset{o}{A}$ (dashed lines). The Magnetic Polaron effect is enhanced by both the magnetic and the Coulomb potentials.

Once the effect of $V_c(z,z_i)$ on electrons is similar to the effect of $V_m(B,T)$, the BMP effect for electrons will not be discussed here.

The free and bound magnetic polaron effects for hole states are both more sensitive to the magnetic and Coulomb potentials. In fig.[10] it is shown the top three energies (solid lines) for valence states in the $Cd_{(0.94)}Mn_{(0.06)}Te - CdTe$ semimagnetic quantum well with $L = 50$ $\overset{o}{A}$ and at temperature $T = 2\ °K$. Here also, the more tightly bound state is weakly affected by the magnetic walls due to the same reasons mentioned for the electrons. However, the two lowest ones feel more the region inside the lateral walls since they are closer to the border of the quantum well and, therefore, their envelope-functions can reach a larger distance from the interface.

The BMP effect is also shown in the same figure. Here we have chosen two different positions to locate the acceptor ion binding the hole. The dotted line shows the energies associated to the BMP where the acceptor is placed in the center of the quantum well and, for the dashed lines, the acceptor is located at 5 $\overset{o}{A}$ from the wall on the right side. Two interesting points should be pointed out: (i) The z-dependent Coulomb potential is magnetic field dependent, as can be seen in eq.(31), through the cyclotron radius. For increasing magnetic field, the cyclotron radius decreases and the magnetic potential increases. Below 8 Tesla, this dependence causes the BMP energy for light-holes (middle lines) to become less tightly bound whereas the first excited heavy-hole state (lowest lines) becomes more tightly bound. Above 8 Tesla the opposite occurs and the light-hole state gets more bounded and the excited heavy-hole becomes more free. (ii) The conjugation of these two effects will cause first an expansion of the effective radius of the Landau cyclotron orbit and, then, followed by the normal contraction of the radius as B_o increases above 8 Tesla.

The overall aspect of this peculiar feature in DMS hetero-structure can be understood since the attractive (negative) z-dependent Coulomb potential competes with the repulsive (positive) magnetic potential up to magnetic fields where the saturation of the thermal average $< S_z(B,T) >$ occurs. Above this magnetic field only the Coulomb potential continues to increases and its effect is enhanced

496

when the impurity is placed closer to the wall. The extra asymmetry in the confining potential produced by impurities closer to the lateral wall certainly will lead to localization of carriers near the interface.

The aspects of the temperature of the sample and the strains associated to the lattice mismatch on the BMP energies will be presented next. Let us first neglect the strains and watch the temperature change. In figure 12-a we show the BMP energies of the same quantum well of figure 10, with the impurity placed at 5 $\overset{o}{A}$ from the right wall, for temperatures 2 °K (solid lines) and 80 °K (dashed lines). Is apparent that a warming up of sample causes two perceptible changes on the energies of the BMP states: (i) A shift in the saturation of the thermal average of spins to lower magnetic fields.

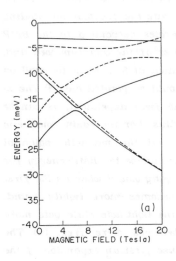 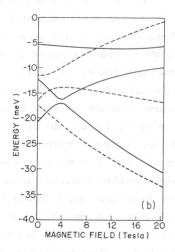

Fig. 11.-Changes caused by temperature and by uniaxial strain on BMP energy of holes in the QW of fig.(10) and bound to an impurity placed in the position $z_i = +20$ $\overset{o}{A}$. Fig.-(a):- The solid lines (dashed lines) show the BMP energies for temperature $T = 2$ °K ($T = 80$ °K) and no strain. Fig. - (b):- The solid lines (dashed lines) show the BMP energies for an uniaxial strain $\delta U = 5.0$ meV ($\delta U = 15.0$ meV) at fixed temperature $T = 2$° K.

(ii) The top BMP heavy-hole level, with spin component $s_z = -\frac{3}{2}$, *is shifted to negative energy (more free) and the BMP light-hole state and the first excited BMP heavy-hole states are shifted upwards (more bounded) for magnetic fields below 5* $\overset{o}{A}$. *Both effects are due to the temperature dependence of the magnetic potential inside the Brillouin function.*

Since the hydrostatic strains, δH, *only affects the fundamental band-gap, let us consider the same quantum well of figure 12-a at the temperature* $T = 2\,^{\circ}K$ *and apply two different uniaxial strains. Figure 12-b shows the BMP energies for 5 meV (solid lines) and 15 meV (dashed lines) uniaxial strains. Here also the mini-gaps between the different BMP states move to lower magnetic fields and the effect is assigned to the same competition between the Coulomb potential and the magnetic plus the strain potentials. If the uniaxial strain is smaller than 7 meV the sequence of BMP states is the same as before. However, above 7 meV the light-hole and the heavy-hole BMP states will crossover and the top state in fig.(12-b), represented by the dashed-line, has its origin in the light-hole state at* $B_o = 0$.

We hope to have gathered in this review several important and peculiar aspects of semimagnetic heterostructures. Also, a deeper discussion on the transport properties of DMS families should be taken in the near future.

VI - ACKNOWLEDGEMENTS

The author acknowledges a partial financial support from Conselho Nacional de Desenvolvimento Científico e Tecnológico (C N Pq) He is also indebted to V. A. Chitta, A. M. Cohen and M. H. Degani for valuable contributions. The calculations were performed in the VAX-780 system at the Instituto de Física e Química de São Carlos-Universidade de São Paulo.

498

VII - REFERENCES

[1] Pajaczkowska, A.; Prog. Cryst. Growth and Charact. 1, 289 (1978).

[2] Furdyna, J. K.; J. Appl. Phys. 53, 7637 (1982);

[3] Awschalom, D. D., Grinstein, G., Yoshino, J., Munekata, H. and Chang, L. L. ; Surf. Sci. 196, 649 (1988).

[4] Novak, M. A., Symko, O. G., Zheng D. J. and Oserroff, O.; Phys. Rev. B33, 6391 (1986)

[5] Brandt, N. B. and Moshchalkov, V. V.; Adv. Phys. 33, 193 (1984).

[6] Warnock, J., Petrou, A., Bicknell, R. B., Giles-Taylor, N. C., Blancks, D. K. and Schetzina, J. F.; Phys. Rev. B32, 8116 (1985).

[7] Zang, X. C., Chang, S. K., Nurmikko, A. V., Kolodziejski, L. A., Gunshor, R. L. and Datta, S.; Phys. Rev. B31, 4056 (1985).

[8] Kane, E. O.; Semiconductors and Semimetals, Vol. 1, ed. by R. K. Williardson and A. C. Beer (Academic Press, N.Y. , 1966), pg. 75.

[9] Dresselhaus, G.; Phys., Rev., 100, 580 (1955).

[10] Löwdin, P. ; J. Chem. Phys. 19, 136 (1951).

[11] Sham, L. J. and Nakayama, M.; Phys. Rev. B20, 734 (1979).

[12] Marques, G. E. and Sham, L. J.; Surf. Sci. 113, 131 (1982).

[13] Luttinger, J. M.; Phys. Rev. ,102, 1030 (1956).

[14] Galaska, R. R. and Kossut, J.; Springer Lectures Notes in Physics, vol. 132 (1986)(Berlin: Springer), p. 245

[15] Weiler, M. H.; Semiconductors and Semimetals, Vol. 16, ed. by R. K. Williardson and A. C. Beer (Academic Press, N.Y. , 1981), p. 119

[16] Pollack, F. H. and Cardona M. ; Phys. Rev. 172, 816 (1986).

[17] Heiman, D., Becla, P., Kershaw, R., Ridgley, R., Dwight, K., Wold, A., and Galaska, R. R.; Phys. Rev. B34, 3961 (1986).

[18] Lawetz, P.; Phys. Rev. B4, 3460 (1971).

[19] Gaj, J. A., Planel, R. and Fishman, G.; Solid State Commun. 29, 435 (1979).

[20] Wu, J., Nurmikko, A. V., Quinn, J. J.; Solid State Commun. 57, 853 (1986)

OPTICAL PROPERTIES OF (Zn,Mn) AND (Cd,Mn) CHALCOGENIDE QUANTUM WELLS AND SUPERLATTICES

W. Heimbrodt and 0. Goede

Humboldt-Universität zu Berlin, Sektion Physik

Invalidenstraße 110, 1040 Berlin, GDR

Contents

1. Introduction

(Zn,Mn) and (Cd,Mn) chalcogenide mixed crystals are typical and the most extensively studied broad-gap diluted magnetic ('semimagnetic') semiconductors (DMS). Their remarkable physical properties are characterized by the combination of a usual semiconductor mixed-crystal behavior with the special properties caused by the half-filled 3d-shell of the Mn^{2+} cations. For both basic and applicative reasons they found a continuously growing interest (see, e. g., survey articles /1-3/), which has been further stimulated by the successful preparation of quantum well structures and superlattices on the basis of these materials. The physical properties of the semimagnetic (Zn,Mn) and (Cd,Mn) chalcogenide mixed crystals are based on the strong s,p-d exchange interaction between electron or hole band states and the Mn^{2+} 3d-electron states. It leads to a giant increase of magneto-optical effects as exciton Zeeman splitting, Faraday rotation and donor spin- flip Raman scattering, which formally can be described by a Mn^{2+} ion-induced enhancement of the effective g-factor of these materials by up to two orders of magnitude. Furthermore, principally new interesting effects occur also in zero magnetic field as bound magnetic polaron formation. A possible application of these materials for magnetic-field tunable opto-electronic devices may be supported by the development or high- T_C superconductivity magnets.

Because of the tunability of their lattice parameters and their energy gaps in a wide range, the (Zn,Mn) and (Cd,Mn) chalcogenides are excellent candidates for the preparation of quantum wells and superlattices with the aim of 'band-gap engineering'. The semimagnetic properties bring new physical effects into the already rich and exciting spectrum of phenomena of quantum wells and superlattices.

In the present article the actual knowledge on the optical properties of such microstructures is reviewed. The first section is dedicated to a small extent to the preparation and structure characterization of (II,Mn)VI microstructures. Then the conventional valence and conduction band related properties are discussed, which are not essentially influenced by the Mn 3d-states. In section 5, the special properties due to the presence of Mn^{2+} are considered which are caused by the exchange interaction between the s-like conduction or p-like valence band states and the Mn 3d-states. Furthermore, the phonon spectra and problems of magnetic phase transitions in multiple-quantum-well (MQW) structures are discussed.

2. Preparation and Structure Characterization

The first successful preparation of II-VI semiconductor super-lattices and multiple-quantum-well structures on the basis of the systems CdTe/(Cd,Mn)Te[4,5] and ZnSe/(Zn,Mn)Se[6] was reported in

1984 and 1985, respectively. A Schematic cross section of a typical CdTe/(Cd,Mn)Te structure is shown in Fig. 1. It consists usually of a CdTe buffer layer grown on the substrate.and the multiple-quantum-well region embedded between (Cd,Mn)Te cladding layers. Usually the structures were grown on (001) or (111) GaAs substrates by molecular beam epitaxy (MBE). Recently CdTe/(Cd,Mn)Te quantum wells were also successfully grown on the better lattice-matched substrates (001) InSb[7], (Zn,Cd)Te[8] and CdTe [11]. Details of the MBE growth technique do not differ significantly from the preparation of II-VI-based structures in general and we shall, therefore, not discuss them here. Also the atomic layer epitaxy (ALE) has been emerging as a new growth technique for (II,Mn)VI microstructures of high quality[9].

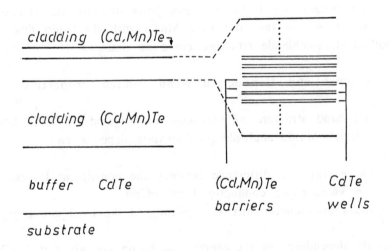

Fig.1. Schematic diagram of a CdTe/(Cd,Mn)Te multiple quantum-well structure.

The above mentioned superlattices and MQW's are of remarkably good structural quality with sharp layer boundaries shown by X-ray diffraction measurements and transmission electron microscopy (TEM)[4-6,10].

An interesting new method to grow doped CdTe/(Cd,Mn)Te-superlattices is the photoassisted MBE (PAMBE), in which the substrate is illuminated during the deposition process[11]. The incident photon beam provides sufficiently high energy to influence surface chemical reactions during the film growth. It is worthwhile to note, that this method allows substrate temperatures Ts< 200°C being considerably lower than the 300-400°C used by conventional MBE. PAMBE obviously can activate dopants such as In by a photochemical reaction, leading to substitutional donors in DMS-superlattices[11].

Another interesting result on the basis of MBE growth technique reported[12] was the preparation of ZnTe/MnTe layer structures on a (001) GaSb substrate, closely lattice-matched to ZnTe. Whereas bulk MnTe crystallizes only in the noncubic NiAs structure, in the case of the MBE layer for the first time MnTe with the, upto this time hypothetical, zincblende structure could be obtained.

3. Valence and Conduction Band related Properties

The real band structure of the most investigated CdTe/(Cd,Mn)Te and ZnSe/(Zn,Mn)Se multiple-quantumwells depends on

(a)   the energy gap difference between the constituent layers,
(b)   the valence-and conduction-band offset,
(c)   the strain-induced shift and splitting of the band gap energies.

From the dependence of the energy gap E_g on x_{Mn} given in Fig.2, it follows that (Cd,Mn)Te and (Zn,Mn)Se act as the barrier layer, whereas CdTe and ZnSe, respectively, are the quantum wells. The valence- and conduction-band offset in the considered systems is always influenced by strain effects. As follows from the dependence of the lattice constants of (Cd,Mn)Te and (Zn,Mn)Se on the Mn concentration x_{Mn} (see Fig.3), substantial strains exist in these structures which seem to accommodate totally the lattice mismatch[13,14]. Taking into account the different lattice constants of quantum well and barrier material, the successive layers in the superlattice are alternately subjected to biaxial tensile and compressive stresses, respectively. Such biaxial stress fields can be considered as a sum of a hydrostatic and a uniaxial stress parallel to the growth direction whose influence on band gap or exciton energies is well known.

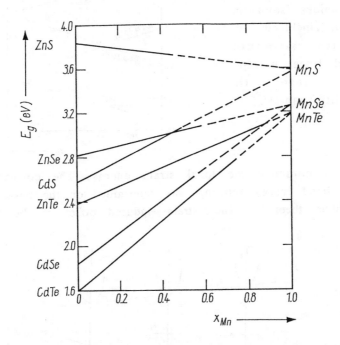

Fig.2. Energy gap, E_g, as a function of the Mn concentration x_{Mn} for the (Zn,Mn) and (Cd,Mn) chalcogenides at low temperatures (T≈4 K) in linear approximation (solid lines). Dashed lines: extrapolation into the miscibility gap/3/.

In Fig. 4 the strain dependence of the band structure is shown schematically for a zincblende-type semiconductor. The valence band degeneracy at k=0 caused by the cubic symmetry is removed. As an example the strain parameters for (111) CdTe/(Cd,Mn)Te superlattices were calculated[15] assuming the distortion of the

504

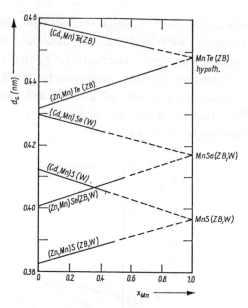

Fig.3. Mean cation-cation distance, d_c, as a function or the Mn concentration x_{Mn} for the (Zn,Mn) and (Cd,Mn) chalcogenides (solid lines). ZB: zincblende, W: wurtzite structure. Dash-ed lines: extrapolation into the miscibility gap-/3/.

layers to minimize the local strain energy. The corresponding valence band energy splitting was determined on the basis of the Bir - Pikus theory[16]. The strain - induced offset of the heavy -

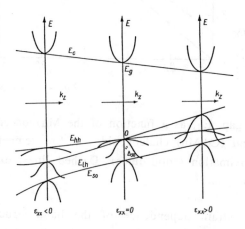

Fig.4. Scheme of the band structure for a zincblende semiconductor under compressive ($\varepsilon_{xx} < 0$) and tensile ($\varepsilon_{xx} > 0$) strain.

hole(hh) band in this case is negligibly small (≈ 1 meV), whereas the corresponding light-hole(lh) band offset is about 50 meV for $x_{Mn} = 0.45$[15].

An interesting method to determine the resulting band offset of a real CdTe/(Cd,Mn)Te quantum-well structure is given in /17/. The magnetic field induced shift of the quantum well emission energies was measured, which is caused by the magnetic field induced change of the barrier band gap (see /3/). Using the known exchange interaction parameters $N_0\alpha$ and $N_0\beta$ of the valence and conduction band, respectively, it was possible to determine the valence band offset by a fitting procedure in the framework of a Kronig-Penney calculation. For $x_{Mn} = 0.25$ it was confirmed that the structures are of type I and a valence band offset in the range 30 - 55 meV was found.

A more sophisticated evaluation of the experimental results on the basis of a variational calculation was given in /25/ including changes of the exciton binding energy and the large Zeeman splitting of the exciton transitions. The calculation was performed for CdTe/(Cd,Mn)Te with $x_{Mn} = 0.24$ in a single-quantum-well limit. The Zeeman splitting is due to the penetration of the exciton envelope function into the (Cd,Mn)Te barrier layer (see chapter 5). Agreement between theory and experiment is obtained and a type I valence band offset of 25 meV is deduced for the heavy-hole. This correspond to a conduction to valence band offset ratio of about 14:1. It should be taken into mind that this value depends on the strain situation and, therefore, the difference to the values between 10:1 and 7:1 obtained in /17/ could be explained.

For ZnSe/(Zn,Mn)Se superlattices the problem of type I or II behavior is not convincingly solved up to now. In any case a small valence band offset should be expected compared to the conduction band offset. As follows from Fig. 3, in ZnSe/(Zn,Mn)Se superlattices the ZnSe layers are subjected to biaxial tensile strain contrary to the CdTe layers in CdTe/(Cd,Mn)Te superlattices and, therefore, bandgap narrowing is expected. A corresponding redshift of the luminescence peaks in comparison to (unstrained) ZnSe epilayers was observed [6,13]. A schematic band diagram for a CdTe/(Cd,Mn)Te multiple-quantumwell structure including strain effects is given in Fig. 5.

The exciton energies of the ground state $E_{exc}^{lh,hh}$ for such a multiple-quantum-well structure are given by

$$E_{exc}^{lh,hh} = E_g(CdTe) + E_1^e + E_1^{lh,hh} - \varepsilon_b^{lh,hh} \qquad (1)$$

The electron (e) or hole (hh,lh) state energies, $E_n(k)$, can be calculated separately by a Kronig-Penney-calculation[18],

$$\frac{a^2-b^2}{2ab} \sinh(aL_b)\sin(bl_w) + \cosh(aL_b)\cos(bL_w) = \cos(kD)$$

$$\text{where} \qquad a^2 = \frac{2m}{\hbar^2}(V-E) \qquad (2)$$

$$b^2 = \frac{2mE}{\hbar^2}$$

$D = L_w + L_B$, m-effective mass, V-effective potential.

As the cosine term on the right-hand side of this dispersion relation varies between +1 and -1, only those values of E are allowed, for which the left-hand side falls in this range.

ε_b in equation (1) is the exciton binding energy. As mentioned above, the valence band offset in CdTe/(Cd,Mn)Te and ZnSe/(Zn,Mn)Se is rather small, leading to the interesting situation that a quasi two-dimensional electron interacts with a quasi three dimensional hole. The exciton binding energy for this case is calculated by generalizing the variational approach-normally used to study excitons in GaAs/(Ga,Al)As quantum wells[26,27]. The binding energy of a Wannier exciton in quasi two-dimensional CdTe quantum wells is enhanced by about a factor of two compared to the bulk.

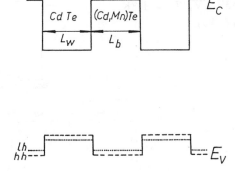

Fig.5. Schematic diagram of the conduction and valence band energies for a CdTe/(Cd,Mn)Te mult-iple-quantum-well struc-ture (-------heavy hole, ... light hole)

An experimental verification of the splitting between the heavy-hole and light-hole valence bands in the case of a CdTe/(Cd,Mn)Te MQW structure with $x_{Mn} = 0.06$ was provided in /19/ The peaks in the emission spectrum shown in Fig. 6a and indicated by X_{hh} and X_{lh} are identified as the heavy- and light-hole exciton bands, respectively. The lower energy peaks are attributed to impurity - bound excitons.

Clear evidence for exciton-confinement effects which are mainly due to electron confinement is given in Fig. 7, showing photoluminescence spectra and free-exciton energies from reflection measurements for a CdTe bulk sample and two (111) CdTe/(Cd,Mn)Te MQW structures with different well thicknesses [14,20,21]. In the case of a large well thickness the same position for

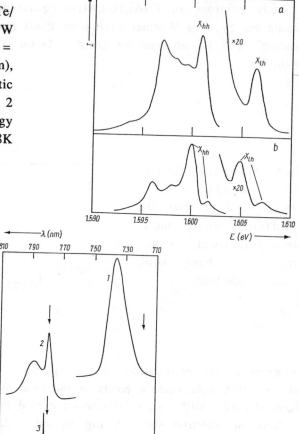

Fig.6. Luminescence spectrum of a CdTe/(Cd,Mn)Te MQW (x_{Mn} = 0.06; L_W = 15.3, L_b = 16.2 nm), a) in zero magnetic field and b) at B= 2 T. Excitation energy 2.539 eV, T = 1.8K (Ref. 19).

Fig.7. Photoluminescence spectra for two (111) CdTe/(Cd,Mn)Te MQW structures (x_{Mn} = 0.26; L_W = 7.1 (1) and 67 nm (2) in comparison with a CdTe bulk sample (3). The arrows denote he free-exciton energies determined from reflection measurements. Excitation energy 1.970 eV, T = 1.8 K (Ref.14).

the free-exciton line is found in the reflection and emission spectra as in the bulk sample. On the other hand, for a well thickness in the order of the magnitude of the exciton Bohr radius in CdTe ($a_B \approx 6.0$ nm) a remarkable confinement-induced shift of the lines to higher energies is observed. In the emission spectrum only a broad band is found which can be attributed to bound exciton transitions. For ZnSe/(Zn,Mn)Se superlattices with very narrow ZnSe wells (2.5 to 5.0 nm; $a_B \approx 2.8$ nm) also a net blue-shift of the exciton lines was found [6,13], because in this case the confinement effect exceeds the strain-induced red shift discussed at the beginning of this article.

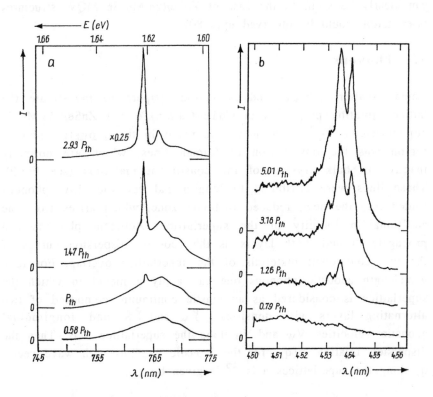

Fig.8. Stimulated emission spectra for MQW structures at four different power levels. a) CdTe/(Cd,Mn)Te (x_{Mn} = 0.45, T = 25 K, P_{th} = 1.35 x 10^4 W/cm^2) (after /23/); b) ZnSe/ (Zn,Mn)Se (x_{Mn} = 0.33 T = 5.5K, P_{th} = 2.0 x 10^5 W/cm^2) (after /24/)

Both in CdTe/(Cd,MnTe)[22,23] and ZnSe/(Zn,Mn)Se[24] MQW's also stimulated emission was observed, indicating the progress in the material preparation technique. After removing the GaAs substrate by a special etching procedure [10,23] the remaining multilayer structure was cleaved into small bar pieces which were pressed onto indium layers on a copper heatsink and covered by a sapphire window. Laser pulses were used for optical pumping. In Fig. 8a,b the emission spectra of CdTe/(Cd,Mn)Te and ZnSe/(Zn,Mn)Se lasers are shown for various pumping intensities. For increasing pump power above the threshold P_{th} the evolution of a mode structure can clearly be seen. In the case of ZnSe/(zn,Mn)Se MQW structures laser action could be observed upto 80K.

4. Phonons

Raman scattering was used in various papers to investigate the lattice vibration properties in CdTe/(Cd,Mn)Te and ZnSe/(Zn,Mn)Se superlattices [28-31]. An interesting result in the <u>acoustic</u> phonon region was the observation of Raman lines, which are normally inactive in bulk crystals of the constituent materials (see Fig.9). These lines were attributed to longitudinal acoustic (LA) phonons folded into the new, reduced Brillouin zone, which arises from the additional periodicity of the superlattice. Acoustic phonons can propagate through both layers as the acoustic dispersion curves of the two constituent materials of the superlattice overlap. For long wavelength acoustic phonons one can apply a model in which the superlattice is considered as an elastic continuum composed of two alternating layers with densities ρ_w and ρ_b and longitudinal acoustic velocities V_w and V_b along the superlattice axis. Then the dispersion relation $\omega(q_z)$ for the acoustic phonons with wave vector q_z along the superlattices axis [32] is given by

$$\cos(q_z D) = \cos\frac{\omega L_w}{V_w} \cos\frac{\omega L_b}{V_w} - (1+\delta) \sin\frac{\omega L_w}{V_w} \sin\frac{\omega L_b}{V_b} \quad (3)$$

with $D = L_w + L_b$; $\delta = \dfrac{1}{2}\dfrac{(\rho_w V_w - \rho_b V_b)^2}{(\rho_w \rho_b V_w V_b)}$

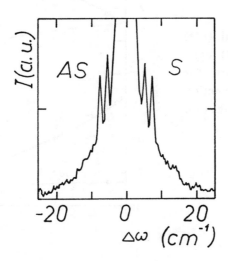

Fig.9. Stokes (S) and anti-stokes (AS) components of the folded longitudinal acoustic phonons in a CdTe/ (Cd,Mn)Te superlattice (x_{Mn} =0.24, L_w = 7.1 nm, L_b = 12.8 nm (Ref. 29)

To discuss the <u>optical</u> phonon behavior of a superlattice it is useful to consider the features of the optical phonons in the constituent bulk materials. (Cd,Mn)Te exhibits a two-mode behavior shown in Fig 10. Besides the CdTe-like zone-centre LO-TO mode, for small x_{Mn} a local mode due to Mn impurities appears which splits into a MnTe-like LO-TO mode with increasing Mn concentration. (Zn,Mn)Se exhibits a mixed-mode behavior due to the smaller mass difference between Zn and Mn (see Ref.3).

512

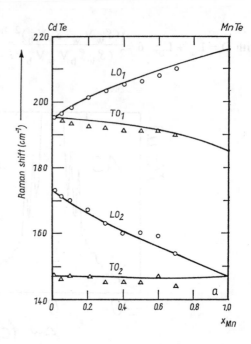

Fig.10. Longitudinal (LO) and transverse (TO) optical-phonon energies for (Cd,Mn)Te as a function of the Mn concentration X_{Mn}.

In the superlattices different kinds of optical phonons were observed in dependence of the overlap of the LO-TO dispersion curves of the constituent layers. If there is no overlap the phonons are 'confined' to their respective layers. In the case of matching, 'propagating' phonon modes can be found. An excellent overview about the confined and propagating phonons in (Cd,Mn)Te/(Cd,Mn)Te and ZnSe/(Zn,Mn)Se superlattices is given in /29/. In Fig. 11 the Raman spectra from LO phonons in a CdTe/(Cd,Mn)Te superlattice are given. Besides the propagating (P) and (Cd,Mn)Te-confined LO phonons (C_b), respectively, the CdTe-confined LO phonon (C_W) can be seen with quantized energies (m = 2,4,6). These phonon energies are equivalent to $\omega_{LO}(q)$ in CdTe-bulk material for wave vectors

$$q_m = m\Pi/(L_w + \frac{a}{2}) \qquad (4)$$

where a is the lattice parameter. It should be mentioned that in a superlattice the intensities of the various LO Raman lines can be selectively enhanced by matching the incident photon energy with electronic transitions in the well or the barrier[29,30].

The existence of interfaces results in new vibrational excitations in addition to the bulk vibrational excitations. These <u>interface</u> modes propagate along the interface planes and are highly localized. The amplitude of an interface mode decays exponentially in the direction perpendicular to the layer plane. In (Cd,Mn)Te/(Cd,Mn)Te superlattices such interface optical phonons are identified in the LO phonon region of Raman spectra[29].

Generally it must be taken into account, that the existence of strain in the considered superlattices lead to a shift of the phonon frequencies. Especially for ZnSe/(Zn,Mn)Se superlattices this shift cannot be neglected, due to the stronger dependence of the lattice constant on x_{Mn} compared to the (Cd,Mn)Te mixed crystal system.

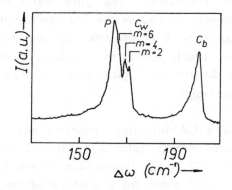

Fig.11. Raman spectra from optical phonons in a CdTe/(Cd,Mn)Te superlattice with $x_{Mn} = 0.25$ ($L_w = 5.7$ nm, $L_b = 9.6$ nm; T = 80K) (Ref. 29)

5. Special Properties due to s,p-Mn^{2+}(3d) Exchange Interaction

As in the diluted magnetic semiconductors the magneto-optical properties of the considered quantum well structures are essentially determined by the Mn 3d-states. The giant Zeeman splitting of the exciton states in the bulk DMS materials is a result of the s,p-d exchange interaction between the electron/hole states and the Mn^{2+} ion 3d states within the exciton orbit. The maximum energy difference ΔE_{exc} between σ^+ and σ^- components as a function of the magnetic field H is given by

$$E_{exc} = x_{Mn} \, N_0 \, (\alpha - \beta) \, |<S_z>|, \qquad (5)$$

with
$$<S_z> = a \, (x_{Mn})S \, B_{5/2} \frac{g\mu_B \, SH}{k(T + \Theta x_{Mn})}$$

where α and β are the exchange interaction parameters for the conduction and valence band, respectively, N_0 is the number of unit cells per cm^3, $<S_z>$ is the mean spin density, $B_{5/2}$ is the Brillouin function for $S = 5/2$, μ_B is the Bohr magnetron, k_B is the Boltzman constant and T the temperature. The Curie-Weiss temperature Θ and the factor a strongly depend on the Mn concentration x_{Mn}. The scaling factor $a(x_{Mn})$ is a consequence of antiferromagnetically ordered Mn clusters which reduce the resulting magnetic moment (see /3/ for further discussion). Typical values of $N_0\alpha$ and $N_0\beta$ are 0.2 and 1 eV, respectively[3].

In the case of CdTe/(Cd,Mn)Te-type quantum well structures the magnetic-field induced energy splitting of the electron/hole states in the semimagnetic (Cd,Mn)Te barrier due to the s,p-Mn^{2+} (3d) exchange interaction also yields to a splitting of the electron/hole states in the well.

This effect can be described by a potential well model (see chapter 3) with different effective potential barriers for the various spin states corresponding to (4). The splitting strongly depends on the penetration depth of the well states into the barrier.

In Fig. 6a the measured luminescence spectrum is shown for a CdTe/(Cd,Mn)Te MQW with small x_{Mn} and, therefore, <u>shallow</u> quantum wells in a magnetic field applied perpendicular to the superlattice layers. A splitting of both the X_{lh} and X_{hh} exciton lines into two components is observed. As shown in Fig. 12 the saturation values of the splitting energies are much smaller than for (Cd,Mn)Te bulk crystals with the same Mn concentration as in the barriers. Although the excitons are mainly confined within the CdTe quantum wells, the exciton envelope function considerably extends into the (Cd,Mn)Te barrier layers leading to exchange interaction with the Mn ions. The value of the splitting energy can be used to determine the part of the exciton wave function in the barriers [8,19].

Similar exciton energy splitting in a magnetic field are found for ZnSe/(Zn,Mn)Se MQW's, too /3/. In CdTe/(Cd,Mn)Te MQW's with deep quantum wells, i. e. sufficiently large x_{Mn}, only a small shift and practically no splitting of the free-exciton energy is observed in an external magnetic field [14,19,21]. In this case the excitons are so strongly confined in the wells that the exchange interaction with the Mn ions in the barriers is negligible and only a diamagnetic exciton energy shift occurs.

As already mentioned in (111) CdTe/(Cd,Mn)Te superlattices with sufficiently narrow wells a broad emission band below the free exciton energy dominates (see Fig. 7). This band was ascribed to localized-exciton transitions as concluded from the temperature dependence of the luminescence intensity[13,20,21]. In a magnetic field this band exhibits a strong shift to lower energies and a circular polarization, obviously due to exchange interaction with the

516

Fig.12. Exciton peak energies as a function of the
applied magnetic field for
a CdTe/(Cd,Mn)Te MQW
(x_{Mn}= 0.06); T = 1.8K (Ref.
19).

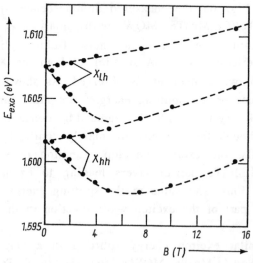

Mn ions in the barriers. As this effect occurs also in the case of deep
quantum wells, a binding of the excitons close to the hetrointerfaces
must be assumed in agreement with the quasi-two-dimensional
character of these bound excitons, concluded from the observed
anisotropic behavior in a magnetic field[20,21,34,43]. Mn
concentration fluctuations in (Cd,Mn)Te were suggested to induce
local fluctuations of the strain and, as a result, of the potential-well
depth in thin CdTe regions near to the interfaces which then bind
excitons.

If (Cd,Mn)Te is used as quantum-well material, the s,p-d exchange
interaction-enhanced exciton-energy shift is expected to be as large

as for bulk crystals. Furthermore, a primary shift of the exciton energy into the visible region can be realized by a sufficiently large x_{Mn}-value. The successful preparation of (Cd,Mn)Te/(Cd,Mn)Te superlattices and the observation of stimulated emission in these structures were reported in[22,35].

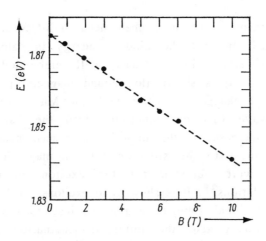

Fig.13. Magnetic-field dependence of the energy of the stimulated emission peak for the MQW laser structure $Cd_{0.81}Mn_{0.19}Te/Cd_{0.64}Mn_{0.36}Te$; T = 1.9 K (Ref.36).

Recently also the possibility of an effective tuning of the energetic position of the laser line by a magnetic field could be demonstrated for these structures (see Fig.13). The magnetic-field-induced line-shift is found to be linear with the slope $\delta E/\delta H = 3.4$ meV/T which is somewhat smaller than for excitons in bulk (Cd,Mn)Te having the same x_{Mn} as the wells. However, no saturation of the shift occurs up to 10T. The bound-magnetic-polaron effects established for bulk crystals of diluted magnetic semiconductors[3] can also be assumed to influence the optical and magneto-optical properties of the considered superlattices or multiple-quantum-well structures. Especially for CdTe/(Cd,Mn)Te MQW structures the role of bound magnetic polarons was discussed in connection with the measured cw and time-resolved luminescence spectra[14,34]. The observed temperature dependence of the exciton energy in a magnetic field, the line-shift in the time resolved spectra, the decrease of the exciton lifetime, and a pronounced linewidth narrowing in the CW luminescence spectra with increasing magnetic field were attributed to bound-magnetic-polaron effects.

The theoretical treatment of these phenomena in MQW structures is quite complicated. In /37/ the binding energy is calculated for shallow-impurity states in a CdTe well embedded between (Cd,Mn)Te barriers. It is shown that bound magnetic polarons are formed by s,p-d exchange interaction with the Mn^{2+} ions because the impurity wave function penetrates into the barriers. The binding energy increases if the impurity is placed nearer to the interface, in contrast to the situation for non-magnetic wells[37]. Similar results were obtained for the exciton energies after variational calculations[38]. It is shown that excitons can be bound in a quasi-two-dimensional state in a region close to the interface. Taking into account also the potential fluctuations near the interface, the exciton-energy shift in a magnetic-field is calculated in good agreement with the experimental results[39].

6. Magnetic Phase Transition

A further subject recently under investigation are the magnetic properties of layered DMS structures. The short-range antiferromagnetic interaction between the neighboring Mn ions leads to a magnetic phase transition from paramagnetism to spin-glass ordering for $x_{Mn} \leq 0.6$ and to antiferromagnetic ordering for $x_{Mn} \geq 0.6$ as the temperature is lowered. An interesting question from theoretical and experimental point of view is, how this behavior is varied as the dimensionality is reduced. In /41,43/ the ac magnetic susceptibility versus temperature was measured in (Cd,Mn)Te layers with various layer thicknesses. In a sample with a (Cd,Mn)Te layer with a thickness of L_b = 8.6 nm between CdTe layers (see Fig. 14a) a well defined sharp peak in the susceptibility curve was found at a temperature close to the known spin-glass transition temperature of the bulk material. For decreasing thickness the peak becomes broadened and for the thinnest layers (Fig. 14b) the peak vanishes, indicating a paramagnetic behavior

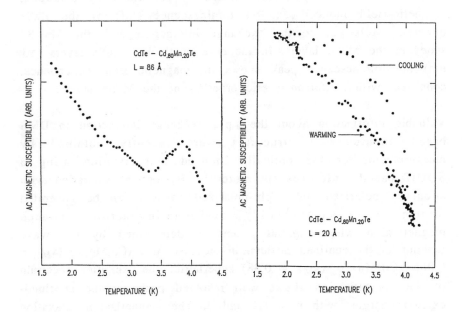

Fig.14. Magnetic susceptibility of CdTe/(Cd,Mn)Te superlattices with 8.6 nm CdTe layers and magnetic layer thicknesses L_b = 8.6 nm (a) and 2.0 nm (b)

even at the lowest temperatures. The results are taken to indicate that spin-glass order can not be found in a two-dimensional system. The appearance of a hysteresis in the thinnest layers was discussed on the basis of trapped short-range spin glass order, not responding to external fields[43].

Different magnetic behavior between a quasi two-dimensional and a three-dimensional case was also reported for MnSe[44,45]. Bulk zincblende MnSe shows an antiferromagnetic ordering of fcc-type III below the Neél temperature $T_N \approx 110K$. Ultrathin MnSe layers $L_W \le 1$ nm embedded between ZnSe layers exhibit paramagnetic behavior at low temperatures, which is confirmed by a strong, magnetic-field induced exciton luminescence shift of the ZnSe qunntum wells due to the exchange interaction with the Mn 3d states in the MnSe layers. In the case of thicker MnSe layers (≥ 3 nm) the luminescence peak shows no magnetic field dependence, consistent with antiferromagnetic ordering of the MnSe layers.

Valuable information about the s,p-d exchange interaction in DMS based quantum-well structures were recently obtained by measurements of the optically induced magnetization using a SQUID-based microsusceptometer[40]. By optical excitation with circularly polarized light spin-polarized carriers can be generated leading to an s,p-Mn^{2+}(3d) exchange-interaction induced magnetizaion whose spatial extent is determined by the wave function of the confined carriers. In the case of a (Cd,Mn)Te ($x_{Mn} = 0.065$)/(Cd,Mn)Te ($x_{Mn} = 0.38$) superlattice the excitation spectrum of the magnetization shows well resolved peaks at the confined-exciton energies with n = 1,2 and 3. The magnetization-relaxation lifetime is found to decrease dramatically for less-confined excitons in excited states or in smaller wells because of the shorter local magnetic relaxation time in the barrier material with the higher Mn concentration.

7. Concluding Remarks

The application possibilities of the considered (Zn,Mn) and (Cd,Mn) chalcogenide quantum wells and superlattices are not the subject of this review. It should be pointed out, however, that the observed high efficiency of the low-temperature exciton luminescence being two or three orders of magnitude larger than for films or bulk samples (see Refs.10 & 46) make such structures interesting for optoelectronic devices as magnetically tunable lasers. Other interesting devices using the outstanding magnetooptical properties were proposed[2,47,52]. Special structures can be used for magnetic-field-induced tuning of resonant tunneling or drastic changes of conductivity using the strong exchange interaction between band electrons and localized Mn^{2+} d-electrons. Furthermore, it should be mentioned, that the possibility of MBE preparation of (II,Mn)VI epitaxial structures on GaAs may be of importance in future attempts for a monolithic integration of layered DMS structures with GaAs based devices.

Although a plenty of papers are published, many basic physical problems are unsolved or must be investigated more systematically. Recent papers give examples for new directions of investigations. In /48/ it has been demonstrated that (Cd,Mn)Se and (Cd,Zn)Se can be stabilized in the cubic zincblende structure on (001) GaAs using MBE. This led to the preparation of novel superlattices on the basis of (Cd,Mn)Se/(Cd,Zn)Se[49]. This material combination allows the preparation of lattice-matched wells and barriers, which is of importance to separate exactly the influence of strain on the excitonic energies, offset etc. Furthermore, ZnSe/(Cd,Mn)Se quantum-well structures would be the first DMS structures with diluted magnetic quantum wells and non-magnetic barriers.

Another exciting idea was realized[50,51] by preparing ZnSe/ (Zn,Mn)Se-MQW's with a special δ-doping of a Te monolayer inside the ZnSe quantum wells The Te isoelectronic centre is known as an effective trapping centre for excitons. The penetration of the bound exciton wavefunction can be measured by the degree of circular polarization in the luminescence in a magnetic field. A systematic shift of the Te-layer in the ZnSe well relatively to the (Zn,Mn)Se barrier allows for example the experimental determination of the radius of the wavefunction in dependence of the position. Growing activities can also be expected in the field of timeresolved spectroscopy of DMS-based microstructures. Recently spin relaxation measurements in (Cd,Mn)Te/(Cd,Mn)Te quantumwell structures were carried out by measurements of the circularly polarized luminescence in a femtosecond time scale[53]. There is no doubt that the research activities, which were largely restricted to a few groups in the past, will attract a broader interest in a wider community.

References

1. N. B. Brandt, V. V.Moshchalkov, Adv. Phys. 33, 193 (1984)
2. J. K. Furdyna, J. Appl. Phys. 64, R 29 (1988)
3   0. Goede, W. Heimbrodt, Phys. Stat.Sol(b) 146, 11 (1988)
4. R. N. Bicknell, R. W. Yanka, N. C. Giles-Taylor, D. E. Blanks, E. L. Buckland, J. F. Schetzina, Appl. Phys. Letters 45, 92 (1984)
5. L. A. Kolodziejski, T. C. Bonsett, R. L. Gunshor, S. Datta, R. B. Bylsma, W. M. Becker, N. Otsuka, Appl. Phys. Letters 45. 440 (1984)
6. L. A. Kolodziejski, R. L. Gunshor, T. C. Bonsett, R. Ven-katasubramanian, S. Datta, R. B. Bylsma, W. M. Becker, N. Otsuka, Appl. Phys. Letters 47, 169 (1985)
7   G. M. Williams, A. G. Cullis, C. R. Whitehouse, D. E. Aahenford, B. Lunn, Appl. Phys. Letters, 55, 1303 (1989)
8. W. Ossau, S. Fischer, R. N. Bicknell-Tassius, Fourth-International Conference on II-VI-Compounds, Berlin (West), 1989
9. M. Pessa, J. Lilja, 0. Jylha, M. Ishiko, W. Asosnen in "Diluted Magnetic (Semimagnetic) Semiconductors", ed. by R. L. Aggarwal, J. K.Furdyna and S.von Molnar (Materials Reaearch Society, Pitsburgh, PA, 1987, Vol. 89, pp. 303)
10. L. A. Kolodziejski, R. L. Gunshor, S. Datta, T. C. Bonsett, M. Yamanishi, R.Frohne, T. Sakamoto, R. B. Bylsma, W. M. Becker, N. Otsuka, J. Vac. Sci. Tech. B3, 714 (1985)
11. R. N. Bicknell, N. C. Giles, J. F. Schetzina, Appl. Phys. Letters 50, 691 (1987)

523

12. N. Pelekanos, Q. Fu, S. Durbin, M. Kobaya8hi, R. Gunshor, A. V. Nurmikko, Fourth International Conference on II-VI Compound, Berlin (West), 1989

13. L. A. Kolodziejski, R. L. Gunshor, N. Otsuka, S. Datta, W. M. Becker, A. V. Nurmikko, IEEE J. Quantum Electronics 22, 1666(1986)

14. A. V. Nurmikko, R. L. Gunshor, L. A. Kolodziejski, IEEE J Quantum Electronics 22, 1785 (1986)

15. D. K. Blanks, R. N. Bicknell, N. C. Giles-Taylor, J. F. Schetzina, A. Petrou, J. Warnock, . Vac. Sci. Technol. A4, 2120 (1986)

16 G. L. Bir, G. E. Pikus, Symmetry and Deformation Effects in Semiconductors, Chap. V, Wiley, New York 1974

17 T. J. Gregory, C. P. Hilton, J. E. Nicholls, W. E. Hagston,. J. J. Davies, B. Lunn, D. E. Ashenford, Fourth International Conference on II-VI Compounds, Berlin (West), 1989

18. R. de Kronig, W. G. Penney, Proc. Roy. Soc. London A 130 , 499(1931).

19. J. Warnock, A. Petrou, R. N. Bicknell, N. C. Giles-Taylor, D. K. Blanks, . F. Schetzina, Phys. Rev. B 32, 8116 (1985).

20. A. V. Nurmikko, X.-C. Zhang, S.-K. Chang, L. A. Kolodziejski, R. L. Gunshor, S.Datta, J. Lum. 34, 89 (1985)

21. X.-C. Zhang, S.-K. Chang, A. V. Nurmikko, L. A. Kolodziejski, R. L. Gunshor, S. Datta, Phys. Rev. B 31, 4056 (1985)

22. R. N. Bicknell, N. C. Giles-Taylor, J. F. Schetzina, N. G. Anderson, W. D. Laidig, Appl. Phys. Lett. 46, 238(1985)

23. R. N. Bicknell, N. C. Giles-Taylor, J. F. Schetzina, N. G. Anderson, W. D. Laidig, J Vac. Sci. Technol. A4, 2126 (1986)

24. R. B. Balsma, W. M. Becker, T. C. Bonsett, L. A. Kolodziejski, R. L. Gunshor, M. Yamanishi, S. Datta, Appl. Phys. Lett. 47, 1039(1985).

25 S.-K Chang, A. V. Nurmikko, J.-W. Wu, . A. Kolodzie jski, R. L Gunshor, Phys. Rev. B 37 1191 (1988)

26 J-W. Wu, A. V. Nurmikko, Phys. Rev. B ,38 1504 (1988)

524

27 J.-W. Wu ,Sol State Comm. 67, 911 (1988)

28 S. Venugopalan, L. A. Kolodziejski, R.L. Gunshor, A. K. Ramdas, Appl. Phys. 45, 974 (1984)

29 E. K. Suh, D. U. Bartholomew, A. K. Ramdas, S. Rodriguez, S. Venugopalan, L. A. Kolodziejski, R. . Gunshor, Phys. Rev. B 36, 4316 (1987)

30 S.-K. Chang, H. Nakata, A. V. Nurmikko, R. L. Gunshor, L. A. Kolodziejski, Appl. Phys. Lett. 51, 667 (1987)

31 S. Perkowitz, S. S. Yom, R. N. Bicknell, J. F. Schetzina, Appl. Phys. Lett. 50, 1001 (1987)

32 S. M. Rytov, Akust. Zh. 2, 71(1956)

33 Y. Hefetz, J. Nakahara, A. V. Nurmikko, L. A. Kolodziejski, R. L. Gunshor, S. Datta, Appl. Phys. Lett. 47, 989 (1985)

34. A. V. Nurmikko, X -C. Zhang, S.-K. Chang, L. A. Kolodziejski, R. L. Gunshor, S. Datta, Surface Sci. 170 665 (1986)

35 R. N. Bicknell, N. C. Giles-Taylor, J. F. Schetzina, N. G. Anderson, W. D. Laidig, Appl. Phys. Letters 46, 1122 (1985)

36 E. D. Isaacs, D. Heimann, J. J. Zayhowski, R. N. Bicknell, J. F. Schetzina, Appl. Phys. Lett. 48 275 (1986)

37. C. E. T. Goncalves da Silva, Pys. Rev. B 32, 6962 (1985)

38 CO.E. T. Goncalvea da Silva, Phys. Rev. B 33, 2923 (1986)

39. Ji-Wei Wu, A. V. Nurmikko, J. J. Quinn, Phys. Rev. B 34, 1080 (1986)

40 D. D. Awschalom, J. Warnock, J. M. Hong, L. L. Chang, M. B. Ketchen, W. J. Gallagher, Phys. Rev. Letters 62 199 (1989)

41 J. M. Hong, D. D. Awschalom, L. L. Chang, A. Segmuller, J. Appl. Phys. 63, 3285 (1988)

42. D. D. Awschalom, J. M. Hong, L. L. Chang, G. Grintein, Phys. Rev. Lett. 59 1733 (1987)

43 L. L. Chang, Superlattice and Microstructures 6 39 (1989)

44 L. A. Kolodziejski, R. L. Gunshor, N. Otsuka, B. P. Gu, Y. Hefetz, A. V. Nurmikko, Appl. Phys. Lett. 48, 1482 (1986)

45 L. A. Kolodzie jski, R. L. Gunshor, N. Otsuka, B. P. Gu, Y. Heetz, A. V. Nurmikko, J. Cryst. Growth 81, 491 (1987)

46 R. N. Bicknell, N. C. Giles-Taylor, D. U. Blanks, R. W. Yanka, E.L. Buckland, J. F. Schetzina, J. Vac, Sci. Technol. B 3, 709 (1985)

47. S. Datta, J. K. Furdyna, R. L. Gunshor, Superlattices and Microstructures 1, 327 (1985)

48 N. Samarth, H Luo, J. K. Furdyna, S. B. Quadri, Y. R. Lee, A. K. Ramdas, N. Otsuka, Appl. Phys. Lett. 54, 2680 (1989)

49. R. G. Alomo, E.-K Suh, A. K. Ramdas, N. Samarth, H. Luo, J. K. Furdyna, Phys. Rev. B 40, 3720 (1989)

50. L. A. Kolodziejski, R. L. Gunshor, Q. Fu, D. Lee, A V. Nurmikko, N. Otsuka, Appl. Phys. Lett. 52, 1080 (1988)

51 Q. Fu, D. Lee, A V Nurmikko, L. A. Kolodziejski, R. L. Gunshor, Phys. Rev. B 39, 3173 (1989)

52. J. Kossut, J. K. Furdyna, Mat. Res. Soc. Symposia Proceedings 89 (1987) 97

53. M. R. Freeman, D. D. Awschalom, J. M. Hong, L. L. Chang, to be published

DMS FOR PHOTOELECTRIC DETECTOR APPLICATIONS

Janusz M. Pawlikowski

Institute of Physics, TUW, Wyspianskiego 27, 50-370 Wroclaw, Poland

1.   INTRODUCTION

This chapter is concerned with device applications of the new class of semiconducting materials, called diluted magnetic semiconductors (DMS) or semimagnetic semiconductors (SMSC), which have rapidly emerged during the last dozen of years[1-3]. It is now well established that these materials, usually the ternary (pseudo-binary) and quaternary alloys, are very interesting for studying the electrical, magnetic, optical, and photoelectric properties and phenomena.

Up to the last days, the two well-developed semiconductor alloy systems, II-VI group $Hg_{1-x}Cd_xTe$ and IV-VI group $Pb_{1-x}Sn_xTe$, have been shown to have the best usefulness for the fabrication of tuned infrared detectors. The great potential for the application of these two alloys has mainly been due to the fact that their spectral sensitivity range can be varied easily along the mole content of the substituting cations. The (Hg,Cd)Te mixed crystals have appeared to be the more versatile - they can cover a wider wavelength range and are suitable for both photoconductive and photovoltaic detectors.

Recently, DMS/SMSC ternary and quaternary compounds based on the "parential" alloys mentioned above have attracted great attention for the three main reasons. First, their fundamental semiconducting properties such as the energy bandgap, the carrier effective mass and mobility, etc. can be varied under control of the molar composition as in non-magnetic ternaries/quaternaries. It can lead to many applications such as those shown, for instance, for graded-bandgap (Hg,Cd)Te epitaxial layers[4]. Secondly, DMS/SMSC exhibit magnetic properties as disordered magnetic alloys. Among these properties are the spin glas transition, formation of antiferromagnetic clusters, etc. The third and the most important feature of these materials is originated in the presence of substitutional magnetic ions which leads to, among other things, to spin-spin exchange interactions between the localized magnetic moments of the ions and the band electrons. This in turn impacts on band structure and impurity states, leading to quite new effects in e.g. magneto-optics and carrier transport, particularly when quantizing magnetic field is present.

Furthermore, the rich spectrum of new physical phenomena ascribed to DMS/SMSC singularities hold promise of new device applications. Infrared detectors, optical non-reciprocal devices and tunable Raman spin-flip lasers are among them. In this paper we have focussed our attention on infrared photoelectric detectors made of DMS/SMSC materials.

2.   DMS MATERIALS

Any semiconducting material, in which a part of the lattice is replaced by magnetic atoms can be treated as a DMS/SMSC. Actually, research interest in both the theory and experiments is focussed almost exclusively on ternary and quaternary compounds, in which one of the cations is replaced by Mn or Fe (sometimes Co, Eu or Yb) ions - see also the other Chapters of this book.

Among the whole class of DMS/SMSC materials the (Zn,Cd,Hg) - (Mn,Fe) - (S,Se,Te) ternary and quaternary compounds of II-VI group are the most frequently studied ones. The type of crystal structure and range of composition of the most interesting DMS/SMSC of this group are listed in Table I after Reference[5]. The mole content ranges given in the Table I correspond to good quality single-phase crystals obtained experimentally. This limited range of x is due to the fact that the crystal structures of Mn-based binaries are different from those of the non-magnetic semiconducting (NMS) copartner. From viewpoint of this structure incompatibility, the high mole content of Mn is remarkable and is presumably due to the natural tendency of Mn to exist in the divalent state.

Table I. The most thoroughly studied DMS of II-VI group

Material	Crystal structure	Range of composition
$Cd_{1-x}Mn_xTe$	zinc blende	$0 \leqslant x < 0.7$
$Cd_{1-x}Mn_xSe$	wurzite	$0 \leqslant x < 0.5$
$Zn_{1-x}Mn_xTe$	zinc blende	$0 \leqslant x < 0.75$
$Hg_{1-x}Mn_xTe$	zinc blende	$0 \leqslant x < 0.3$
$Hg_{1-x}Mn_xSe$	zinc blende	$0 \leqslant x < 0.3$

In general, the band structure of zinc-blende II-VI DMS/SMSC are similar to those of zinc-blende NMS. They have a direct bandgap at the Γ point increasing, as a rule, with increase of molar concentration of magnetic ions[1].

The second important group of DMS/SMSC materials is a subgroup of IV-VI compounds, the lead salts. Replacing Pb by magnetic ions, mixed ternary and quaternary crystals of Pb-(Mn,Eu,Yb)-(S,Se,Te) have been obtained. The crystal structure and range of composition of the most interesting IV-VI DMS/SMSC are listed in Table II along References[6-8]. However, an important difference as compared to the tetra-

Table II. The most thoroughly studied DMS of IV-VI group

Material	Crystal structure	Range of composition
$Pb_{1-x}Mn_xTe$	rock salt	$0 \leqslant x < 0.12$
$Pb_{1-x}Eu_xTe$	rock salt	$0 \leqslant x < 0.15$

hedrally bonded zinc-blende II-VI compounds has been found. In PbTe, the electronic configuration of Pb is $6s^2 6p^2$ and Te is $5s^2 5p^4$ and the s states of both components form a deep valence levels whereas the highest valence level exhibits mainly p-type character due to p electrons of Te[9]. Ions of Mn, having $4s^2 3d^5$ configuration and introduced here as Mn_{Pb}, do not fit into rock-salt structure as smoothly as they do it into II-VI compounds. Therefore, the amount of hybridization between $3d^5$ states and the p-electrons might be different[6] in zinc-blende and rock-salt type structures. In general, the band structures of cubic IV-VI DMS/SMSC with small content of magnetic ions are similar to those of cubic NMS. They have a direct bandgap at the L point which increases with molar composition of magnetic ions[10].

There are also other DMS/SMSC materials investigated recently. Among them are II-V binaries with by-manganese-replaced cations: (Zn,Cd)-Mn-(P,As). Tetragonal $(Cd_{1-x}Mn_x)_3As_2$ and $(Zn_{1-x}Mn_x)_3As_2$ were studied in References 11 and 12, respectively. It is suggested that Mn ions interact antiferromagnetically in $(Cd,Mn)As$[13].

The physical properties of DMS/SMSC materials are extensively described elsewhere. Let us make here only two fundamental remarks. In general, the spin-independent properties of DMS/SMSC such as the crystallographic, electrical, and optical are qualitatively similar to those of "parential" non-magnetic compounds. For instance, the partial replacement of the cation with magnetic ion in binaries leads to composition-dependent variations of almost all the properties of the material, including the fundamental energy gap, like in the NMS ternaries. However, e.g. Mn substitution of e.g. Hg in HgTe and for Cd in CdTe widens the energy gap more remarkably. The same effect was found in the lead salts. These spin-independent properties of DMS/SMSC are just utilized in the infrared detector applications . On the other side, there are two groups of phenomena which distinguish DMS/SMSC materials from their NMS copartners. The first group is connected with magnetic properties of these materials. They are originated in the partially-filled 3d shell ($3d^5, 3d^6$, and $3d^7$ in Mn, Fe and Co, respectively) highly localized at the site in the cation sublattice. The knowledge of magnetic properties is gathered from the measurements of specific heat, neutron and Raman scattering, low-field magnetic susceptibility, high-field magnetisation, and electron paramagnetic resonance. The second group of the phenomena mentioned above is connected with spin-spin exchange interaction between the localized electrons of the 3d shell and the band electrons. This exchange interaction is manifested experimentally in magneto-optical and magneto-transport measurements. It also leads to a splitting of the top of the valence band in zinc-blende DMS/SMSC[14,15] (i.e. the four-fold degenerated Γ_8 level) as well as to a splitting of the impurity states.

3.   MATERIAL PREPARATION

Almost all the preparation techniques have been applied to grow DMS/SMSC materials in the form of both the single crystals, epitaxial layers and thin polycrystalline films. The most representative collection of these technological methods has been found for II-VI based DMS/SMSC, particularly the ternary $Cd_{1-x}Mn_xTe$ and $Hg_{1-x}Mn_xTe$ alloys as well as quaternary $Hg_{1-x-y}Cd_xMn_yTe$. These semiconductors have been recognized as possessing the most considerable potential for infrared applications[16] and their preparation techniques will be described extensively as an example.

Single crystals of the materials mentioned above were grown by a number of techniques: the Bridgeman method[17-19], the two-phase liquid method[20], the traveling-solvent method[21], the liquid-phase epitaxy, LPE[22], and the isothermal vapor-phase epitaxy, VPE[23]. A new technique which combines the modified Bridgman method and the recrystallization method was applied[24] and a method called modified evaporation-condensation-interdiffusion was also used[25]. Epitaxial layers of these materials were obtained by various LPE techniques, i.e. the tipping, dipping and sliding ones[26-30] and by the pressure-controlled LPE[31,32]; as well as by the standard VPE process[33] and by the pressure-controlled VPE[34]. Some effort to form the thin film of ternary (Hg,Mn)Te was also carried out[35]. From the viewpoint of the detector applications, the two techniques among all the mentioned above are particularly important. They are briefly described below.

3.1.   LPE Technique

This technique has been succesfully applied for the growth of ternary (Cd,Mn)Te and (Hg,Mn)Te epitaxial layers from Te-rich solution[30]. Both the standard LPE growth system and the pressure-

controlled LPE growth set-up have been used. The schematic outlines
of the systems are compared in Fig. 1.

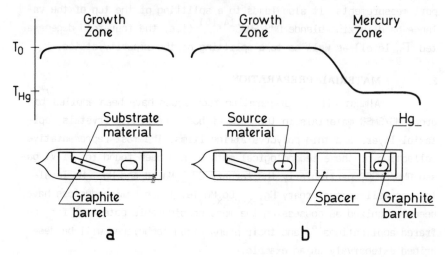

Fig. 1. *Experimental arrangement used for epitaxial growth
in the standard (a) and in the pressure-controlled (b) LPE
growth systems.*

Both set-ups shown in Fig. 1 consist of graphite (tipping)
boats positioned in evacuated and sealed quartz ampules having the substrate plates and the source materials placed inside. In both technological variants, temperature has to be controlled and maintained very
precisely. Usually, sodium heat pipes with extremaly high thermal conductivity are applied to the furnace, keeping the temperature zones
of several tens of cm long within 0.1°C range of temperature variations.

The standard LPE growth was carried out under isothermal conditions, with temperature profile maintained carefully as it is shown
schematically in Fig. 1a. The (Hg,Mn)Te layers were usually grown on
(Cd,Mn)Te substrates at T_0 = 550 to 670°C from Te-rich (Hg,Mn)Te
source. The (Cd,Mn)Te layers were grown on CdTe or (Cd,Mn)Te substrates at T_0 = 710 to 800°C, from Te-rich (Cd,Mn)Te source. Temperature

T_0 was adjusted at the liquidus point for the Te-rich homogeneity range of the source materials. After a few tens of min. of equilibration,
the source solution was tipped onto the substrate (usually by tilting
the furnace) and growth was developed. Pouring the source solution
off terminated the growth.

The pressure-controlled two-zone LPE growth was also carried
out under isothermal conditions with the substrate-zone temperature
controlled independently of the mercury-zone one (Fig. 1b). The Hg zone was included to provide: (1) precize and continuous control of the
equilibrium mercury pressure during the growth process, and (2) continuous in-situ annealing of the growing layers. The role of the mercury pressure is described below. Temperature of the substrate zone, T_0,
was adjusted at the liquidus point for the Te-rich homogeneity range
of the source material. Conditions of the growth were complemented by
adjusting the mercury-zone temperature, T_{Hg}, accordingly. After the
source material equilibration, growth was performed as mentioned above.

3.2. VPE Technique

This technique has been succesfully employed to grow quaternary (Hg,Cd,Mn)Te epitaxial layers[33,34]. Both the standard VPE
growth system[33] and the pressure-controlled one[34] have been used.
It should be mentioned that a very similar method has been used for
years to prepare epitaxial (Hg,Cd)Te layers - the well known material
for infrared p-n detectors - see e.g. references[36-38].

The standard VPE growth system is shown in Fig. 2a. Single
(Cd,Mn)Te crystals were used as the substrate and stoichiometric HgTe
was applied as the source material. Growth took place at 600°C for 250
hours, approximately. The idea of two-zone isothermal VPE configuration sketched in Fig. 2b was developed for (Hg,Cd,Mn)Te layers in papers[23,34]. Controlled mercury vapor pressure was found to be a crucial parameter in this method. The substrate and the source material
are positioned in a crucible (usually made of pyrolitic boron nitride)
that is used - among other reasons - to reduce the free volume and to
define the deposition distance (usually a few mm). This assembly is
finally mounted together with a mercury drop in an evacuated quarz

534

ampule and sealed.

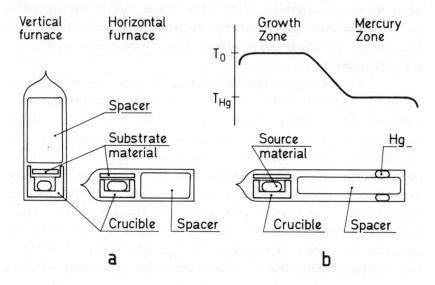

Fig. 2. Experimental arrangements used for epitaxial growth in the standard (a) and in the pressure-controlled (b) VPE growth systems.

The (Hg,Cd,Mn)Te layers were grown from Te-rich HgTe source onto CdTe or (Cd,Mn)Te substrates at T_o = 600 to 640°C, typically, and for several hours. The mercury pressure (via T_{Hg}) was adjusted accordingly within the range of 1-6 atm., approximately. The Hg-to-Te pressure ratio was an important parameter for the isothermal VPE growth, governing both the rate of the transport of material from the source to the substrate and the interdiffusion coefficient of Cd and Mn from the substrate to the growing layer. Therefore, it controled the overall electronic parameters of the layers.

This effect has been known for years (see e.g. papers[39,40]). The role of the mercury pressure[41] during the VPE growth of the materials mentioned above is schematically shown in Fig. 3 (see also Section 4). It has been widely accepted that the optimum of growing process (its rate and output parameters of the layer) is reached when T_o

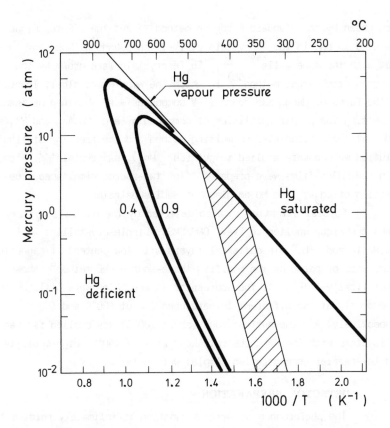

Fig. 3. *Schematic diagram of mercury pressure over binary and ternary Hg-based solid solutions. The p-n transition region is marked as shaded area. Numbers indicate mole content of Hg in (Hg, Cd)Te.*

line and p_{Hg} line (defined by T_{Hg}) cross closely to the Te-rich boundary line. There is also shown in Fig. 3 the region of p ⇌ n type transitions (shaded area) suggested in earlier papers[42-45] and combined by Jones et al.[46].

Other DMS/SMSC materials were also grown by similar techniques; e.g. (Hg,Fe)Se and (Hg,Cd,Fe)Se were prepared by a modified Bridgman method[47], single (Pb,Mn)Te crystals with Mn up to 12 per cent

were grown by the standard Bridgman method[48] and the VPE technique
was used to obtain large crystals of IV-VI compounds without any contact with the tube walls[49]. $Pb_{1-x}Eu_xTe$ crystals were grown by the
Bridgman-Stockbarger method[50] in which the ampule was slowly lowered
in the furnace. Manganese-doped II-V compounds were obtained by means
of melting the proper quantities of components (e.g. Cd_3As_2 and Mn_3As_2)
and next the horizontal-zone melting method and the freezing of liquid
solution method were applied to get $(Cd_{1-x}Mn_x)_3As_2$ crystals[13]. Also,
thin (Cd,Mn)As films were prepared[51] by two-source simultaneous deposition of Cd_3As_2 and Mn on glass or NaCl substrates.

Special attention is also deserved by the hot wall epitaxy
(HWE) technique applied to IV-VI DMS/SMSC. Single crystalline
(Pb,Mn)Te and (Pb,Eu)Te epitaxial layers with low content of magnetic
atoms were prepared by the modified three-source HWE method[6] shown
schematically in Fig. 4. The sources of Te and PbTe were utilized independently at two different temperatures (about 400°C and 600°C,
respectively) and the hottest zone (about 700°C) was applied for magnetic atoms with the BaF_2 substrate kept at $T_o \cong 450°C$. High-mobility
and low-carrier concentration samples were obtained.

4.   DETECTOR PREPARATION

The performance of any p-n junction is intimately related to
the electrical characteristics (stoichiometry and purity of the semiconductor, level of intentional doping, etc.) of both materials constituting the junction. These characteristics are, in turn, determined
both by introduced impurities and by native defects. The latter are a
product of the growth process carried out.

Both in DMS/SMSC II-VI compounds containing Hg and in DMS/SMSC
IV-VI lead salts the native defects (vacancies and interstitials) play
an important role in setting the electrical conductivity type. Therefore, the local deviation from stoichiometry is the best and easy way
to settle the conductivity type and to make p-n junction. This technological process is widely known for II-VI mercury-containing compounds
and shall be described below as an example.

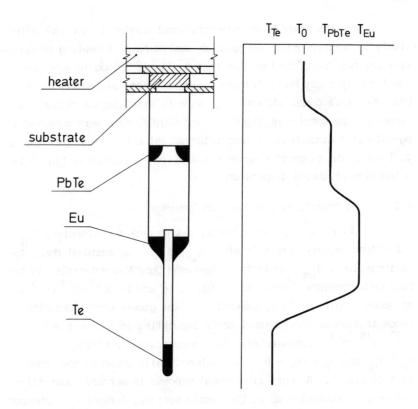

Fig. 4. Experimental arrangement used for epitaxial growth in the three-source HWE growth system.

There is a wide consensus that mercury vacancies and interstitials are the most numerous native defects in Hg-based ternary and quaternary compounds. Furthermore, the rapid escape of mercury atoms that occurs when the material is heated is much more important than the escape of others. Therefore, the control of mercury partial pressure during both growth and anneal process is fundamental; and the reasons are twofold. First, free mercury must be supplied to avoid destruction of the sample (see section 3). Second, adjusting of the mercury partial pressure provides the best means to settle the electrical properties of the material (see Fig.3) and to make p-n junction, particularly.

On the other hand, substitutional donors do not work effec - tively in semiconductor compounds discussed here. Annealing in mercury vapor has therefore been the best available method to make p-n junctions in every Hg-containing compound. Then, p-n junctions in DMS/SMSC II-VI compounds were also made by using such a method. As- -grown p-type samples of (Hg,Mn)Te and (Hg,Cd,Mn)Te were annealed in Hg-saturated atmosphere at temperatures just above 300°C for time up to 1 hour. The range of p $\rightleftharpoons$ n transitions is shown in Fig. 3 in a function of anneal temperature.

4.1.   A Profile of p-n Junction Detector

As it was mentioned above, p-type mercury vacancies, V_{Hg}, and n-type mercury interstitials, Hg_i, as well as residual donor-type impurities, D_{imp}, dominate in Hg-based DMS/SMSC materials. In the compounds considered here, N_A-N_D typically equals 10^{17}-10^{18} cm^{-3} in as-grown samples at 77 K, depending on the growth conditions. The concentration of the residual donor impurities is in the range of 10^{15}-10^{18} cm^{-3}, approximately. The concentration profiles of $N_A \cong V_{Hg}$ and $N_D \cong Hg_i + D_{imp}$ are schematically shown in the upper part of Fig. 5. A short-time anneal process in mercury vapor fills the mercury vacancies up at the sample surface, leaving N_D unchanged substantially. Therefore, it leads to the formation of a near-surface layer with $N_D > N_A$ as shown in the lower part of Fig. 5. This proce- dure gives the net donor concentration at the surface up to 10^{16} cm^{-3} at 77 K.

These concentration profiles as well as the thickness of both the n-type layer and the p-n junction are a function of the an- nealing time and temperature. Under technological conditions discus- sed above as typical, the n-type layer comes out to be about 5 μm thick and the thickness of the p-n junction is equal to about 0.05 - - 0.1 μm. The built-in potential barrier height is usually in the range of 0.1 - 0.2 eV.

A corresponding procedure has also been used to prepare p-n junctions in DMS/SMSC lead salts. $Pb_{1-x}Eu_xTe$, as an example, is n-type material for very low Eu content whereas it is the p-type

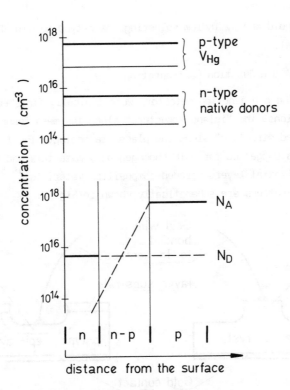

Fig. 5. *The concentration profiles of native defects before the annealing in Hg vapor (top) and after the anneal process (below).*

one for $x \geqslant 0.01$. The p-n junctions were prepared[50], as in II-VI compounds containing Hg, by the deviation from stoichiometry. This was done by diffusion of Te from $Pb_{0.97}Te$ source into n-type material and by diffusion of Pb from $PbTe_{0.97}$ source into p-type (Pb,Eu)Te. The junction dimensions were 0.5x0.5x0.3 mm typically.

Similarly, the p-n junctions in as-grown p-type $Pb_xMn_ySn_zSe$ with $x = 1-y-z$, $y \leqslant 0.04$ and $z \leqslant 0.12$ have been obtained by the annealing procedure performed in a closed system in presence of $(Pb_{1-z}Sn_z)_{0.51}Se_{0.49}$ at 500°C for a few days[52]. The diodes with dimensions of about 0.4x0.4x0.2 mm were cleaved from the samples and ohmic contacts were obtained by electrolytic covering the p-type

side with gold and by indium soldering the n-type to cold finger of
the cryostat.

4.2. A p-n Junction Configuration

The p-n junction detectors were typically finished in two
configurations: the "island structure" along the mesa technology and
the "drowned structure" along the planar technology. The first configuration is suggested for bulk (homogeneous) materials and the second
one for epitaxial layers (graded-composition materials)[36,53]. Both
the configurations are schematically shown in Fig. 6.

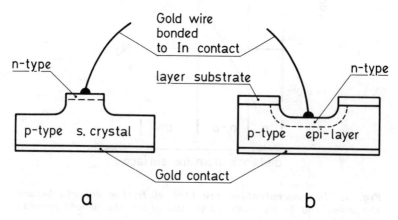

Fig. 6. *The configurations of the p-n junctions discussed in
the text: the mesa structure (a) and the planar structure
(b).*

Typical process of detector preparation in II-VI DMS/SMSC is described
in an earlier paper[53].

The mesa-structure junctions were made in the single crystals with the lithography techniques by using a black wax as a masking material and 5% bromine in methanol as the etching solution. After etching, the black wax was removed in trichloroethylene and the
sample was repeatedly rinsed in methanol and acetone, revealing a well-
-defined mesa structure of about 0.2 mm in diameter. A solution of
gold chloride in water was used to form contacts with the p-type re-

gion and silver paint was eventually employed to attach the junction in a standard TO-transistor mounting. Pure indium was used to form contacts to the n-type region, to which a thin gold wire was subsequently bonded.

The planar-structure junctions were made in the graded-composition epitaxial layers. The back side of the layers was used, where the molar composition of Cd and Mn is very large. After thinning the substrate to about 20 μm, the hole of about 0.5mm in diameter was etched in the substrate to reach the p-type epitaxial layer. The uncovered region of the layer was doped to n type by Hg annealing mentioned above. Gold and indium contacts were finally made to p-type and n-type regions, respecively, according to their properties[54].

The planar technique has two distinct advantages over the mesa one[53]: (i) better contact to p-type region and lower series resistivity due to lower molar composition of Cd and Mn on the surface at Au contact; and (ii) very low surface leaking current due to a high resistivity protection ring formed by the substrate material on the edges of the p-n junction.

5.   DETECTOR PROPERTIES

Both the photoconductive, PC, and photovoltaic, PV, detectors have been made from DMS/SMSC. Photoconductivity of (Hg,Mn)Te is reported in paper[55], PC of (Cd,Mn)Te is described in papers[56-59] and that of (Zn,Mn)Te is also discussed[60]. Infrared PV detectors made of DMS/SMSC materials have been described in a few papers. Both II-VI PV detectors[25,33,53,61-67] and IV-VI ones[50,52] were prepared and measured. In this section the electrical properties of p-n junctions are described on the base of II-VI DMS/SMSC materials as they are relatively well known actually. Instead, the photovoltaic spectra are presented for both II-VI and IV-VI DMS/SMSC diodes. Also, luminescence spectra for p-n diodes are finally shown for both types of compounds.

5.1. Electrical Properties of DMS Detectors

To analyze the main mechanisms of current transport in the detectors, the dark current-voltage (I-V) and differential resistivity (dV/dI) versus voltage and magnetic field characteristics were measured in the temperature range of 2K - 300 K. Also, the dark differential resistivity at zero bias, $R_o = (dV/dI)_{V=0}$, was measured as a function of temperature and magnetic field. The measurements were performed by the standard d.c. and a.c. modulation techniques.

The exemplary I-V and dV/dI-vs-voltage curves for one of the mesa-type (Hg,Cd,Mn)Te detectors[62] are shown in Fig. 7.

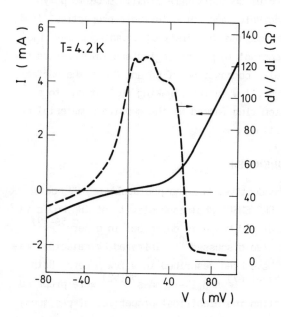

Fig. 7. The I-vs-V and dV/dI-vs-V plots measured for one of the (Hg,Cd,Mn)Te detectors with $E_g = 0.09$ eV at 77 K.

It was found[33,62,63] that the temperature coefficient of the reverse-bias current changed its sign within the 50 - 100 K temperature range, while in the case of forward bias the sign remained positive. The singularities observed on the dV/dI-vs-voltage curves disappeared as the temperature increased above 10 K[33,62,63]. For certain mesa-type (Hg,Cd,Mn)Te detectors the singularities were observed for forward

bias only (cf. Fig. 7) for the other ones for both forward and reverse biases[33,62,63].

The peak observed in $dV/dI = f(V)$ dependences and positioned
close to the zero-voltage bias (the so-called zero-bias anomaly,ZBA)
appeared in the mesa-type (Hg,Cd,Mn)Te detectors[62,63] within the low
temperature range (T < 10 K) and decreased with temperature increase.
The ZBA has been explained in the NMS (Hg,Cd)Te detectors (without Mn)
with the help of the resonant tunneling effect[64,65].

The presence of Mn ions in DMS/SMSC materials has led to the
hypothesis that the ZBA could be sensitive to a magnetic field. Therefore, the measurements of the dV/dI-vs-V characteristics were also
performed in a magnetic field up to 6 T[63]. Figure 8 presents the exemplary curves for one of the detectors, without and in the magnetic

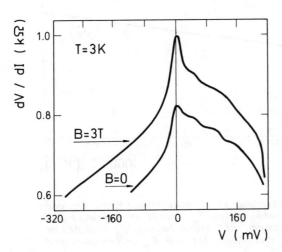

Fig. 8. The dV/dI-
vs-V plots measured
for one of the
(Hg,Cd,Mn)Te
detectors (E_g = 0.1
eV at 77 K.) without
and with the
magnetic field.

field. It can be concluded from Fig. 8 and other studies[33,62,63] that
the presence of Mn ions does not influence the ZBA visibly - no shift
or energy splitting of the ZBA caused by the magnetic field has been
found. Therefore, the origin of the ZBA is probably the same for both
(Hg,Cd)Te and (Hg,Cd,Mn)Te p-n junctions; i.e. it can be ascribed to
a resonant tunneling process in the carrier current.

Figure 9 shows the exemplary characteristics of R_oA product versus reciprocal temperature in the absence of magnetic field for three exemplary mesa-type (Hg,Cd,Mn)Te detectors[62,63], having diffe-

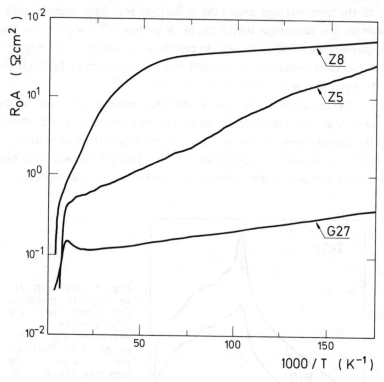

Fig. 9. The RoA product along reciprocal temperature for one of the HgCdMnTe detectors having E_g = 0.34 eV (denoted Z8) and for two narrow-bandgap detectors having E_g = 0.14 eV (detectors denoted Z5 and G27). All the E_g values measured at 77 K.

rent molar composition/energy bandgap. The behaviour of the R_oA product along temperature differed from one diode to another, being a function of energy bandgap of the material. For the diodes with $E_g > 0.3$ eV, approximately, the R_oA increased as temperature decreased. The slope of the R_oA-versus-1/T curves is characteristic for the dominance of the thermal G-R mechanism of the dark current at high temperatures (above 100 K, approximately). For the diodes with $E_g < 0.2$ eV most of the R_oA-vs-1/T characteristics exhibited minimum at low temperatures ranging from 20 to 100 K approximately, (e.g. G27 detector in Fig. 9). However, there were also some mesa-type (Hg,Cd,Mn)Te diodes having similar bandgaps[62,63] for which this minimum was not observed (e.g. Z5 detector in Fig. 9). The presence of the minimum accompanied the inverse of the sign of temperature coefficient of the I-V curves under reverse bias, mentioned above. This behaviour is evident for the tunneling current. The temperature measurements of R_o in the range of 140 K - 300 K showed that the R_oA-vs-1/T dependences for planar-type detectors[53] made in (Hg,Cd,Mn)Te epitaxial layers did not exhibit any minimum of these plots.

It seemed interesting to look for an oscillation effect in R_o-vs-magnetic field plots in narrow-bandgap DMS/SMSC. Typical R_oA-vs-B characteristics of diodes made from (Hg,Mn)Te and (Hg,Cd,Mn)Te[61] are shown in left side of Fig. 10. These characteristics are quite similar for the p-n junctions in ternary and quaternary DMS/SMSC. No visible oscillation, however was found of $R_o = f(B)$ for both the materials. Instead, right-side part of Fig. 10 shows the exemplary characteristics of R_o along a magnetic field for the far-infrared (Hg,Cd,Mn)Te detector[63] measured at two temperatures. One can notice the strong temperature-dependent oscillation effect which is the quantum one in nature. The Landau quantization of the degenerated conduction band levels causes the observed non-monotonic behaviour of R_o (the Shubknikov-de Haas effect). The presence of the magneto-oscillation in some of the (Hg,Cd,Mn)Te detectors[63] can be connected with the relatively weak built-in junction electric field in the narrow gap diodes. In the case of non-magnetic (Hg,Cd)Te p-n junction, the R_o-vs-B oscillations have not been found. Therefore, there was the strong suggestion[63] that this effect is due to the exchange interaction between free electrons and magnetic Mn ions.

Summarizing all these electrical measurement results, we can make some general conclusions on transport mechanism, listed below.

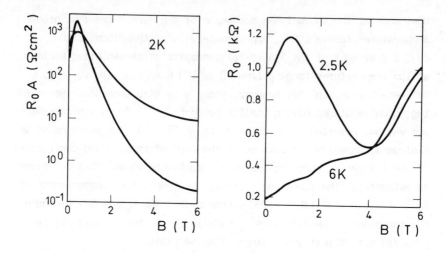

Fig. 10. Left: The R_0A-vs-B characteristics for $Hg_{.78}Cd_{.20}Mn_{.02}Te$ detector (upper curve) and for $Hg_{.87}Mn_{.13}Te$ detector (lower curve).
Right: The R_0-vs-B characteristics for (Hg,Cd,Mn)Te detector with $E_g = 0.1$ eV at 77 K.

For the planar-type near-infrared (NIR) detectors[53] made in (Hg,Cd,Mn)Te epitaxial layers an analysis showed that the major source of the dark current at high temperatures was minority-carrier diffusion and that the I-V curves fitted the standard diode equation

$$I = I_{sat}exp\left[(qV/bkT) - 1\right] \qquad (1)$$

with factor b = 1.05 at 140 K and b = 1.15 at 300 K. Reverse I−V curves of these detectors showed very low saturation current and relatively high breakdown (reverse) voltage. The temperature dependence of R_0A was described by the well-known equation[66] for a dominance of the minority-carrier diffusion

$$R_{o(diff)} = kTq^{-2}n_i^{-2}(D_h/L_hN_D + D_e/L_eN_A)^{-1} \qquad (2)$$

(the notation used has the usual meaning) which transforms (for narrow-gap DMS and under the condition $N_D < N_A$, the minority hole lifetime, τ_h, in the n-type region is much longer than minority electron lifetime, τ_e, in the p-type region) into

$$R_{o(diff)} = (kT)^{1/2} q^{-3/2} n_i^{-2} N_A (\tau_e/\mu_e)^{1/2} \qquad (3)$$

The $R_o = f(n_i^{-2})$ dependence is a feature of this type of current mechanism.

For the mesa-type NIR detectors[62,63] made in (Hg,Cd,Mn)Te epitaxial layers the generation-recombination current (originated within depletion layer with thickness W) was found to be important at high temperatures equally to the diffusion current mentioned above. The $R_o A$ product was fitted to the generation-recombination equation, either in the form of[66]

$$R_{o(G-R)} = V_b \tau_o (q n_i W)^{-1} \qquad (4)$$

(where τ_o is an average carrier lifetime and V_b is the built-in junction potential) or in the form of[67]

$$R_{o(G-R)} = \tau_o (E_g N_D)^{1/2} (2 \varepsilon_o \varepsilon_s)^{-1/2} (q n_i)^{-1} \qquad (5)$$

In any case, $R_o A$ was proportional to n_i^{-1}. At low temperatures the resonant-tunneling current mechanism (both the elastic and phonon-assisted tunneling) came to the fore as the bandgap decreased. There was the impurity-assisted tunneling rather than the band-to-band one. The mesa-type bulk (Hg,Mn)Te detectors suited for midle-infrared (MIR) range[53] showed soft breakdown at a relatively low voltage (similarly to (Hg,Cd)Te detectors). The breakdown voltage decreased with increasing dopant concentration and decreasing temperature. These effects suggested the impurity-assisted tunneling mechanism. The diode factor b for these detectors was 1.5 - 2.7 at 77 K. The I-V and R_o-vs-1/T curves fitted the respective equations characteristic for generation--recombination current. The same behaviour, i.e. the strong contribution from the tunneling current at low temperatures and the generation--recombination current at high temperatures, was also found in far-infrared (FIR) mesa-type bulk (Hg,Mn)Te detectors[53].

Generally, the tunneling current was found to dominate at low temperatures in the NIR, MIR and FIR detectors, both in the mesa-

and planar-type. The strength of this dominance varied from one detector to another being a function of the molar composition/energy bandgap and the technological conditions. Usually, the lower T and the narrower E_g, the higher probability of tunneling. For the MIR and FIR mesa-type detectors made of both the epitaxial (Hg,Cd,Mn)Te layers[33,62,63] and the bulk (Hg,Mn)Te[53] an analysis showed that the generation-recombination current is a major part of the dark current at high temperatures.

5.2.   Photoelectric Properties of DMS Detectors

The spectral characteristics of photoconductivity, PC, of DMS/SMSC materials were usually measured in the standard way, with a lock-in amplifier and were computed by means of reducing the photoresponse with photon flux density. The PC spectra of open-bandgap n-type $Hg_{1-x}Mn_xTe$ ($0.14 < x < 0.17$) with $n = 10^{16}$ cm^{-3} were determined[55] in the temperature range 4.2 - 300 K. Figure 11 shows PC spectra measured

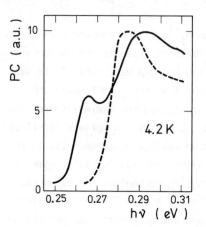

Fig. 11. The photo - conductivity spectra for sample $Hg_{0.85}Mn_{0.15}Te$ with 6.8 T magnetic field (solid line) and without it (dashed line).

with magnetic field and without it[55]. The effect of exchange interaction mentioned in the section 2 above is clearly manifested as the PC edge shift towards lower photon energies. This closing of the bandgap with magnetic field is due to the shift of the highest Landau level of the valence band and can be of practical importance as a means of tuning of photoresponse peak position. PC in (Cd,Mn)Te samples were measured by a few authors (see e.g. papers[56-59]). The imperfection PC was found[57] in p-type $Cd_{0.85}Mn_{0.15}Te$ as connected with acceptor levels and the role of Ag and/or Cu in an annealing procedure of the sample was shown to be crucial. The strong and sharp peak of PC was observed[59] in $Cd_{1-x}Mn_xTe$ ($0.05 \leqslant x \leqslant 0.4$) in the temperature range 63 - 110 K. The energy position of the peak as well as its sharpness suggest its relation to via-exciton transitions. It can be utilized in narrow-band tuned photoconductors.

The spectral sensitivity, S_λ, of photovoltaic DMS/SMSC detectors was determined from the standard photoelectric measurements as a ratio of the photo-voltage response of a detector and the incident illumination power

$$S_\lambda = PV/P_i \tag{6}$$

whereas the normalized detectivity, $D*_\lambda$, was computed as

$$D*_\lambda = (A\Delta f)^{1/2}S_\lambda/V_N \tag{7}$$

where Δf and V_N are the noise bandwidth and the noise voltage in this bandwidth, respectively. Assuming that under zero bias the noise of the p-n junction detector is limited both by a thermally fluctuated carrier concentration and by a background radiation, the normalized detectivity is expressed as

$$D*_\lambda = (Qq \lambda/hc)(2q^2Q\emptyset_B + 4kT/R_0A)^{-1/2} \tag{8}$$

where Q is the quantum efficiency and $\emptyset_B$ is the background photon flux density. The importance of the R_0A product discussed above is evident, particularly in the case of thermally limited performance of a detector.

By controling the mole content of the materials, the DMS/SMSC detectors were fabricated with the photoresponse peak varied in the wide photon energy range from approximately 1.5 to 15 μm, at 77 K. For instance, the normalized detectivity of MIR and FIR detectors made of (Hg,Mn)Te and (Hg,Cd,Mn)Te and described in[53] was close to the background radiation limit, having the quantum efficiency within the 0.2 - 0.4 range without an antireflection coating.

Figure 12 shows the spectral characteristics of normalized detectivity, D^*_λ of (Hg,Mn)Te and (Hg,Cd,Mn)Te detectors[53]. The highest mea-

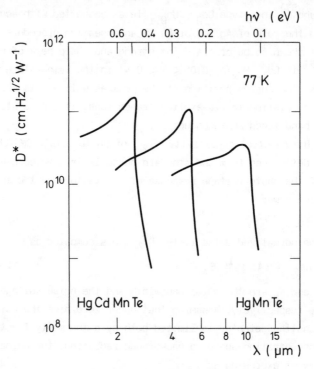

Fig. 12. *The spectral detectivity characteristics of (Hg,Mn)Te and (Hg,Cd,Mn)Te detectors.*

sured detectivities were 1.05×10^{11} cm $Hz^{1/2}W^{-1}$ for 5.3 μm and
3.6×10^{10} for 10.6 μm wavelength at 77 K. The sharp long-wavelength
cut-off of D^*_λ suggests a good compositional homogeneity achieved inside the mixed crystals.

The spectral characteristics of photovoltaic response of
p-n junctions prepared in (Hg,Cd,Mn)Te crystals with various composition (and obtained in the different way[25]) than the materials discussed above) were also measured. The PV spectra are presented in Fig.
13 for $Hg_{0.712}Cd_{0.27}Mn_{0.018}Te$ p-n diode. The temperature shift of

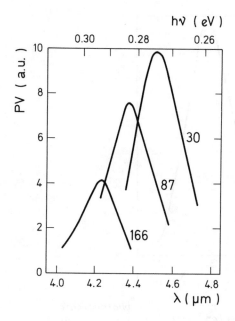

Fig. 13. *The spectral characteristics of PV response of (Hg,Cd,Mn)Te diode. Numbers indicate measurement tempera - ture.*

the PV maxima clearly shows the bandgap variations. The high detectivity values were observed in FIR (Hg,Cd,Mn)Te detectors described in [33,62,63]. An exemplary result of photovoltage response of the FIR (Hg,Cd,Mn)Te mesa-type detector[33] is shown in Fig. 14. The normalized detectivity of the uncoated FIR detectors[33] was also close to the background radiation limit with quantum efficiency of 0.3, approximately.

Spectral photoresponse characteristics of DMS detectors (made of HgMnTe and HgCdMnTe) and of NMS detectors (made of HgCdTe and HgCdZnTe) were measured in magnetic fields[61]. Figure 15 shows a comparison between the spectral responses (at 0 and 6 T magnetic field) of a DMS $Hg_{0.78}Cd_{0.20}Mn_{0.02}Te$ detector and a NMS $Hg_{0.78}Cd_{0.21}Zn_{0.01}Te$ detector[61]. By ascribing the long-wavelength cut-off values of photoresponses to energy bandgaps, the dE_g/dB could be estimated. The DMS detectors showed a negative dE_g/dB coefficient, up to 30 meV/T, whereas this coefficient was positive for the NMS detectors and of similar values. The exchange interaction of Mn ions

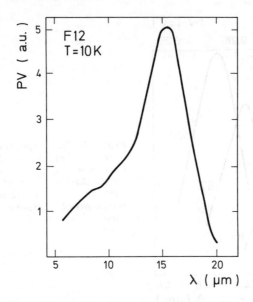

Fig. 14. *The spectral characteristic of photovoltaic response of FIR(Hg,Cd,Mn)Te detector with E_g = 0.99 eV at 77 K.*

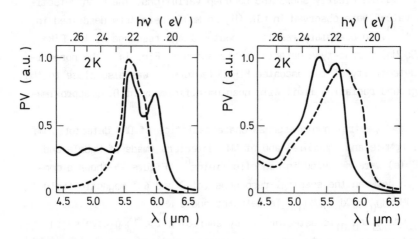

Fig. 15. *The spectral characteristics of photovoltaic responses of DMS(Hg,Cd,Mn)Te detector (left) and of NMS(Hg,Cd,Zn)Te detector (right) with 6 T magnetic field (solid lines) and without it (dashed lines).*

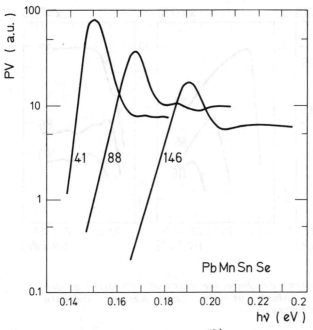

Fig.16. *The spectral PV characteristics of p-n $Pb_{0.9}Mn_{0.02}Sn_{0.08}Se$ junction. Numbers indicate measurement temperature.*

with band electrons was suggested[61] to be responsible for this influence on the photovoltaic properties of detectors.

Spectral characteristics of photovoltaic response were also measured for IV-VI DMS/SMSC detectors. For instance, Fig. 16 shows the PV spectra for p-n junction prepared in $Pb_{0.9}Mn_{0.02}Sn_{0.08}Se$[52]. Relatively sharp low-energy cut-off of PV response suggests good sample homogeneity. PV spectral responses of backside illuminated three p-n diodes prepared in (Pb,Eu)Te with different molar compositions are presented in Fig.17 at various temperatures[50]. The PV spectra possess two maxima; the lower-energy one was ascribed to the energy bandgap of the material.

Finally, let us say a few words about light-emission effects found in DMS/SMSC materials.

A strong electroluminescence was observed from MIR (Hg,Cd,Mn)Te detector below 77 K[61,68]. The typical electroluminescence plots at two temperatures[61] are shown in Fig. 18. The shorter--wavelength peak was ascribed to band-to-band transitions and it was

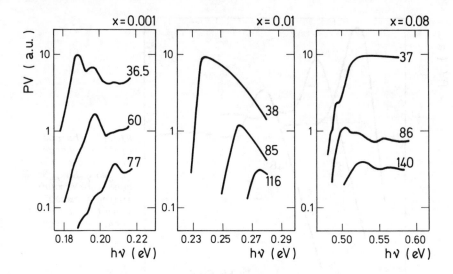

Fig. 17. *The spectral PV characteristics of (Pb,Eu)Te diodes with Eu content shown at the top. Numbers indicate measurement temperature.*

suggested that the longer-wavelength bump originated from band-to-acceptor state transitions. After corrections, the internal electroluminescence efficiencies were estimated[61] to be equal to several percents.

The cathodoluminescence spectra of (Cd,Mn)Te samples were measured at 80 K[69]. Two luminescence bands/peaks shown in Fig. 19 have been found. The weaker one has been ascribed to band-to-band recombination (via the free direct excitons) and therefore its energy position is dependent upon the manganese mole content. The stronger band/peak has been suggested to be a result of transitions of the d-electrons of Mn ions between the first excited state and the ground state.

Light emitting diodes (LED's) and lasers have also been fabricated employing group II-VI and IV-VI DMS/SMSC materials. The

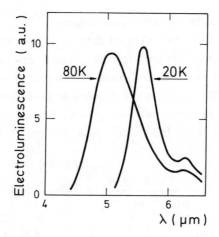

Fig. 18. *The spectral electroluminescence plots at two temperatures, taken from $Hg_{.78}Cd_{.20}Mn_{.02}Te$ detector.*

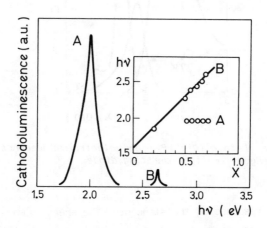

Fig. 19. *The spectral cathodoluminescence plot for $Cd_{0.3}Mn_{0.7}Te$ sample at about 80 K. Insert shows energy positions of the luminescence peaks (points) and the fundamental absorption edge (line) as the function of Mn mole content.*

(Hg,Mn)Te and (Hg,Cd,Mn)Te LED´s were prepared[70] by a method similar to the one used earlier for infrared detectors (see above). For the preparation of (Hg,Mn)Te heterostructural laser, new techniques were developed[70], combining the isothermal VPE to create a graded-band-gap substrate and the two-step LPE process to grow both active and passive layers of the laser heterostructure. Typical electrolumines-cence plots for (Hg,Mn)Te LED[70] are shown in Fig. 20 together with

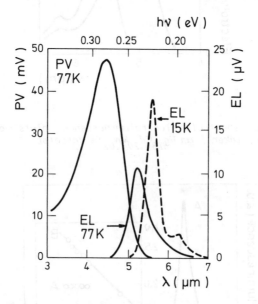

Fig. 20. *The photovoltaic (PV) and electroluminescence (EL) spectra of (Hg,Mn)Te light emitting diode.*

a photovoltaic response of the structure. The energy positions of the electroluminescence peaks correspond to the long-wavelength cutoff of PV response and therefore the electroluminescence resulted undoubted-ly from band-to-band recombinations.

Laser emission has also been obtained from $Pb_{0.9}Mn_{0.02}Sn_{0.08}Se$ sample at 4.2 K[52]. It has been suggested that the emission resulted from band-to-band recombination.

6. FINAL COMMENTS

Desired infrared detector wavelength range of response is well satisfied by the ternary and quaternary DMS/SMSC alloys of both II-VI and IV-VI groups since they provide continuously variable cut-off wavelength over a wide spectrum. There are two additional limits imposed upon the maximum operating temperature and upon the background noise in the detectors[71]. The first limit requires (for PC detector only) the concentration of thermally generated carriers contributing to the g-r noise to be negligible compared to that of photo-carriers generated by the background radiation. From this point of view, the II-VI DMS/SMSC are the very good materials. The second limit demands Johnson noise to be negligible compared to background noise. For PV detectors it requires a good enough p-n junction barrier (chiefly technological problem) to yield a large enough R_oA product for a given background radiation flux density. Therefore, it means in consequence that the advanced technology is required to yield high photon collection efficiency and high carrier efficiency in the junction. For intrinsic PC detectors it requires the best compromised combination of high electron-to-hole mobility ratio and low majority carrier concentration and mobility. These conditions for PC detectors are quite well satisfied in II-VI DMS/SMSC whereas IV-VI DMS/SMSC materials show very poor performance. Concluding, II-VI DMS/SMSC materials are very good for both PC and PV modes while IV-VI DMS/SMSC ones are potential for PV mode application, although strong Auger recombination predicted in lead salts is a performance limiting factor. Moreover, response time is usually shorter in II-VI than in IV-VI DMS/SMSC alloys. These properties are similar to those characterising the NMS "parential" materials.

Narrow-gap (Hg,Mn)Te and (Hg,Cd,Mn)Te alloys have already been used successfully for the fabrication of infrared photovoltaic detectors. Despite some differences, the combined results let us make some general comments. Epitaxial DMS/SMSC layers of well controlled composition, purity and morphology were obtained. A better lattice match of DMS detectors (made of HgCdMnTe and HgMnTe) versus NMS detectors (made of HgCdTe) provided a reduction of the alloy scattering effect, at the substrate-epitaxial layer especially. Both the current--voltage characteristics and the $R_o A$ products were found to be superior in DMS detectors[33,53,61-63] to those in NMS HgCdTe detectors[72,73] with similar bandgaps. Particularly, the R_o values were generally higher suggesting better potential for high detector performance. The measured photovoltaic responses confirmed these expectations, giving the normalized detectivity values close to the background radiation limit, for the MIR and FIR detectors especially[33,53].

Experimental evidences of exchange interaction between Mn ions and band electrons were found. The exchange interaction affected both the $R_o A$ product and the bandgap value and, therefore, influenced on the photovoltaic properties of DMS/SMSC detectors. Then, by varying the cation mole content the zero-magnetic field bandgap can be tuned (as in the NMS detectors[72,73]) while the photoresponse in magnetic field can be changed independently along the varying magnetic ion concentration.

7.   REFERENCES

1. R. R. Galazka, Proc. 14th Intern. Conf. Phys. Semiconductors, Edinburgh (1978) IOP Conf. Ser. 43, 133 (1978).

2. J. K. Furdyna, J. Appl. Phys. 53, 7637 (1982).

3. see e.g. papers presented at the 18th and the 19th Intern. Conf. Phys. Semiconductors held in Stockholm (1986) and in Warsaw (1988) respectively.

4. J. M. Pawlikowski, "Electrical and Photoelectric Properties of Graded-Gap Epitaxial $Cd_x Hg_{1-x} Te$ Layers", Wydawnictwo Politechniki Wroclawskiej, Wroclaw, (1978)

5. A. Pajaczkowska, Prog. Crystal Growth and Charact. 1, 289 (1978)

6. G. Bauer, Proc. 12th Conf. Phys. Semicond. Compounds, Jaszowiec 1983, Ossolineum, Wroclaw (1983) p.62

7. Z. Korczak and M. Subotowicz, as Ref. 6, p. 392

8. I. I. Zasavitskii and A. V. Sazonov, Sov. Phys.-Solid State 30, 962 (1988).

9. G. Martinez, M. Schlüter and M. L. Cohen, Phys. Rev. B11, 651 (1975).

10. J. Niewodniczanska-Zawadzka and A. Szczerbakow, Solid State Commun. 34, 887 (1980).

11. Z. Celinski, W. Zdanowicz, K. Kloc and A. Burian, as Ref. 6, p. 468.

12. C. J. M. Denissen, S. Dakun, K. Kopinga, W. J. M. de Jonge, H. Nishibara, T. Sakakibara and T. Goto, Phys. Rev. B36,5316 (1987).

13. W. Zdanowicz, W. Lubczynski, J. C. Portal and E. Zdanowicz, Acta Phys. Pol. A67, 203 (1985).

14. M. Jaczynski, J. Kossut and R. R. Galazka, Phys. Status Solidi b88, 73 (1978).

15. J. A. Gaj, J. Ginter and R. R. Galazka, Phys. Status Solidi b89, 655 (1978).

16. J. M. Pawlikowski, Infrared Phys. 30, no 4 (1990).

17. R. T. Delves, Br. J. Appl. Phys. 16, 343 (1965).

18. R. T. Holm and J. K. Furdyna, Phys. Rev. B15, 844 (1977).

19. P. Becla, D. Heiman, J. Misiewicz, P. A. Wolff and D. Kaiser, SPIE vol. 796, "Growth of Compound Semiconductors" pp. 108-114, (1987).

20. I. E. Lopatynski, Inorg. Mater. 12, 296 (1976).

21. R. Triboulet, D. Triboulet and G. Didier, J. Crystal Growth 38, 82 (1977).

22. U. Debska, M. Dietl, G. Grabecki, E. Janik, E. Kierzek-Pecold and M. Klimkiewicz, Phys. Status Solidi a64, 707 (1981).

23. P. Becla, J. Lagowski, H. C. Gatos and L. Jedral, J. Electrochem.

Soc. <u>129</u>, 2855 (1982).

24. A. Mycielski, J. Wrobel, and M. Arciszewska, as Ref. 6, p. 404.

25. T. Piotrowski and J. Niewodniczanska-Zawadzka, Acta Phys. Pol. <u>A67</u>, 353 (1985).

26. D. A. Nelson, W. R. Higgins and R. A. Lancaster, Proc. Soc. Photo-Optical Instrum. Eng. <u>225</u>, 48 (1980).

27. J. A. Mroczkowski and H. R. Vydyanath, J. Electrochem. Soc. <u>128</u>, 655 (1981).

28. C. C. Wang, S. H. Shin, M. Chu, M. Lauir and A. H. B. Vanderwych, J. Electrochem. Soc. <u>127</u>, 275 (1980).

29. T. C. Harman, J. Electron. Mater. <u>9</u>, 945 (1980).

30. P. Becla, P. A. Wolff, R. L. Aggarwal and S. Y. Yuen, J. Vac. Sci. Technol. <u>A3</u>, 116 (1985).

31. T. C. Harman, J. Electron. Mater. <u>10</u>, 175 (1981).

32. H. Ruda, P. Becla, J. Lagowski and H. C. Gatos, J. Electrochem. Soc. <u>130</u>, 228 (1983).

33. E. Placzek-Popko, L. Jedral, E. Dudziak and J. M. Pawlikowski, Proc. Intern. Conf. Semicond. Mater., New Delhi (1988) in the press.

34. P. Becla, P. A. Wolff, R. L. Aggarwal, S. Y. Yuen and R. R. Galazka, J. Vac. Sci. Technol. <u>A3</u>, 119 (1985).

35. A. Milczarek and S. A. Ignatowicz, Electron Technol. 5, 47 (1972).

36. J. M. Pawlikowski and P. Becla, Phys. Status Solidi <u>a 32</u>, 639 (1975).

37. J. M. Pawlikowski, Thin Solid Films <u>44</u>, 241 (1977).

38. P. Becla, E. Dudziak and J. M. Pawlikowski, Mater. Sci. <u>3</u>, 27 (1977).

39. H. Rodot and J. Henoc, C. R. Acad. Sci. Paris. <u>256</u>, 1954 (1963).

40. F. Bailly, G. Cohen-Solal and Y. Marfaing, C. R. Acad. Sci. Paris

261, 103 (1963).

41. T. Tung, Ch.-H. Su, P.-K. Liao and R. F. Brebrick,
    J. Vac. Sci. Technol. 21, 117 (1982).

42. H. Rodot, J. Phys. Chem. Solids 25, 85 (1964).

43. L. Schmit and E. L. Stelzer, J. Electron. Mater. 7, 65 (1968).

44. R. A. Farrar, C. J. Gillham, B. Barlett and M. Quelch, J. Mater.
    Sci. 12, 836 (1977).

45. G. Dittmar, B. Forbig and F.Schubert, Krist. Tech. 13, 817 (1978).

46. C. L. Jones, M. J. T. Quelch, P. Capper and J. J. Gosney, J. Appl.
    Phys. 53, 9080 (1982).

47. A. Lewicki, Z. Tarnawski and A. Mycielski, Acta Phys. Pol.
    A67, 357 (1985).

48. Z. Korczak and M. Subotowicz, as Ref. 6, p. 392.

49. Z. Golacki, Z. Furmanik, M. Gorska, A.Szczerbakow and W. Zaho-
    rowski, J. Crystal Growth 60, 150 (1982).

50. Z. Golacki and M. Gorska, Acta Phys. Pol. A67, 379 (1985).

51. G. Pocztowski, G. Zielinski, E. Szibel, B. Rzepa, B. Jarzabek,
    J. Jarusik and L. Zdanowicz, Acta Phys. Pol. A67, 207 (1985).

52. L. Kowalczyk and A.Szczerbakow, Acta Phys. Pol. A67, 189 (1985).

53. P. Becla, J. Vac. Sci. Technol. A4, 2014 (1986).

54. J. M. Pawlikowski, Phys. Status Solidi a40, 613 (1977).

55. J. Wrobel and A. Mycielski, as Ref. 6, p. 440.

56. B. S. Sundersheshu, Phys. Status Solidi, a61, K155 (1980).

57. A. J. Szadkowski and A. Lubomirska-Wittlin, as Ref. 6, p. 432.

58. K. Yamada, M. Lindström, J. heleskivi and R. R. Galazka, Jap.
    J. Appl. Phys. 19, Suppl. 19-3, 361 (1980).

59. A. Zareba and A. J. Nadolny, as Ref. 6, p. 444.

60. J. Stankiewicz and M. Di Lorenzo, Proc. 19th Intern. Conf. Phys.

Semiconductors, Warsaw 1988, WDN Wrclaw (1988) p. 1562.

61. S. Wong and P. Becla, J. Vac. Sci. Technol. $\underline{A4}$, 2019 (1988).

62. E. Placzek-Popko and L. Jedral, Infrared Phys. $\underline{28}$, 249 (1988).

63. E. Placzek-Popko, E. Dudziak, L. Jedral, J. F. Kasprzak and J. M. Pawlikowski, Infrared Phys. $\underline{29}$, 903 (1989).

64. E. Placzek-Popko and J. M. Pawlikowski, IEEE Trans. Electron Devices $\underline{ED-32}$, 842 (1985).

65. E. Placzek-Popko and J. M. Pawlikowski, J. Crystal Growth $\underline{72}$, 485 (1985).

66. S. M. Sze, "Physics of Semiconductor Devices", Wiley, New York (1969) p. 108.

67. M. R. Johnson, R. A. Chapman and J. S. Wrobel, Infrared Phys. $\underline{15}$, 317 (1975).

68. P. Becla, R. L. Aggarwal, P. A. Wolff and S. Y. Yuen, presented at the U.S. Workshop on the Physics and Chemistry of Cadmium Mercury Telluride, San Diego, 1984, unpublished.

69. J. Petryk, as Ref. 6, p. 424.

70. P. Becla, J. Vac. Sci. Technol. $\underline{A6}$, 2725 (1988).

71. D. Long, in "Optical and Infrared Detectors", Topics in Applied Physics, Springer, Berlin (1980), p. 101.

72. J. M. Pawlikowski and P. Becla, Infrared Phys. $\underline{15}$, 331 (1975).

73. P. Becla and J. M. Pawlikowski, Infrared Phys. $\underline{16}$, 457 (1976).

DEVICE APPLICATIONS OF $Hg_{1-x-y}Cd_xMn_yTe$ (0< x,y <1)

G.N. Pain

Telecom Australia Research Laboratories

770 Blackburn Road, Clayton, Victoria,

Australia 3168

ABSTRACT

The ternary diluted magnetic semiconductors $Cd_{1-x}Mn_xTe$, $Hg_{1-y}Mn_yTe$ and the quaternary $Hg_{1-x-y}Cd_xMn_yTe$ are used in a variety of optoelectronic devices. The recent literature is reviewed with an emphasis on operation at or near room temperature.

1.   INTRODUCTION

Manganese-doped tellurides of cadmium and mercury have received intensive study due to their intrinsic scientific interest and potential exploitation in a wide variety of optoelectronic devices. Many studies of these materials are performed at liquid helium temperatures and often magnetic fields obtainable only in the laboratory. In this review the emphasis is directed to reports of devices which can function at room temperature or temperatures which can be achieved with thermoelectric or liquid nitrogen cooling, and thus are more likely to find wide application. Thin-film devices utilizing epitaxial material grown by liquid phase epitaxy (LPE), ionized cluster beam deposition (ICBD), molecular beam epitaxy (MBE) or metal organic chemical vapour deposition (MOCVD), are more commonly reported. This reflects the difficulties associated with bulk growth and the desire to be able to economically produce large numbers of devices by exploiting potential for optoelectronic integration.

Doping of II-VI semiconductors with manganese, which occupies the cation sites, can be achieved to relatively high levels compared to III-V compounds. Although not discussed here in any detail, the interested reader will find low level manganese doping reported for $Ga_{0.47}In_{0.53}As$[1], InP[2,3], $GaAs$[4], $InAs_xP_{1-x}$[5] and $In_{1-x}Ga_xAs_yP_{1-y}$[6]. The low-level doping is mainly used to achieve p-type or semi-insulating material. Most attempts to achieve high doping levels of III-V compounds have resulted in phase segregation, but recently a new class of metastable diluted magnetic semiconductors, $In_{1-x}Mn_xAs$ (x < 0.18) has been prepared by low-temperature molecular beam epitaxy[7,8]. So far no III-V DMS have been reported as the active element of a device, although they have been used as current blocking layers.

This review will focus on three of the more commonly studied II-VI compounds which are already finding commercial application in optoelectronic devices.

2. $Cd_{1-x}Mn_xTe$

By far the most widely studied DMS, this alloy forms the active element of numerous devices which will be considered below. As discussed in a recent review[9], DMS device possibilities can be grouped into three classes. The first depend only on the semiconductor bandgap, lattice constant or other composition-dependent properties. For example, $Cd_{1-x}Mn_xTe$ can be used as a lattice matched substrate, or heteroepitaxial buffer layer, for $Hg_{1-x}Cd_xTe$ or quaternary epilayers.

The second class of devices utilize optical transitions of the Mn^{2+} ion such as electroluminescent devices discussed elsewhere in this volume. The third class are based on the sp-d exchange interaction. For economy and convenience, it is desirable to miniaturize devices and reduce touch labour costs. Early devices reported from $Cd_{1-x}Mn_xTe$ employed bulk single crystal. The preparation of $Cd_{1-x}Mn_xTe$ bulk crystal from the melt can involve temperatures in excess of $1000^\circ C$. It is particularly difficult to grow from the melt due to small stacking fault energy, tendency to twin formation, small thermal conductivity, and low critical shear stress energy leading to high dislocation density[10]. Further details of the bulk crystal growth, physico-chemical and optical properties of $Cd_{1-x}Mn_xTe$ can be found in several excellent reviews and monographs [11-13] and in Chapter 1 of this book.

LPE growth has been reported[14] from Te-rich solutions at temperatures as low as $700^\circ C$, but this severely restricts the range of substrates which can be used for epitaxy. The lowest growth temperatures of thin films have been achieved with MBE[15], pulsed laser evaporation and epitaxy[16] or ionized cluster beam techniques[17]. Unfortunately manganese is not available commercially more pure than 99.99% and it has been necessary for researchers to vacuum distil it prior to MBE or solution growth. For mass production, MOCVD growth would be preferred since chemical precursors are more readily refined. The high cracking temperature of commonly employed organometallic manganese feedstocks limits the growth of $Cd_{1-x}Mn_xTe$ to temperatures of $410-450^\circ C$[18-20]. Operating at the lower temperature, the maximum Mn fraction achievable was limited by lack of feedstock volatility as well as inefficient cracking.

Through careful selection of precursors[30,31], $Cd_{1-x}Mn_xTe$ has recently been grown by MOCVD at 320-350°C[32]. The new feedstocks enabled growth of superlattices[33-35] at temperatures comparable to MBE and Mn fraction up to 0.8.

2.1 Solar cells from $Cd_{1-x}Mn_xTe$

Perhaps the largest scale use which can be envisaged for $Cd_{1-x}Mn_xTe$ is as the top, high band gap, component of multilayer, or cascade solar cells. It has been estimated[21] that a top cell of potential efficiency 17.5%, combined with a $Hg_{1-x}Cd_xTe$ bottom cell efficiency of about 15%, could yield overall efficiency of 24%. Allowing for non-optimum performance[24], a top cell with >10% efficiency and about 80% subgap transmission, coupled with a bottom cell efficiency of 12-15%, can produce tandem efficiency of 15-20%. This would provide electricity at a potential cost of <10 cents $kW^{-1}h^{-1}$. There is commercial pilot scale development of these cells, and various methods are being explored to reduce production costs versus efficiency[22].

A very interesting application for solar cells, is the recent demonstration of efficient power delivery over optical fibre using small coupled solar cells[23]. This would eliminate the need to supply electric power supply to fibre optic control systems, or even repeaters. Coupling in series can provide up to 12 volt electrical power. The efficiency of the power transfer can be as high as 50%, because only a limited range of wavelengths is transmitted. II-VI materials can extend the useful wavelength range for power delivery considerably, and could be used, for example, in fluoride or chalcogenide fibre systems. The availability of high power laser diodes of many watts continuous power have made this application a practical reality. Many military uses for micro solar cells have been forecast, including fly-by-light aircraft control systems, electromagnetic interference-free devices and systems, phased array radar with completely fibre-fed antennae, bomb detonators for nuclear weapons and sensors for harsh environments[23].

Polycrystalline $Cd_{1-x}Mn_xTe$ grown by MOCVD has been used to construct solar cells with efficiencies of up to 6.7% using the non-optimum Mn fraction of 0.05[24]. Further work was considered necessary to control doping, ohmic contact formation and adhesion to the substrate before these cells become a commercial reality[24]. The use of the new precursors[32-35] mentioned above to grow solar cell structures at lower temperatures can be anticipated to dramatically improve solar cell characteristics through improved surface morphology, decreased thermal and chemical degradation of substrates, and improvement of doping characteristics. Efficient use of organometallic feedstocks will reduce production costs and waste generation. Large area MOCVD of $Cd_{1-x}Mn_xTe$ has already been demonstrated[33].

2.2 Magnetic field sensors from $Cd_{1-x}Mn_xTe$

When placed in a magnetic field, $Cd_{1-x}Mn_xTe$ experiences a very large Zeeman splitting of energy levels and very large Faraday rotation is observed at room temperature. This leads to interesting applications such as magnetic and electric field sensors[37-46]. Sumitomo Electric developed magnetic field (current) sensors in 1982 using bismuth silicon oxide (BSO) single crystal for

use in current measurement in power transmission lines[37].
In 1983 they employed ZnSe as well as BSO for this
purpose. In 1986 Sumitomo began to develop a more
sensitive device, employing Liquid Encapsulated
Czochralski (LEC) bulk single crystalline $Cd_{1-x}Mn_xTe$, for
use in power distribution lines that have lower currents
than transmission lines. Horizontal Bridgman growth was
used for volume production of these sensors[37]. For
$Cd_{0.75}Mn_{0.25}Te$, at 730 nm wavelength, the Verdet constant
of 5 min/Oe.cm was measured[37], far superior to ZnSe or BSO
crystals. However optical transmission loss varied from 1-
8 dB/mm. This was attributed to poor homogeneity of the
crystal.

Reduction of growth rate reduced transmission loss to
1 dB/mm[37]. This enabled production of fibre optic magnetic
field sensors with signal to noise ratio more than 10
times that of ZnSe or BSO and less than 1% variation with
temperature from -10 to 80°C. The sensors can be used in
watthour meters and non-destuctive inspection of steel
structures. In order to optimize sensor design, the
dispersion of Faraday rotation with composition,
temperature, and wavelength is required[38,39,42]. The
Faraday effect can be used to frequencies of several GHz,
so that a sensor's operating band is determined by the
other system components, such as the detector. Speckle
noise of laser diodes can reduce sensitivity and signal to
noise ratio but LEDs prove satisfactory[39].

2.3 Optical isolators and modulators from $Cd_{1-x}Mn_xTe$

Nonreciprocal components for optical communications
are important to reduce optical feedback and noise in the
light source. Reflections can lead to frequency pulling,

changes in phase, amplitude and polarization[40]. Isolation of 60 dB or more can be required. Coherent transmission systems require extreme stability of the laser to <1 MHz. Other important applications of isolation are multiport circulators for bidirectional use of fibres, separation of forward and backward waves in reflection measurements such as optical time domain reflectometry, or stability improvements in laser gyros[40]. It would be highly desirable to incorporate isolation in integrated optoelectronic chips. Isolators are fabricated from DMS by aligning a polarizer, the DMS Faraday rotator and an analyzer. Circulators are formed by coupling the isolator to separate ports using polarization-dependent beam splitters. A large Verdet constant enables size reduction of the chip. The integration of magnetic layers with III-V technology is an active area of research.

Recently $Cd_{1-x}Mn_xTe$ superlattices were grown by low temperature MOCVD on GaAs and InP substrates[33,34]. Using thin (~1 um) films grown by ICBD on transparent conductive indium tin oxide, it has been shown that electric fields applied perpendicular or parallel to the film surface can induce large changes in the absorption edge and Faraday rotation[41]. This was attributed[43] to the Franz-Keldysh effect, in which an external field induces wave functions to leak into the forbidden band, decreasing the apparent band gap. The electric fields were applied by use of 30 nm thin, transparent, aluminium contacts and voltages of 0-20 V perpendicular to the films, or 0-1 kV for electrode spacings of 1 mm parallel to the surface[41]. The high resistivity of $Cd_{1-x}Mn_xTe$ enables the use of relatively high fields. The change in Verdet constant was found to increase with manganese fraction. The material is considered satisfactory for fabrication of optical modulators, even on glass subtrates[43,45]. An extension of the films to multilayer structures has shown that quantum size effects enhance the disperion of Faraday rotation[46]. The superlattices, deposited by ICBD at 300°C, were 40-60 period stacks of alternating layers of $Cd_{0.9}Mn_{0.1}Te$ and $Cd_{0.5}Mn_{0.5}Te$. Because there is little difference in the valence band energy levels between the two compositions, the $Cd_{0.9}Mn_{0.1}Te$ forms deep quantum wells with a conduction band discontinuity of 0.56 eV at 300K[46].

The multilayer structure provides another method of tuning the properties of the material through variation of the well width. The Verdet constant was observed[46] to drastically increase when the well width decreased below the Bohr radius of an exciton (~6 nm), suggesting the enhancement is caused by confinement of excitons in the well layer, as observed in AlAs-GaAs quantum wells.

2.4   MESFETs and Schottky diodes from $Cd_{1-x}Mn_xTe$

The close lattice match of $Cd_{1-x}Mn_xTe$ and $Hg_{1-x}Cd_xTe$ should make it possible to monolithically integrate devices such as transistors, lasers and even infrared focal plane arrays in the II-VI or II-VI/III-V heteroepitaxial material systems. The first electronic devices reported in a DMS material were metal-semiconductor field-effect transistors and Schottky diodes fabricated from n-type indium-doped $Cd_{0.94}Mn_{.06}Te$ grown by photoassisted MBE[47] at 230°C. The doping density was 1 X 10^{17} cm^{-3}. Indium was used for ohmic contacts and gold for Schottky contacts. The diodes had turn-on voltages of 0.8 V and reverse breakdown voltages of 5.5-10.5 V. Transistor action was limited by large gate length and the thick epilayer prevented pinch-off, but the measured characteristics were encouraging for the prospect of monolithic integration[47]. The possibility of modulation of electronic devices, such as those described above, by an external magnetic field, opens interesting possibilities not currently available in III-V technology.

2.5 Visible and near-infrared lasers from $Cd_{1-x}Mn_xTe$

Stimulated emission by optical pumping at low temperatures has been observed in $Cd_{1-x}Mn_xTe$-$Cd_{1-y}Mn_yTe$ and $Cd_{1-x}Mn_xTe$-CdTe multiple quantum well structures grown on GaAs (111B) substrates[48-50]. With CdTe wells, near infrared emission from 760-790 nm was obtained and addition of manganese to the well enabled visible red operation at 665-670 nm. The threshold pumping power was comparable to that required for optical pumping of III-V laser structures. Magnetic tuning has been demonstrated[48] with the spectral peak of emission shifting to lower energy with increasing field at the rate of 3.4 meV/T. This was slower than in bulk material of comparable manganese fraction to the wells, possibly due to the high carrier temperature compared with the liquid helium bath temperature. In order to obtain injection lasers it will be necessary to produce doping heterostructures in $Cd_{1-x}Mn_xTe$ and there has been considerable success in this area using indium or antimony in photoassisted MBE[50]. It appears that difficulties in achieving good ohmic contacts to the material might be the reason for lack of reports to date on light emission by carrier injection.

The compound MnTe is particularly difficult to grow[11] from the bulk and crystallizes in the nickel arsenide structure, which is incompatible with the zincblende CdTe. Recently however, it has been found that metastable cubic

MnTe can be incorporated in heterostructures grown by MBE[51]. While the optical gap of bulk hexagonal MnTe is 1.3 eV (p-d) transition), the useful gap in the cubic epilayers is about 3.2 eV at liquid helium temperatures[51]. There is a 3% lattice mismatch with CdTe, but good pseudomorphic material has been obtained. Efficient photoluminescence was obtained from single quantum wells in the infrared, red, yellow-green and blue (2.6 eV) by varying the well thickness. If these structures can be made to emit at higher temperatures, they will become attractive for light sources across the visible spectrum. The possibility exists that such structures could be made by MOCVD. Recently a novel organometallic precursor incorporating a manganese-tellurium bond was shown to decompose to MnTe at 300°C in hydrogen[52]. Unfortunately the compound lacked sufficient volatility, but work is continuing to find similar, more satisfactory precursors.

2.6 Waveguide structures in $Cd_{1-x}Mn_xTe$

The dispersion of refractive index versus composition and wavelength has been reported for $Cd_{1-x}Mn_xTe$ films grown on glass substrates by hot wall epitaxy[53]. The refractive index decreases with an increase in manganese content or decrease in photon energy. Multilayer structures would offer the possibility of both optical and electrical confinement. The addition of mercury to make the quaternary alloy gives further flexibility to this system. Multilayer stacks are of interest also as filters, reflectors, surface emitting lasers and modulators. The effect of electric and magnetic fields on such films remains an active and fruitful area for study.

2.7 Nonlinear devices from $Cd_{1-x}Mn_xTe$

Theoretical studies of Hg-based diluted magnetic semiconductor superlattices show that the optical cutoff wavelength can be substantially tuned by a magnetic or electric field at 77 K[36]. A superlattice of HgTe (7.3 nm) and $Cd_{0.87}Mn_{0.13}Te$ (7.5 nm), chosen for lattice match, showed a variation of cutoff wavelength from 16 to 26 microns at 77 K for fields up to 6 T. Decreasing the well width to 5.8 nm shifted the cutoff range to 10-14 microns. Applications of such films include high-speed infrared modulators, bandpass filters and high-speed nonlinear devices. The latter would respond to the electric field induced by a powerful short pulsed laser, and be faster than devices dependent on changes in carrier concentration or thermoeffects[36].

Recently two-beam coupling at 1.5 microns was reported for CdTe:V, with higher efficiency than for GaAs at 1.06 microns[54]. High resistivity layers of $Cd_{1-x}Mn_xTe$ can probably be expected to yield interesting photorefractive effects, with the possibility of electric or magnetic field tuning. Multiple quantum well structures of DMS have been found to have anisotropic optical propeties for light polarized parallel and perpendicular to the layers[55], leading to new possibilities in nonlinear behaviour. Recently laser beam self-focusing was observed in bulk $Cd_{0.4}Mn_{0.6}Te$[56]. Thermally induced change in nonlinear refractive index is the main mechanism and this can be exploited in such practical applications as light power limiters and optically bistable devices. A magnetic field was found not to affect self-focusing[56].

3. $Hg_{1-x}Mn_xTe$

Theoretical and photoemission studies indicated that $Hg_{1-x}Mn_xTe$ should exhibit more stable Hg-Te bonding than is found for $Hg_{1-x}Cd_xTe$[57]. The energy gap varies roughly two times faster with x for manganese than for cadmium, making cutoff wavelength control more difficult. An empirical formula for the energy gap has recently been reported[59], E_g = -0.275 + 3.47x + $6.45X10^{-4}T$ $-3X10^{-3}xT$. Growth by MBE gives n-type or p-type material depending on the mercury overpressure[58], whereas the Bridgman method from Te-rich solution or other bulk techniques yields p-type material[59,61]. It is also made by LPE[14] and recently was grown polycrystalline for the first time by low temperature MOCVD[32]. The first epitaxial MOCVD of $Hg_{1-x}Mn_xTe$ was achieved at 350-400°C directly on GaAs substrates[60].

3.1 Infrared detectors from $Hg_{1-x}Mn_xTe$

Detectors are already commercially available in this material and a recent excellent review[61] provides comprehensive information on the properties of the ternary, theoretical performance and the fabrication technology of photoconductive, photovoltaic and photoelectromagnetic detectors made from it. It was concluded that at present $Hg_{1-x}Mn_xTe$ photodiodes show inferior performance to $Hg_{1-x}Cd_xTe$ ones[61]. Possible advantages of the manganese material for device applications include lattice stability mentioned above and reduced disorder scattering since less manganese is required than cadmium to produce the equivalent gap. Further details on diode fabrication and near room temperature operation have been reported[62-66]. At liquid helium temperatures magnetic tuning of the band gap has been observed[62]. Although some authors consider[65] there are no decisive advantages of $Hg_{1-x}Mn_xTe$ over $Hg_{1-x}Cd_xTe$, calculations show that HgMnTe superlattices could be magnetically tuned at 77K[36]. Low temperature epitaxy will be required for such structures.

3.2 Optical isolators from $Hg_{1-x}Mn_xTe$

Faraday rotation and absorption of $Hg_{1-x}Mn_xTe$ has been studied in the wavelength region 1.3 to 1.55 microns, corresponding to silica optical fibre communications[67]. There is intense interest in high-quality inexpensive magneto-optical isolators to protect semiconductor lasers from reflected noise at these wavelengths. Since Faraday rotation in DMS is largest just below the band gap, Hg provides a means of moving to longer wavelength than $Cd_{1-x}Mn_xTe$. It was found[67], using bulk material, that room-temperature operation of $Hg_{1-x}Mn_xTe$ isolators is feasible, although specific rotation was one tenth that of bismuth-doped ferrimagnetic garnets. The fact that the ternary can be grown epitaxially on GaAs, and the likely improvement of material properties by low temperature MOCVD, make this material one of only a few which can be considered for monolithic integration of waveguides and nonreciprocal devices.

3.3 Light emitting diodes from $Hg_{1-x}Mn_xTe$

Pulsed current light emitting diodes and laser heterostructures have been fabricated by a two step LPE growth of $Hg_{1-x}Mn_xTe$ layers on graded $Hg_{1-x-y}Mn_xCd_yTe$

which was obtained by isothermal vapor phase epitaxy of HgTe onto $Cd_{1-x}Mn_xTe$ doped with indium[68]. This reduced the series resistivity, enabling higher luminescence efficiency than reported earlier, where junctions were achieved by mercury annealing p-type substrates[63,69]. The LEDs emitted in the 5.3-5.7 micron region. Laser action was achieved at 5.33 microns by current injection at operating temperatures of 77 K with a threshold current density of about 1.2 kA/cm^2. The application of a transverse magnetic field at 15 K increased the saturation current density. This was attributed to a high negative magnetoresistivity. The emission peak shifted to lower energy and the shape changed to one characteristic of quantized interband emission[68]. Spin-flip Raman lasers are considered possible in this material[9].

3.4 Nonlinear devices from $Hg_{1-x}Mn_xTe$

Free-carrier-induced optical non-linearity with saturation power densities of 100 kW cm^{-2} to 1 MW cm^{-2} has been studied in narrow gap semiconductors, including $Hg_{1-x}Mn_xTe$[70], with possible application in phase conjugation elements, tunable filters and power limiters. Free carriers enable picosecond speeds and can be controlled electrically or magnetically. The non-linear optical coefficient of zero-gap $Hg_{0.97}Mn_{0.03}Te$ is independent of laser intensity[70]. The use of superlattices incorporating $Hg_{1-x}Mn_xTe$ could enable tunable saturable absorbers for generation of ultrashort light pulses.

4. $Hg_{1-x-y}Mn_yCd_xTe$

This homogeneous quaternary alloy can be grown by bulk or LPE methods. Other techniques such as isothermal vapour phase epitaxy inevitably lead to graded material. The lattice constant of $Cd_xHg_yMn_zTe$ varies as reported recently, $a = 0.6481 - 0.0019y - 0.0144z$ and the room-temperature bangap $E_g = 1.46 - 1.62y + 1.33z$[71]. The quaternary alloys are expected to become the most important II-VI alloys for optoelectronic applications[72].

Thermodynamic analysis shows that $Hg_{1-x-y}Mn_yCd_xTe$ is the II-VI quaternary least hindered by solid miscibility problems, being completely miscible over the entire composition range for temperatures above 35 K[72]. This is particularly important for solution or vapour phase growth. The quaternary was recently grown by LPE from Hg rich solution for the first time on CdTe substrates[73]. This inevitably leads to a composition gradient region of more than 10 microns due to interdiffusion with the substrate at 390°C over the growth time (up to 50 hours). There is great interest in growth of the alloy from the vapour phase but so far there have been no reports of MBE or direct alloy MOCVD growth of the quaternary. Recently however, it has been shown that the quaternary can be grown by interdiffusion of $Cd_{1-x}Mn_xTe$ and HgTe multilayers grown on substrates such as GaAs, InP or sapphire by low temperature MOCVD[74]. Such layers can be completely homogenized by careful control of annealing.

4.1 Infrared detectors from $Hg_{1-x-y}Mn_yCd_xTe$

In order to make sensitive room-temperature detectors for wavelengths in the range 1.3 to 1.8 microns, one approach has been to achieve a resonantly enhanced ionization ratio of holes to electrons such that the bandgap equals the spin-orbit splitting. This occurs at 1.3 microns for $Hg_{1-x}Cd_xTe$ and 1.8 microns for $Hg_{1-x}Mn_xTe$. Calculation[75] for the quaternary shows the room-temperature resonance to occur at 1.55 microns and a composition of $Hg_{0.49}Mn_{0.19}Cd_{0.32}Te$. Variation of $x + y$ allows tuning of peak response in the range 1.5-1.65 microns. Using LPE grown epilayers on CdTe (lattice mismatch 0.6%), at about 550°C, p-type quaternary has been grown and fabricated into boron-implanted mesa diodes with breakdown voltages of ~50V and leakage currents similar to Ge or InGaAs avalanche photodiodes[75]. It was concluded that the quaternary gives equal or superior performance to III-V or IV devices. Longer wavelength detectors have been reported[63,76,77] and further details can be found in the chapter by Pawlikowski in this volume[78].

4.2   Light emitting diodes from $Hg_{1-x-y}Mn_yCd_xTe$

Light emitting diodes have been reported in conjunction with mercury manganese telluride LEDs[63,68,69] discussed above in section 3.3. The quaternary allows more flexibility in tuning of lattice constant, band gap, optical confinement and electrical properties. Control of material quality and doping through techniques such as low temperature MOCVD should lead to exciting developments in tunable infrared sources from this alloy system.

4.3   Optical isolators from $Hg_{1-x-y}Mn_yCd_xTe$

Recently it has been found that the quaternary alloy demonstrates larger Verdet constants for material with the same energy gap as can be achieved with the corresponding ternary, $Cd_{1-x}Mn_xTe$[79]. This allows fabrication of non-reciprocal devices only half the size required with the ternary alloy and has important economic and performance consequences for optoelectronic integrated circuits, as well as larger hybrid components.

CONCLUSION

The intensive research activity in $Hg_{1-x-y}Mn_xCd_yTe$, including end members, has been largely sponsored by military requirements to date. However the fundamental science supporting the material and device technology has shown that broader markets for these materials, and systems derived from them, exist, often limited only by imagination. The involvement of major commercial players such as AT&T[67,72], IBM, Mitsubishi Electric Corporation[39], NASA[19], NTT[79], Rockwell International[75] and Sumitomo Electric[37] shows that these materials are more than a scientific curiosity. Smaller organizations such as Ametek[22,29], Brimrose Corporation[54], Boston Optronics[64] and Vigo Ltd.[64,65], are also actively involved in bringing optoelectronics based on manganese-doped narrow band gap II-VI semiconductors to the market place for a wide range of applications.

REFERENCES

1. Kunzel, H., Bochnia, R., Gibis, R, Harde, P. and Passenburg, W., Appl. Phys. A51, 508-514 (1990)

2. Huang, K. and Wessels, B.W., J. Appl. Phys. 67, 6882-5 (1990)

3. Laiho, R., Lahderanta, E., Supponen, E. and
    Vlasenko, L., Phys. Rev. B41, 7674-7 (1990)

4. Fabre, F., Bacquet, G., Frandon, J., Bandet, J.
    Taouint, R. and Paget, D., Sol. State. Comm. 71,
    717-719 (1989)

5. Huang, K. and Wessels, B.W., Appl. Phys. Lett. 52,
    1155-7 (1988)

6. Ishikawa, J., Takahashi, N.S., Ito, T., Sube, M. and
    Kurita, S., J. Electrochem. Soc. 137, 343-5 (1990)

7. Munekata, H., Ohno, H., von Molnar, S., Segmuller,
    A., Chang, L.L. and Esaki, L., Phys. Rev. Lett. 63,
    1849-52 (1989)

8. Munekata, H., Ohno, H., von Molnar, S., Harwit, A.,
    Segmuller, A. and Chang, L.L., J. Vac. Sci. Technol.
    B 8, 176-80 (1990)

9. Samarth, N. and Furdyna, J.K., Proc. IEEE 78, 990-
    1003 (1990)

10. Kotani, T., Nakanishi, F., Yasuo, H., Shibata, M.
    and Tada, K., Sumitomo Electric Technical Review 27,
    166-173 (1988)

11. Pajaczkowska, A. Prog. Crystal Growth Charact. 1,
    289-326 (1978)

12. Goede, O. and Heimbrodt, W., Phys. Stat. Sol. 146,
    11-62 (1988)

13. Aggarwal, R.L., Furdyna, J.K. and von Molnar, S.
    (editors) "Diluted Magnetic (Semimagnetic)
    Semiconductors" (Vol. 89, Materials Research
    Society Symposia Proceedings, Pittsburgh, PA,
    (1987))

14. Becla, P., Wolff, P.A., Aggarwal, R.L. and Yuan,
    S.Y., J. Vac. Sci. Technol. A 3, 116 (1985)

15. Lunn, B. and Davies, J.J., Semicond. Sci. Technol.
    5, 1155-60 (1990)

16. Wrobel, J.M. and Dubowski, J.J., Appl. Phys. Lett.
    55, 469 (1989)

17. Koyanagi, T., Matsubara, K., Takaoka, H. and Takagi,
    T., J. Appl. Phys. 61, 3020 (1987)

18. Nouhi, A. and Stirn, R.J., Appl. Phys. Lett. 51, 2251
    (1987)

19. Nouhi, A. and Stirn, R.J., US Patent Application
    (National Aeronautics and Space Administration)
    No. US 248019 (1988)

20. Mazur, J.H., Grodzinski, P., Nouhi, A. and Stirn,
    R.J., Mater. Res. Soc. Symp. Proc. 102, 337 (1988)

21. Landis, G.A., Solar Cells 25, 203 (1988)

22. Meyers, P.V., Solar Cells 24, 35-42 (1988)

23. Henderson, B.W., Aviation Week & Space Technology
    August 13 (1990) p74

24. Rohatgi, A., Ringel, S.A., Welch, J., Meeks,
    Pollard, K., Erbil, A., Meyers, P.V. and Liu, C.H.,
    Solar Cells, 24, 185-94 (1988)

25. Feng, Z.C., Sudharsanan, R., Perkowitz, S., Erbil,
    A., Pollard, K.T. and Rohatgi, A., J. Appl. Phys. 64,
    6861 (1988)

26. Feng, Z.C., Perkowitz, S., Sudharsanan, R., Erbil,
    A., Pollard, K.T., Rohatgi, A., Bradshaw, J.L. and
    Choyke, W.J., J. Appl. Phys. 66, 1711-16 (1989)

27. Sudharsanan, R., Feng, Z.C., Perkowitz, S.,Rohatgi,
    A., Pollard, K.T. and Erbil, A., J. Electronic Mater.
    18, 453-5 (1989)

28. Nouhi, A., Stirn, R.J., Meyers, P.V. and Liu, C.H.,
    J. Vac. Sci. Technol. A 7, 833-36 (1989)

29. Rohatgi, A., Ringel, S.A., Sudharsanan, R., Meyers,
    P.V., Liu, C.H. and Ramanathan, V., Solar Cells 27,
    219-30 (1989)

30. Pain, G.N., Christiansz, G.I., Dickson, R.S.,

582

Deacon, G.B., West, B.O., McGregor, K. and Rowe,
R.S., Polyhedron 9, 921-29 (1990)

31. Christiansz, G.I., Elms, T.J., Pain, G.N. and
Pierson, R.R., J. Crystal Growth 93, 589-93 (1988)

32. Pain, G.N., Bharatula, N., Christiansz, G.I.,
Kibel, M.H., Kwietniak, M.S., Sandford, C.,
Dickson, R.S., Rowe, R.S., McGregor, K., Deacon,
G.B., West, B.O., Glanvill, S.R., Hay, D.G.,
Rossouw, C.J. and Stevenson, A.W., J. Crystal Growth
101, 208-10 (1990)

33. Pain, G.N., Russo, S.P., Elliman, R.G., Wielunski,
L.S., Gao, D., Glanvill, S.R., Rossouw, C.J.,
Stevenson, A.W., Rowe, R.S., Deacon, G.B., Dickson,
R.S. and West, B.O., Materials Forum 15, 35-43 (1991)

34. Pain, G.N., Warminski, T., Sulcs, S., Kwietniak,
M.S., Gao, D., Glanvill, S.R., Rossouw, C.J.,
Stevenson, A.W., Russo, S.P., Elliman, R.G.,
Wielunski, L.S., Rowe, R.S., Deacon, G.B., Dickson,
R.S. and West, B.O., Applied Surface Science 1991, in
press

35. Gao, D., Stevenson,A.W., Wilkins, S.W., and Pain,
G.N., J. Crystal Growth 1991, in press.

36. Yang, Z., Schetzina, J.F. and Furduyna, J.K., J. Vac.
Sci. Technol. A 7, 360-64 (1989)

37. Katsuda, M., Hosoe, K. and Nakaseko, M., Sumitomo
Electric Technical Review 30, 84-7 (1990)

38. Aksionov, D.A., Konov, V.I., Nikitin, P.I.,
Prokhorov, A.M., Savchuk, A.I., Savitski, A.V. and
Ulyanitski, K.S., Sensors and Actuators, A21-A23,
875-878 (1990)

39. Mikami, N., Nagao, C., Sawada, T., Takahashi, H.,
Furakawa, Y. and Aikawa, E., J. Appl. Phys. 69, 433-8
(1991)

40. Schmitt, H.J. and Dammann, H., J. Instn. Electronics & Telecom. Engrs. 34, 286-297 (1988)

41. Yamano, K., Sota, T., Koyanagi, T., Nakamura, K., and Matsubara, K., IEEE Translation J. Magnetics in Japan, 5, 342-7 (1990)

42. Koyanagi, T., Yamano, K., Sota,T. and Matsubara, K., IEEE Translation J. Magnetics in Japan 5, 306-312 (1990)

43. Nakamura, K., Koyanagi, T., Yamano, K. and Matsubara, K., J. Appl. Phys. 65, 1381-3 (1989)

44. Nakamura, K., Koyanagi, T., Yamano, K. and Matsubara, IEEE Translation J. Magnetics in Japan 3, 597-8 (1988)

45. Koyanagi, T., Yamano, K., Sota, T., Nakamura, K. and Matsubara, K., Jap. J. Appl. Phys. 28, L669-671 (1989)

46. Koyanagi, T., Watanabe, T., Nakamura, K., Yamano, K. and Matsubara, K., Nucl. Instrum. Meths. Phys. Res. B37/38, 878-81 (1989)

47. Dreifus, D.L., Kolbas, R.M., Harper, R.L., Tassitino, J.R., Hwang, S. and Schetzina, J.F., Appl. Phys. Lett. 53, 1279-81 (1988)

48. Isaacs, E.D., Heiman, D., Zayhowski, J.J., Bicknell, R.N. and Schetzina, J.F., Appl. Phys. Lett. 48, 275-7 (1986)

49. Bicknell, R.N., Giles-Taylor, N.C., Schetzina, J.F., Anderson, N.G. and Laidig, W.D., J. Vac. Sci. Technol. A 4, 2126 (1986)

50. Bicknell, R.N., Giles, N.C. and Schetzina, J.F., Mater. Res. Symp. Proc. Vol. 90, 163-70 (1987)

51. Ding, J., Pelekanos, N., Fu, Q, Walecki, W., Nurmikko, A.V., Han, J., Durbin, S., Kobayashi, M. and Gunshor,R.L., Abstract MoP-134, 20th

International Conference on the Physics of
Semiconductors, Thessaloniki, Greece, August 6-10
(1990)

52. McGregor, K., Deacon, G.B., Dickson, R.S., Fallon,
    G.D., Rowe, R.S. and West, B.O., J.C.S. Chem. Comm.
    1293-4 (1990)

53. Miotowski, I. and Miotowska, S., Thin Solid Films,
    165, 91-7 (1988)

54. Partovi, A., Millerd, J., Garmire, E.M., Ziari, M.,
    Steier, W.H., Trivedi, S.B. and Klein, M.B., Appl.
    Phys. Lett. 57, 846 (1990)

55. Gunshor, R.L., Kolodziejski, L.A., Nurmikko, A.V.,
    and Otsuka, N., Ann. Rev. Mater. Sci. 18, 325-50
    (1988)

56. Dai, X.D., Ito, Y. and Ja, Y.H., Aust. J. Phys. 43,
    303-10 (1990)

57. Wall, A., Caprile, C., Franciosi, A., Reifenberger,
    R. and Debska, U., J. Vac. Sci. Technol. A 4, 818-22
    (1986)

58. Faurie, J.P., Reno, J., Sivananthan, S., Sou, I.K.,
    Chu, X., Boukerche, M. and Wijewarnasuriya, P.S.,
    J. Vac. Sci. Technol. A 4, 2067-71 (1986)

59. Piotrowski, T., Tomm, J.W. and Puhlmann, N., Phys.
    Stat. Sol. 117, K181 (1990)

60. Clifton, P.A., Brinkman, A.W. and Al Allak, H.M.,
    Semicond. Sci. Technol. 5, 1067-9 (1990)

61. Rogalski, A., Infrared Phys. 31, 117-166 (1991)

62. Wong, S. and Becla, P., J. Vac. Sci. Technol. A 4
    2019-23 (1986)

63. Becla, P., J. Vac. Sci. Technol. A 4, 2104-18 (1986)

64. Becla, P., Grudzien, N. and Piotrowski, J., Abstracts
    1990 U.S. Workshop on the Physics and Chemistry of
    Mercury Cadmium Telluride and novel IR detector

materials, San Francisco, Oct. 1-4 1990

65. Piotrowski, J., Galus, W. and Grudzien, M., Infrared Phys. 31, 1-48 (1991)

66. Rogalski, A. and Rutkowski, J.,Infrared Phys. 29, 887-93 (1990)

67. Dillon, J.F.Jr., Furdyna, J.K., Debska, U. and Mycielski, A., J. Appl. Phys. 67, 4917-9 (1990)

68. Becla, P., J. Vac. Sci. Technol. A 6, 2725-7 (1988)

69. Becla, P., MIT Materials Research Review, September 1988, p489

70. Wolff, P.A. and Auyang, S.Y., Semicond. Sci. Technol. 5, S57-67 (1990)

71. Manhas, S., Khulbe, K.C., Beckett, D.J.S., Lamarche, G. and Woolley, J.C., Phys. Stat. Sol. 143, 267 (1987)

72. Kisker, D.W., J. Crystal Growth 98, 127-139 (1989)

73. Takita, K., Uchino, T. and Masuda, K., Semicond. Sci. Technol. 5, S277-80 (1990)

74. Pain, G.N., Rossouw, C.J., Glanvill, S.R., Rowe, R.S. Dickson, R.S., Deacon, G.B. and West, B.O., J. Crystal Growth 107, 632-36 (1991)

75. Shin, S.H., Pasko, J.G., Lo, D.S., Tennant, W.E., Anderson, J.R., Gorska, M., Fotouhi, M. and Lu, C.R. Mater. Res. Symp. Proc. Vol. 89, 267-74 (1987)

76. Placzek-Popko, E., Dudziak, E., Jedral, L., Kasprzak, J.F. and Pawlikowski, J.M., Infrared Phys. 29, 903-5 (1989)

77. Placzek-popko, E. and Jedral, L., Infrared Phys. 28, 249-53 (1988)

78. Pawlikowski, J.M., Chapter 14, this volume.

79. Inukai, T. and Ono, K., Jap. J. Appl. Phys. 30, L198-201 (1991)

ELECTROLUMINESCENT DEVICES
USING ZnS: Mn FOR THE PHOSPHOR LAYER

Atsushi Abe

Matsushita Electric Industrial Co., Ltd.
3-15, Yagumo-Nakamachi, Moriguchi, Osaka
JAPAN

ABSTRACT

Characteristics of ZnS: Mn phosphor films for ac thin-film electroluminescent
(TFEL) devices were examined in detail ; the relationships between the
crystallographic, microstructural, chemical, and electroluminescent
characteristics and the growth mechanism, postannealing temperature, and
effects of an oxygen and some metal oxide impurities introduced into the
ZnS:Mn phosphor films were clarified.

1. INTRODUCTION

Figure 1 shows the device structure of the double-insulator-
layer-type ac thin-film electroluminescent (TFEL) device laminated
with the following thin films ; indium-tin oxide thin-film (a
transparent electrode film) / the first insulator layer / ZnS:Mn thin-
film (a phosphor layer) / the second insulator layer / Al thin-film (
a counter electrode film). Our investigation has been directed to the
development of new dielectric thin films having high permittivity
and figure-of-merit (permittivity multiplied by field strength : $\varepsilon \times E_b$)
and to the optimum design of the ac TFEL device structure[1]. A
figure-of-merit for insulator layers in ac TFEL devices is important
for realizing both sufficient brightness and stable operation without
breakdown. The value of figure-of-merit is the same as the
maximum charge density which can be stored on the surfaces of
insulator layers.

In a process used to produce a TFEL device, the annealing of a ZnS:Mn film after deposition is generally imposed to raise luminance of the EL device. Redistribution of Mn ions, improved crystallinity of the ZnS, and reduction of nonradiative recombination centers by the annealing process are generally considered to account for the improved luminance. The improved luminescence as a move of the Mn atoms / ions from the interstitial to the Zn lattice site is interpreted. The temperature of this heat treatment is, as a rule, highest among the processes for producing the EL device. In this annealing process an interdiffusion of ions between the ZnS:Mn and previously deposited films, such as dielectric and ITO (indium-tin oxide) films, may also take place in addition to the effects described above. This phenomenon should affect the luminescent characteristics of the EL device, although it is considered to depend on the dielectric materials.

In the postannealing of the phosphor film, the diffusions of an oxygen and various metal oxide impurities are possible from an annealing atmosphere and adjacent other films, respectively. This paper describes the experimental results of the influences of these

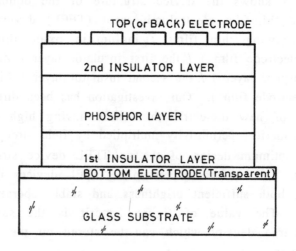

Fig. 1. Schematic illustration of the ac TFEL device.

impurities in the phosphor film on the EL characteristics such as luminance / voltage, charge density / voltage and emission threshold voltage.

2. STUDY OF THE CRYSTALLOGRAPHIC AND LUMINESCENT CHARACTERISTICS OF ZnS:Mn FILMS

Characteristics of ZnS:Mn films prepared by rf magnetron sputtering method (SP ZnS:Mn films) for ac TFEL devices were examined[2]. We have studied the crystallographic, microstructural, chemical, and electroluminescent characteristics of the SP ZnS:Mn films regarding the deposition conditions, and compared these characteristics with those of EB methods (EB ZnS:Mn films) in order to clarify features of SP ZnS:Mn films. To determine the direction of further improvement of the characteristics, the growth mechanism of SP ZnS:Mn film was considered, qualitatively, based on the results described above and in accordance with preliminary experimental results concerning incident species to the substrates during sputtering. It was found that the incident species to the substrate were composed of ZnS, Zn_2, Zn_2S, MnS and ZnMn in addition to Zn and S atoms ; in the case of the electron-beam method, they were composed of Zn and S_2.

2.1. Preparation of ZnS:Mn Ceramic and Powder Targets

The preparation of ZnS:Mn ceramic targets was carried out as follows. ZnS, MnS and sulfur powders were used as ingredients. These powders were mixed in a dry state by using a mill containing resin-coated iron balls. Sulfur powder was included to 2 wt.% to the total weight of ZnS and MnS mixtures after adjusting the ratios of MnS to ZnS. After mixing, the powders were hot-pressed at 1100°C for 0.5 h at a pressure of 25 kg/cm^2 using a finely machined carbon die. ZnS:Mn ceramics produced by hot-pressing have an apparent density of about 80% and those X-ray diffraction patterns exhibited hexagonal crystal symmetry and no evidence of a ZnO compound. The ZnS:Mn ceramics discs were coated by copper metal on one side of the major surface for metal bonding to a backing plate.

Powder-type targets of ZnS:Mn were also prepared by the following method. Ingredient ZnS powder and a $MnSO_4$ aqueous solution were mixed in a wet state with pure water. After drying at 150°C, the mixed powder was fired at 1200°C for 2 h in an atmosphere of Ar gas. While the temperature was being raised up to 500°C during firing, the vacuum atmosphere was maintained in order to remove any residual water in the mixed powder. Then, the atmosphere was changed to Ar gas. By this method, the yield of the ZnO compound by oxidization of the powder due to the water was prevented. After firing, the loosely sintered powder was pulverized in a mortar. Using the ZnS:Mn ceramic and powder targets discussed above, ZnS:Mn films were prepared by an rf magnetron sputtering method under various deposition conditions.

2.2. Preparative Examination of SP and EB ZnS:Mn Films

The deposition rate as a function of the sputtering (SP) gas pressure (P_s) was examined at a substrate temperature (T_s) of 200°C using the ceramic target and Ar sputtering gas. The rate was roughly constant below 3×10^{-2} Torr and decreased rapidly at higher pressures, showing about half at 8×10^{-2} Torr, compared to that at 3×10^{-2} Torr.

The dependences of the rate on T_s at $Ps = 3 \times 10^{-2}$ Torr and a power density of 0.8 W/cm^2 are shown in Fig. 2 for both ceramic and powder targets. The curves show average values of many data, since these data exhibited considerable fluctuation at each T_s. For a comparison, the result of the EB method is also described in Fig. 2. It was found that in the case of a powder target the deposition rate decreased more rapidly with T_s, though it was larger below 300°C, compared with the case of the ceramic target. A much stronger thermal quenching of the deposition rate was observed in the EB method, indicating almost zero rate at 270°C, as shown in Fig. 2. Generally, the SP ZnS:Mn film could be formed even in the T_s region where the sticking coefficient of the EB film was zero.

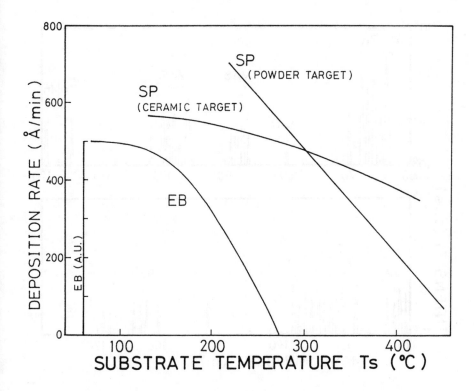

Fig. 2. Dependence of the deposition rates on the substrate temperature.

2.3. Growth Mechanism of SP and EB ZnS:Mn Films

The difference between the SP and EB methods with respect to the T_s dependence of the deposition rate implies an obvious difference in the incident species to the substrate in each case. In order to examine this difference, the sputtering deposition was simulated by using SIMS (A-DIDA 3000) equipment. Although the energies of Ar$^+$ ions are comparatively high (3 keV), compared with that of the actual sputtering, the mass spectrum of sputtered species

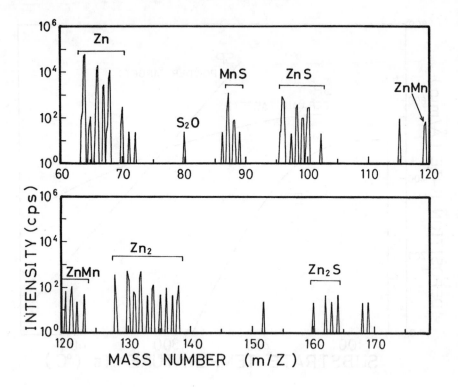

Fig. 3. Full mass range scanning of species sputtered from the ZnS:Mn target.

by incident Ar^+ ions into a ZnS:Mn target was measured by full-mass-range scanning, as shown in Fig. 3. S, Mn and SO were detected as relevant species at mass numbers lower than 60. As shown in Fig. 3, complex MnS, ZnS, MnZn, Zn_2 and Zn_2S were observed as sputtered species in addition to S and Zn atoms. On the other hand, Ban and White revealed Zn and S_2 as the species in thermal evaporation[3].

Film growth during thermal evaporation is considered to proceed by a formation of alternate layers of Zn and S atoms, resulting in an automatic achievement of stoichiometry. However, this mechanism may not be feasible in an SP film, as predicted by the sputtered species designated above, resulting in different deposition rate dependences on T_s between the SP and EB methods.

The higher values of the rate in a powder target than those in a ceramic target at a T_s below about 300°C were considered to be due to thermally assisted sputtering, due to a poorer thermal contact to cathode in the power target. However, a T_s higher than 300°C the deposition rates in the. powder target were rather smaller than those in the ceramic target. As one possible explanation, this may be attributed to the approach of the film growth to the thermal evaporation behavior to some extent, since the sputtered species from the heated powder target are inferred to include more Zn and S atoms compared with those from a cooled ceramic target.

The Mn concentration change in the film with the preparative conditions was examined. The results are summarized qualitatively in the following.

At low SP power, the Mn concentrations in the films were not sensitive to the T_s in both the ceramic and powder targets, being similar to those of the target concentration. But in the high SP power, the Mn concentration became sensitive to the T_s in the powder target. This may be due to heating of the powder target, approaching an evaporation behavior similar to the EB method, in which the Mn concentration tends to increase with T_s.

2.4. Crystallographic Analyses of SP and EB ZnS:Mn Films

Crystallographic analyses were carried out by X-ray diffraction on SP and EB ZnS:Mn films. The SP films at 8000Å thickness were prepared at a T_s of room temperature, 200°C and 350°C in Ar gas of 3×10^{-2} Torr using a ceramic target containing 1.2 at.% Mn.

It was noted that the SP film prepared at room temperature exhibited weak (100) and (101) peaks, which were characteristic to hexagonal symmetry, in addition to strong (002) and weak (112), (004) peaks by hexagonal indexing. Since (002), (112) and (004) peaks can also be indexed as cubic symmetry, only the hexagonal phase does not necessarily take place. Other SP films prepared at 200 and 300°C showed a cubic symmetry with the strongest peak of (111) in addition to weak (222) and (220), (311), respectively. The (111) peak intensity was highest in the film prepared at 200°C, and rather weaker at 350°C. A (111) orientation is also highest in the film prepared at 200°C. The EB films prepared at a T_s of room temperature and 200°C using a ZnS:Mn (0.8 at.%) ceramic evaporation source showed only a cubic phase, with the strongest peak of (111) in addition to weak (220), (311) and (311), respectively. The appearance of the hexagonal phase in the SP film is considered essentially to be due to previously indicated complex incident species such as ZnS, Zn_2 and Zn_2S to the substrate.

2.5. Microstructure of SP and EB ZnS:Mn Films

The microstructures of the films prepared on 7059 glass and glass / ITO substrates at a thickness of about 1.2 μm by the SP method (ceramic target of 1.2 at.% Mn content, T_s=200°C, and P_s=3x10^{-2} Torr) were observed.

A disordered columnar structure was observed up to 6000Å from the film / substrate interface. In this region the columns were shorter and had smaller diameters, compared to those in the upper part. In the upper region, above 6000Å, fairly thick columns of 1000-2000Å in diameter were observed. In general, however, the columnar structure was more ambiguous, compared with the one by the EB method. This columnar structure disappeared upon an annealing at 600°C, being different from the previous reports concerning films prepared by both the SP[4] and EB[5] methods. The films prepared on 7059 glass did not show the columnar structure. On the other hand, the EB films exhibited a distinct columnar structure at sizes of 600-1000Å in diameter, just growing from the surface of a 7059 glass substrate. The columnar structure by the EB method seemed to show no change after annealing at 600°C by a microscopic observation, confirming a previous report[5].

2.6. Chemical characteristics of SP and EB ZnS:Mn Films

Contamination by impurities and diffusion phenomenon through the interface were examined. The SP ZnS:Mn film is presumed to be contaminated by target impurities and an oxygen component in the deposition chamber. On the other hand, in an EB film, a purification effect for heavy-metal elements is expected, due to a difference in the vapor pressure between Zn and those elements. This is predicted from evidence that Mn tends to remain in the ceramic evaporation source, exhibiting a greenish surface color after EB evaporation. Table 1 shows one instance of a relative comparison with respect to each impurity by full mass range scanning of SIMS in

Table 1.   Impurities in ZnS:Mn films.

Impurity / Film	Na	B	OH	Mg	Ca	Cu	Fe Al Si Cr
EB ZnS : Mn	3 - 5	10	1	1	1	1	1
SP ZnS : Mn	1	1	6	20	50	300	3 - 5

the ZnS:Mn films prepared by the SP and EB methods. Both of the films were prepared by the same ZnS powder ingredient. Concerning light metals, Na for instance, EB film contains more by factors of 3-5, compared to SP film. As shown in the Table, the SP film contains more impurities, except for Na and B, than the EB film. These impurities were confirmed to be also included in the SP target and could be pursued to each root by emission spectroscopy.

Figure 4 shows one instance of the concentration profile by SIMS measurement in the neighborhood of an interface between ZnS:Mn and $Sr(Zr_{0.2}Ti_{0.8})O_3$ films. A substrate glass / ITO / $Sr(Zr_{0.2}Ti_{0.8})O_3$ was divided into two pieces, and films were deposited on each substrate by either SP or EB methods. After depositing, both

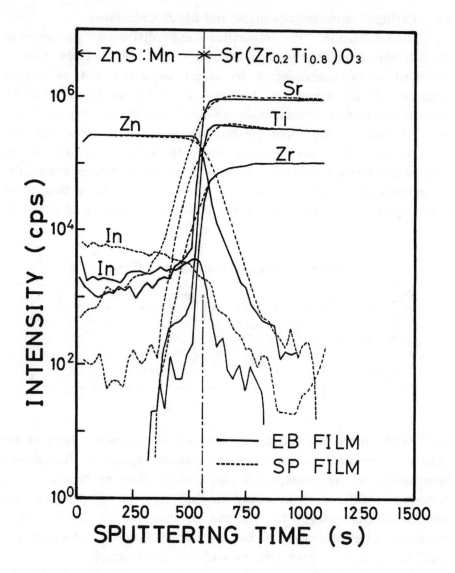

Fig. 4. In-depth concentration profiles of the composing ions in the neighborhood of an interface between the ZnS:Mn and Sr(Zr$_{0.2}$Ti$_{0.8}$)O$_3$ films.

the films were simultaneously annealed at 600°C for 1 h in a vacuum atmosphere. Each concentration profile of the composing elements (Zn, Sr, Ti, and In) was measured in the neighborhood of the ZnS:Mn and $Sr(Zr_{0.2}Ti_{0.8})O_3$ interface. In the SP ZnS:Mn film more pronounced counterdiffusion was observed near the interface. The In ion, which composes the ITO film, piles up more through the $Sr(Zr_{0.2}Ti_{0.8})O_3$ film in the SP ZnS:Mn film than in the EB film.

2.7. Comparison of the Luminances of SP and EB ZnS:Mn Films Deposited at Various T_s and P_s

Using a few kinds of EL thin-film structures, the luminances were appreciated at an effective charge density of 1.8 $\mu C/cm^2$, a half cycle driven at a 5 kHz sine wave voltage. A comparison of the luminances among the EL devices was conducted by using normalized values corresponding to a 1 μm thickness of the ZnS:Mn films. The ZnS:Mn films were deposited at various T_s and P_s using ceramic and powder targets ; they were then annealed at about 600°C for 1 h in a vacuum.

The best luminance of 5000 cd/m^2 (normalized to 1 μm ZnS:Mn thickness) was observed in the EL structure ITO / Y_2O_3 / ZnS:Mn / Y_2O_3 / Al_2O_3 / Al, in which ZnS:Mn film was prepared using a powder target, as shown in Fig. 5. This luminance corresponds to the one by the EB method, which shows 5000 cd/m^2 and higher in the same EL structure and driving conditions. It is noted that the optimum Mn concentration was found to be located at 1.2-1.6 at.% in the EL device, being different from the previously reported values [6,7] related to the EB films and larger by factors of 2-4 compared to those of the EB films.

In the case of the ceramic target, the luminance was not so high as that obtained in the powder target, being about three fifths of of that in the powder target. The highest luminance was also obtained at Mn concentrations of 1.2-1.6 at.%. The higher luminance by the powder target seems to be one reason, since a higher-purity target

598

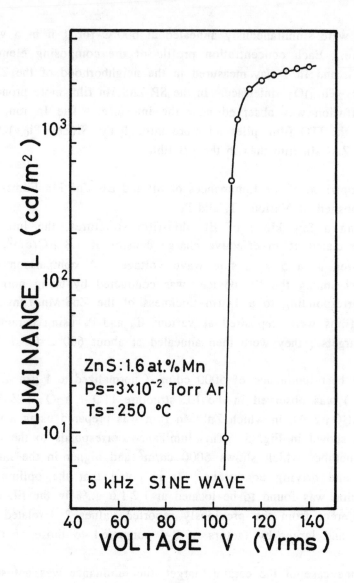

Fig. 5. Luminance-voltage (L-V) characteristics of the EL
device with the sputtered ZnS:Mn film.

can be obtained compared with the ceramic case. The higher optimum Mn concentration, such as 1.2-1.6 at.%, and comparatively low luminance in the SP films were considered, basically, to originate from the insufficient crystallinity, additional impurities and stronger counterdiffusion at the interfaces. In such ZnS:Mn film, the ineffective Mn emission center is presumed to increase.

3. INTERDIFFUSION PHENOMENA CAUSED BY POSTANNEALING OF ZnS:Mn FILMS IN TFEL DEVICE

In the EL device laminated with the following thin films, ITO / $Sr(Zr_{0.2}Ti_{0.8})O_3$ / ZnS:Mn / Y_2O_3 / Al_2O_3 / Al, diffusions of In, Sr and Ti ions to the ZnS:Mn film were observed at high postannealing temperature of the ZnS:Mn films. We confirmed that an optimized annealing temperature needed to obtain the highest luminance was located at 450°C, where pronounced diffusions did not take place. The diffusion of the Ti ion into the ZnS:Mn film was considered to affect the performance of the EL device.

3.1. Experimental Procedure

An EL device structure of ITO / $Sr(Zr_{0.2}Ti_{0.8})O_3$ / ZnS:Mn / Y_2O_3 / Al_2O_3 / Al was constructed to examine the interdiffusion phenomena by the postannealing of the ZnS:Mn film. ITO is an indium-tin oxide transparent electrode film. The $Sr(Zr_{0.2}Ti_{0.8})O_3$ (SZT) dielectric film was deposited on an ITO-coated glass substrate by an rf magnetron sputtering method using a ceramic target[1]. The ZnS:Mn (Mn concentration 0.8 at.%) film in thickness of 4200Å was superposed on the SZT film by an electron beam (EB) evaporation method. The substrate temperature was kept at 180°C. After deposition of the ZnS:Mn film, the substrate glass was divided into several pieces, and each piece was annealed at a different temperature up to 600°C for 1 h in a vacuum atmosphere. One piece remained as a nonannealed sample. Using the reactive EB evaporation method, all samples were then simultaneously coated with Y_2O_3 and then Al_2O_3 dielectric films. Finally, the Al counter-electrode was also simultaneously formed by the EB method to complete the EL devices. As explained above, all samples were constructed simultaneously in the same chamber except for the process of postannealing.

The luminance of the EL device was measured by a Pritchard photometer (1980A). To examine the interdiffusion between the ZnS:Mn and other films, concentration profiles of several composing ions were measured using SIMS (ATOMIKA, A-DIDA 3000). X-ray diffraction was applied to evaluate the crystallinity of the ZnS:Mn film.

3.2. Dependences of L-V and L-Q Characteristics of the EL Devices on the Postannealing Temperature

Figures 6 and 7 show the dependences of luminance-voltage (L-V) and luminance-charge density (L-Q) on the postannealing temperature, respectively. As shown in Fig. 6, emission threshold voltages (V_{th}) changed with the annealing temperature. V_{th} was defined as a voltage from which an effective charge started to flow through the ZnS:Mn film. At V_{th} obtained in this way, driven by a 5 kHz sine wave voltage, the luminance showed about 100 cd/m^2. As the effective charge began to flow through the ZnS:Mn film in each polarity of an ac voltage, an onset of the EL emission could be observed from an intersection of the L-Q curve and the abscissa, as shown in Fig. 7. An increment of the charge density Q above the intersection is the effective charge density. From the data in Fig. 7, the luminance (L_{std}) was obtained as a value at a constant 1.8 $\mu C/cm^2$ effective charge density and was compared. That is, the charge density plus 1.8 $\mu C/cm^2$ added to the value of the intersection of L-Q curve with the abscissa in Fig. 7 was calculated, and then the luminance at that charge density was obtained from the L-Q curve for the measurement of L_{std}, implying a correspondence to a quantum efficiency of the ZnS:Mn phosphor film. The effective charge density about 1.8 $\mu C/cm^2$ flows through the phosphor film of the EL device in an ordinary driving. From Q values when the effective charge began

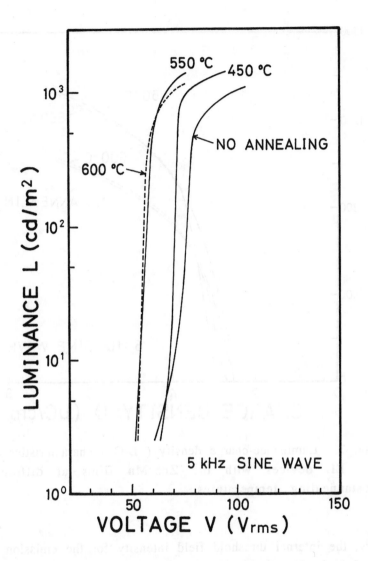

Fig. 6. Luminance-voltage (L-V) characteristics of the EL devices with the ZnS:Mn films at different postannealing temperatures.

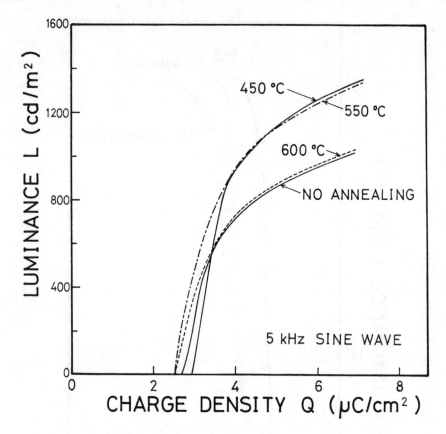

Fig. 7. Luminance-charge density (L-Q) characteristics of the EL devices with the ZnS:Mn films at different postannealing temperatures.

to flow, the internal threshold field intensity for the emission (E_{th}) in the ZnS:Mn films is calculated according to the following equation.

$$Q / 2 = \varepsilon_0 \varepsilon_r S E_{th}$$

ε_0 : dielectric constant of vacuum
ε_r : relative dielectric constant of ZnS
S : electrode area

The dependences of V_{th}, L_{std} and E_{th} on the annealing temperature are summarized in Fig. 8. The data at the 180°C annealing temperature correspond to the ZnS:Mn film without

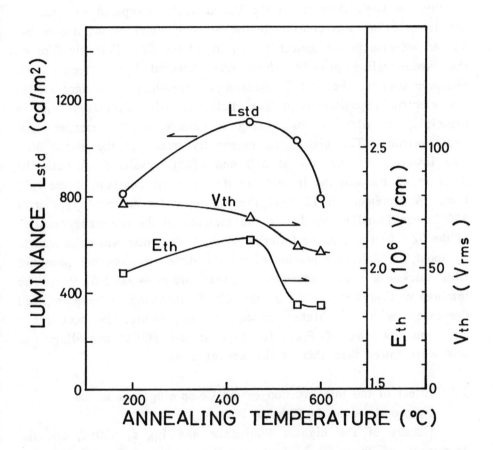

Fig. 8. Dependences of emission threshold voltage (V_{th}) internal threshold field intensity for emission in the ZnS:Mn film (E_{th}) and luminance (L_{std}) on postannealing temperature of the ZnS:Mn film.

annealing. As shown in Fig. 8, the highest E_{th} is observed at 450°C, while the lowest value is observed at both 550 and 600°C annealings. Although the E_{th} at the 450°C annealing became higher than that at the nonannealing, the V_{th} at the 450°C, on the contrary decreased slightly. A slight decrease of the V_{th} at 450°C compared with that at the nonannealing was considered due to competition of the higher E_{th} and an increase in the dielectric constant of the SZT dielectric film by the postannealing process which was detected by a steeper L-V characteristics at the 450°C annealing temperature. An increase in the electric capacitance of the dielectric film decreases V_{th}, in principle, in addition to causing a change to the steeper L-V characteristic. The higher E_{th} essentially results in the higher V_{th}. The lowering of the V_{th} at 550 and 600°C results from both the decrease of E_{th} and the increase in the dielectric constant of the SZT film. A decrement of 13 V of the V_{th} from that at 450°C to that at 550°C was found to be due to the increase in the dielectric constant of the SZT film and to the lowering of E_{th}. The latter was calculated to be about 7.5 V from a product of the E_{th} difference and the phosphor film thickness, and consequently, the former was 5.5 V. The luminance L_{std} was highest at the 450°C annealing temperature, and decreased with the higher annealing temperatures in accordance with the lowering of E_{th}. The L_{std} at the 600°C annealing was somewhat lower than that at the nonannealing.

3.3. Effect of the Interdiffusion of the Composing Ions in the Neighborhood of the ZnS:Mn Film

Causes of the highest luminance and E_{th} at 450°C, and the decrement of those at higher annealing temperatures of 550 and 600°C were examined as explained below. The crystallinity of the ZnS:Mn films was improved at the higher annealing temperature, as compared relatively by a ratio of the (111) peak intensity of the ZnS:Mn film and the (101) peak of the ITO in X-ray diffraction patterns. Since the (101) peak intensity of the ITO did not change as a result of the annealing, it was used as a standard of the crystallinity for the ZnS:Mn film. The highest luminance at 450°C cannot be explained from the results of the crystallinity. An emission symmetry according to the polarity of the voltage was measured by applying the triangle wave at 66 Hz[8]. The EL device with the ZnS:Mn at the annealing temperature of 450°C showed the equivalent emission intensities in each polarity of the voltage. But in the EL devices with the ZnS:Mn annealed at 550°C and 600°C, the symmetries were degraded ; that is, an intensity ratio of the emission at different polarities was 1 : 2 at 600°C. The emission intensity was smaller when electrons flowed through the ZnS:Mn film from the ZnS:Mn / SZT to the ZnS:Mn / Y_2O_3 interfaces by applying positive voltage to the Al electrode. This implies a change in electronic properties of the ZnS:Mn / SZT interface. One possible explanation to

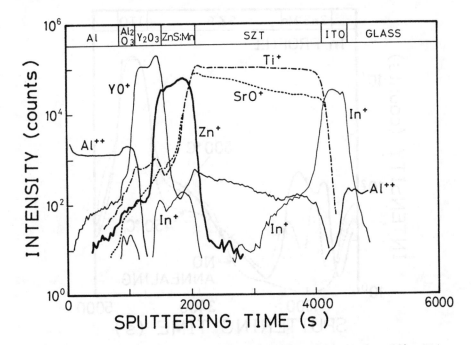

Fig. 9. In-depth profiles of the composing ion concentrations in the EL device with the ZnS:Mn film postannealed at 600°C.

the results described above is as follows. An interface trap state of the ZnS:Mn / SZT may change to a shallower state by high annealing temperatures, such as 550 and 600°C, as detected by the lowered E_{th} and the degraded emission intensities when a positive voltage is applied to the Al electrode as described above. In such a condition, trapped electrons are released at a smaller internal electric field from the ZnS:Mn / SZT interface compared with the opposite ZnS:Mn / Y_2O_3 interface. Some relatively excess electrons are stored at the ZnS:Mn / Y_2O_3 interface through the driving, that is, an imbalance of polarizations between the two interfaces takes place in each polarity of the driving voltage. As a result, a smaller E_{th} is observed compared with one of the devices with a normal deep interface trap state in both interfaces by the L-Q measurement of the EL device.

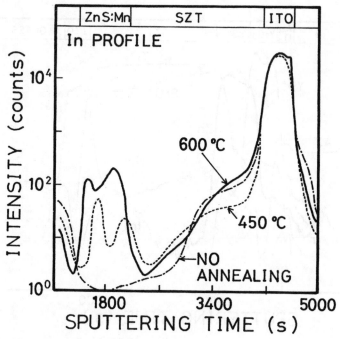

Fig. 10. In-depth profiles of In ion concentration in the EL devices with the ZnS:Mn films postannealed at different temperatures.

The E_{th} is measured as an intermediate value of two internal electric fields which start the electron releasing from the respective interfaces. Although the number of the electrons flowing through the ZnS:Mn film is the same in each polarity of the voltage, an excitation probability for the Mn emission centers becomes smaller when the electrons flow from the ZnS:Mn / SZT interface because of the lowered electric field. This results in the imbalanced emission intensities according to the polarity of the voltage, and the resulting lowered luminance of the EL device as observed in the 550 and 600°C postannealing temperatures. The same E_{th} is observed in the

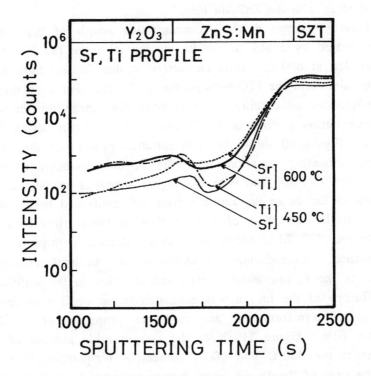

Fig. 11. In-depth profiles of Sr and Ti ion concentrations in the EL devices with the ZnS:Mn films postannealed at 450 and 600°C.

EL devices with ZnS:Mn films annealed at 500°C and 600°C. But the luminance at the same effective charge density is much lower in the EL device at 600°C, as shown in Fig. 7. This may indicate degraded emission efficiency inside the ZnS:Mn phosphor film in addition to the interface state transformation.

As a basic cause of the results described so far, we considered interdiffusion of the composing ions in the neighborhood of the ZnS:Mn / SZT interface which took place at the postannealing of the ZnS:Mn film.

3.4. In-depth Profiles of the Composing Ion Concentrations in the EL Devices with the ZnS:Mn Film

Figure 9 shows a whole concentration profile of the composing ions measured by SIMS in the EL device with the the ZnS:Mn film postannealed at 600°C. It is characteristic that In accumulates in the ZnS:Mn film from the ITO through the SZT film. For a comparison of the differences concerning the effect of the annealing temperature, the concentration profiles of In, Sr and Ti were drawn out in Figs. 10 and 11. Figure 10 shows the concentration profiles of the In ion at different annealing temperatures. The In ion obviously accumulates more in the ZnS:Mn film with the higher annealing temperature. This diffusion of the In to the ZnS:Mn film was confirmed to be harmless, since a further insertion of a very thin In metal film between the ZnS:Mn and SZT films before the ZnS:Mn deposition improved the EL performance. Considering the shape of the in-depth concentration profiles of the In ion which exhibit accumulation in the ZnS:Mn film, the influence of the In ion was estimated to be small with respect to the interface transformation, and to be on properties of the ZnS:Mn phosphor film. Figure 11 shows the concentration profiles of Sr and Ti ions at the 450°C and 600°C annealing temperatures. Differing from the case of the In diffusion, higher concentrations Sr and Ti ions in the ZnS:Mn films were observed in the neighborhood of the ZnS:Mn / SZT interface, implying an influence on electronic properties of that interface. Both Sr and Ti have similar profile curves at each annealing temperature. The concentration profiles for the nonannealing almost coincide with those for 450°C. Also, the profiles for 550°C are close to those for 600°C. From the results, Sr and Ti ions were found to begin to diffuse to the ZnS:Mn film at temperatures above 550°C. An observation of the pronounced diffusions of Sr and Ti ions at 550-600°C is qualitatively in agreement with the results of the EL characteristics. Because of this diffusion, the interface trap state between the ZnS:Mn and SZT films was considered to be transformed ; in particular, at 600°C the emission efficiency in the ZnS:Mn phosphor further degraded in addition to the interface state transformation. One other experiment which used rf-sputtered TiO_2 film (oxide target) as the first dielectric instead of the SZT showed a similar result, implying that the Ti ion was responsible for the results of the EL characteristics at high annealing temperatures.

4. INFLUENCE OF OXYGEN AND METAL OXIDE IMPURITIES IN ZnS:Mn FILM ON CHARACTERISTICS OF TFEL DEVICES

The effects of an oxygen and some metal oxide impurities introduced into the ZnS:Mn film at its postannealing process were examined with respect to the characteristics of a TFEL device. Depending on the oxygen pressure at the postannealing, the oxygen either decreases the emission threshold field strength in the ZnS:Mn film or forms an additional insulating film such as $ZnSO_4$. The metal oxide impurities Fe, V, and Co caused a decrease in luminance via a killer effect on the Mn emission center. On the other hand, Ti, In, Na, and V decreased the emission threshold voltage by lowering the emission threshold field strength in the ZnS:Mn film.

4.1. Experimental Method to Examine the Impurity Effects for the ZnS:Mn Phosphor Film

In order to examine the effects of impurities in the phosphor film, the EL device shown in Fig. 12 was constructed. The first and second insulating films of $Sr(Zr_{0.2}Ti_{0.8})O_3$ (6000Å), $BaTa_2O_6$ (1500Å)

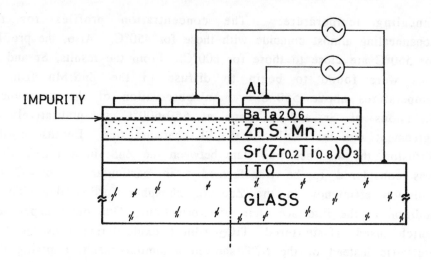

Fig. 12. EL device structure constructed to examine the impurity effects for ZnS:Mn film.

were prepaed by an rf magnetron sputtering method using oxide ceramic targets with the same compositions as the films. The ZnS:Mn phosphor film (Mn concentration 0.8 at.%, thickness 4200Å) was deposited at a substrate temperature of 180°C using an electron-beam evaporation method.

For an examination of the oxygen impurity, a glass substrate with glass / ITO / $Sr(Zr_{0.2}Ti_{0.8})O_3$ / ZnS:Mn film layers was divided into several small pieces, and each piece was annealed at an atmosphere with a different oxygen pressure and temperature for 1 h. After the annealing of all pieces, the $BaTa_2O_6$ insulating film was deposited simultaneously on the ZnS:Mn film, and then Al electrode was also deposited to complete the EL devices. The luminance / voltage (L-V) and charge density / voltage (Q-V) characteristics of each devices were measured to examine the effect of the oxygen impurity. The charge density Q is determined as the charge which

Table 2. Impurity sources for ZnS:Mn films.

COMPOUND	SOLVENT
Nai – OC$_3$H$_7$	(CH$_3$)$_2$CHOH
Sr (i – OC$_3$H$_7$)$_2$	"
Ti (n–OC$_3$H$_7$)$_4$	"
VO (tert – OC$_4$H$_9$)$_3$	"
Fe (i – OC$_3$H$_7$)$_3$	"
Sn (i – OC$_3$H$_7$)$_4$	"
Sb (i – OC$_3$H$_7$)$_3$	"
Mg (OCH$_3$)$_2$	CH$_3$OH
B (OCH$_3$)$_3$	"
Cr (C$_6$H$_6$)$_2$	C$_6$H$_6$
In (CH$_3$COCHCOCH$_3$)$_3$	"
Si (OC$_2$H$_5$)$_4$	C$_2$H$_5$OH
Co (i – OC$_3$H$_7$)$_2$	C$_6$H$_6$ CH$_3$COCH$_2$COCH$_3$

flows in a half cycle of the voltage (5 kHz, sine wave) and is divided by the Al electrode area.

To examine the effects of the various metal oxide impurities, mainly alcolate compounds shown in Table 2 were used as impurity sources. These compounds were supposed to become active diffusion sources after those thermal decompositions. The substrate glass just after the ZnS:Mn film deposition was also divided into small pieces for the respective impurities. Each alcolate compound was coated on half of the ZnS:Mn film in the small piece, as shown by dashed line in Fig. 12. By controlling concentrations of the alcolate solutions, the addition of all impurities to the ZnS:Mn was adjusted to be almost 1 wt.% as the respective oxide form. After the coating, all small pieces were dried in air and then annealed simultaneously at 450°C for 1 h in vacuum atmosphere to diffuse the impurities to the ZnS:Mn film. The $BaTa_2O_6$ insulating film and Al electrode were then deposited simultaneously in sequence on the ZnS:Mn film by the same method as in the case of the oxygen impurity. The luminances of both parts with and without the impurity in each piece of EL device were measured at an effective charge density $\Delta Q = 1.8$ $\mu C/cm^2$ (charge density flowing through the ZnS:Mn film) using 5 kHz sine wave voltage, and were compared with each other by calculating a ratio (L(I) / L(S)) of those luminances. L(I) indicates the luminance at the part having the impurity, and L(S) at the standard part with no impurity at the previous driving condition. The emission threshold voltage (V_{th}) were also measured in the two parts of each piece of the EL device, obtaining the differences $V_{th}(I)-V_{th}(S)$. I and S in the parentheses mean the same as in the case of the luminance described above. V_{th} was defined as a voltage of a turning point of the Q-V curve, from which the effective charge started to flow through the ZnS:Mn film. The luminance at V_{th} was about 100 cd/m^2 by the driving of 5 kHz sine wave voltage.

4.2. L-V and Q-V Characteristics of the EL Devices Annealed at Various $P(O_2)$ and T_a

The L-V and Q-V characteristics of the EL devices in which the ZnS:Mn films were annealed at various oxygen pressures and temperatures of 450°C and 550°C as shown in Figs. 13 and 14. In these figures, the oxygen pressure ($P(O_2)$, Torr) and the annealing

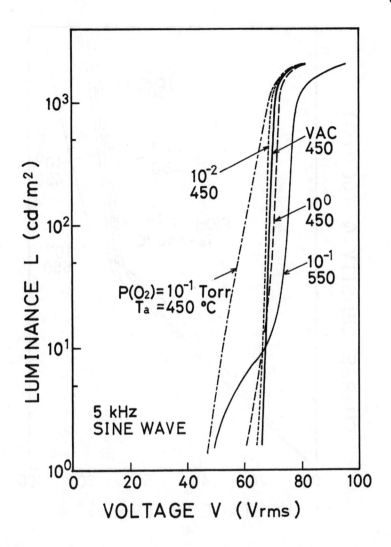

Fig. 13. Luminance-voltage (L-V) characteristics of the EL devices with ZnS:Mn films postannealed at various oxygen pressures P (O_2) and temperatures T_a after the deposition.

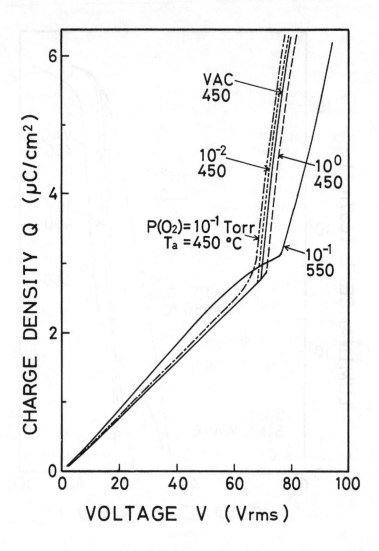

Fig. 14. Charge density-voltage (Q-V) characteristics of the EL devices with ZnS:Mn films postannealed at various oxygen pressures P (O_2) and temperatures T_a after the deposition.

temperatures (T_a, °C) are indicated, and VAC means vacuum atmosphere of an order of 10^{-6} Torr usually used for the annealing. No changes of the L-V and Q-V characteristics were observed at 10^{-2} Torr oxygen pressure and 450°C annealing temperature, compared with the characteristics of the ordinary vacuum atmosphere. In the case of higher pressure of 10^{-1} Torr, the emission threshold voltage once lowered, and then rose in the higher 10^0 Torr oxygen pressure. But saturated luminances were unchangeable according to the oxygen pressure. In connection with the data of the Q-V characteristics, these results could be explained as follows. The smaller threshold voltage in the oxygen pressure of 10^{-1} Torr was considered to be ascribed to the formation of a shallower ZnS:Mn / $BaTa_2O_6$ interface trap density due to oxygen diffusion to the ZnS:Mn surface. The increase of the oxygen concentration at the surface of the ZnS:Mn films with the annealing temperature was confirmed by secondary-ionization mass spectroscopy (SIMS) measurements of the films annealed even in a vacuum of the order of 10^{-6} Torr. The shallower interface trap was predicted by a more gradual slope in the neighborhood of a turning point (the effective charge starts to flow), and also lowered Q and V values of the turning point in the Q-V characteristic. The lowered Q means a smaller emission threshold field strength of the ZnS:Mn phosphor film. When the ZnS:Mn film was annealed in the even higher oxygen pressure of 10^0 Torr, the $ZnSO_4$ layer might be formed by an oxidization at the surface of the ZnS:Mn film. The formation of the $ZnSO_4$ by heating the ZnS, although heated in an air atmosphere, was confirmed by thermogravimetric analysis. Therefore, an additional insulating layer of $ZnSO_4$ might be inserted into the EL thin-film device structure, resulting in a rise in the emission threshold voltage. In the case of the annealing at 10^{-1} Torr oxygen pressure and the higher temperature of 550°C, even thicker $ZnSO_4$ might be formed by the oxidization, and the ZnS:Mn film might become slightly leaky. These were predicted by a larger gradient up to the turning point (gradient indicates total electric capacitance of the EL device) and a smaller gradient above the turning point (gradient indicates electric capacitance of the insulating film layers only) in the Q-V characteristic. The highest Q and V values of the turning point were considered also due to the formation of the additional $ZnSO_4$ film. The EL device with the ZnS:Mn film annealed at high oxygen pressure and temperature such as 10^{-1} Torr and 550°C tended to break down, maybe due to the previously described leaky properties of the ZnS:Mn film.

4.3. Effects of the Metal Oxide Impurities Introduced to the ZnS:Mn Film

The effects of the metal oxide impurities on the EL characteristics are summarized in Fig. 15. The luminance ratio L(I) / L(S) and the difference of the emission threshold voltage ΔV_{th} =V_{th}(I)—V_{th}(S) obtained from both parts with impurity-doped ZnS:Mn and that nondoped were plotted as one point for the respective impurity in Fig. 15. The ΔV_{th} was obtained as the difference of zero-peak voltages of the two emission thresholds. Since the preparation processes, dimensions and properties of the constituent films are the same in each piece of EL device except for the impurity doping, L(I), L(S) and V_{th}(I), V_{th}(S) should indicate differences only in the impurity doped and nondoped ZnS:Mn films. The annealing temperature of 450°C of all samples was chosen as the one at which no influence by diffusion of the composing ions such as Ti and Sr from the first insulating film of $Sr(Zr_{0.2}Ti_{0.8})O_3$ was confirmed by the measurements of the SIMS and the EL characteristics in other experiments. Such diffusion was observed above 550°C annealing temperature. As shown in Fig. 15, Fe, V, and Co oxide impurities decreased the luminance. On the other hand, Ti, In, V, and Na lowered the emission threshold voltage. The lowering of the emission threshold voltage is ascribed to the smaller emission threshold field strength in the ZnS:Mn phosphor film because of the same constituent films and preparation processes, as described previously. The Fe and Co decreased the luminance without lowering the emission threshold voltage. This was considered to be due to the killer effect of the Fe

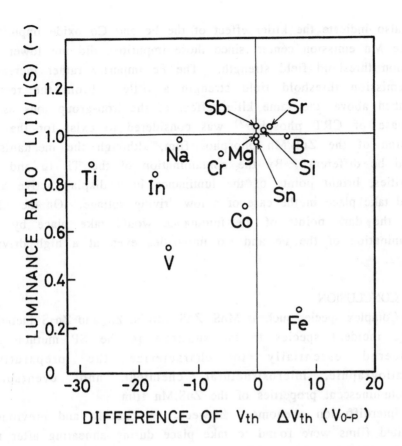

Fig. 15. Luminance ratios (L(I) / L(S)) and differences of
the emission threshold voltage (ΔV_{th}) of the EL devices
with metal oxide impurities doped and not doped ZnS:Mn
films.

and Co on the Mn emission center. From our other experiments, the
luminance of the same EL device structure tended to be higher in the
EL devices having larger emission threshold field strength. This fact

618

may also indicate the killer effect of the Fe and Co oxide impurities on the Mn emission center, since those impurities did not lower the emission threshold field strength. The Fe impurity rather increased the emission threshold field strength a little. From the results described above, the same killer effect of the iron-group ions as in the case of CRT phosphor[9] was considered to exist for the EL emission of the ZnS:Mn phosphor film, although the mechanisms should be different. By the contamination of the Ti, In and Na impurities, bright points of the luminance in a display active area should take place in the case of a low driving voltage. On the other hand, the dark points of the luminance would take place by the contamination of the Fe and Co impurities even at a high driving voltage.

5. CONCLUSION

Complex species such as MnS, ZnS, MnZn, Zn_2 and Zn_2S included in the incident species to the substrate in the SP method are considered essentially to characterize the preparative, crystallographic, microstructural, chemical, and, eventually, electroluminescent properties of the ZnS:Mn film.

Interdiffusion phenomena between the ZnS:Mn and previously deposited films were found to take place during annealing after the ZnS:Mn film deposition, although these phenomena are supposed to depend on the composing materials of the dielectric films. Therefore, it is necessary to optimize the annealing temperature in such a case.

The effect of the oxygen impurity on the EL characteristics included when the ZnS:Mn film is postannealed were considered fundamentally by the diffusion of the oxygen to the ZnS:Mn surface layer and the formation of the additional insulating layer such as $ZnSO_4$. The diffusion of the oxygen decreased the emission threshold field strength in the ZnS:Mn phosphor film, and the additional insulating layer increased the emission threshold voltage. However, the saturated luminances were independent of the oxygen pressure at the postannealing.

Among the examined metal oxide impurities in the ZnS:Mn film, Fe, V, and Co were found to degrade the luminance of the EL device, and Ti, In, Na, and V to lower the emission threshold voltage. The Fe and Co are considered to have a killer effect on the Mn luminescent center. The lowering of the emission threshold voltage by Ti, etc. was confirmed to be due to the smaller emission threshold field strength in the ZnS:Mn film.

REFERENES

1. Fujita, Y., Kuwata, J., Nishikawa, M., Tohda, T., Matsuoka, T., Abe, A. and Nitta, T., Proc. SID 25, 177 (1984).

2. Matsuoka, T., Kuwata, J., Nishikawa, M., Fujita, Y., Tohda, T. and Abe, A., Jpn. J. Appl. Phys. 27, 592 (1988).

3. Ban, V. S. and White, E. A. D., J. Cryst. Growth 33, 365 (1976)

4. Cattel, A. F. and Cullis, A. G., Thin Solid Films 92, 211 (1982).

5. Venghaus, H., Theis, D., Oppolzer, H. and Schild, S., J. Appl. Phys. 53, 4146 (1982).

6. Marrelo, V. and Onton, A., IEEE Trans. Electron Devices ED-27, 1767 (1980).

7. Sasakura, H., Kobayashi, H., Tanaka, S., Mita, J., Tanaka, T. and Nakayama, H., J. Appl. Phys. 52, 6901 (1981).

8. Matsuoka, T., Nishikawa, M., Kuwata, J., Fujita, Y., Tohda, T., and Abe, A., Jpn. J. Appl. Phys. 27, 1430 (1988).

9. Tabei, M., Shionoya, S., and Ohmatsu, H., Jpn. J. Appl. Phys. 14, 240 (1975).

ON Eu ACTIVATED II-VI SEMICONDUCTING COMPOUNDS

M. Godlewski and K. Świątek

Institute of Physics, Polish Academy of Sciences
02-668 Warsaw, Al. Lotników 32/46, Poland

In this monograph we review the optical and magnetic properties of Eu doped II-VI semiconducting compounds. The main emphasis is placed on analysis of the wide bandgap sulphides (ZnS, CaS, SrS) due to their possible application in multicolour thin film electroluminescence displays. A brief review of EuX (X = S, Se, Te, O) magnetic compounds and their solid solutions with wide bandgap II-VI compounds, such as CaS, BaS, SrS, is also given. The magnetic properties of narrow bandgap EuTe - HgTe mixed crystals are explained as being due to EuTe solid solutions in HgTe and to the low Eu solubility in this material. The nature of the solid solutions and Eu rich precipitates in HgTe and ZnS is clarified.

1. Introduction

In recent years, wide bandgap II-VI compounds, such as ZnS and the alkaline earth sulphides (CaS, BaS, SrS) activated with rare earth (RE) ions, were studied intensively as potential materials for multicolour cathodes and electroluminescence (EL) devices [1,2,3]. The main emphasis of these studies is on obtaining multicolour thin film EL displays in the visible light range. Wide bandgap II-VI compounds, such as ZnS (3.83 eV for zinc blende phase), CaS (4.41 eV), SrS (4.30 eV), and BaS (3.78 eV), are the most promising candidates for such applications. Among the various crystals and RE ions checked, $ZnS:TbF_3$ and CaS:Ce for green colour [1,6], CaS:Eu for red colour [4,5], SrS:Ce for blue colour [6] and SrS:Pr,K, SrS:Ce,K,Eu for white colour [7], are the most promising ones.

In this monograph, we review the physical aspects of Eu activated wide bandgap II-VI compounds. The main topic discussed will be the Eu-related properties that may be utilized in constructing efficient displays either for red colour or, in the case of Ce codoping, for white colour. For ZnS, the RE solubilities are rather low, well below those leading to pronounced interactions. This will be discussed in detail, e.g., on the basis of the electron spin resonance (ESR) investigations of Eu doped ZnS, and microanalysis of ZnS: Eu and HgTe: Eu.

Eu doped ZnS: Pb, Cu powder was recently applied in nonradiative night-luminous phosphors [9]. The very long glow after the initial uv illumination (12 hours [9]) make this material an attractive one for applications in luminous paints in, e.g., watches. The very long afterglow of Eu codoped powders is related to the deep energy level introduced in the ZnS forbidden gap. The exact position of the Eu^{2+} energy level in ZnS will be discussed in Section 3.3.

Some physical problems related to the magnetic interactions in Eu chalcogenides will be discussed briefly in Section 4. The solid solutions of $Eu_{1-x}M^{2+}_x X$ for X=S, 0, Te were studied previously, and the main emphasis of these studies was on optimizing the EuX doping in order to increase the magnetic ordering temperatures [8]. The problems arising in connection with the EuS, EuTe solid solutions in heavily Eu doped ZnS and HgTe will be discussed separately in Section 2.

2. Some comments on Eu solubility in thin film and bulk II-VI compounds

An important constraint on development of more efficient EL devices for RE activated phosphors is the rather low solubility of the RE ions in most semiconducting materials. This was shown for Tm in GaAs to be in the range of $4*10^{17} cm^{-3}$ [10], or not more than 0.01 mol% for binary chalcogenides [11]. This is certainly less that the optimum RE concentration of between 0.1 and 10 mol% required for thin film electroluminescence devices [2, 17, 24]. This indicates the role which can be played by, e.g., grain boundaries in polycrystalline thin film layers, which can be decorated with RE ions.

The first successful EL device, constructed by Inoguchi et al. [16], was based on a thin film ZnS: Mn phosphor. For practical applications in TFEL (thin film EL), a range of different growth methods was developed for RE doped EL phosphor films. These include the most widely used method of deposition due to sputtering, vacuum evaporation, and the recently introduced chemical deposition methods such as MOCVD (metal organic chemical vapour deposition) and ALE (atomic layer epitaxy), based on a CVD process. More details on the technologies used to obtain TFEL structures may be found in recent reviews (e.g. [3]).

To obtain RE doped TFEL devices, similar growth methods were used, modified by using different starting sources of the RE ions. EL phosphors can be obtained by using RE doped ZnS as the starting material, or by evaporating RE compounds together with ZnS, or by rf sputtering of ZnS and RE compounds, or by epitaxial growth of ZnS and sublimation of RE chelates, as reviewed in reference 3.

Alkaline earth sulphides (CaS, SrS, BaS) are more difficult to obtain in the form of thin films. This is because of their more refractory and hydroscopic nature, chemical instability, and an increased tendency to decompose thermally.

As has been mentioned above, RE doping level in polycrystalline films may be much larger, and several attempts were made to optimize RE concentration in order to obtain more efficient EL. Different concentrations were tried e.g. from 1.4 mol% [2] to 7 mol% [17] for TbF_3 in ZnS, and as much as 10 wt% for Tb in CaS [24]. The higher doping levels turned out to be unprofitable. At increased concentrations (4.5 mol% for TbF_3 [2]), the Tb^{3+} emission is no longer observed, due to nonradiative cross-relaxation processes. In CaS:Eu, the optimum Eu concentration is fairly low (0.2 wt%) [24]. For EuF_3 doped SrS, Eu^{2+} EL could be observed even for the relatively high doping level of 1 mol%, but under the condition that the substrate temperature during deposition was high (~450 °C) [18]. For lower substrate temperatures the interstitial F^- ions were stabilized, and for Eu concentrations of over 0.6 mol% Eu^{3+}, emission becomes predominant. Here we touched one of the most fundamental questions of the nature of the RE sites in II-VI compounds. Several remarkable differences were observed, depending on the RE doping procedure. The charge state of Eu can be controlled by choosing a suitable doping procedure. ZnS doped with EuS and EuF_2 show EL related only to Eu^{2+} [19,20]. ZnS and SrS films doped with EuF_3 show Eu emission from the 3+ charge state. Several differences not only in the charge state but also in excitation and recombination efficiencies, indicate that, by intentional codoping, we do not only control the valence of Eu and other RE impurities, but that more complex RE centers can be formed in the materials. The "lumocen" concept was introduced to explain the much larger brightness of the films doped with the RE ions in the form of molecules (REF_3) [21,22]. Formation of extended RE centers is responsible for the increased cross sections for impact excitation with hot carriers.

Even the simple codoping required for charge compensation of the RE in the 3+ charge state may result in quite distinctly different features of TFEL devices. For example, different EL excitation mechanisms were observed for Ce compensated with K ions (SrS:Ce,K) and with F ions (ZnS:Ce,F) [23].

Table I.   RBS analysis of Eu doped bulk ZnS crystals

crystal	initial Eu conc. (mol/mol)	concentr. from RBS (mol%)	remarks
ZnS:Eu,Li	10^{-2}	>5	inhomogeneous, highest concentration at the surface
ZnS:Eu,Li	10^{-3}	<4	inhomogeneous, better homogeneity in the depth profile, different concentrations in different parts of the sample
ZnS:Eu	10^{-2}	0.05	homogeneous sample
ZnS:Eu	10^{-2}	4	a relatively homogeneous sample with the maximum of concentration at the surface of 8.6 mol%
ZnS:Eu	10^{-2}	8	prepared identically
ZnS:Eu	10^{-3}	0.09	homogeneous sample
ZnS:Eu_2O_3	10^{-3}	7.5	peak at the surface
ZnS:Eu_2O_3, Li	10^{-3}	27	no relevant data, due to inclusions

The data collected in Table I show the high inhomogeneity of Eu doping in bulk ZnS crystals. RBS, whose penetration depth in ZnS is rather limited (250 to 300 nm), detects high Eu concentrations at the surface. The ESR measurements performed on the same samples show that only about $10^{17} cm^{-3}$ of Eu enters ZnS lattice substitutionally on the zinc sites, being in the 2+ charge state. It is interesting to note the correlation between ESR and X-ray examinations [12]. X-ray studies indicated that only low Eu concentration samples are true bulk materials, with fairly homogeneous Eu spread. Crystals with low europium concentrations showed strong Eu^{2+} ESR signals due to isolated Eu^{2+} centers in the hexagonal phase of ZnS, as verified by X-ray studies confirming the hexagonal structure of these samples, or cubic one, with 4H and 6H polytypes. With an increase in the doping level (above 0.1 mol%), even macroscopic layer-like growth was observed.

The up to two orders in magnitude difference between Eu concentrations, as concluded from RBS and ESR experiments, can be explained by using an electron probe microanalyser [13]. Two extreme II-VI compounds were tested. The first, wide bandgap ionic ZnS crystals, and the second, narrow bandgap HgTe. The microanalysis proved that a considerable part of the europium in HgTe and ZnS forms solid solutions of EuTe and EuS, or precipitates enriched with the dopant elements [13, 14]. The precipitates in HgTe consist mainly of a phase close or identical to Eu_4Te_7, with Eu in the 3+ charge state. In ZnS, the precipitates consist mainly of a solution of ZnS and $ZnEu_2S_4$, the valence of Eu in these precipitates is 3+. The real solubility level of Eu as an impurity was found to be fairly low, lower than 10^{-3} at.% for HgTe crystals, and about 10^{-2} at.% for ZnS crystals [13]. It should be stressed that the presence of EuS or EuTe type solid solutions in ZnS and HgTe may, most probably, be responsible for some magnetic properties of these crystals. For example, the magnetic properties of $Hg_{1-x}Eu_xTe$ ($0 \leq x \leq 0.07$) and $Hg_{1-x}Eu_xSe$ ($0 \leq x \leq 0.01$) reported recently by Krylov et al. [15] may be explained in this way. On the other hand, Eu^{3+}-rich precipitates may be the source of the Eu^{3+}-related emission in ZnS [14], which would thus be related to a substitutional impurity.

3. Eu charge states and energy structure

RE ions enter the cation sites of II-VI compounds in their 3+ charge states, i.e., charge compensation is necessary [25, 26]. Until now, only four RE ions were observed in the 2+ charge state in ZnS. These are: Eu [28, 58], Yb [66], Sm, and Tm [27]. The increased band gaps of CaS, MgS, SrS mean that for few other RE ions can an excited 2+ state enter the forbidden gap, and be populated under illumination [27]. In this sense, the 2+ charge states of Eu and Yb are the exceptions. Both of them are deep states with energy levels close to midgap positions in ZnS and other wide bandgap sulphides, and can be observed as stable charge states without prior excitation.

Eu^{3+} has the $4f^6$ electronic configuration with a spin singlet ($^7F_0$) being the ground state. This is a diamagnetic configuration and, hence, could not be detected via ESR experiments. In this monograph, we turn our attention to the optical and magnetic properties of the Eu^{2+} charge state.

u^{2+} has 4f^7 electron configuration, with a $^8S_{7/2}$ ground state. This state s easily detected with the ESR technique, as described in detail in the ollowing section.

i.1 Eu^{2+} energy structure

Eu^{2+} has a half filled 4f shell with the $^8S_{7/2}$ ground state. The first excited configuration requires excitation of one of the 4f electrons to the 5d shell.

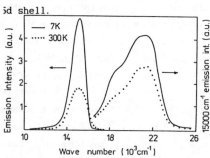

Fig.1. The PL and PL excitation bands of Eu^{2+} in a CaS lattice. The bright red emission observed is due to $4f^6 5d^1 \Rightarrow 4f^7$ recombination. Both PL and PLE ($4f^7 \Rightarrow 4f^6 5d^1$) bands show a characteristic broad featureless structure, further broadened at increased temperatures due to strong electron-phonon coupling.

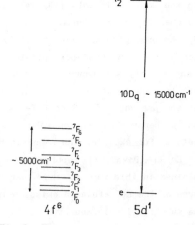

Fig.2. The energy structure of the first $4f^6 5d^1$ excited state of Eu^{2+}.

Eu-based EL devices utilize the intense emission from the lowest-lying $\Gamma_8(4f^6 5d^1)$ level of Eu^{2+} to the $^8S_{7/2}$ (4f^7) ground state. The $4f^n \Rightarrow 4f^{n-1} 5d^1$ transition is quite different from the intra-shell 4f^n transitions. Its energy depends strongly on lattice covalency and the relevant absorption and PL bands are broadened due to the strong electron-phonon coupling of the "external" 5d electron (see Fig.1). The crystal field splitting is much larger for the 5d electron than the spin orbit splitting of the 4f^6 "core" electrons (see Fig.2) [29]. The important question which was examined by many authors was the strength of the coupling between the 5d spin and the 4f^6 (^7F_J) multiplets. This problem was solved by Chase [29], who studied the optically detected magnetic resonance (ODMR) on the Eu^{2+}-related emission in CaF$_2$, SrF$_2$, and BaF$_2$. The studies of Chase led to rejection of the model of strong Coulomb coupling of the 5d electron to the 4f^6 configuration, resulting in ^8P_J multiplets, proposed to explain the results of studies of Faraday rotation [30,31]. An analysis of the ODMR data indicates that the $^7F_o - ^7F_6$ spin-orbit multiplets of the $4f^6$ core levels are weakly coupled to the e and t_2 orbitals of the external $5d^1$ electron, giving a lowest-lying Γ_8 quartet [29]. The large average g value of 3.8 for this level can be explained if exchange coupling between the 5d spin and the $4f^6$ (^7F_J) spin components is comparable to the 400 cm^{-1} spin-orbit coupling parameter for the ^7F_J multiplets. The shape of the resonance spectrum also indicates the presence of dynamic Jahn-Teller distortion of the e orbital of the 5d electron.

The "external" $5d^1$ electron of the Eu^{2+} excited state is sensitive to the crystal field. A quite noticeable change of the $4f^7 \Rightarrow 4f^6 5d^1 (t_2)$ energy separation was observed for $Ca_x Sr_{1-x}S$:Eu mixed crystals (wurtzite structure). The characteristic Eu^{2+} red emission in CaS shifts to orange in SrS [32].

3.2 ESR study of Eu^{2+} in II-VI and europium chalcogenides

In semiconductor and DMS systems, magnetic impurities mostly conserve the electronic structures of the free ions, though the effects related to delocalization of the electronic orbitals and hybridization with the atomic orbitals of the neighboring ions cannot be neglected. These delocalization phenomena indicate, through a significant reduction of the hyperfine interaction between the electronic spin and the nuclear spin of the impurities, that the ligand bonds are more covalent [33]. Therefore, it seems reasonable that superhyperfine interactions with neighboring nuclei may be sensitive to the type and distance of the ligands in the lattice.

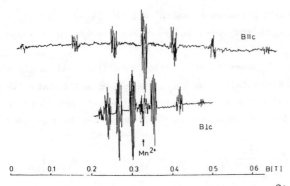

Fig. 3 ESR spectrum of a ZnS:Eu,Li single-crystal at 114 K, for two different orientations of the magnetic field vs the crystallographic axes in the ($1\bar{1}00$) plane.

In Fig. 3, we show the ESR spectrum of Eu^{2+} in the wurtzite phase of

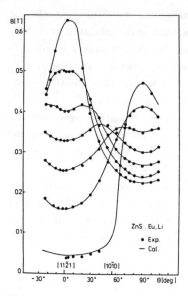

B[T]

ZnS : Eu, Li
• Exp.
— Cal.

[11$\bar{2}$1]   [10$\bar{1}$0]

-30°   0°   30°   60°   90°   θ[deg]

Fig. 4. Angular dependence of the Eu^{2+} ESR spectrum upon rotation in the (1$\bar{1}$00) plane.

ZnS, with the applied magnetic field parallel to the [11$\bar{2}$1] and [$\bar{1}\bar{1}$22] axes [12]. This consists of two overlapping spectra, 7 x 6 lines each, because of a superposition of two sets of lines corresponding to fine and hyperfine splittings caused by two odd isotopes (^{151}Eu and ^{153}Eu) of europium (47.77 % and 52.23 % of natural abundance), both with 7/2 electron spin and nuclear spin I=5/2.

In a crystalline axial field, the $^8S_{7/2}$ level is resolved into four doublets, split further by the external magnetic field, and observed in an ESR experiment as the fine structure of the spectrum. The z-axis of the defect center is along the crystal c-axis [11$\bar{2}$1]. The angular dependence of the ESR spectrum in the (1$\bar{1}$00) plane, containing the c-axis of the ZnS:Eu crystals, is shown in Fig. 4.

The ESR spectrum observed in this situation can be analyzed with the spin Hamiltonian of the form:

$$\mathcal{H}_s^{\alpha} = g_{\parallel}\mu_B B_z S_z + g_{\perp}\mu_B(B_x S_x + B_y S_y) + \frac{1}{3}b_2^0 O_2^0 + \frac{1}{16}b_4^0 O_4^0 + \frac{1}{3}b_4^3 O_4^3 +$$
$$+ \frac{1}{1260}b_6^0 O_6^0 + \bar{S}\cdot\tilde{A}_{151}\cdot\bar{I}_{151} + \bar{S}\cdot\tilde{A}_{153}\cdot\bar{I}_{153}$$

where b_n^m are the crystal field parameters, and O_n^m are the Stevens operator equivalents. The $\tilde{A}_{151}$, $\tilde{A}_{153}$ are the hyperfine coupling tensors of the two isotopes.

In a cubic field (e.g., T_d symmetry of a cation substitutional position in a zinc blende structure crystal), the $^8S_{7/2}$ level is split into two doublets and one quartet which are further split by the external magnetic field. In this situation, the appropriate spin Hamiltonian can be expressed as follows:

$$\mathcal{H}_s^c = g\mu_B \bar{B}\cdot\bar{S} + \frac{b_4}{60}(O_4^0 + 5\, O_4^4) + \frac{b_6}{1260}(O_6^0 - 21\, O_6^4) + A_{151}\bar{S}\cdot\bar{I}_{151} + A_{153}\bar{S}\cdot\bar{I}_{153} \; .$$

In other types of crystal field symmetries, we should first construct an appropriate spin Hamiltonian, as described in [34], which can then be fitted to the experimental data. The spin Hamiltonian parameters obtained for Eu^{2+} as an impurity in wide bandgap II-VI compounds, are summarized in Table II. All these results suggest that the Eu^{2+} ions simply substitute for the cations in II-VI compounds.

Table II. Spin Hamiltonian parameters for Eu^{2+} in II-VI semiconductors.

Comp.	g	b_2^{0*}	b_4^{0*}	b_6^{0*}	b_4^{3*}	b_6^{3*}	b_6^{6*}	A_{151}^*	A_{153}^*	Ref.
ZnS	$g_\parallel$=1.989 ∓0.002 $g_\perp$=1.993 ∓0.002	-422.3 ∓0.3	-12.4 ∓0.3	-0.8 ∓0.1	8.3 ∓0.3	1.2 ∓0.3	0.0 ∓0.1	-21.8 ∓0.6	-9.4 ∓0.4	[12]
CdS	1.991	-34.2	-12.0	1.0						[35]
CdSe	1.991	240.0	-2.7	0.2	11.7		2.4			[36]
ZnSe	1.993 ∓0.001		-9.87 ∓0.15	0.28 ∓0.25				-21.8 ∓0.2	-9.8 ∓0.1	[37]
CdTe	1.9917 ∓.0007		-7.66 ∓0.10	-0.12 ∓0.14				-23.19 ∓0.15	-10.25 ∓0.10	[36] **

* in 10^{-4} cm^{-1}
** a more extensive study of the transferred hyperfine interaction in the ESR spectra of Eu^{2+} in CdTe was given by M. Tovar and M.T. Causa [38].

Table III. ESR data on Eu^{2+} in Sr-chalcogenides.

Comp.	a(A)	g	$\|A_{151}\|$ $(10^{-4}cm^{-1})$	$\|A_{153}\|$ $(10^{-4}cm^{-1})$	T(K)	Ref.
SrTe	6.47	1.9890 ∓.0005	28.2 ∓0.1	12.5 ∓0.1	300	
		1.9920 ∓.0005	28.3 ∓0.1		77	
SrSe	6.23	1.9917 ∓.0005	29.1 ∓0.1	13.0 ∓0.1	300	[43]
		1.9921 ∓.0005	29.3 ∓0.1		77	
SrS	5.87	1.9917 ∓.0005	29.9 ∓0.1	13.2 ∓0.1	300	
		1.9918 ∓.0005	30.0 ∓0.1		77	
SrO	5.14	1.991	29.9	13.2	4.2	[44]

In Table III, the g values and hyperfine coupling constants A of Eu^{2+} in SrS are listed. The values of A_{151} in the alkaline earth chalcogenides are summarized in Table IV.

Table IV. Values of $|A_{151}|$ (in units of 10^{-4} cm^{-1}) for Eu^{2+} in the alkaline earth chalcogenides at room temperature.

	Ca^{2+}	Sr^{2+}	Ba^{2+}	Ref.
O^{2-}	30.16	29.9	29.6	[45]
S^{2-}	30.6	29.9	29.6	
Se^{2-}	29.7	29.1	28.9	[44]
Te^{2-}	28.5	28.2	28.1	

For a given anion, the value of A_{151} decreases with increasing Eu-X bond distance in going along the CaX - BaX series. Reduction of A_{151} for Eu^{2+} with covalency is rather complicated. It may be partly explained by the mechanism which involves an electron transfer from the p orbitals of the anion to the 6s orbitals of the impurity. There also occurs an electron transfer to the 5d orbitals. The spin polarization of the 5d electrons produces a negative field at the nucleus, and the relative amounts of the 5d and 6s bondings is expected to determine the magnitude of the reduction of A. In addition, the outermost $5s^2 5p^6$ electron shells, exchange-polarized by the 4f electrons, are affected by covalent bonding with the anion orbitals [45]. The contribution of the $5s^2 5p^6$ shells to the hyperfine field becomes more positive with increasing covalency.

Europium chalcogenides EuX (X= O, S, Se, Te) crystallize in the NaCl structure. EuO and EuS are ferromagnetic, EuSe is ferrimagnetic, and EuTe is antiferromagnetic, as will be described in more details further on. These magnetic properties are phenomenologically described by the exchange parameters J_1 and J_2 of Eu, corresponding to the interactions between the nearest neighbors (n.n.) and the next nearest neighbors (n.n.n.), respectively.

The hyperfine fields at the Eu nuclei in EuX have been investigated by many workers. Several attempts were made to correlate the hyperfine data with physical properties such as the Eu-X bond character, the exchange interaction, and others [39, 40]. The hyperfine field at an Eu nucleus in EuX arises mainly from the core polarization due to the 4f electron spins of the ion, and is partly transferred from the neighboring Eu^{2+} ions.

The effective magnetic field at an Eu nucleus in EuX is given by

$$H_{eff} = H_{hf} + H_d + H_L + H_D + H$$

where H_{hf} is the hyperfine field, H_d the dipole field, H_L the Lorentz field, H_D the demagnetizing field, and H the applied field. H_{hf} consists of the core polarization field H_{hf}^a, and the transferred hyperfine field H_{hf}^t:

$$H_{hf} = H_{hf}^a + H_{hf}^t$$

In an NaCl structure, Eu^{2+} ion has six n.n. Eu^{2+} spins within its own (111) plane, six n.n. and six n.n.n. Eu^{2+} spins in adjacent (111) planes. The transferred hyperfine field in the ferromagnetic spin arrangement is given by

$$H_{hf}^t = 12A_1 <S> + 6A_2 <S>$$

where the first and second terms represent contributions of the twelve n.n. and six n.n.n. spins <S>, respectively.

In the ferromagnetic state of EuX, the dipole lattice sum H_d is zero, and, using the saturation magnetizations M, the Lorentz field $H_L = \frac{4\pi M}{3}$ is estimated to be 8.0, 4.7, 4.6, and 3.8 kGs for EuO, EuS, EuSe, and EuTe, respectively. In EuO and EuS, values of H_{eff} are obtained from zero-field NMR frequencies [41,42].

3.3 Eu^{2+} energy level position

Until now, only for two II-VI compounds was the energy level position of Eu^{2+} determined. This includes the early studies of Keller and Pettit [46], and Title [47] of Eu and Sm doped SrS, and the more recent photo-ESR experiments on Eu-doped ZnS [58]. The former results require some corrections, since the model of Keller and Pettit for PL and thermoluminescence processes in SrS:Eu,Sm was based on the incorrect assumption that the 3+ ⇒ 2+ ionization energies of Eu and Sm are identical. The appropriate 3+ ⇒ 2+ ionization energies for a range of RE ions were calculated recently by us, using the refined electron-spin-pairing theory (RESPET) of Jörgensen [48]. In Fig.5, we show the calculated ionization energies for $3+ \Rightarrow 2+$ transitions of the RE ions in sulphides [49,27,50]. For $Eu^{3+} \Rightarrow Eu^{2+}$, a transition energy of ~ 12000 cm^{-1} (1.5 eV) is predicted. This estimation is consistent with the $9700 - 12500$ cm^{-1} band in the photo-ESR experiments of Title [47].

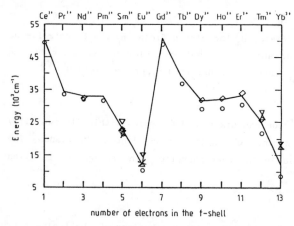

Fig.5. Energy of RE ($3+ \Rightarrow 2+$) ionization transition in CaGa$_2$S$_4$, SrS, and ZnS. Theoretical data: (——), ($\circ$) fit to the maximum and the edge of PL excitation band in CaGa$_2$S$_4$, respectively. Experimental data: ($\lozenge$) for CaGa$_2$S$_4$ [59], ($\triangle$) for ZnS [66,58,53,54], ($\times$) for SrS [47], and ($\triangledown$) for hexaiodides [60].

A much more refined photo-ESR experiment was performed for Eu in ZnS [58]. The spectral dependence of the Eu^{2+} ESR signal for the $Eu^{2+} \Rightarrow Eu^{3+} + e_{CB}$ transition, shown in Fig.6, was analyzed with the theoretical formulae describing center photoionization. A fit to the experimental results gave the following energies of the $Eu^{2+} \Rightarrow Eu^{3+} + e_{CB}$ transitions: $E_{opt} = 2.60$ eV, $E_{th} = 2.20-2.25$ eV, $E_{rel} \sim 0.35-0.40$ eV [58]. In the fit, the lattice relaxation concomitant with the change of the center charge state was taken into account. This results in an optical ionization energy (E_{opt}) larger than the thermal ionization energy (E_{th}) by the lattice relaxation energy (E_{rel}).

The energy values obtained mean that the Eu^{2+} energy level is localized at $\sim 1.5-1.6$

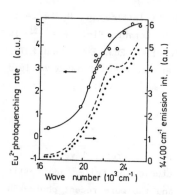

Fig.6. The spectral dependence of the Eu^{2+} ESR signal photoquenching (empty circles) and the excitation band of the Eu-related 14400 cm^{-1} emission in ZnS:Eu (dashed curve) and ZnS:Eu,Li (dotted curve) crystals, respectively.

eV above the edge of the ZnS valence band and, in accordance with the basic assumption of the RESPET theory [48,49,27,50], should occupy similar positions for other wide bandgap II-VI sulphides. Consequently, Eu^{2+} should be localized at about 2.7-2.75 eV below the conduction band edge in CaS and 2.6-2.65 eV in SrS. This, as will be explained further on, allows for Eu^{2+} emission in this materials not observed in ZnS due to the resonant character of the Eu^{2+} excited state in this lattice.

3.4 The concept of RE bound excitons

The possibility of change of the charge state of some of the RE ions created hope of uncovering a more efficient EL excitation mechanism. The most common impact excitation mechanism of RE excitation [51] is rather inefficient, due to the small optical cross sections for such transitions. Allen [52] observed that higher efficiencies are expected when impact excitation is replaced by impact ionization of a RE center by a hot carrier. This requires efficient recapture of the ionized carrier via the RE excited state, as explained in Fig. 7.

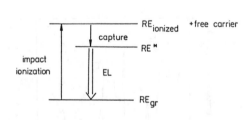

Fig.7. A three level system for efficient RE intra-ion excitation via impact ionization of a RE center by a hot carrier.

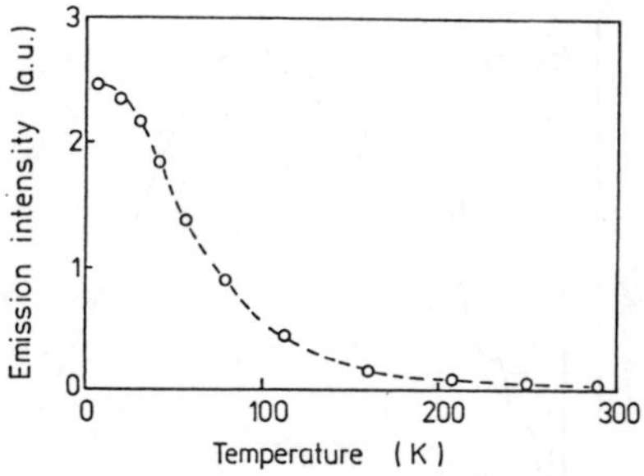

Fig.8. The temperature dependence of the Eu-based exciton emission in ZnS.

The theoretical calculations presented in the previous section showed that 3+ ⇔ 2+ ionization transitions are possible for four (in ZnS) or more (for CaS and SrS) RE ions. Moreover, Ce and, probably, Pr can change their charge state from 3+ to 4+. This makes the proposition of Allen an attractive one. The first experimental verification of the high efficiency of RE intra-ion emission under the ionization transition produced, however, strong evidence that the model shown in Fig. 7 should be complemented by introduction of an intermediate state for free carrier recapture. This state was attributed to an exciton bound to a RE site, and was observed for Yb [66], Eu[64], Sm[53], and Tm [54] in ZnS, and postulated for Sm in SrS [46].

There are several important consequences connected with an appearance of such an "external excited" state of a RE ion. The first is the strong reduction of PL excitation efficiency, due to temperature-induced dissociation of an RE bound exciton (BE) [64,66]. The very high efficiency of Yb^{3+} intra-shell excitation via Yb photoionization is rapidly reduced with a rise in temperature [66]. Similar experimental results are shown in Fig.8, in which the temperature dependence of the Eu-bound exciton PL is shown, indicating BE dissociation at increased temperatures [64].

Another consequence to the BE formation is the increased probability of energy transfer from an RE excited center to some other centers, including energy transfer to PL deactivators. In Fig.9, the relevant experimental results are presented, proving high efficiency of such transitions. Efficient energy transfer from the Eu bound exciton in ZnS to a Fe "killer" center was observed [55].

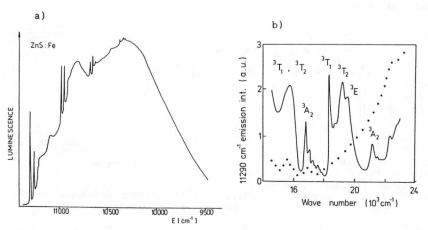

Fig.9. PL band for the $^3T_1 \Rightarrow {}^5E$ transition of the isolated Fe^{2+} center in ZnS (a). In (b), the excitation band for this emission is shown for ZnS:Fe (solid curve) and ZnS:Eu,Fe (dotted curve). The latter band is of $Eu^{2+} \Rightarrow Eu^{3+} + e_{CB}$ photoionization nature, i.e., identical to that occurring during formation of an Eu-bound exciton.

The third consequence is the possibility of BE ionization by the strong electric field present in the active layer of an EL device. This has not been documented experimentally until now.

The high efficiency of the impact ionization mechanism is, however, evidenced by the experimental data [62,56,32,57]. The experimental results clearly imply that Eu^{2+} and Ce^{3+} centers are ionized in CaS:Eu, SrS:Eu, and SrS:Ce,K under the impact of hot carriers. This proceeds via the $4f^{n-1}5d^{1}$ excited state, with the carrier being first excited to this state, then ionized and finally undergoing recapture. Formation of an RE bound exciton has not been observed in this case, and their probable formation must be verified in experiments performed at low temperatures.

3.5 Relation between Eu energy level positions and PL, EL mechanism

As explained previously, europium enters II-VI compounds in two stable charge states Eu^{3+} with a $4f^{6}$ electron configuration, and Eu^{2+} with a $4f^{7}$ configuration. The photo-ESR studies presented in the previous section allowed us to localize the 2+ charge state in the forbidden gap of wide bandgap sulphides, e.g., 2.25 eV below the edge of the conduction band in ZnS [58] and, as estimated in a previous section, ~2.75 eV in CaS.

In consequence, the Eu^{2+} $4f^{6}5d^{1}$ excited state is a resonant state in a ZnS lattice and the $4f^{6}5d^{1}(t_{2}) \Rightarrow 4f^{7}$ Eu^{2+} PL is not observed. This is due to fast autoionization from the $4f^{6}5d^{1}$ state, if populated, as demonstrated previously is a direct photo-ESR experiment for Eu in the CdF_{2} lattice [61]. In Fig.10, we show the relevant experimental results, proving directly the fast autoionization from the Eu^{2+} excited state to the continuum of the conduction band states. In the ESR experiment performed for Eu doped CdF_{2}, photo-quenching of the Eu^{2+} ESR signal was measured. The data shown in Fig.10 prove that the photo-quenching band of the Eu^{2+} ESR signal ($Eu^{2+} \Rightarrow Eu^{3+} + e_{CB}$) coincides with the $Eu^{2+}(4f^{7}) \Rightarrow Eu^{2+}(4f^{6}5d^{1}(e))$ transition.

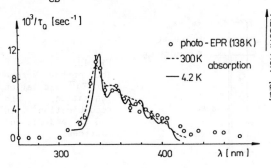

Fig. 10. Optical absorption for the Eu^{2+} $4f^{7} \Rightarrow 4f^{6}5d^{1}(e)$ transition in CdF_{2}. The spectral dependence of the Eu^{2+} photoquenching rate $1/\tau_{Q}$ is also shown (o).

636

For CaS and SrS the ground $4f^7$ state is deep and the first excited $4f^6 5d^1$ state lies close to the conduction band minima at the X point of CaS and SrS. This may result in a strong mixing of the d-like conduction electron wave function and the wave function of the 5d excited electron of Eu^{2+} which, in turn, should affect the efficiency of the Eu^{2+} $4f^6 5d^1 \Rightarrow 4f^7$ emission [62, 63]. Further experiments supporting this model are still necessary.

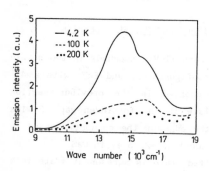

Fig. 11. Eu-connected "anomalous" emission in ZnS:Eu, excited by a $Eu^{2+} \Rightarrow Eu^{3+} + e_{CB}$ photoionization transition.

In section 3.4, we introduced the concept of an exciton bound at an RE center. The evident consequence of the near midgap Eu^{2+} energy level position in ZnS is the absence of Eu^{2+} $4f^6 5d^1 \Rightarrow 4f^7$ emission, as explained above, and appearance of Eu-connected "anomalous" emission [64]. The "normal" Eu^{2+} red emission is replaced by 14400 cm^{-1} PL (Fig. 11), with PL excitation band of $Eu^{2+} \Rightarrow Eu^{3+} + e_{CB}$ nature (Fig. 6). Such "anomalous" Eu^{2+} or Yb^{2+} and Yb^{3+} emissions were observed previously for a range of different crystals ($SrF_2:Yb^{2+}$, $BaF_2:Eu^{2+}$ [65], ZnS:Yb^{3+} [66]), and were explained as being due to direct radiative recombination of excitons bound at RE centers. Fig. 12 presents the energy scheme for Eu^{2+} excitation, ionization, and deexcitation in ZnS vs. CaS and explains the nature of Eu-related emissions in these two materials.

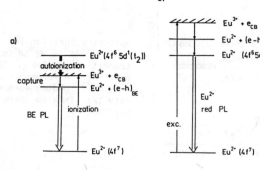

Fig. 12.
The energy scheme for Eu^{2+} excitation, ionization, and recombination in ZnS (a) and CaS (b) lattices.

3.6 <u>Energy transfer mechanism of the Eu^{2+} excitation in SrS:Ce,K</u>

There is another approach to obtaining multicolour EL panels. This can be realized by applying a colour filter to a white light EL display. Three different classes of white-colour EL displays were constructed recently [7]. Two of them include Eu doped SrS. White EL can be obtained by a superposition of blue-green Ce^{3+}-related and red Eu^{2+}-related emissions, satisfying a complementary colour relationship, and giving white colour sensation to a human eye. The physical mechanisms responsible for white EL are explained in Fig. 13.

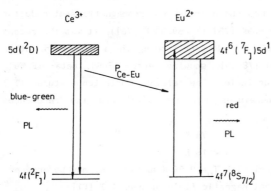

Fig. 13. The mechanism responsible for white colour EL in SrS:Ce,Eu (~0.1 mol% of Ce and ~0.03 mol% of Eu). Nonradiative energy transfer from the Ce^{3+} centers reduces the blue-green emission of Ce^{3+}, resulting in Eu^{2+} excitation followed by red $4f^6 5d^1 \Rightarrow 4f^7$ emission.

It is important to notice that high brightness of the EL is related to ionization of Ce, and probably also Eu, under the impact of hot carriers [7, 67]. First, a Ce^{3+} center is ionized by the impact of hot carriers, changing its charge state from 3+ to 4+. Subsequent hole recapture proceeds via the Ce^{3+} excited state, resulting in either direct Ce^{3+} intra-ion emission, or in energy transfer to Eu and Eu^{2+} excitation. It is not clear at the moment if Eu^{2+} centers are induced only indirectly, via the energy transfer from Ce. The theoretical estimation given in section 3.3 indicates that Eu$^{2+} \Rightarrow$ Eu^{3+} + free electron ionization requires about 2.75 eV, which is sufficient for recapture via an Eu^{2+} $4f^6 5d^1$ excited state. Experimental results indicate, however, that such direct Eu^{2+} excitation is less efficient than the indirect process via energy transfer from Ce^{3+} centers. This can be explained by a higher threshold electric field for Eu^{2+} excitation to the $4f^6 5d^1$ state, followed by ionization of this state, than is the case for the Ce^{3+} centers [69]. Efficient Ce – Eu energy transfer coupling means that codoping with a small amount of Ce is often used for improving the brightness and slow response characteristics of the CaS:Eu displays [68].

4. Towards Eu-based diluted magnetic compounds

The initial interest in EuX X=O, S, Se, Te compounds was motivated by their possible applications as magneto-optic memories and magneto-optic modulators. A comprehensive review of the magnetic properties Eu chalcogenides was given in a number of reviews and original papers [see, e.g., 70-73]. The Eu chalcogenides crystallize in a rock salt structure. When pure, they are insulating. The dominant exchange mechanism in insulators and semiconductors is the Kramers antiferromagnetic superexchange mechanism [74]. However, EuO was shown to be a ferromagnetic semiconductor with a high Curie temperature ~69K [75] (T_c=69.33 K [76]). It was the second magnetic semiconductor identified, $CrBr_3$ [77] being the first. Soon, other Eu chalcogenides were studied [78-80], showing ferro- (EuS), meta- (EuSe), and antiferromagnetic behaviour (EuTe). The relevant Curie temperatures (T_c) and paramagnetic Curie temperatures (ϑ) are ([72] and the references therein)

EuO T_c = 69.33 K ϑ = 74 K

EuS T_c = 16.57 K ϑ = 18 K

EuTe which is an antiferromagnet with T_N= 9.58 K and ϑ=-4 K, may be ordered ferromagnetically by an external magnetic field of over 1 T [81].

The main emphasis in the studies of EuX materials was on increasing the magnetic ordering temperature, with the room temperature as the final goal, to make practical application of Eu chalcogenides realistic.

To increase the magnetic ordering temperature, several doping (mixing) procedures were used. The main idea was to utilize the very strong dependence of the magnetic interactions on the presence of free carriers' and on modification of lattice parameters. The latter was demonstrated for $Eu_{1-x}Ca_xS$ and $Eu_{1-x}Ca_xO$ crystals. The decrease of the lattice parameter due to the smaller ionic radius of Ca^{2+} as compared to that of Eu^{2+}, leads to an increase in the ferromagnetic interaction [70] and, consequently, could result in a rise in the magnetic ordering temperature.

Different diluted magnetic compounds were studied. Solid solutions of $Eu_{1-x}M_x^{2+}X$ for X = O,S and M = Ca,Ba,Sr or Gd [82-86], were prepared for the whole range of x (0≤x≤1). The experiments performed showed that doping with nonmagnetic M^{2+} ions, which dilutes spins and reduces ϑ and T_c, prevails over the above mentioned mechanism of increase of ϑ and T_c, due to reduction of the lattice parameters. Until now, the highest magnetic ordering temperature was obtained for $Eu_{1-x}Gd_xO$ mixed crystals, with the Curie temperature being 137 K for x = 0.034, and 136 K for x=0.078 [87]. These temperatures are still far too low to realize practical applications of Eu based magnetic compounds. The main interest in II-VI doped Eu crystals was, thus, turned to optimizing Eu doping with a view to obtaining efficient EL displays, as described in some details in previous sections of this article. For example, efficient red thin film EL displays are obtained by deposition from a mixed powder of CaS and 0.25 mol% of EuS [88].

For low doping levels, realized for in ZnS or HgTe due to limited Eu solubility, magnetic properties of the crystals can be related to the presence of solid solutions of EuS and EuTe form. Such solutions were identified in ZnS and HgTe in microanalysis [13], and their influence on magnetic properties was visualized by the recent studies of Krylov et al. [15].

5. Conclusions

Eu doped or Eu mixed II-VI semiconducting compounds were considered for different applications. Eu-based diluted magnetic compounds attract less attention at the moment, due to their low magnetic ordering temperatures. Initial hopes of their practical applications, as magneto-optic memories in computer systems or magneto-optic modulators turned out to be unrealistic.

The main interest in Eu doped (mixed) II-VI compounds is now motivated by their utilization in thin film electroluminescence displays or, as shown recently, as nonradiative night-luminous phosphors (ZnS:Cu, Pb, Eu). To obtain multicolour EL displays, different RE doped wide bandgap II-VI compounds were prepared and studied, with ZnS and the alkaline earth sulphides (CaS, SrS, BaS) being the most promising materials.

Such focus of practical interest in Eu doped II-VI compounds is reflected in the contents of the above article. In this review, we stressed and discussed in more detail all the aspects connected with performance and optimization of TFEL structures. Much less attention was given to analysis of the magnetic interactions related to the highly localized $4f^7$ (S=7/2) electronic configuration of Eu^{2+}. The properties and energy structure of Eu^{3+} was not analyzed for two reasons. First, due to its nonmagnetic ground state and, second, since till now, Eu^{2+} emission was utilized in tested CaS:Eu, SrS:Eu red-orange and SrS:Eu,Ce, CaS:Eu,Ce white EL displays. The

640

question of whether such displays will find practical applications is still open. The brightness of the EL from these displays is a few to ten times too low for mass production. Till now, only Mn and TbF_3 doped ZnS TFEL are manufactured, with a range of other II-VI:RE systems prospective, but still requiring improvements.

Acknowledgements

This work has been supported by the Polish Academy of Sciences Program CPBP 01.12.

References

1. H. Ohnishi, Y. Yamamoto, Y. Katayama, in Conference Record 1985 International Display Research Conference, San Diego (IEEE, New York and Society for Information Display, Los Angeles, 1985), p.159.
2. H. Kobayashi, S. Tanaka, V. Shanker, M. Shiiki, T. Kunou, J. Mita, H. Sasakura, Phys. Stat. Sol. (a), **88**, 713 (1985).
3. M. Leskelä, M. Tammenmaa, Materials Chemistry and Physics **16**, 349 (1987).
4. S. Tanaka, H. Deguchi, Y. Mikami, M. Shiiki, H. Kobayashi, 1986 Digest SID Int. Symp., San Diego, 1986, p. 29.
5. M. Yoshida, A. Mikami, T. Ogura, K. Tanaka, K. Taniguchi, S. Nakajima, 1986 Digest SID Int. Symp., San Diego, 1986, p. 41.
6. S. Tanaka, J. of Luminescence **40/41**, 20 (1988).
7. S. Tanaka, H. Yoshiyama, J. Nishiura, S. Ohsio, H. Kawakami, H. Kobayashi, SID 1988 Digest **29**, 305.
8. A. Mauger, C. Godart, Physics Reports **141**, 51 (1986).
9. X. Mao, Z. Wu, Y. Feng, J. of Luminescence **40/41**, 891 (1988).
10. H.C. Casey, G.L. Pearson, J. Appl. Phys. **35**, 3401 (1964).
11. M.R. Brown, W.A. Shand, J. Mater. Sci. **5**, 790 (1970).
12. K. Świątek, M. Godlewski, D. Hommel, H. Hartmann, Phys. Stat. Sol. (a) **114**, 127 (1989).
13. G. Jasiołek, Z. Gołacki, M. Godlewski, J. Phys. Chem. Sol. **50**, 277 (1989).
14. Y. Charreire, H. Dexpert, J. Loriers, J. Mater. Res. Bull. **15**, 657 (1980).
15. K.R. Krylov, A.I. Ponomarev, I.M. Tsidilkovski, N.P. Gavalishko, V.V. Homiak, Physics and Technics of Semicond. **23**, 429 (1989).
16. T. Inoguchi, M. Takeda, Y. Kakahara, Y. Nakata, M. Yoshida, SID 1974 Digest **5**, 84 (1974).
17. R.T. Tuenge, R.E. Coovert, W.A. Barrow, 159th Electrochem. Soc. Meeting, Extended Abstracts p. 422 (1981).
18. H. Xion, G. Zhong, S. Tanaka, H. Kobayashi, Japanese J. Appl. Phys. **28**, L1019 (1989).
19. E.W. Chase, R.T. Hepplewhite, D.C. Krupka, D. Kahng, J. Appl. Phys. **40**, 2512 (1969).
20. D.C. Krupka, D.M. Mahoney, J. Appl. Phys. **43**, 2314 (1972).
21. D. Kahng, Appl. Phys. Lett. **13**, 910 (1968).

22. J. Bendit, P. Benalloul, B. Blanzat, J. of Luminescence **23**, 175 (1981).
23. S. Tanaka, H. Yoshiyama, Y. Mikami, J. Nishiura, S. Ohshio, H. Kobayashi, SID Digest **29**, 77 (1988)
24. M. Leskelä. M. Mäkelä, L. Niinistö, M. Tammenmaa, Chemtronics **3**, 113 (1988).
25. M.R. Brown, A.F. J. Cox, W.A. Shand, J.M. Williams, Adv. Quantum Electronics **2**, 69 (1974).
26. R. Boyn, Phys. Stat. Solidi (b) **148**, 11 (1988).
27. K. Świątek, A. Suchocki, M. Godlewski, Appl. Phys. Lett. **56**, 195 (1990).
28. Z.A. Trapeznikova, V.V. Suchaenko, Dokl. Akad. Nauk SSSR **106**, 230 (1956).
29. L.L. Chase, Phys. Rev. **B2**, 2308 (1970).
30. Y.R. Shen, N. Bloembergen, Phys. Rev. **133**, A515 (1963).
31. Y.R. Shen, Phys. Rev. **134**, A661 (1964).
32. S. Tanaka, H. Deguchi, Y. Mikami, M. Shiiki, H. Kobayashi, SID Digest **28**, 21 (1987).
33. O. Matamura, J. Phys. Soc. Jpn. **14**, 108 (1959).
34. A. Abragam, B. Bleaney, *Electron paramagnetic resonance of transition ions*, Clarendon Press, Oxford 1970.
35. R.R. Schehl, P. Wigen, Solid State Commun. **8**, 669 (1970).
36. R.S. Title, Phys. Rev. **133**, A198 (1964).
37. S. Ibuki, H. Komiya, M. Nakada, H. Masui, H. Kimura, J. of Luminescence **1/2**, 797 (1970).
38. M. Tovar, M.T. Causa, Phys. Rev. **B36**, 44 (1987).
39. A.A. Gomes, R.M. Xavier, J. Danon, Chem. Phys. Lett. **4**, 239 (1969).
40. W. Zinn, in Proc. Int. Conf. Magnetism, Grenoble 1970, J. Phys. **32**, Suppl. C I, p. 724 (1971).
41. E.L. Boyd, Phys. Rev. **145**, 174 (1966).
42. S.H. Charap, E.L. Boyd, Phys. Rev. **133**, A811 (1964).
43. K. Kojima, T. Komaru, T. Hihara, Y. Koi, J. Phys. Soc. Jap. **40**, 1570 (1976).
44. A. Calhoun, J. Overmeyer, J. Appl. Phys. **35**, 989 (1964).
45. R.E. Watson, A.J. Freeman, Phys. Rev. **133**, A1571 (1964).
46. S.P. Keller, G.D. Pettit, Phys. Rev. **111**, 1533 (1958).
47. R.S. Title, Phys. Rev. Lett. **3**, 273 (1959).
48. C.K. Jörgensen, Molecular Physics **5**, 3 (1962).
49. K. Świątek, M. Godlewski, Acta Phys. Polonica (1990) (in press).
50. K. Świątek, A. Suchocki, H. Przybylińska, M. Godlewski, J. Crystal Growth (1990) (in press).
51. D.C. Krupka, J. Appl. Phys. **43**, 476 (1972).
52. J.W. Allen, in *Electroluminescence*, Springer Proc. in Physics **38**, eds. S. Shionoya and H. Kobayashi (Springer Verlag, Berlin, Heidelberg, 1989) p. 10.
53. K. Świątek, M. Godlewski, D. Hommel, unpublished results.
54. H. Zimmermann, R. Boyn, Phys. Stat. Sol. (b) **135**, 379 (1986).
55. K. Świątek, M. Godlewski, unpublished results.
56. W. Lehmann, F.M. Ryan, J. Electrochem. Soc., **118**, 477 (1971).
57. S. Tanaka, J. Crystal Growth (1990), in press.
58. M. Godlewski, D. Hommel, Phys. Stat. Sol. (a) **95**, 261 (1986).
59. A. Garcia, F. Guillen, C. Fouassier, J. of Luminescence **33**, 15 (1985).
60. J.L. Ryan, Inorg. Chem. **8**, 2053 (1969).
61. M. Godlewski, D. Hommel, J.M. Langer, H. Przybylińska, J. of Luminescence **24/25**, 217 (1981).
62. S. Tanaka, J. of Luminescence **40/41**, 20 (1988).

63. H. Yoshiyama, S.H. Sohn, S. Tanaka, H. Kobayashi, in *Electroluminescence* Springer Proc. in Physics **38**, eds. S. Shionoya and H. Kobayashi (Springer Verlag, berlin, Heidelberg, 1989), p. 48.

64. K. Świątek, M. Godlewski D. Hommel, unpublished results.

65. D.S. McClure, C. Pedrini, Phys. Rev. **B32**, 8465 (1985).

66. H. Przybylińska, K. Świątek, A. Stąpor, A. Suchocki, M. Godlewski, Phys. Rev. **B40**, 1748 (1989).

67. S. Tanaka, S. Ohshio, J. Nishiura, H. Kawakami, H. Yoshiyama, H. Kobayashi, Appl. Phys. Lett. **52**, 2102 (1988).

68. M. Ando, Y.A. Ono, K. Onisawa, H. Kobayashi, in *Electroluminescence*, Springer Proc. in Physics **38**, eds. S. Shionoya and H. Kobayashi (Springer Verlag, Berlin, Heidelberg, 1989), p. 171.

69. S. Tanaka, H. Yoshiyama, J. Nishiura, S. Ohshio, H. Kawakami, H. Kobayashi, Appl. Phys. Lett. **51**, 1661 (1987).

70. A. Mauger, C. Godart, Physics Reports **141**, 51 (1986).

71. A. Mauger, Phys. Stat. Sol. (b) **84**, 761 (1977).

72. W. Notting, Phys. Stat. Sol. (b) **96**, 11 (1979).

73. P. Wachter, Physics Reports **44**, 159 (1978).

74. H.A. Kramers, Physica **1**, 182 (1934).

75. B.T. Matthias, R.M. Bozorth,, J.H. van Vleck, Phys. Rev. Lett. **7**, 160 (1961).

76. P. Schwob, Phys. Kondens. Materie **10**, 186 (1969).

77. I. Tsubokawa, J. Phys. Soc. Japan **15**, 1664 (1960).

78. G. Busch, P. Junod, M. Risi, O. Vogt, Proc. Intern. Conf. Semicond., Exeter 1962, Inst. Phys. and Phys. Soc., London 1962, p. 727.

79. T.R. McGuire, B.E. Argyle, M.W. Schafer, J.S. Smart, Appl. Phys. Lett. **1**, 17 (1962).

80. S. van Houten, Phys. Letters **2**, 215 (1962).

81. J. Schoenes, unpublished, reference given in [72].

82. S. Methfenel, D.E. Eastman, F. Holtzberg, T.R. McGuire, T. Penney, M.W. Schafer, S. Von Molnar, Colloque Inter. Terres Rares, C.N.R.S., Paris, 1969 (Edition du C.N.R.S.), p. 565.

83. A.A. Smokhvalov, N.N. Loshkareva, V.G. Bamburov, Sov. Phys. Solid State **13**, 1574 (1972).

84. A.A. Smokhvalov, Yu.N. Morozov, N.V. Volkenshtein, T.D. Zotov, Izv. Akad. Nauk Fiz. ser. **36**, 1401 (1972).

85. A.A. Smokhvalov, N.V. Volkenshtein, V.G. Bamburov, T.D. Zotov, A.A. Ivakin, Yu.N. Morozov, M.I. Sinonova, Sov. Phys. Solid State **9**, 555 (1967).

86. A.A. Smokhvalov, T.I. Arbuzova, A.Yu. Afanasev, V.A. Babushkin, N.N. Loshkareva, Yu.N. Morozov, M.I. Simonova, V.G. Bamburov, N.I. Lobachevskaya, Sov. Phys. Solid State **17**, 26 (1975).

87. M.W. Schafer, T.R. McGuire, J. Appl. Phys. **39**, 588 (1969).

88. D. Yebdri, P. Benalloul, J. Benoit, in *Electroluminescence*, Springer Proc. in Physics **38**, eds. S. Shionoya and H. Kobayashi (Springer Verlag, Berlin, Heidenberg, 1989), p. 167.

Author Index

Subject index

650